Volumes already published

Volumes in preparation

Biotechnology in Agriculture and Forestry 21

Medicinal and Aromatic Plants IV

Edited by Y.P.S. Bajaj

With 250 Figures

Springer-Verlag
Berlin Heidelberg New York
London Paris Tokyo
Hong Kong Barcelona
Budapest

Professor Dr. Y.P.S. Bajaj
A-137
New Friends Colony
New Delhi 110065, India

ISBN-13:978-3-642-77006-7 e-ISBN-13:978-3-642-77004-3
DOI: 10.1007/978-3-642-77004-3

Library of Congress Cataloging-in-Publication Data. (Revised for volume 4). Medicinal and aromatic plants. (Biotechnology in agriculture and forestry ; 4, 15, 21) Includes bibliographies and index. 1. Medicinal plants—Biotechnology. 2. Aromatic plants—Biotechnology. 3. Plant cell culture. 4. Materia medica, Vegetable. I. Bajaj, Y.P.S., 1936– . II. Series: Biotechnology in agriculture and forestry ; 4, etc. TP248.27.P55M43 1988 660'.62 88-3059
ISBN-13:978-3-642-77006-7 (U.S.: v. 1)

Softcover reprint of the hardcover 1st edition 1993

Production Editor: Ingrid Samide, Heidelberg

Typesetting: Macmillan India Limited, Bangalore
31/3145-5 4 3 2 1 0 – Printed on acid-free paper

Dedicated to
Andrea Bajaj

Preface

Recently, the pharmaceutical industry has shown considerable interest in the en masse micropropagation of elite plants, and the large-scale production of secondary metabolites in plant cell cultures through the use of bioreactors. This is now being realized by the selection of high-yielding somaclones or through the production of hairy roots by transformation with *Agrobacterium rhizogenes*. These new developments need to be high-lighted and brought to the attention of workers dealing with medicinal, aromatic, and other plants of industrial importance. A series of books on the biotechnology of medicinal and aromatic plants is therefore being compiled to provide a survey of the literature focusing on recent information and the state of the art.

This book, *Medicinal and Aromatic Plants IV*, like the previous three volumes (Medicinal and Aromatic Plants I, II, and III, published in 1988, 1989, and 1991, respectively) is unique in its approach. It comprises 28 chapters dealing with the distribution, economic importance, conventional propagation, review of tissue culture studies, and the in vitro production of important medicinal and pharmaceutical compounds in various species of *Ammi*, *Bergenia*, *Canavalia*, *Capsicum*, *Cassia*, *Cephaelis*, *Cornus*, *Cucurbita*, *Elettaria*, *Eupatorium*, *Genipa*, *Gentiana*, *Gypsophila*, *Hygrophila*, *Leontopodium*, *Nerium*, *Picrasma*, *Polygonum*, *Ptelea*, *Rheum*, *Scopolia*, *Silene*, *Solanum*, *Strophanthus*, *Tagetes*, *Thymus*, and *Uncaria*.

This book is tailored to the needs of advanced students, teachers, and research workers in the fields of plant biotechnology and chemical engineering, pharmacy, botany, and phytochemistry.

New Delhi, October 1992

Professor Y. P. S. Bajaj
Series Editor

Contents

XIV *Hygrophila erecta:* In Vitro Culture and the Production of Verbascoside
M. Henry, J.-L. Roussel, and C. Andary (With 4 Figures)

XV *Leontopodium alpinum* Cass. (Edelweiss): In Vitro Culture, Micropropagation, and the Production of Secondary Metabolites
I. L. I. Hook (With 13 Figures)

XVI *Nerium* spp.: In Vitro Culture and the Production of Secondary Metabolites
D. H. Paper and G. Franz (With 12 Figures)

XVII *Picrasma quassioides* Bennet (Japanese Quassia Tree): In Vitro Culture and Production of Quassin
A. H. Scragg and E. J. Allan (With 20 Figures)

XVIII *Polygonum hydropiper* L. (Water Pepper): In Vitro Culture and the Production of the Aphid-Antifeedant Polygodial
D. V. Banthorpe (With 4 Figures)

XXIII *Solanum dulcamara* L. (Bittersweet):
Accumulation of Steroidal Alkaloids in the Plant
and in Different in Vitro Systems
A. Ehmke and U. Eilert (With 4 Figures)

XXIV *Solanum glaucophyllum* Desf. (Duraznillo Blanco):
In Vitro Culture and the Production of Steroidal Secondary
Metabolites
M. Weissenberg, A. Levy, and R. H. Wasserman
(With 4 Figures)

XXV *Strophanthus* Species (Members of the Dogbane Family):
In Vitro Culture and the Production of Cardenolides
K. Kawaguchi, M. Hirotani, and T. Furuya
(With 10 Figures)

XXVI *Tagetes* spp. (Marigolds): In Vitro Culture
and the Production of Thiophenes
H. Breteler and D. H. Ketel (With 11 Figures)

List of Contributors

ALLAN, E. J., Department of Agriculture, University of Aberdeen, 581 King Street, Aberdeen AB9 1UD, Scotland

ANDARY, C., Laboratoire de Botanique, Phytochimie et Mycologie, Faculté de Pharmacie, Avenue Charles Flahault, 34060 Montpellier Cedex, France

BAJAJ, Y. P. S., Former Professor of Tissue Culture, Punjab Agricultural University, Ludhiana, India. (Present address: A-137 New Friends Colony, New Delhi 110065, India)

BANTHORPE, D. V., Chemistry Department, University College London, 20 Gordon St., London WC1H 0AJ, United Kingdom

BATTERMAN, S., University of Groningen, University Center for Pharmacy, Department of Pharmacognosy, A. Deusinglaan 2, 9713 AW Groningen, The Netherlands

BISTER-MIEL, F., University of Paris-Sud, Faculty of Pharmacy, 3 rue J.B. Clément, 92296 Châtenay-Malabry Cedex, France

BOTTA, B., Centro Chimica dei Recettori, Istituto di Chimica, Università Cattolica del S. Cuore, Lagro F. Vito 1, 00168 Roma, Italy

BRETELER, H., Center of Agrobiological Research (CABO), P.O. Box 14, 6700 AA Wageningen, The Netherlands

CHENIEUX, J. C., Plant Biotechnology Laboratory, Faculty of Pharmacy, University of Tours, 2 bis Bd. Tonnellé, 37042 Tours Cedex, France

DAS, N. P., Laboratory of Flavonoid Research, Department of Biochemistry, Faculty of Medicine, National University of Singapore, Singapore 0511

DELLE MONACHE, G., Centro Chimica dei Recettori, Istituto di Chimica, Università Cattolica del S. Cuore, Lagro F. Vito 1, 00168 Roma, Italy

EHMKE, A., Institut für Pharmazeutische Biologie der Technischen Universität Braunschweig, Mendelssohnstraße 1, W-3300 Braunschweig, FRG

EILERT, U., Institut für Pharmazeutische Biologie der Technischen Universität Braunschweig, Mendelssohnstraße 1, W-3300 Braunschweig, FRG

EKIERT, H., Nicolaus Copernicus Academy of Medicine, Department of Pharmaceutical Botany, 16 Krupnicza Str., 31-123 Krakow, Poland

ESAKA, M., Faculty of Applied Biological Science, Hiroshima University, Kagamiyama, Higashi-Hiroshima 724, Japan

FRANZ, G., Institute of Pharmaceutical Biology, University of Regensburg, Universitätsstraße 31, W-8400 Regensburg, FRG

FURMANOWA, M., Department of Biology and Pharmaceutical Botany, Medical Academy, Banacha 1, 02-097 Warsaw, Poland

FURUYA, T., School of Pharmaceutical Sciences, Kitasato University, Minato-ku, Tokyo 108, Japan

GODOY, G., Visión de Biologia Vegetal, Centro de Investigación Cientifica de Yucatán Apado, Postal 87 Cordemex, 97310 Yucatán, Mexico

GOPALAKRISHNAN, P. K., Department of Plantation Crops and Spices, College of Horticulture, Kerala Agricultural University, Trichur 680654, India

GUIGNARD, J. L., University of Paris-Sud, Faculty of Pharmacy, 3 rue J.B. Clément, 92296 Châtenay-Malabry Cedex, France

HENRY, M., Laboratoire de Botanique, Faculté des Sciences Pharmaceutiques, 31, Allées Jules Guesde, 31000 Toulouse, France

HIROTANI, M., School of Pharmaceutical Sciences, Kitasato University, Minato-ku, Tokyo 108, Japan

HOOK, I. L. I., Department of Pharmacognosy, School of Pharmacy, Trinity College, Dublin 4, Ireland

KAWAGUCHI, K., Medicinal Plant Garden, School of Pharmaceutical Sciences, Kitasato University, Sagamihara, Kanagawa 228, Japan

KETEL, D. H., Center of Agrobiological Research (CABO), P.O. Box 14, 6700 AA Wageningen, The Netherlands

LAW, K. H., Laboratory of Flavonoid Research, Department of Biochemistry, Faculty of Medicine, National University of Singapore, Singapore 0511

LEVY, A., Department of Medicinal, Spice and Aromatic Plants, Agricultural Research Organization, The Volcani Center, Bet Dagan 50250, Israel

LOYOLA-VARGAS, V. M., División de Biologia Vegetal, Centro de Investigación Cientifica de Yucatán Apado, Postal 87 Cordemex, 97310 Yucatán, Mexico

MONTAGU, M., Plant Biotechnology Laboratory, Faculty of Pharmacy, University of Tours, 2 bis Bd. Tonnellé, 37042 Tours Cedex, France

OKUDA, T., Faculty of Pharmaceutical Sciences, Okayama University, Tsushima, Naka 1-1, Okayama 700, Japan

OROPEZA, C., División de Biologia Vegetal, Centro de Investigación Cientifica de Yucatán Apado, Postal 87 Cordemex, 97310 Yucatán, Mexico

PAPER, D. H., Institute of Pharmaceutical Biology, University of Regensburg, Universitätsstraße 31, W-8400 Regensburg, FRG (Present address: St. Georgs-Platz 8, W-8451 Rieden, FRG)

PETIT-PLAY, G., Plant Biotechnology Laboratory, Faculty of Pharmacy, University of Tours, 2 bis Bd. Tonnellé, 37042 Tours Cedex, France

PROKSCH, P., Institut für Botanik und Pharmazeutische Biologie, Universität Würzburg, Mittlerer Dallenbergweg 64, W-8700 Würzburg, FRG

QUIROZ, J., División de Biologia Vegetal, Centro de Investigación Cientifica de Yucatán Apado, Postal 87 Cordemex, 97310 Yucatán, Mexico

RAPCZEWSKA, L., Department of Biology and Pharmaceutical Botany, Medical Academy, Banacha 1, 02-097 Warsaw, Poland

REGHUNATH, B. R., Department of Plantation Crops and Spices, College of Horticulture, Kerala Agricultural University, Trichur 680654, India

RIDEAU, M., Plant Biotechnology Laboratory, Faculty of Pharmacy, University of Tours, 2 bis Bd. Tonnellé, 37042 Tours Cedex, France

ROUSSEL, J.-L., Laboratoire de Botanique, Phytochimie et Mycologie, Faculté de Pharmacie, Avenue Charles Flahault, 34060 Montpellier Cedex, France

SCHOLTEN, H. J., Agricultural University, Department of Horticulture, Haagsteeg 3, 6708 PM Wageningen, The Netherlands

SCRAGG, A. H., Department of Applied Science, The University of the West of England, Cold Harbour Lane, Frenchay, Bristol BS16 1QY, United Kingdom

SHIBATA, S., Laboratory of Natural Medicinal Materials, c/o Minophagen Co, 3rd Tomizawa Bldg, 4th floor, Yotsuya 3-2-7, Shinjuku-ku, Tokyo 160, Japan

SHIMOMURA, K., Tsukuba Medicinal Plant Research Station, National Institute of Hygienic Sciences, Ministry of Health and Welfare, 1 Hachimandai, Tsukuba Ibaraki, 305, Japan

SIEBERTZ, R., Institut für Pharmakologie der Universität Düsseldorf, Moorenstraße 5, W-4000 Düsseldorf, FRG

SKRZYPCZAK, E., Department of Pharmaceutical Botany, K. Marcinkowski Medical Academy, Wieniawskiego 1, 61-712 Poznán, Poland

SKRZYPCZAK, L., Department of Pharmaceutical Botany, K. Marcinkowski Medical Academy, Wieniawskiego 1, 61-712 Poznán, Poland

SUGISAWA, H., Department of Bioresource Science, Faculty of Agriculture, Kagawa University, Miki-cho, Kagawa-ken, Japan 761-07

TAKEBAYASHI, T., Department of Bioresource Science, Faculty of Agriculture, Kagawa University, Miki-cho, Kagawa-ken, Japan 761-07

TAMURA, H., Department of Bioresource Science, Faculty of Agriculture, Kagawa University, Miki-cho, Kagawa-ken, Japan 761-07

TRÉMOUILLAUX-GUILLER, J., Plant Biotechnology Laboratory, Faculty of Pharmacy, University of Tours, 2 bis Bd. Tonnellé, 37042 Tours Cedex, France

UEDA, S., Faculty of Pharmaceutical Sciences, Kyoto University, Sakyo-ku, Kyoto 606-01, Japan

VIEL, C., University of Tours, Faculty of Pharmacy, 2 bis Bd. Tonnellé, 37042 Tours Cedex, France

VISSER, J. F., Agricultural University, Department of Horticulture, Haagsteeg 3, 6708 PM Wageningen, The Netherlands

WASSERMAN, R. H., New York State College of Veterinary Medicine, Cornell University, Ithaca, NY 14853, USA

WEISSENBERG, M., Department of Chemistry of Natural Products and Pesticides, Agricultural Research Organization, The Volcani Center, Bet Dagan 50250, Israel

WESOLOWSKA, M., Department of Pharmaceutical Botany, K. Marcinkowski Medical Academy, Wieniawskiego 1, 61-712 Poznán, Poland

WILLIAMS, P. D., Department of Chemical Engineering, University of Manchester, Institute of Science and Technology, Manchester, United Kingdom

YAZAKI, K., Faculty of Pharmaceutical Sciences, Okayama University, Tsushima, Naka 1-1, Okayama 700, Japan

YOSHIMATSU, K., Tsukuba Medicinal Plant Research Station, National Institute of Hygienic Sciences, Ministry of Health and Welfare, 1 Hachimandai, Tsukuba, Ibaraki 305, Japan

I *Ammi majus* L. (Bishop's Weed): In Vitro Culture and the Production of Coumarin Compounds

H. Ekiert[1]

1 Introduction

1.1 Distribution and Importance

Ammi majus L. (Bishop's weed) (Fig. 1) is a subtropical species belonging to the family Apiaceae. It is a widely distributed species in the Mediterranean region, from the Canary Islands to Iran. Its range covers North Africa (Morocco, Egypt, Ethiopia) and all of southern Europe. The species also occurs on other continents under similar climatic conditions: in Argentina, southern United States, and less commonly in Australia and New Zealand (Hegi 1906–1931; Tutin et al. 1968; Hoppe 1975; Nowiński 1980).

A. majus L. is regarded as the richest, natural source of linear furanocoumarins called psoralens. These compounds are found mostly in the fruits of this species (Nielsen 1970, 1971; Beyrich 1981).

The psoralens are successfully applied in photochemotherapy of numerous dermatological diseases, e.g., in treating vitiligo, psoriasis, mycosis fungoides, atopic eczema, pityriasis lichenoides, urticaria pigmentosa, alopecia areata, and others. The therapy mostly makes use of photosensitizing and antiproliferating properties of psoralens. These properties are particularly enhanced in the presence of long-wavelength UV, called UV-A ($\lambda = 320–400$ nm), hence the treatment is often referred to as PUVA therapy (psoralens + UV-A) (Pathak et al. 1981; Ben-Hur and Pill-Soon 1984; Roenigh 1984; Turjanmaa et al. 1985; Plewig et al. 1986).

The use of fruits of *A. majus* in treating vitiligo has a long tradition. Medicinal virtues of this species were known to the ancient Egyptians (and were first mentioned in Ebers' papyrus) (Fahmy and Abu-Shady 1947; Ramadan 1982).

Most of the preparations whose photosensitizing properties are currently used in dermatological treatment methods have been developed on the basis of *A. majus*. These are, among others, such preparations as Meladinine (Memphis Chemical Co., Promedica, Basotherm, Pharma-Medica); Oxsoralen (Elder, Gerot); and Ammifurin (Medexport). The main components of these are bergapten, xanthotoxin, isopimpinellin, and imperatorin (Podlewski and Chwalibogowska-Podlewska 1986; Wolff and Tessa 1986).

---

[1] Nicolaus Copernicus Academy of Medicine, Department of Pharmaceutical Botany, 16 Krupnicza Str., 31-123 Kraków, Poland

Biotechnology in Agriculture and Forestry, Vol. 21
Medicinal and Aromatic Plants IV (ed. by Y.P.S. Bajaj)

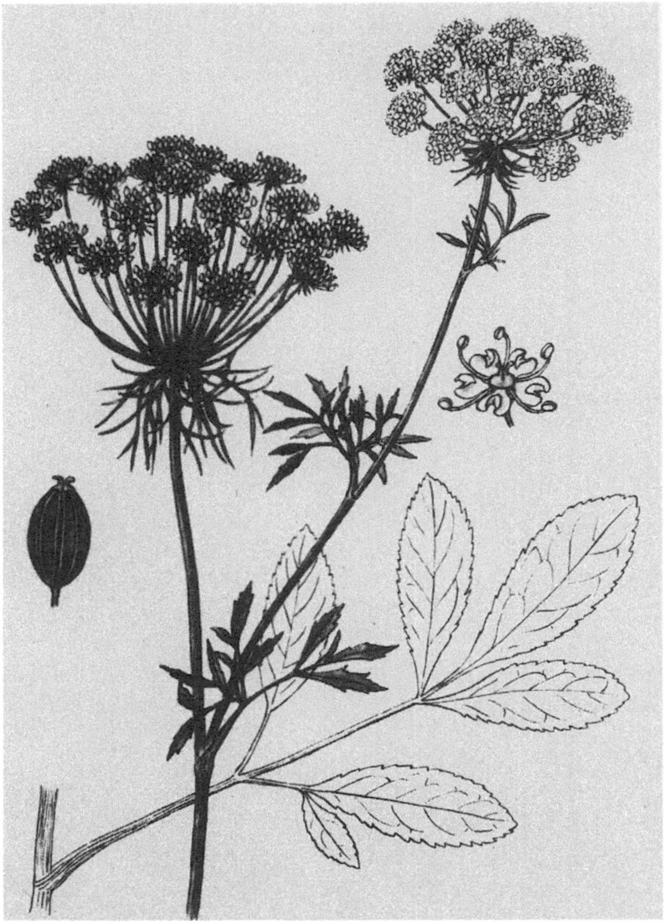

Fig. 1. *Ammi majus* L. (Korbelar and Endris 1981)

The biological properties of psoralens have been intensively studied in recent years. Apart from their photosensitizing and antiproliferating properties (Ben-Hur and Pill-Soon 1984; Rodigiero et al. 1984; Rodigiero 1985; Rodigiero and Dall'Acqua 1986), it has been recently proved that these compounds might show calcium-antagonistic action (Vuorela et al. 1988). This discovery opens new fields for their therapeutic applications.

1.2 Chemistry

Linear furanocoumarins, which determine the therapeutic properties of *Ammi majus* L., constitute the principal group of secondary metabolites in this species. They are a subgroup of coumarin compounds with a benzo-α-pyrone (coumarin) structure condensed with the furan ring at positions 6 and 7 (among furanocoumarins of angular structure, at positions 7 and 8). Psoralen is the principal compound in this group, hence the entire group is referred to as psoralens (Soine 1964; Tandon and Rastogi 1979).

In *A. majus* of various origin, about 20 different derivatives of psoralen were found. The first furanocoumarins isolated from fruits of *A. majus* of Egyptian origin were xanthotoxin (0.5%), imperatorin (0.3%), and bergapten (0.04%) (Fahmy and Abu-Shady 1947, 1948; Fahmy et al. 1947; Schönberg and Sina 1948, 1950). The Egyptian material provided several more psoralens: marmesin and its glucoside marmesinin (Abu-Mustafa et al. 1958; Starkowsky and Badran 1958; Abu-Mustafa and Fayez 1961); isopimpinellin (Abdel Hay et al. 1966); isoimperatorin (Abu-Mustafa et al. 1968); alloimperatorin (Abu-Mustafa et al. 1975a); and ammirin (Abu-Mustafa et al. 1975b) (Fig. 2). In plant material of American origin, eight more linear furanocoumarins were identified (Ivie 1978). The analysis of plant raw material of various origin indicates marked quantitative and qualitative differences in the composition of furanocoumarins (Akačić and Kuštrak 1958; Trojanek et al. 1961; Kohlmünzer and Wacław 1962; Kohlmünzer and Wacław-Rozkrutowa 1963; Nikonov 1965; Tarpo et al. 1965; Ivie 1978; Eisenreichová et al. 1980). The plant material of Egyptian origin is regarded as the most valuable, since it provides components of therapeutic value such as xanthotoxin, bergapten, imperatorin, and isopimpinellin (Fahmy and Abu-Shady 1948; Abdel Hay et al. 1966).

Psoralens occur mainly in the fruits of *A. majus* L. (about 1%). Apart from fruits, flowers and axes of umbels also contain high concentrations of furanocoumarins. Leaves and stems contain much less of these compounds, while least quantities occur in roots (Kibalchich et al. 1960; Blažek and Stary 1965, 1966; Gheorghiu et al. 1965; Tarpo et al. 1965; Blažek 1966).

Because of the high therapeutic value of the species, in many countries of Central Europe and in the Soviet Union, attempts have been made to acclimatize it

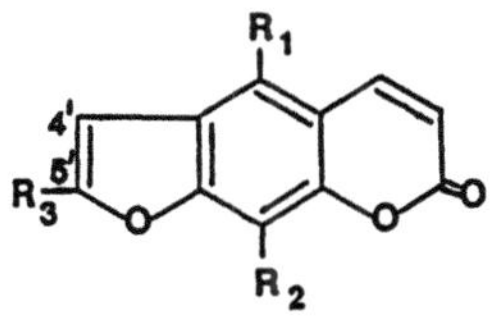

	R_1	R_2	R_3
BERGAPTEN	$-OCH_3$	H	H
XANTHOTOXIN	H	$-OCH_3$	H
ISOPIMPINELLIN	$-OCH_3$	$-OCH_3$	H
IMPERATORIN	$-O-CH_2CH=C/CH_3/_2$	H	H
MARMESIN /4' H, 5' H/	H	H	$-C(OH)/CH_3/_2$

Fig. 2. Chemical structure of basic furanocoumarins occurring in *Ammi majus* L.

(Trojanek et al. 1961; Kohlmünzer and Wacław 1962; Gheorghiu et al. 1965; Nikonov 1965). The efforts have not always been completely successful. Within a moderate climatic zone, it is not always possible to obtain fully mature fruits. The plants have been also attacked by diseases.

Taking this into account, and also due to the fact that high chemical variability related to climatic and edaphic conditions affects the quality of plant material, studies on the biosynthesis of secondary metabolites in in vitro cultures were initiated in this project. Newly emerging reports on secondary metabolites biosynthesis in tissue cultures of medicinal plants gave yet another impulse to undertake this project (e.g., Butcher 1977; Czygan 1984; Staba 1985; Bajaj 1988, 1989).

2 In Vitro Approaches

2.1 Review of Biotechnology Studies

The first brief mention of tissue culture of *Ammi majus* L. pertained to the possible biotransformation of ^{14}C-tagged visnagin (Chen et al. 1969). It was found that visnagin was metabolized to unidentified products having higher polarity than the substrate. Tissue culture of *A. majus*, as well as that of the principal object of the study, *Ammi visnaga*, did not show the expected ability to carry out the biotransformation of visnagin into khellin.

Two reports that followed referred to attempts of polyembryos induction and regeneration of plants in the in vitro culture of *Ammi majus* L. (derived from the ovary or hypocotyl) as a potential means for micropropagation of the species (Sehgal 1972; Grewal et al. 1976).

In the 1980s, intensive studies started on the biosynthesis of secondary metabolites, characteristic for the native plant, under various conditions of in vitro culture (Ekiert 1986a, b; 1989a, b; 1990), followed later by a research project on the biosynthesis of these metabolites induced by elicitors (skleroglucan from *Sclerotium sclerotiosum*, elicitors from *Phytophthora megasperma* f. sp. *glycinea* and *Alternaria carthami*) (Hamerski and Matern 1988a, b; Hamerski et al. 1990a, b).

2.2 Experimental In Vitro Culture

2.2.1 Establishment of In Vitro Culture

The tissue culture was derived from fruits of *A. majus* L. obtained from the Botanical Garden of the Justus Liebig University at Giessen, FRG. The fruits sown under sterile conditions germinated after 2–6-week periods in about 90% of the culture (Fig. 3). On LS medium (Linsmaier and Skoog 1965) containing 2 mg/l NAA and 2 mg/l BAP, callus initiation was observed in hypocotyl parts of sterile seedlings. The tissue formed there gave rise to the tissue culture (Fig. 4).

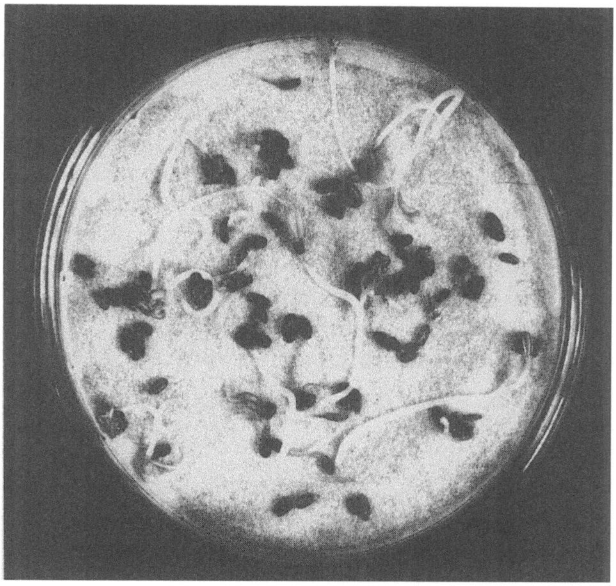

Fig. 3. Fruits of *Ammi majus* L. germinating under aseptic conditions from which in vitro tissue cultures were derived. Medium: solution of kinetin and gibberellic acid ($c = 10^{-4}$ g/l)

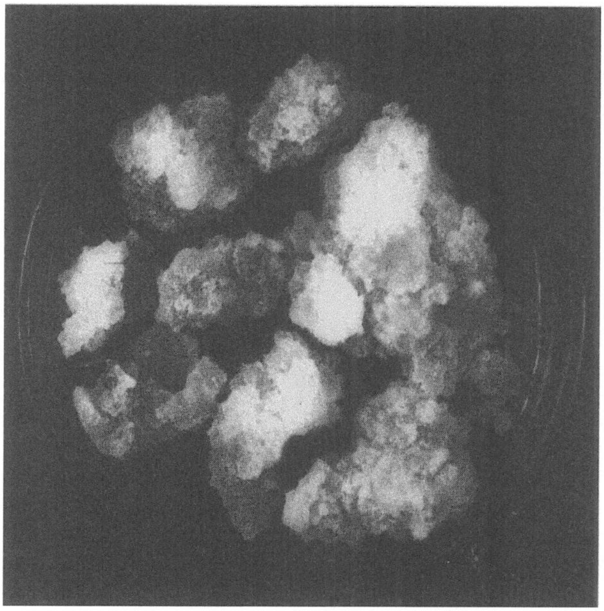

Fig. 4. Typical callus tissue in agar culture. LS medium containing 1 mg/l BAP and 1 mg/l IBA

2.2.2 Tested Culture Conditions

To determine the most suitable conditions for tissue mass increments, the effects of chemical composition of culture media (variants of LS medium), light regime, and culture types were studied.

Media. The media of the LS type were modified by adding cytokinins (BAP or kinetin) and auxins: 2,4-D, IAA, IBA or NAA in the variable quantities of 0.1, 1, 5, and 10 mg/l, respectively. Nutritive complex additions were also applied: coconut water (10000 mg/l); maize milk (10000 mg/l); casein hydrolysate (100 mg/l); vitamin complex according to modified MS medium (Murashige and Skoog 1962). In some media, an addition of umbelliferone was also applied as a precursor in the following quantities: 25, 50, 100, and 200 mg/l (Table 1).

The cultures were maintained as agar, stationary liquid, and suspension cultures (on selected variants of media).

The cultures were kept under constant artificial light of 900 lx intensity. Some cultures were kept in absolute darkness.

Tissues were subcultured every 4 weeks.

Table 1. Variants of Linsmaier and Skoog medium tested in tissue cultures of *Ammi majus* L.

Variants	Components (mg/l)			
	Auxins	Cytokinins	Other components	
	2,4-D, IAA, IBA, NAA	BAP kinetin		
1	1.0	10.0		
2	5.0	5.0		
3	10.0	1.0		
4	1.0	1.0		
5	0.1	0.1		
6	5.0	1.0		
7	1.0	5.0		
8	10.0	10.0		
9	10.0	5.0		
10	5.0	10.0		
11	1.0	–		
12	5.0	–		
13	10.0	–		
14	–	1.0		
15	–	5.0		
16	–	10.0		
17	–	–		
18[a]	0.1	0.1	Coconut water	– 10000
19	0.1	0.1	Maize milk	– 10000
20	0.1	0.1	Casein hydrolysate	– 100
21	0.1	0.1	Vitamin complex[c]	
22[b]	0.1	0.1	Umbelliferone	– 25
23	0.1	0.1	Umbelliferone	– 50
24	0.1	0.1	Umbelliferone	– 100
25	0.1	0.1	Umbelliferone	– 200

[a] Variants 18–21 – IAA-BAP and IBA-BAP combinations.
[b] Variants 22–25 – IBA-BAP combinations.
[c] Composition as in Murashige and Skoog medium (1962).

2.2.3 Results

Effect of Growth Substances on Biomass Increments. The best increments of fresh mass of tissue were obtained on media containing BAP in combination with IAA (Fig. 5), and BAP with IBA, but slightly less in media with BAP-NAA combinations. Combinations of BAP and 2,4-D (Fig. 6) were decisively poorest in respect to growth enhancement. The variants of media containing kinetin in combinations with IAA and IBA supported much lesser increments than media containing identical combinations of BAP with these auxins.

The tissues of *Ammi majus* L. in in vitro cultures showed, in further passages, the ability to grow on media containing only cytokinins or auxins. A phenomenon of simultaneous anergization was found in respect to both auxins and cytokinins.

Effect of Nutritive Complexes on Biomass Increments. Clear positive effects of cytokinins present in coconut water, and of the mixture of amino-acids in

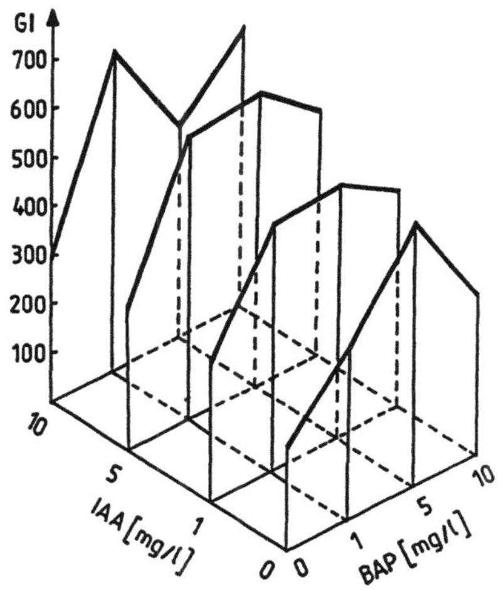

Fig. 5. Tissue mass increments in agar cultures on LS media containing various BAP and IAA combinations (*GI* Growth increment, relative mass increment after 4 weeks)

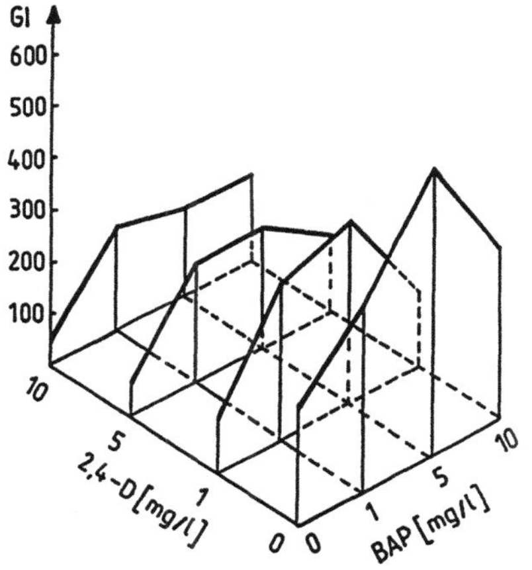

Fig. 6. Tissue mass increments in agar cultures on LS media containing various BAP and 2,4-D combinations (*GI* as in Fig. 5)

casein hydrolysate on the increments of tissue mass, without alteration of their appearance, was observed. The changes in appearance occurred on media containing maize milk.

The results of our study also indicate that the tissues of *A. majus* L. have limited requirements of vitamin contents in the culture media. The tissues grew much better when only thiamine hydrochloride was present than in the presence of the entire complex of vitamins (Fig. 7).

Effect of Precursor on Biomass Increments. In cultures with umbelliferone, a gradual reduction in dry mass increment was seen along with the increase in concentration of this precursor in the medium. The cultures, however, showed no significant differences between increments in number of cells in the culture cycles.

Effect of Light Conditions on Biomass Increments. It was found that the tissues of *A. majus* L. grow well both under constant light of about 900 lx intensity and under absolute darkness. Under those utterly different light conditions, the increments were equal. At the same time, it was proved that the effect of light on biomass increase was linked to concentration of growth substances in the culture media (Fig. 8). In media variants containing BAP/2,4-D combinations, a stimulating effect of light on tissue increments was evident. Elimination of light slightly increased the increments on the media with BAP-NAA combinations. The tissues growing on media with BAP-IAA and BAP-IBA combinations were variably affected by light conditions, depending on the proportions between kinin and auxin concentrations in the media. Such a variable effect of light conditions on growth of *A. majus* tissues, depending on composition and concentrations of growth substances in the media, is perhaps a combined result of both factors which simultaneously affect growth processes of tissues.

Effect of Culture Type on Biomass Increments. The biomass increments were high in agar cultures, as well as in stationary liquid and suspension cultures. The increments found in the first two types of cultures were identical for cultures growing on media of the same composition and kept under the same light conditions.

In suspension cultures, the cell-number increments (Fig. 9) and dry-mass percentages were also similar to the average values normally obtained in culture cycles.

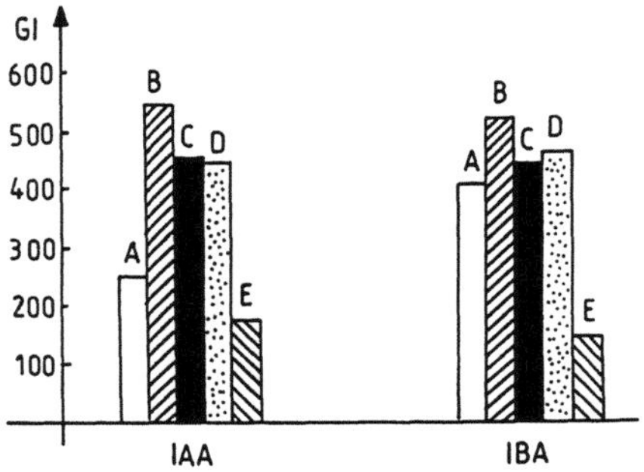

Fig. 7. Tissue mass increments in agar cultures on LS media with nutritive complex addition (*GI* as in Fig. 5). *A* basic medium (v.5); *B* with coconut water (v.18); *C* with maize milk (v.19); *D* with casein hydrolysate (v.20); *E* with vitamin complex (v.21); v. = variant of LS medium (cf. Table 1)

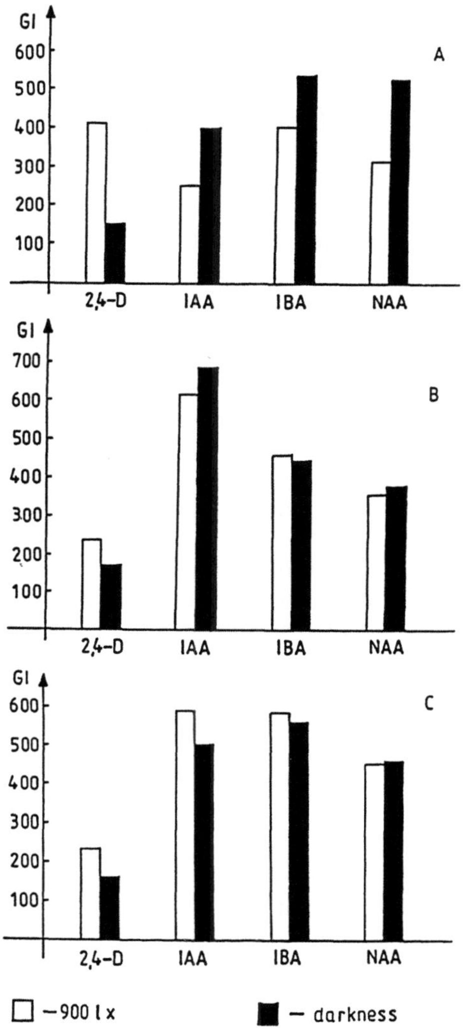

Fig. 8. Effect of light on tissue mass increments in agar cultures on LS media (*GI* as in Fig. 5). *A* v.5; *B* v.2; *C* v.6; v. = variant of LS medium; (all variants with BAP; cf. Table 1)

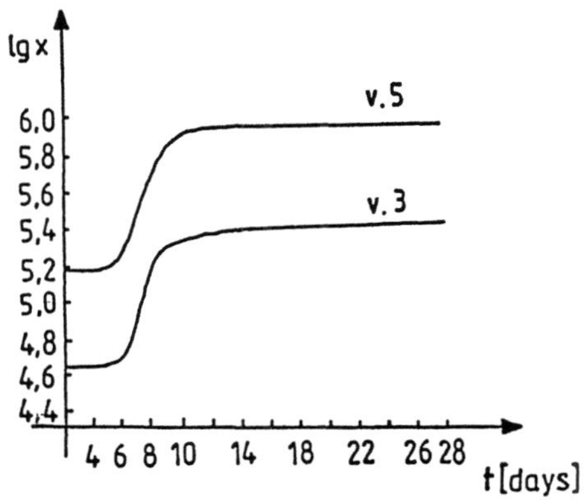

Fig. 9. Suspension cultures – relationship between logarithm of cell number (x) in 1 ml and the duration of culture (t), $\lg(x) = f(t)$; v. = variant of LS medium, v.3-BAP and IAA, v.5-BAP and IBA (cf. Table 1)

2.3 Biosynthesis of Secondary Metabolites

The studies of biosynthesis of secondary metabolites included the quantitative analysis of their contents in material from cultures and also covered structural studies of the metabolites thereby isolated.

2.3.1 Quantitative Analysis

Material. The analysis covered material obtained under various culture conditions to determine their effect on the biosynthesis of secondary metabolites. The following materials were studied:

1. Tissue material obtained from agar cultures under light condition; on 32 variants of LS medium containing the combinations of BAP with 2,4-D, IAA, IBA, NAA (variants 1–8) – passages 9–13, and also 2 variants of LS medium with BAP and IBA (variant 5), and with BAP and IAA (variant 2) – passages 23–25;
2. Tissue material obtained from agar cultures (passages 9–13) under darkness; on 12 variants of medium containing combinations of BAP with 2,4-D, IAA, IBA, NAA (variants 2, 5, 6);
3. Tissue material obtained from suspension cultures (passages 9–13) on 3 variants of LS medium containing combinations of BAP and IAA (variants 2, 3), and BAP and IBA (variant 5);
4. Media from the above suspension cultures;
5. Tissue material obtained from suspension cultures with umbelliferone addition (passages 10–12) on LS medium variants containing BAP and IBA (variants 22–25);
6. Media from the above suspension cultures with umbelliferone;
7. The fruits of *Ammi majus* L. from which the tissue culture was derived (as a comparison material);
8. Seedlings obtained from these fruits under sterile conditions (as a comparison material).

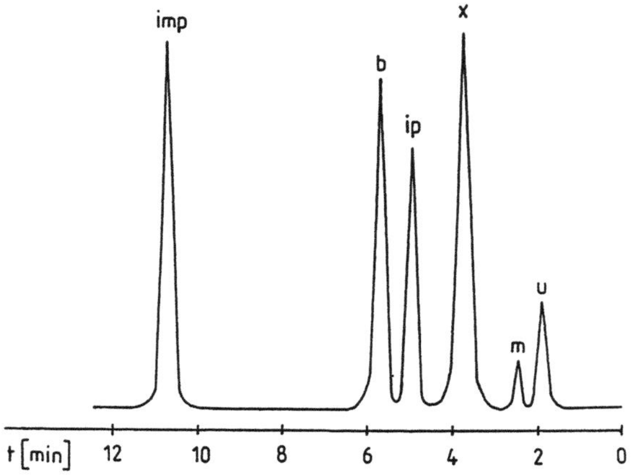

Fig. 10. Separation of 0.1% ethanol solutions of standards in HPLC technique. RP-18, water: methanol 30:70 v/v, $\lambda = 254$ nm; *u* umbelliferone; *m* marmesin; *x* xanthotoxin; *ip* isopimpinellin; *b* bergapten; *imp* imperatorin

Method. Dried, pulverized tissue material was thoroughly extracted with 96% ethanol. After purification at the latest stage, ethanol solutions suitable for direct analysis by HPLC method with reversed phases were obtained. The media were extracted with chloroform. Dry residue after evaporation was dissolved in 96% ethanol. Analyses were carried out by high-pressure liquid chromatography under the following conditions: Pump Liquochrom OE 307 (Labor MIM, Budapest); UV detector $\lambda = 253.7$ nm; column (Labor MIM, Budapest), Chromsil C-18; water: methanol (30:70 v/v) (Ekiert and Ekiert 1987). The analyses covered six compounds: bergapten, xanthotoxin, isopimpinellin, imperatorin, marmesin and umbelliferone (Fig. 10).

2.3.2 *Quantitative Analysis Results*

Effect of Growth Substances on Biosynthesis of Metabolites. High variability, both qualitative and quantitative, was found among metabolites in the material studied (Fig. 11), associated with various content of growth substances in the media, from trace quantities

> up to 22 mg% for bergapten, 6.67 mg% for xanthotoxin; 28.57 mg% for isopimpinellin; 22.11 mg% for imperatorin; 34.48 mg% for marmesin; and 34.90 mg% for umbelliferone (Table 2).

Among the compounds studied, bergapten was the only metabolite present in the tissue material obtained from all the variants of media studied. The best biosynthetic medium for all furanocoumarins studied under this project was an LS medium with 10 mg/l BAP and 10 mg/l 2,4-D.

Effect of Light on Biosynthesis of Metabolites. Light conditions exerted variable effects on biosynthesis of various metabolites. In the case of bergapten, the effect produced was evident activation, while for marmesin, isopimpinellin, and imperatorin, the biosynthesis was markedly suppressed in darkness. Umbelliferone and xanthotoxin synthesis was affected in a variable way, depending on the chemical composition of the culture media.

Effect of Type of Culture on Biosynthesis of Metabolites. Again, both quantitative and qualitative diversification of metabolites contents were found in the material analyzed. When compared with identical agar media, suspension cultures on media containing BAP and IAA combinations (Fig. 11), had tissue fractions containing greater quantities of furanocoumarines; their composition also included xanthotoxin and imperatorin. On media with BAP and IBA, the range of metabolites was reduced. It was, however, a universal observation that the conditions in suspension cultures supported the accumulation of bergapten. The tissue fractions of these cultures contained 1.7–5.6 times higher contents of this metabolite than in corresponding agar cultures.

The degree of release of metabolites into medium was also variable. A high proportion of this fraction was noted for marmesin and isopimpinellin (57.9–100%), and much more variable for umbelliferone and bergapten (in the

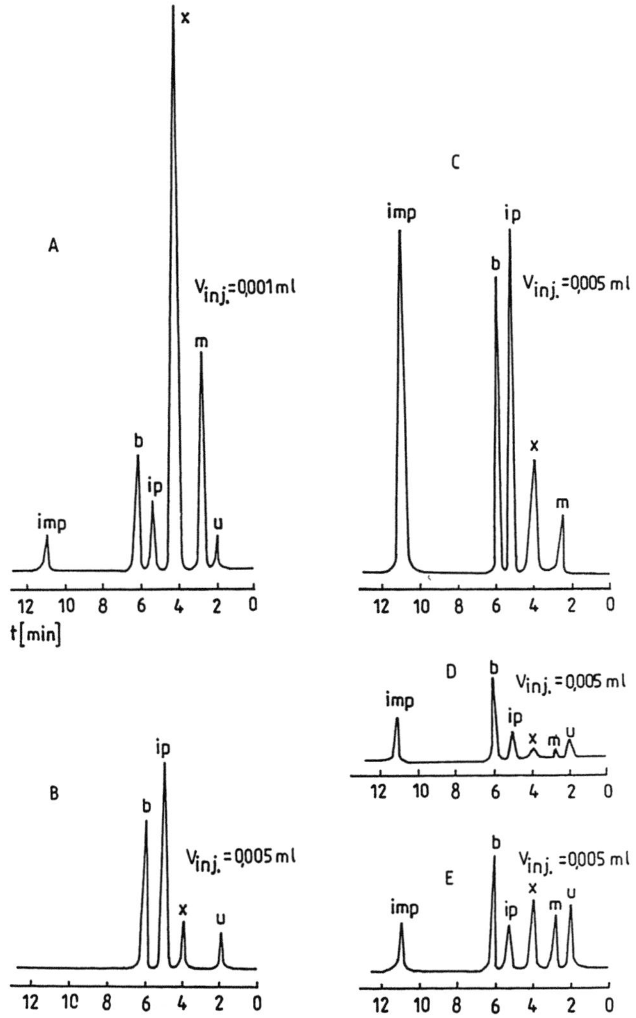

Fig. 11A–E. Examples of separations by HPLC technique. Extracts: **A** fruits; **B** seedlings; **C–E** tissue material from in vitro culture (**C** agar culture, BAP and 2,4-D, v.8; **D** agar culture, BAP and IBA, v.3; **E** suspension culture, BAP and IAA, v.2); v. = variant of LS medium (cf. Table 1); $V_{inj.}$ injection volume; *u*, *m*, *x*, *ip*, *b*, *imp* as in Fig. 10

range 34.0–82.5%). Imperatorin and xanthotoxin were not found at all in media fractions.

Effect of Umbelliferone on Biosynthesis of Metabolites. Increase in the content of bergapten was found in both tissue and media fractions along with the increase in concentration of precursor in the media. Xanthotoxin and marmesin were also found among metabolites of media fractions.

Table 2. Contents of metabolites in fruits of *Ammi majus* L., in seedlings, and their maximum contents obtained in in vitro culture

Metabolites	Content (mg %)		
	Fruits	Seedlings	Tissue material[a]
Bergapten	43.30	10.90	25.16
Imperatorin	10.90	–[b]	22.11
Isopimpinellin	29.67	17.18	28.57
Marmesin	610.69	–	34.48
Xanthotoxin	172.06	2.90	6.67
Umbelliferone	40.30	8.40	35.83

[a] Tissue material obtained under various culture conditions.
[b] Content < 0.001 mg %.

Effect of Age of Culture on Biosynthesis of Metabolites. The analysis of material from the cultures kept for about 2 years (passages 23–25) showed retained ability for biosynthesis of the metabolites studied.

2.3.3 *Structural Studies of Metabolites*

Five metabolites occurring in greater quantities in the extracts from cultures were isolated by using preparative TLC technique. These compounds were designated as A, B, C, D, and E.

Isolated metabolites were then subjected to spectral analysis, by obtaining UV, IR, and MS spectra. UV spectra in 96% ethanol were obtained in a Specord UV-VIS spectrophotometer, IR spectra in paraffin oil in a Specord 75-IR spectrophotometer (both VEB Zeiss, Jena), and MS spectra at 15 and 70 eV in a LKB-2091 mass spectrophotometer (Sweden).

Spectral analysis confirmed the identity of A compound with umbelliferone, and the identity of B compound with bergapten. It was also proved that C, D, and E compounds were furanocoumarins of 386, 314, and 342 molecular masses, respectively. The spectral data compared with those known from literature indicated that C, D, and E compounds were not identical with any of the about 20 furanocoumarin compounds known to occur in native plants. These compounds, along with umbelliferone and bergapten, were present in culture material obtained under almost any conditions.

3 Conclusions and Prospects

1. Tissues of *Ammi majus* L. grow satisfactorily under various conditions of in vitro cultures, so they can be used, for instance, in studies on the biosynthesis of secondary metabolites.

2. Secondary metabolism of *A. majus* tissues is markedly changed under conditions of in vitro culture. Both qualitative and quantitative deviations from the secondary metabolites content in comparison to that of native plants are found.
3. The tissues in in vitro cultures show the ability for endogenic biosynthesis of those coumarin compounds characteristic for native plants: bergapten, imperatorin, isopimpinellin, xanthotoxin, marmesin, and umbelliferone, although their contents differ from those found in native plants.
4. Under in vitro culture conditions, some new furanocoumarin compounds are produced (compounds C, D, and E) that are not known in the metabolism of the native plant.
5. Tissue culture of *A. majus* L. can offer a potential source of furanocoumarins: bergapten, isopimpinellin, and imperatorin, the maximum contents of these furanocoumarins found in the tissue culture material, equal or exceed those found in fruits from which tissue cultures were derived (Table 2).
6. Tissue culture of *A. majus* L. may provide a source for new furanocoumarin compounds (compounds C, D, and E), hence they are also synthesized in great quantities.
7. The conditions in cultures (composition of growth substances, light conditions, type of culture) affect the level of biosynthesis of various metabolites. There is a possibility of stimulation of biosynthesis of particular furanocoumarins by changing the conditions of tissue culture.

4 Protocol

Hypocotyl parts of sterile seedlings constituted the best explants for establishment of culture. Linsmaier and Skoog's (1965) medium, i.e., LS, containing NAA (2 mg/l) and BAP (2 mg/l) was the best for callus initiation. For the growth of tissue culture, LS medium containing IBA (1 mg/l), and BAP (1 mg/l) or IAA (10 mg/l) and BAP (1 mg/l) (Fig. 5), agar culture, and light regime (900 lx) provided the most effective conditions. LS medium containing 2,4-D (10 mg/l) and BAP (10 mg/l), agar culture, and light regime (900 lx) provided the best conditions for the production of furanocoumarin compounds.

Acknowledgements. I would like to express my sincere thanks to Prof. S. Brown (Trent University, Peterborough), Prof. G. Rodigiero (Universitá di Padova), and Dr. B. Rozkrutowa (Polish Academy of Sciences, Kraćow) for the standards of furanocoumarins; Dr. H. Bartoń, Dr. L. Ekiert, and Dr.hab. J. Mokrosz (Medical Academy, Kraćow) for their assistance in that part of the research involving chemistry; and Dr. R. Tertil (Jagiellonian University, Kraćow) for translating this chapter into English.

References

Abdel Hay FM, Abu-Mustafa EA, Fayez MBE (1966) Isolation of isopimpinellin from the fruits of *Ammi majus* L. Natural coumarins. Part IV. Naturwissenschaften 53: 406–406

Abu-Mustafa EA, Fayez MBE (1961) Natural coumarins I. Marmesin and marmesinin, further products from the fruits of *Ammi majus* L. J Org Chem 26: 161–166

Abu-Mustafa EA, Badran N, Fayez MBE, Starkowsky NA (1958) Isolation of marmesin from the fruits of *Ammi majus* Linn. Nature 182: 54–55

Abu-Mustafa EA, El-Bay FKA, Fayez MBE (1968) Isoimperatorin, a further constituent of the fruits of *Ammi majus* L. Recl Trav Chim Pays-Bas 87: 925–928

Abu-Mustafa EA, El-Bay FKA, Fayez MBE (1975a) Ammirin, a new coumarin constituent from *Ammi majus* L. fruits. Naturwissenschaften 62: 39–40

Abu-Mustafa EA, El-Bay FKA, Fayez MBE (1975b) Alloimperatorin from *Ammi majus* fruits. Naturwissenschaften 62: 40–40

Akačić B, Kuštrak D (1958) Pokusne kulture *Ammi visnaga* (L.) Lam. i *Ammi majus* L. Acta Pharm Jugosl 8: 101–112

Bajaj YPS (ed) (1988) Biotechnology in agriculture and forestry, vol 4. Medicinal and aromatic plants I. Springer, Berlin Heidelberg New York Tokyo

Bajaj YPS (ed) (1989) Biotechnology in agriculture and forestry, vol 7. Medicinal and aromatic plants II. Springer, Berlin Heidelberg New York Tokyo

Ben-Hur E, Pill-Soon S (1984) The photochemistry and photobiology of furocoumarins (psoralens). Adv Radiat Biol 11: 131–171

Beyrich Th (1981) Furocumarine, Wiss Z E- M- Arndt-Univ Greifswald 30: 25–28

Blažek Z (1966) Farmakognosie rostlinných časti *Ammi majus* L. Farm Obz 35: 495–504

Blažek Z, Stary F (1965) Qualitative und quantitative Veränderungen der Furocumaringehalte bei *Ammi majus* L. im Laufe der Vegetation. Arch Pharm 35: 117–127

Blažek Z, Stary F (1966) Qualität und Quantität der Furocumarine der Art *Ammi majus* L. im Laufe der Blüte und Reifezeit. Sci Pharm 34: 97–106

Butcher DN (1977) Secondary products in tissue cultures. In: Reinert J, Bajaj YPS (ed) Applied and fundamental aspects of plant cell, tissue, and organ culture. Springer, Berlin Heidelberg New York, pp 668–693

Chen M, Stohs SJ, Staba EJ (1969) The biosynthesis of visnagin from 2 ^{14}C-acetate by *Ammi visnaga* suspension cultures and the metabolism of ^{14}C-visnagin and ^{14}C-khellin by *Ammi visnaga* and *Ammi majus*. Lloydia 32: 339–348

Czygan FC (1984) Pflanzliche Gewebe- und Zellkulturen als Arzneistoffproduzenten. In: Czygan FC (ed) Biogene Arzneistoffe. Vieweg, Braunschweig Wiesbaden, pp 85–105

Eisenreichová E, Bučková L, Kosáková Ĺ, Tomko J (1980) Obsahové látky *Ammi majus* L. Farm Obz 49: 503–506

Ekiert H (1986a) Tissue culture of *Ammi majus* L. and its metabolites. Acta Pol Pharm 43: 634–636

Ekiert H (1986b) Metabolity hodowli tkankowej *Ammi majus* L. In: Mat Zjazd XIII Naukowego Zjazdu PTFarm, Katowice Poland, p 132

Ekiert H (1989a) The influence of some growth substances on biosynthesis of coumarin compounds in tissue culture of *Ammi majus* L. In: Proliferation, differentiation and maturation of cells and tissues. Abstr 22nd Symp Polish Histochem Cytochem Soc, Szczecin Poland, p 70

Ekiert H (1989b) Wpływ warunków hodowlanych na biosyntezę bergaptenu w kulturach tkankowych *Ammi majus* L. In: Mat Zjazd XIV Naukowego Zjazdu PTFarm, Wrocław Poland, p 32

Ekiert H (1990) Furanocoumarins in tissue culture of *Ammi majus*. Planta Med 56: 572–572

Ekiert H, Ekiert L (1987) Rozdziały pochodnych psoralenu w materiale roślinnym techniką TLC i HPLC. In: Chromatograficzne metody badania związków organicznych. Mat Zjazd XI Środowiskowego Seminarium Naukowego, Katowice Poland, pp 60–61

Fahmy IR, Abu-Shady H (1947) *Ammi majus* Linn. – pharmacognostical study and isolation of a crystalline constituent ammoidin. Q J Pharm Pharmacol 20: 281–281

Fahmy IR, Abu-Shady H (1948) Isolation and properties of ammoidin, ammidin and majudin. Q J Pharm Pharmacol 21: 499–503

Fahmy IR, Abu-Shady H, Schönberg A, Sina A (1947) A crystalline principle from *Ammi majus* L. Nature 160: 468–469

Gheorghiu M, Pavel M, Tarpo E, Contz O, Grintescu A, Georgescu V (1965) Contributii la studiul speciei *Ammi majus* L. aclimatizată. Farmacia (Bucharest) 13: 541–546

Grewal S, Sachdeva U, Atal CK (1976) Regeneration of plants by embryogenesis from hypocotyl cultures of *Ammi majus* L. Indian J Exp Biol 14: 716–717

Hamerski D, Matern U (1988a) Biosynthesis of psoralens. Psoralen 5-monooxygenase activity from elicitor-treated *Ammi majus* cells. FEBS Lett 239: 263–265

Hamerski D, Matern U (1988b) Elicitor-induced biosynthesis of psoralens in *Ammi majus* L. suspension cultures. Microsomal conversion of demethylsuberosin into (+)marmesin and psoralen. Eur J Biochem 171: 369–375

Hamerski D, Beier RC, Kneusel RE, Matern U, Himmelspach K (1990a) Accumulation of coumarins in elicitor-treated cell suspension cultures of *Ammi majus*. Phytochemistry 29: 1137–1142
Hamerski D, Schmitt D, Matern U (1990b) Induction of two prenyltransferases for the accumulation of coumarin phytoalexins in elicitor-treated *Ammi majus* cell suspension cultures. Phytochemistry 29: 1131–1135
Hegi G (1906–1931) Illustrierte Flora von Mittel-Europa, vol V/2. A Pichler's Witwe und Sohn, Wien, pp 1170–1173
Hoppe A (1975) Drogenkunde, vol I, Walter de Gruyter, Berlin New York, pp 66–69
Ivie WG (1978) Linear furocoumarins (psoralens) from the seed of Texas *Ammi majus* L. (Bishop's weed) J Agric Food Chem 26: 1394–1403
Kibalchich PN, Nikonov GK, Chernobai NKh, Ivashin DS (1960) O kulture ammi bolshoi kak istochnika poluchenya furokumarinov. Med Promst' SSSR 14: 23–26
Kohlmünzer S, Wacław R (1962) Badania fitochemiczne owoców *Ammi majus* L. Diss Pharm 14: 69–81
Kohlmünzer S, Wacław-Rozkrutowa B (1963) Izolowanie ksantotoksyny i marmezyny z owoców *Ammi majus* L. Acta Pol Pharm 20: 321–327
Korbelar J, Endris Z (1981) Naše rostliny v lékarštvi. Zdravotnicke Nakladatelnosti, Avicenum Praha, pp 314–315
Linsmaier EM, Skoog F (1965) Organic growth factor requirements of tobacco tissue cultures. Physiol Plant 18: 100–127
Murashige T, Skoog F (1962) A revised medium for rapid growth and bioassays with tobacco tissue cultures. Physiol Plant 15: 473–497
Nielsen BE (1970) Coumarins of umbelliferous plants. Kandrup and Wunsch, Copenhagen
Nielsen BE (1971) Coumarin patterns in the Umbelliferae. In: Heywood VH (ed) Biology and chemistry of Umbelliferae. Acad Press, London, pp 325–336
Nikonov GK (1965) Furokumarini ammi bolshoi. Med Promst' SSSR 19: 21–24
Nowiński M (1980) Dzieje upraw i roślin leczniczych. PWRL Warszawa, p 108, 282
Pathak MA, Parrish JA, Fitzpatrick TB (1981) Psoralens in photochemotherapy of skin diseases. Farm Ed Sci 36: 479–491
Plewig G, Hölzle E, Lehmann P (1986) Phototherapy for photodermatoses. Curr Probl Dermatol 15: 254–264
Podlewski JK, Chwalibogowska-Podlewska A (1986) Leki współczesnej terapii, PZWL, Warszawa, pp 384–385
Ramadan S (1982) The *Ammi majus* plant. Hamdard 25: 32–35
Rodigiero G (1985) Hyperpigmentation induced by furocoumarins. Farm Ed Prat 40: 172–186
Rodigiero G, Dall'Acqua F (1986) Present aspects concerning the molecular mechanisms of photochemotherapy with psoralens. Drugs Exp Clin Res 12: 507–515
Rodigiero G, Dall'Acqua F, Pathak MA (1984) Photobiological properties of monofunctional furocoumarin derivatives. In: Smith KC (ed) Topics in photomedicine. Plenum, New York, pp 319–398
Roenigh HH Jr (1984) Effectiveness of psoralens in mycosis fungoides. In: Photochemotherapeutic aspects of psoralens. Natl Cancer Inst Monogr 66: 179–183
Schönberg A, Sina A (1948) Xanthotoxin from the fruits of *Ammi majus*. Nature 161: 481–482
Schönberg A, Sina A (1950) Experiments with xanthotoxin and imperatorin obtained from the fruits of *Ammi majus* L. J Am Chem Soc 72: 4826–4828
Sehgal CB (1972) In vitro induction of polyembryony in *Ammi majus* L. Curr Sci 41: 263–264
Soine TO (1964) Naturally occurring coumarins and related physiological activities. J Pharm Sci 53: 231–264
Staba EJ (1985) Milestones in plant tissue culture systems for the production of secondary products. Lloydia 48: 203–209
Starkowsky NA, Badran N (1958) Ammajin, a new constituent of *Ammi majus* (L.). J Org Chem 23: 1818–1820
Tandon S, Rastogi RP (1979) Recent advances in naturally occurring coumarins. Ind J Sci Res 38: 428–441
Tarpo E, Gheorghiu M, Contz O (1965) Studiul dozarii furanocumarinelor din fructele de *Ammi majus* L. Farmacia (Bucharest) 13: 331–335
Trojánek J, Hodková J, Čekan Z (1961) Über die Furocoumarine der Früchte *Ammi majus* L. Planta Med 9: 200–211

Turjanmaa K, Salo H, Reunala T (1985) Comparison of trioxsalen bath and oral methoxsalen PUVA in psoriasis. Acta Dermatol Venereol (Stockh) 65: 86–88
Tutin TG, Heywood VH, Burges NK, Moore DM, Valentine DH, Walters JH, Webb DA (ed) (1968) Flora Europaea, vol II, Cambridge Univ Press, Cambridge, p 353
Vuorela H, Törnquist K, Nyiredy Sz, Sticher O, Hiltunen R (1988) Calcium antagonistic activity of the main furocoumarins from *Peucedanum palustre*. In: Abstr 36-th Yearly Congr Soc Med Plant Res, Freiburg, Thieme, Stuttgart, pp 44–45
Wolff FA, Tessa TV (1986) Clinical pharmacokinetics of methoxsalen and other psoralens. Clin Pharmacokinet 11: 62–75

II *Bergenia crassifolia* (L.) Fritsch (Bergenia): Micropropagation and Arbutin Contents

M. Furmanowa and L. Rapczewska[1]

1 Introduction

Bergenia crassifolia (L.) Fritsch (Fig. 1), a species in the *Bergenia* genus belongs to the family *Saxifragaceae*, the order Rosales. For more than 100 years the plant has been known in Asia as a valuable raw material, a source of tannins and pigments.

Fig. 1. Intact plants of flowering *Bergenia crassifolia* (L.) Fritsch growing in the Botanic Garden of the Institute of Medicinal Plants in Poznań (third year of vegetation)

[1] Department of Biology and Pharmaceutical Botany, Medical Academy, Banacha 1, 02-097 Warsaw, Poland

Biotechnology in Agriculture and Forestry, Vol. 21
Medicinal and Aromatic Plants IV (ed. by Y.P.S. Bajaj)

Apart from that, *Bergenia crassifolia* has been used as a medicinal and ornamental plant. Due to its rich and varied chemical composition (arbutin, tannins, bergenin) the species continues to be the object of pharmaceutical and pharmacological studies (Borissowa 1956).

In the light of research confirming the usefulness of this plant as a source of chemical compounds, it has become increasingly obvious that plant tissue culture should be employed to provide a method of rapid multiplication of *Bergenia crassifolia* as an alternative to propagation from seeds. The second part of this chapter deals with arbutin determination in regenerated plants. The observations are based on the results of the experiments carried out by the authors.

1.1 Systematics and Distribution of *Bergenia* Plants

The genus *Bergenia* Moench (Meth. pl. 1794, 664) which is also known in the literature under the synonymous *Geryonia* Schrank, *Megarea* Haw., *Eropheron* Tausch., *Piarophylla* Raf. and *Saxifraga* L. is said to consist of 11 species (Borissowa 1956) although the Index Kewensis (1895–1974) lists 19 species and Engler (1964) 8 species. According to Yeo (1966) the genus *Bergenia* was named in memory of Carl August von Bergen, author of *Flora Francofurtana.*

Bergenia crassifolia (L.) Fritsch is also known by the following names:

- *Bergenia bifolia* Moench.,
- *Bergenia cordifolia* (Haw.) A.Br.,
- *Bergenia media* (Haw.) Engl.,
- *Bergenia orbicularis* Stein,
- *Bergenia sibirica* Hort.,
- *Bergenia aemula* Tausch.,
- *Saxifraga crassifolia* L.,
- *Saxifraga aemula* Tausch.,
- *Saxifraga cordifolia* Haw.,
- *Saxifraga cuneifolia* Hort., non L.,
- *Saxifraga orbicularis* Stein,
- *Saxifraga sibirica* Hort., non L.,
- *Saxifraga macrophylla* Clement. Marot,
- *Megasea crassifolia* Haw.,
- *Megasea cordifolia* Haw.,
- *Megasea media* Haw.,
- *Megasea orbicularis* Hort.,
- *Geryonia crassifolia* Schrank,
- *Geryonia cordifolia* Schrank,
- *Piarophylla cordifolia* Rafin.,
- *Piarophylla elliptica* Rafin.

The detailed systematics of the genus *Bergenia* was compiled by Borissowa (1956). She described 11 species, their various names, morphology, distribution, and economic uses. Yeo (1966) offered a revision of the genus *Bergenia* and gave the key to the determination of only six *Bergenia* species and eight hybrids. The systematics

Table 1. Classification of *Bergenia* species

Series	After Borissowa (1956)	After Yeo (1966)
1. Crassifoliae Boriss. (1954)	1. *B. crassifolia* (L.) Fritsch[a] (= *B. sibirica* Hort.) *B. crassifolia* var. *cordifolia* (Haw.) Boriss. = *B. cordifolia* (Haw.) A.BC *B. crassifolia* var. *baicalensis* Boriss. 2. *B. pacifica* Kom.	1. *B. crassifolia* (L.) Fritsch *B. crassifolia* (L.) Fritsch var. *pacifica* Kom. 2. *B. cordifolia* (Haw.) Sternb.
2. Intermediae Boriss. (1954)	3. *B. purpurascens* (Hook.f.et Thoms.) 4. *B. Delavayi* (Franchet) Engler	3. *B. purpurascens* (Hook.f.et Thoms.) Engl. 3a. *B. ugamica* V. Pavlov.
3. Ciliatae Boriss. (1954)	5. *B. himalaica* Boriss. (nomen novum) 6. *B. ciliata* (Haw.) Sternb. 7. *B. stracheyi* (Hook.f.et Thoms.) Engl. 8. *B. Gorbunovii* B. Fedtsch. 9. *B. hissarica A. Boriss.* 10. *B. orbata* Stein. 11. *B. spathulata* (Nagels nomen) ex quillaumin	4. *B. stracheyi* (Hook .f.et. Thoms.) Engl. 5. *B. hissariaca* Boriss. 6. *B. ciliata* (Haw.) Sternb.

[a] The species are classified under other names; for more information on the systematics and distribution of the species, see Borissowa (1956).

of the *Bergenia* species is outlined in Table 1. Apart from the three series presented in Table 1, there are many hybrids cultivated in botanic gardens. A mixed series of hybrids, obtained from crossing many species, was introduced in 1895 under the name *Bergenia* hybrida by Georg Arends of Wuppertal-Ronsdorf, Germany.

Bergenia is a typical East-Asian genus distributed in two regions, one to the north of and one to the south of the Central Asian Plateau. The series Crassifoliae, to which *Bergenia crassifolia* belongs, is confined to the northern region of the Central Asian Plateau, which is divided into three separate parts; the area extending from the Tomsk region to Transbaikal; the mountains of the Maritime Territory; and the mountains of North Korea. The plant is widely distributed throughout the USSR and has many garden forms. In its natural condition it grows even at an altitude of 2500 m. In many European countries it is cultivated as an ornamental plant, especially in rock gardens. *Bergenia* is a perennial herb; the leaves of most species last two seasons. The leaves aciliate petiolar sheath does not form a ligule and is 15–30 cm thick, with shining, broadly elliptic-obovate, petiole sheathing at the base. The flowering stems are up to 30 cm long, and are leafless, with numerous hanging flowers in a rather dense panicle. The flowers in each inflorescence are opened over a long period; a few usually appear before the elongation of the scape. Seeds take about 2 months from the time of pollination to ripen. The creeping rhizomes are ca. 2 cm in diameter. Webb (1964) mentions only one species, *Bergenia crassifolia*, stressing the fact that *Bergenia* is like *Saxifraga* in floral structure, but of different habit. In the USSR, the common name of Bergenia is *badan* (Borissowa 1956).

1.2 Conventional Propagation and Importance

Bergenia plants are propagated by seeds or by rhizome cuttings. Propagations by seeds is time- and labor-consuming, as this species has a slow rate of natural propagation. The plants developed from seeds can be transferred from the pot to the field in the second year after sowing. The leaves can be gathered starting from the second or third year of plant cultivation in the field. The leaves are gathered in autumn, when the arbutin content is at its highest, and should be dried indoors. Raw material dried in the open air has a lower arbutin content and turns brown. The yield of dried leaves from a several years plantation in the USSR is 3–4 t/ha (Tyszyńska-Kownacka 1979).

In folk medicine, the *Bergenia* plants have been used to prepare bacteriostatic and anti-inflammatory drugs. Traditional Chinese medicine claims to effectively treat coughs and other respiratory disorders with *B. sibirica* (= *B. crassifolia*). The major antitussive constituent is bergenin (Lewis 1977). In some parts of the USSR the dry leaves of the second year are used for making a beverage, named "Siberian tea". The rhizomes of *B. crassifolia* and *B. pacifica* have been used for years in Siberian tanneries. Some fractions of tannins are used for dying cotton and woolen fabrics green and black. Borissowa (1956) in her review of industrial uses of *Bergenia* plants concentrated on the importance of these plants as a source of tannins and arbutin.

1.3 Phytochemical Studies

Of the numerous species of the genus *Bergenia*, most phytochemical studies have been carried out on *B. crassifolia* and *B. cordifolia*. They have been found to contain various chemical compounds, the most important ones being the glucoside of hydroquinone, arbutin, tannins, and bergenin (Fig. 2). The detailed chemical composition of the *Bergenia* species was described by Ostrowska and Gorecki (1988). From the chemotaxonomical point of view, the family *Saxifragaceae* is characterized by the presence of phenols and polyphenols. Only the *Bergenia* species can synthesize arbutin, which is very important in pharmacy and medicine, as an antiseptic used in urinary tract infections. The majority of phytochemical investigations concern the arbutin content of *Bergenia* plants, since this compound, in great demand due to its medicinal properties, is not frequently found in plants. The leaves of *Arctostaphylos uva-ursi* (Ericaceae) are the typical arbutin-containing raw material, while those of *Vaccinium vitis-idaea*, another species of the same family, demonstrate less activity. These natural drugs are very much in demand, but the supply, i.e:, the number of plants containing arbutin, is limited. *Arctostaphylos uva-ursi*, which grows on heaths, rocky ground, and in open woods, is often under partial environmental protection. Therefore, it is very important to find some other plant material which could be used in place of *Arctostaphylos uva-ursi* and *Vaccinium vitis idaea*.

The leaves of the *Bergenia* species are an especially important source of arbutin. The content of this compound in old leaves may reach up to 22% (Borissowa 1956). Arbutin was first isolated from the fresh *Bergenia crassifolia* leaves by Tschitschibabin (1930). In contrast to the leaves of *Arctostaphylos uva-ursi* and *Vaccinium vitis idaea*, the yield of arbutin in the leaves of *Bergenia crassifolia* depends on the process of drying and storage. Elbanowska (pers comm) found that the best temperature for drying raw material is 60–50 °C for 5 h in a drying chamber of the *Leśniczanka* type, but the leaves of plant should be cut into long pieces (ca. 4 cm wide).

The isolation and quantitative determination of arbutin and tannins in *Bergenia crassifolia* have been studied by a number of authors. Friedrich and Wehnert (1973a) investigated the distribution of arbutin and bergenin in the *Bergenia* plants. Previously it had been generally believed that in *Bergenia*, arbutin was present only in the leaves, while bergenin was found solely in the rhizomes. Using the TLC method, the two authors isolated small amounts of arbutin from the *B. cordifolia* rhizomes and bergenin from the leaves. In their subsequent study, Friedrich and Wehnert (1973b) elaborated a new method of separation and

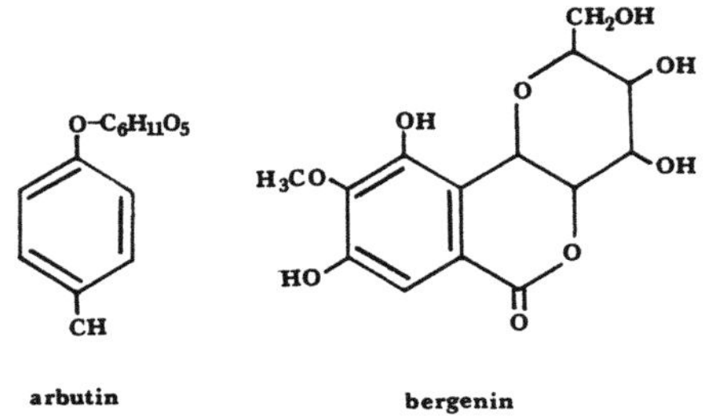

Fig. 2. Structure of arbutin and bergenin

quantitative determination of arbutin and bergenin, the two compounds which are present in the extracts of raw material. The highest yield of arbutin was found in the leaves of *B. crassifolia* gathered in June (14.10%) although a small amount (mean range 1.62–2.31%) was observed also in the rhizomes bergenin content was higher in the rhizomes (6.73–10.6%) as the leaves contained only 1.1–2.14%. Srepel et al. (1978) found 8.6% arbutin in the leaves and 5.2% in the stems.

The second important group of chemical compounds found in the *Bergenia* plants in considerable amounts are tannins, of which the species is also considered one of the main sources. The investigations carried out by Ostrowska and Gorecki (1988) allow us to compare the raw material derived from four species in which arbutin and tannins were analyzed. Their results show that the hybrid *B. crassifolia* × *B. cordifolia* is the best source of arbutin (9.1%) and tannins (17.5%). *Folium bergeniae* and *Folium uvae-ursi* contain similar amounts of arbutin (6.0–6.9%) and tannins (16.2–17.5%). However, possible introduction of *Bergenia* raw material to therapy will depend on the results of toxicological and pharmacological studies of bergenin. The chemical structure of bergenin was described by Hay and Haynes (1958), Pasternak and Durr (1958), Fujise et al. (1959), and Adamanis and Bederska-Plotkowiakowa (1960). The authors suggested to use raw material containing arbutin, e.g. the leaves of *Bergenia* plants, rather than *Folium uvae-ursi*.

The modified method of Friedrich and Wehnert (1973a, b) was used by Ostrowska et al. (1989) for the quantitative determination of bergenin in *B. crassifolia* and *B. cordifolia* leaves and rhizomes. For determination of bergenin in raw material and extracts, the spectrophotometric method elaborated by Ostrowska et al. (1989) was used. The bergenin content in *B. crassifolia* leaves was 1.03%, in *B. cordifolia* leaves 1.28%. In the rhizomes of *B. crassifolia* × *B. cordifolia*, 10.07% bergenin was found. If pharmacological and toxicological studies of this raw material confirm its medicinal properties, *B. crassifolia* and *B. cordifolia* could replace rare plants containing arbutin and tannins. *B. purpurascens* Engl., which grows in the mountains of southwest China at an altitude of 2000 m, is also a plant rich in tannin. This species contains bergenin and its gallates, 6-O-galloyarbutin and other phenolic compounds (Xin-Min et al. 1987).

Other chemical compounds found and isolated from the *Bergenia* species were described by Thieme et al. (1969), Bluhme and Frohne (1973), and Bohm et al. (1986)

In recent years, arbutin production by plant tissue culture has been investigated, Duŝkova and Dušek (1989) analyzed the secondary metabolites in *Arctostaphylos uva-ursi* tissue culture.

2 In Vitro Approaches

To the authors' knowledge, *Bergenia crassifolia* is at present the only species of the genus *Bergenia* studied in vitro. The other species of the *Saxifragaceae* family propagated in vitro include: *Ribes nigrum and Ribes rubrum* Zatyko and Simon 1988. Axillary buds were used as the explants for the micropropagation of these species. Preliminary reports on *B. crassifolia* micropropagation were presented by Furmanowa et al. (1981), Furmanowa and Rapczewaska (1982) and Bajaj et al. (1988).

2.1 Plant Propagation in Vitro

Clonal propagation of *B. crassifolia* plants in vitro was carried out using two methods:

1. Induction of adventitious shoots via the callus:
2. Organogenesis directly from sterile rhizome cuttings, taken from plantlets.

2.1.1 Micropropagation Via Callus

The culturing process was carried out in three stages:

1. Initiation of the callus culture:
2. Adventitious shoots formation:
3. Rooting of the shoots and their transfer to "in vivo" growth (Fig. 3).

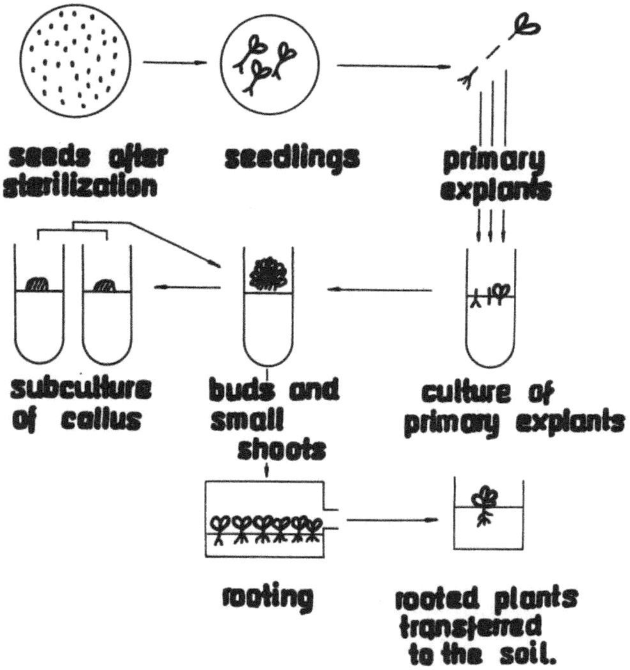

Fig. 3. Schematic representation of in vitro propagation of *Bergenia crassifolia* (L.) Fritsch via callus

Seeds of *Bergenia crassifolia* obtained from the Institute of Medicinal Plants in Poznań were surface-sterilized with 75% alcohol for 2 min, followed by treatment with 5% calcium hypochlorite solution for 10 min, and rinsed four times with sterile distilled water. Seeds placed in the Petri dishes germinated after 7–11 days at the temperature of 26 °C. The seedlings were cut into three parts: (1) cotyledons with shoot tip and part of hypocotyl; (2) hypocotyl; (3) root tip.

The explants were used for the callus induction. For the primary culture, three media were prepared in various modifications; Murashige and Skoog, MS (1962); Murashige's Lili Multiplication Medium, ML (Gibco Bio-Cult.); Murashige Cattleya, MC (Gibco Bio-Cult.). For rooting of the shoots three media were also employed; Miller (1967); Nitsch and Nitsch, NN (1969:); and MS medium.

Kinetin, BAP, NAA, IBA, IAA, and AS at various concentrations were used as growth regulators; 100 media combinations were prepared. The cultures were maintained at 25 ± 1 °C, 60% relative humidity under photoperiod 12 h/day and light intensity of 800–1000 lx, subculture every 4 weeks. Among primary explants used for initiation of the culture, the hypocotyl proved the best source of callus. The hypocotyl placed on the majority of the media produced a friable, light, yellow, or green callus after three weeks. Light stimulated callus growth. The effect of hormones on the growth of callus and its differentiation varied. Some of numerous combinations of MS, MC, ML and MN media demonstrate markedly stimulating effect on the growth of tissue.

A vigorous growth of callus was obtained on MS medium with 0.3 mg/l NAA, 1 mg/l BAP and 80 mg/l AS. On this medium the callus was friable and light-green.

Fig. 4. Adventitious shoots of *Bergenia crassifolia* (L.) Fritsch after 1 month of cultivation in MS liquid medium with 5 mg/l BAP and 2.5 mg/l IBA. Small leaves with long petioles are visible

The adventitious shoots developed well after transferring the callus onto two MS media containing: 3 mg/l BAP, 0.1 mg/l NAA, 80 mg/l AS; and 5 mg/l BAP, 2.5 mg/l IBA.

The shoots were rooted on the following four media: MS supplemented with 1 mg/l IAA and sucrose 5%; NN supplemented with 0.3 mg/l IBA and 10 mg/l AS; NN supplemented with 0.1 mg/l kinetin and 0.3 mg/l NAA; and NN supplemented with 0.1 mg/l kinetin and 0.5 mg/l IBA. NN medium with 0.1 mg/l kinetin and 0.3 mg/l NAA, proved very convenient since two processes of regeneration could be observed, i.e., adventitious shoot development after 1–3 weeks, and rooting of the shoots after 3–6 weeks of the same passage. The plantlets obtained on this medium could be transferred to the pots. Very good results with rooting of the shoots were obtained on the NN medium supplemented with 0.5 mg/l IBA, 0.1 mg/l kinetin, 10 mg/l AS and 40 g/l sucrose. This medium was used by Olszowska and Furmanowa (1986) for rooting all shoots of *Coluria geoides*. In an attempt to improve

Table 2. Composition of Murashige's lily multiplication medium used for micropropagation of *Bergenia* plants

Component	mg/l	Component	mg/l
NH_4NO_3	1650	KJ	0.83
KNO_3	1900	Na_2MoO_4	0.25
$CaCl_2$	332	$CuSO_4 \cdot 2H_2O$	0.025
$MgSO_4 \cdot 7H_2O$	370	$CoCl_2 \cdot 6H_2O$	0.025
KH_2PO_4	170	Adenine sulphate	80
$NaH_2PO_4 \cdot H_2O$	170	2IP[a]	3
Na_2EDTA	37.25	Inositol	100
$FeSO_4 \cdot 7H_2O$	27.85	NAA[b]	0.30
$MnSO_4 \cdot 4H_2O$	22.30	Thiamin HCl	0.40
$ZnSO_4 \cdot 7H_2O$	8.60	Sucrose	30000
H_3BO_3	6.20	Agar	70000

[a] BAP was used instead of 2IP.
[b] NAA and BAP were used at various concentrations.

Table 3. Effect of NAA and BAP in ML medium on growth and differentiation of *Bergenia crassifolia* (L.) Fritsch callus in primary culture after 4 weeks[a]

NAA (mg/l) \ BAP (mg/l)	0	0.5	1	2	4
0	–	–	–	–	–
0.5	–	±↑	+↑	+↑	±
1	–	±	+↑	+↑	±
2	±	±	±	±	±
4	±	±	±	±	±

[a] –, No callus induction; ±, callus induction on part of explant; +, callus covered whole explant; ↑, bud formation after 5 weeks of culture.

Fig. 5. Adventitious shoot formation from callus of hypocotyl origin cultured on ML medium containing 3 mg/l BAP, 0.1 mg/l NAA, and 80 mg/l AS, 6 weeks of culture in vitro

Fig. 6. Rooted plantlets of *Bergenia crassifolia* (L.) Fritsch after 1 month of cultivation on solid NN medium with 0.1 mg/l kinetin, 0.5 mg/l IBA, 10 mg/l AS and sucrose 40 g/l

the penetration of the growth regulators into the plant organ, MS liquid medium with 5 mg/l BAP and 2.5 mg/l IBA was prepared, but the results were not satisfactory. Although adventitious shoots proliferated well, after some days abnormal symptoms appeared such as the physiological phenomenon of vitrification, manifested as elongated leaves with a glassy appearance (Fig. 4). The plantlets ceased to grow after transferring to the pot.

Very interesting results were obtained using a ML medium supplemented with 3 mg/l BAP, 0.1 mg/l NAA. The ML medium contains AS 80 mg/l and 170 mg/l $NaH_2PO_4 \cdot H_2O$, which are not added to the MS medium (Table 2). In our modification, BAP was used instead of 2IP in the ML medium. On this medium the callus was friable and light green after 4 weeks. After a longer culture of 4–6 weeks,

Fig. 7. *Bergenia crassifolia* (L.) Fritsch plants regenerated in vitro 2 years after transferring to the soil

Table 4. Rooting of the shoots of *Bergenia crassifolia* (L.) Fritsch on various media

Medium	Kinetin (mg/l)	NAA (mg/l)	IAA (mg/l)	IBA (mg/l)	AS (mg/l)	Sucrose (mg/l)	Percentage of shoots rooted
Miller	0.1	1	–	–	–	30	58.0
Miller	0.2	2	–	–	–	30	95.5
Miller	0.2	–	2	–	–	30	64.3
Miller	10.0	2	–	–	–	30	0
NN	–	–	–	0.2	10	30	60.0
NN	–	–	–	0.3	10	30	79.0
NN	0.1	1	–	–	10	30	65.0
NN	0.1	1	–	–	–	30	76.0
NN	0.5	1	–	–	–	30	59.0
NN	0.1	–	–	0.5	10	40	100
NN	0.1	0.3	–	–	–	30	80.2
MS	–	–	1	–	–	50	81.5

the callus differentiation was observed as early as the first passage (Table 3). The callus inoculated on the same medium differentiated very fast and numerous adventitious bud shoots could be observed (Fig. 5). For the rooting of these shoots, a NN medium with 0.1 mg/l kinetin, 0.5 mg/l IBA, 10 mg/l AS, and 4% sucrose was the best (Fig. 6; Table 4). Using the media described, more than 5000 plantlets can be easily obtained from one seedling after four passages (5 months). These plantlets have four to five leaves and are transferred to the pots. After 2 months they can grow in the soil.

When the plant-tissue culture method via the callus is employed to propagate *Bergenia crassifolia*, the period of time necessary for plant regeneration is shorter than in the case of plant development from seeds (Fig. 7).

2.1.2 Organogenesis Directly Within Rhizome Cuttings

In the second culturing process, plantlets developed in vitro from the callus of hypocotyl origin, and after being cultured in Erlemeyer flasks for 2 months, were used as the source of explants. The microrhizome cuttings (0.5 cm) were taken from well-developed plantlets growing on the NN medium with 0.5 mg/l IBA, 0.1 mg/l kinetin, 10 mg/l AS and 40 g/l sucrose. From one plantlet, ca. two to five explants could be obtained. The explants were inoculated onto the ML medium supplemented with 3 mg/l BAP and 0.1 mg/l NAA. Adventitious bud development was observed after 3 weeks. One month later the shoots were rooted on the same medium. From one explant, five to eight plantlets were obtained. Apart from plantlets, a small amount of callus was observed, which differentiated into bud shoots (Table 5). Upon transferal to flower-pots, the recovered plantlets required high humidity which was provided by covering the plants with glass beakers for 2 weeks. Thus, production of the next generation of plantlets can be repeated, providing a new, clean source of explants. Using this method, this number of regenerated plantlets is smaller than that of propagation via callus.

On the basis of the experiments concerning micropropagation of *Bergenia crassifolia*, one can say that for mass propagation of this species the regeneration process via callus is fast and the propagation rate using this method is high, but in those cases when the *Bergenia* seeds are infected and seedlings cannot be used as

Table 5. Development of adventitious shoots of *Bergenia crassifolia* (L.) Fritsch on ML medium with BAP – 3 mg/l and NAA – 0.1 mg/l in relation to the explant used

Source of explant	Explant	Growth response in first passage	Growth response in second passage	Plants which survived after transfer to pots (%)
Seedling	Hypocotyl	Callus	Callus differentiation	50
Plantlet	Rhizome cuttings	Callus differentiation	Adventitious shoots	60
		Adventitious shoots developing on explant	Adventitious shoots with small roots	80

a source of explants, a direct method within the parent tissue such as rhizome cuttings should be applied. Intact plants could also be used as the source of explants, but then the material must be sterilized.

2.2 Production of Arbutin

Arbutin production in the intact plants of the *Bergenia* species, *Arctostaphylos uva-ursi* and *Vaccinium vitis-idaea* has been reported by many authors (see Sect. 1.3). This compound has also been studied in tissue culture. Duškova and Dušek (1989) found that the addition of 4-hydroxybenzoic acid to the medium as a potential precursor of arbutin in tissue culture of *Arctostaphylos uva-ursi* did not influence its biosynthesis; it stimulated proliferation of the culture.

2.2.1 Laboratory Protocol

Material and Methods. Arbutin content was determined in the following organs:

1. Leaves and rhizomes of *Bergenia crassifolia* plants propagated in vitro via callus, growing for 1, 2 and 3 years in soil, gathered at different times (from March until December);
2. Leaves and rhizomes of *Bergenia crassifolia* plants propagated by conventional methods growing for 3 years in soil, gathered from March until December.

Chemical Analysis. Arbutin content was determined according to the method presented in the Polish Pharmacopoea (1970) used by Walewska and Wyrzykowska (1967). The leaves and rhizomes were dried at room temperature and powdered. The powder was extracted with boiled water. Lead acetate was added and the mixture was left in the refrigerator for 24 h. The precipitated impurities were filtered off and sodium hydrogen phosphate was added to the filtrate and then filtered. Determination of arbutin was assayed by the colorimetric method with 4-aminophenazone. Specol Carl Zeiss Jena was used, cuvette 1 cm, at 535 nm.

Results and Discussion. The results presented in Table 6 indicate that arbutin was found in all the organs of the investigated plants which were regenerated in vitro and propagated by a conventional method. The highest arbutin content was determined in leaves gathered in autumn (11.6–11.9%). The method of propagation has no influence on arbutin biosynthesis. Even the leaves of plants propagated in vitro and transferred to the soil, showed in the autumn after 1 year the same yield of arbutin (11.6%) as the plants growing for 2 or 3 years in the soil. The plants propagated by a conventional method show a very similar content of this compound, i.e., 11.8–11.9%.

A lesser arbutin content was found in rhizomes (4.3–5.4%). In this organ, the content of arbutin varies less than that in leaves. In plants propagated in vitro and in vivo in all vegetation periods, the yield of arbutin in rhizomes is similar (Table 6).

3 Conclusions and Prospectives

The application of in vitro culture for the propagation of *Bergenia crassifolia* is useful and promising. This method can help to solve two problems, as it permits (1) a high propagation rate of plants, ca. 5000 plantlets from one seedling via

Table 6. Content of arbutin in leaves and rhizomes of *Bergenia crassifolia* (L.) Fritsch propagated in vitro and in vivo from rhizome cuttings

Month of harvest	Content of arbutin (%) in							
	Plants propagated in vitro via callus						Plants propagated by rhizome cuttings growing only in the soil for 3 years	
	1 year after transfer to the soil		2 years after transfer to the soil		3 years after transfer to the soil			
	Leaves	Rhizomes	Leaves	Rhizomes	Leaves	Rhizomes	Leaves	Rhizomes
March	5.6	No study	9.3	4.5	9.4	4.9	10.1	5.1
June	6.0	No study	11.3	4.8	11.0	5.2	11.4	5.3
August	11.5	4.7	11.6	4.8	11.4	5.3	11.8	5.3
October	11.6	5.1	11.7	4.9	11.8	4.9	11.9	5.4
December	11.6	4.3	11.6	4.4	11.7	4.6	11.8	5.3

callus, in the period of 6 months; and (2) a shorter period necessary for plant development. Plant regeneration from the callus and directly within rhizome cuttings is faster than the development of the plants from seeds.

These conclusions were based on the following results:

1. *Bergenia crassifolia* can be propagated in vitro through adventitious shoots from the intermediate callus or directly via rhizome cuttings.
2. Hypocotyl is the best source of callus.
3. Murashige and Skoog (1962) medium supplemented with 1 mg/l BAP, 0.3 mg/l NAA and 80 mg/l AS is suitable for callus induction.
4. Murashige Lily Multiplication Medium (Gibco Bio-Culture) supplemented with 3 mg/l BAP, 0.1 mg/l NAA (also containing 80 mg/l AS) is the best for adventitious shoot development.
5. Nitsch and Nitsch medium (1969) supplemented with 0.5 mg/l IBA, 0.1 mg/l kinetin, 10 mg/l adenine sulphate (AS) and 40 g/l sucrose is suitable for rooting of the shoots.

Studies on the micropropagation of *B. crassifolia* plants seem to confirm, that the arbutin content in plants propagated in vitro is within the same range as in plants propagated by conventional methods.

The research done so far has demonstrated the presence of bergenin in *B. crassifolia*. If future toxicological tests confirm its application in medicine, *Bergenia* will become an important raw material. In the near future, *Bergenia* leaves could possibly replace other raw material containing arbutin and tannin, which are scarce and therefore in great demand.

Acknowledgements. We would like to thank Prof. Dr. J. Kozčowski (Institute of Medicinal Plants, Poznań) for seeds of *B. crassifolia*, Dr. E. Walswska (Department of Pharmacognosy, Medical Academy, Warsaw) for discussion of arbutin determination, J. Ciscierska (Medical Academy, Warsaw) for her help with the preparation of this manuscript, I. Rudnicki and A. Purzycka (Dept. of Biology and Pharm. Botany, Medical Academy, Warsaw) for technical assistance.

References

Adamanis F, Bederska-Płotkowiakowa Z (1969) The chromatographical analysis of *Bergenia* sp. leaves. Acta Pol Pharm 17: 7–11

Bajaj YPS, Furmanowa M, Olszowska O (1988) Biotechnology of the micropropagation of medicinal and aromatic plants. In: Bajaj YPS (ed) Biotechnology in Agriculture and Forestry vol 4. Medicinal and Aromatic Plants I. Springer, Berlin Heidelberg New York Tokyo pp 60–103

Von Bluhme H, Frohne D (1973) Zur Extraktion und Charakterisierung von Phenoloxidase-aktivitäten aus gerbstoff-haltigen Blattmaterial. Planta Med 24: 219–233

Bohm BA, Donevan LS, Bhat UG (1986) Flavonoid of some species of *Bergenia*, *Francoa*, *Parrnassia* and *Lepuropetalon*. Biochem Syst Ecol 14: 75–77

Borissowa AG (1956) Badan (*Bergenia* Moench) ego sistematics i chosjajstwennoe znaczenije. (*Bergenia* Moench – its systematics and economic meaning). Tr Bot Inst im A Komarowa Akad Nauk SSSR, ser V wyp 4: 297–339

Duŝkova J, Dušek J (1989) A study of the production of selected metabolites by the tissue culture of *Arctostaphylos uva-ursi* (L.) spr. and its possible influencing. Cesk Farm 38: 257–260

Engler A (1964) Syllabus der Pflanzenfamilien. Gebrüder Borntraeger, Berlin Nikolassee, p 203
Engler E, Prantl KAE (1930) Nat Pflanzenfam 111, 2a/189, 51
Friedrich H, Wehnert NU (1973a) Zur Verteilung von Arbutin und Bergenin in *Bergenia*- Pflanzen. Arch Pharm 306 (10): 757–764
Friedrich H, Wehnert NU (1973b) Die quantitative Verteilung von Arbutin und Bergenin in Blättern und Rhizomen von *Bergenia cordifolia* (Haw) Sternb. Sci. Pharm 41: 141–149
Fritsch K (1889) Verh Zool Bot Ges Wien 34: 587
Fujise S, Suzuki M, Watanabe Y, Matsueda S (1959) Structure of bergenin. Bull Chem Soc Jpn 32: 97–98 cf CA 1960, 54: 1511
Furmanowa M, Rapczewska L (1982) Effects of bioregulators on micropropagation of *Bergenia crassifolia* (L.) Fritsch. Natl Conf Med Plants, Sofia, October 1982, Abstr 25
Furmanowa M, Olędzka H, Olszowska O, Rapczewska R, Józefowicz J (1981) Vegetative propagation of some medicinal plants by tissue culture. Bull Breed Acclim Plants 145: 99–105
Hay JE, Haynes LJ (1958) Bergenin a c-glycopyranosyl derivative of 4-O-methylgallic acid. J Chem Soc 6: 2231–8
Index Kewensis Plantarum Phanerogamarum (1895) Supl I–XV (1906–1974), Oxford
Lewis WH (1977) Medical Botany – Plants Affecting Man's Health. J Wiley and Sons, New York, London, pp 300
Miller CO (1967) Cytokinins in *Zea Mays.* Ann NY Acad Sci 144: 251
Murashige T, Skoog F (1962) A revised medium for rapid growth and bioassays with tobacco tissue cultures. Physiol Plant 15: 473–497
Nitsch JP, Nitsch C (1969) Haploid plants from pollen grains. Science 163: 85–87
Olszowska O, Furmanowa M (1986) Micropropagation of *Coluria geoides* by axillary shoots. Planta Med 52: 521
Ostrowska B, Gorecki P (1988) The investigation on possibility of utilization of *Bergenia* leaves to therapeutics in place of arbutin and tannin raw materials deficiency. I. Phytochemical investigation. Herba Pol 34: 21–25
Ostrowska B, Gorecki P, Wolska D (1989) Investigations of possibility of utilization of *Bergenia* leaves to therapeutics in place of arbutin and tannin raw materials deficiency. II. Isolation of bergenin and a method for its quantitative determination. Herba Pol 35: 117–122
Pasternak Th, Durr K (1958) Constitution of bergenin. Helv Chim Acta 41: 1159–62, cf CA 1959, 53, 428
Srepel B, Apostoloski R, Simic Z, Coric L (1978) Arbutin und Gerbstoffgehalt in den Blättern einiger *Bergenia* Arten. Acta Pharm Jugosl 28: 71–73
Sticher O, Soldati F, Lehmann D (1979) Hochleistungsflüssigchromatographische Trennung und quantitative Bestimmung von Arbutin, Methylarbutin, Hydrochinon und Hydrochinon-monomethylaether in *Arctostaphylos*-, *Bergenia*-, *Calluna* und *Vaccinium* Arten. Planta Med 35: 253–261
The Polish Pharmacopoea IV ed (1970) PWN, Warszawa
Thieme H, Walewska E, Winkler NJ (1969) Isolierung von Rhododendrin aus *Bergenia* – Arten. Pharmazie 10: 468
Tschitschibabin AE (1950) Notannins compounds from extracts of *Saxifraga* (=*Bergenia*) *crassifolia.* DAS USSR, w2
Tyszyńska- Kownacka D (1979) *Bergenia* grubolistna – roślina przydatna dla przemysłu zielarskiego (*Bergenia crassifolia* – plant useful for the herb industry). Wiadomości Zielarskie Herb News 5: 2–3
Walewska E, Wyrzykowska J (1967) Annual variation of content of arbutin and pyroside in the leaves of *Vaccinium vitis-idaea* L. Dissert. Pharm Pharmacol 19: 351–355
Webb DA (1964) *Saxifragaceae.* In: Tutin TG, Heywood VH, Burges NA, Valentine DH, Walters SM, Webb DA (eds) Flora Europaea vol I. *Lycopodiaceae* to *Platanaceae.* Cambridge Univ Press, Cambridge, p 380
Xiu-Min Ch, Yoshida J, Hatano T, Fukushima M, Okuda T (1987) Gallcylarbutin and other polyphenols from *Bergenia purpurascens.* Phytochemistry 26: 515–517
Yeo PF (1966) A revision of the genus *Bergenia* Moench (*Saxifragaceae*) Kew Bull 20: 113–148

III *Canavalia ensiformis* L. DC (Jackbean): In Vitro Culture and the Production of Canavanine

C. Oropeza, G. Godoy, J. Quiroz, and V.M. Loyola-Vargas[1]

1 Introduction

Canavalia ensiformis (L.) DC commonly known as jackbean is a legume native to Central and South America from México south to Brazil and Peru, and to the West Indies (Duke 1981). The earliest known record of its existence is from about 900 A.D., from Oaxaca in central Mexico (Sauer and Kaplan 1969). It is cultivated on a small scale throughout the tropics, but is of relatively little economic importance in world trade (Smartt 1985).

C. ensiformis is an annual plant, growing bushy and erect from 0.6–1.6 m in height (Fig. 1), becoming woody with age. It can also grow as a climber, especially if shaded. Branches are normally found singly at the lower nodes. However, as many as three can develop at one node. Leaves arise alternately and are trifoliate. Leaflets, commonly up to 12 cm long, are ovate-elliptic. The center leaflet is slightly elongated. Petioles are as long as the leaflets. The flowers are typically papillionaceous, rose to purple in color, rounded, about 2 cm long; up to 20 may develop in groups of 2–3 on each peduncle. Pods are linear, slightly curved, and commonly 30 × 2.5 cm in size. Seeds (Fig. 2) are white or ivory with a brownish mark near the grayish hilum, about 2.1 × 1.5 × 1 cm in size; 12–20 develop in each pod (Piper 1925; Sauer 1964).

The jackbean has been more or less employed as feed for domestic animals, but both the herbage and seeds are eaten with reluctance by most animals (Piper 1925). The dry mature seeds are used as foodstuff, but are not popular because of their unacceptable flavor and texture (Babar et al. 1989). The jackbean is presently cultivated throughout the tropics as a cover crop, a soil cover for erosion control, and forage for ruminants. As a green manure, it is intercropped with cacao, coffee, and sugarcane. Young pods and immature seeds are used as vegetables. Flowers and young leaves are steamed as a condiment, in Indonesia (Duke 1981). Roasted beans are sometimes used as a substitute for or filler in coffee.

The plant is reported to tolerate disease, drought, fungus, insects, low pH, salts, sand shade, slope virus, and waterlogging (Duke 1981). Seed yield reports vary considerably (Table 1). However, this may be the result of different growing conditions. High yields in Puerto Rico and Venezuela were obtained under experimental conditions. In the case of the lowest yields, not enough information on the environmental conditions was reported, but these yields may have resulted from

[1] División de Biología Vegetal, Centro de Investigación Científica de Yucatán, Apado. Postal 87 Cordemex 97310 Yucatán, México

Biotechnology in Agriculture and Forestry, Vol. 21
Medicinal and Aromatic Plants IV (ed. by Y.P.S. Bajaj)

Fig. 1. *Canavalia ensiformis*

restricted rainfall conditions. Nevertheless, the figures show the survival potential of the jackbean. This is more evident when compared with other crops. Herrera (1983) reported a jackbean seed yield in Yucatán of 1800 kg/ha, whereas average yields for maize and the common bean under similar conditions were 600 and 1500 kg/ha, respectively. This was carried out in an area too rocky for mechanization, where the range of crops was already limited by the alkaline soil and the tropical environment. Additionally, the jackbean being a legume with a nitrogen-fixing ability, it would not be expected to need fertilizer nitrogen. However, to what extent this is correct is not yet clear. In a review of the nitrogen-fixing ability of 13 species of pulse, legumes, the jackbean was recorded as having the lowest value (Nutman 1976).

In contrast with the soybean, the jackbean forms seeds which accumulate starch rather than lipids (De la Vega and Sotelo 1982; Ellis and Belmar 1985; Babar et al. 1989). Its carbohydrates can be efficiently used by ruminants and poultry, therefore it is a good source of energy. The seed is also a good source of minerals with a high phosphorus content (3.5 g/kg dry wt.) (Ellis and Belmar 1985). The protein content is high, i.e., 24–31% (Duke 1981; De la Vega and Sotelo 1982; Babar et al. 1989). The amino-acid composition is very similar to that of the soybean seed, although it has a lesser lysine content, and similar to other legumes, is poor in amino acids containing sulfur (De la Vega and Sotelo 1982; Ellis and

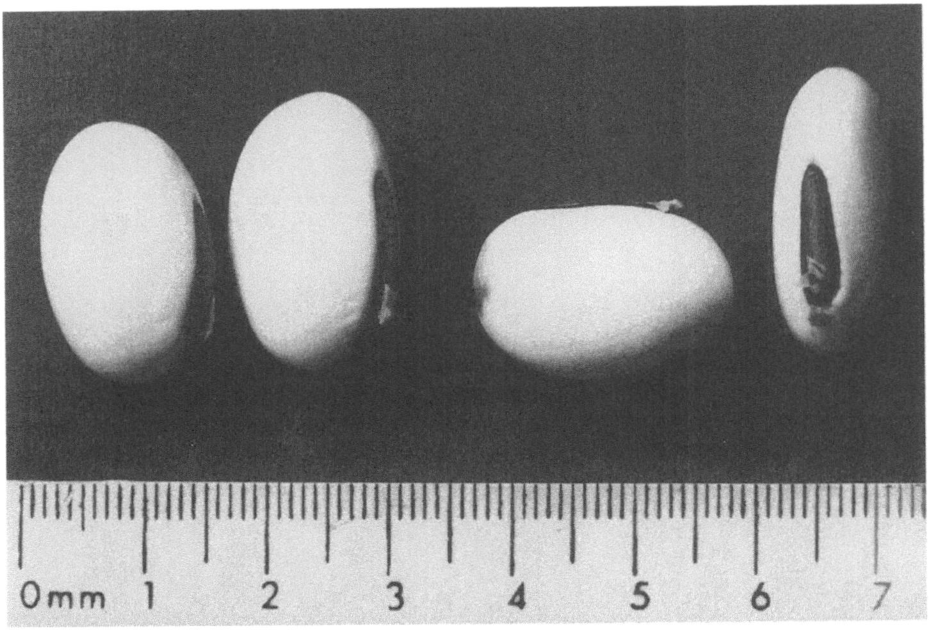

Fig. 2. *Canavalia ensiformis* seeds

Table 1. Jackbean seed yields. (Kessler 1987)

Yield (kg/ha)	Location
800–1200	Brazil
1800	Mexico
2000–2500	Rhodesia
2876	Nigeria, Ibadan
2400–3950	Hawaii
3531	Dominican Republic
4580	Nigeria, Ibadan
4300–5400	Puerto Rico
6000	Venezuela

Belmar 1985; D'Mello et al. 1985). Molina et al. (1977) reported a product extracted from jackbean seed which was rich in protein and starch, with a high nutritive value.

Since the jackbean has a high protein content in seeds and foliage, high seed yields without substantial management inputs, and probably grows in soil independent of nitrogen, the plant has been identified by the United States National Academy of Sciences (NAS 1979) as a potential food source for the tropics, which could help agricultural production in developing countries.

2 Toxic Components of the Seed

The potential of jackbean, both agronomically and in nutritive value, is unfortunately restricted by the presence of several antimetabolites. It contains trypsin and quimotrypsin inhibitors (Duke 1981; Babar 1987); lectins (concanavalin A, B and conatoxine) (Carlini and Guimaraes 1981); and L-canavanine (Fig. 3), a toxic nonprotein amino acid analogous to L-arginine (Kitagawa and Tomiyama 1929). Of these factors, the protease inhibitors and the lectins have been shown to be heat-labile (Liener 1980; Babar et al. 1989), whereas L-canavanin (Fig. 3) is heat-stable (it decomposes at 180 °C). Therefore, cooking or autoclaving can eliminate the protein antimetabolites, but not L-canavanine, which still limits the usefulness of *C. ensiformis*.

L-Canavanine is accumulated in the seeds of several legumes (Bell et al. 1978). Some species contain over 12% dry wt. of the seed, representing more than 95% of the free amino acid nitrogen (Rosenthal 1977a). The L-canavanine content of dry jackbean seeds ranges from 25–31 g/kg (dry wt.) (Fearon and Bell 1955; De la Vega and Sotelo 1982; Natelson 1985; Oropeza et al. 1988). It has been reported that L-canavanine is toxic to animals, insects, and bacteria (Rosenthal 1977b). Some organisms which have been found to be resistant to L-canavinine have enzymes which can break down the amino acid into L-homoserine and guanidine (Kalyankar et al. 1958). Addition of L-arginine to diets containing L-canavanine reduces its toxic effects (Wright and Srb 1950), therefore, it has been suggested that at least part of its toxicity could be explained by competition with L-arginine (Hegdekar 1970; Vanderzant and Chremos 1971). For instance, protein structure and function could be affected if L-canavanine, with a lower isoelectric point (8.2), is incorporated instead of L-arginine (10.8) (Greenstein and Winitz 1961). This could result in the formation of structurally aberrant proteins with altered properties (Pines et al. 1981). L-Canavanine toxicity has also been explained on the basis of its conversion to L-canaline, which is also a toxic amino acid because it is an effective inhibitor of pyridoxal enzymes (Rahiala et al. 1971).

The biosynthesis of L-canavanine has been suggested to take place in the pericarp of the fruit (Warren and Hunt 1971). It occurs in a pathway similar to the Krebs-Henseleit ornithine-urea cycle, where each component has the terminal methylene group replaced by oxygen. The corresponding enzymes have been found in jackbean (see Rosenthal 1982). One such enzyme, arginase (EC 3.5.3.1), cleaves

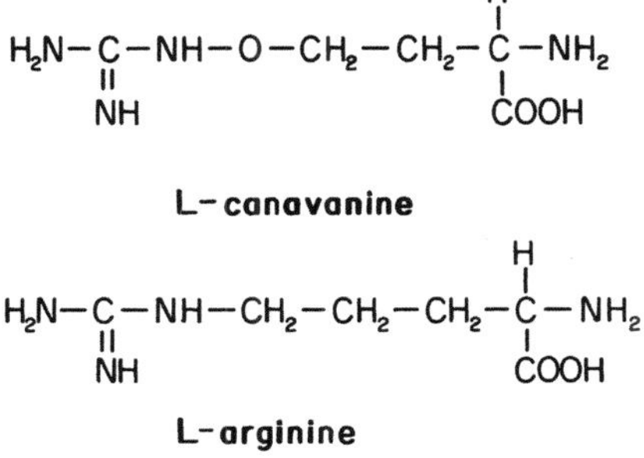

Fig. 3. Chemical structures of L-canavanine and L-arginine

L-canavanine into L-canaline and urea (Damodaran and Narayanan 1940). Urea is then hydrolyzed by urease (EC 3.5.1.5) to CO_2 and NH_3 (Rosenthal and Rhodes 1984). These two reactions are important in the mobilization of nitrogen from L-canavanine. During germination and the first days of growth, the cotyledonous L-canavanine content notably decreases, whereas it increases in different parts of the developing plantlet, suggesting a translocation from cotyledons to plantlet (Rosenthal 1970). This is supported by experiments in which ^{14}C-L-canavanine was fed to cotyledons and the labeled amino acid was recovered from the plantlet organs (Rosenthal and Rhodes 1984). This L-canavanine mobilization from the seed, together with its preponderance in the seed and the high nitrogen content and nitrogen-to-carbon ratio, led to the proposal of a nitrogen-storing role for L-canavanine (Rosenthal 1977b), similar to that of L-arginine in other members of the leguminosae family (Thompson 1980; Ruiter and Kollöffel 1983; Shargool et al. 1988; Micallef and Shelp 1989). In addition, based on its toxicity, L-canavanine is considered important as a defense against predator insects (see Rosenthal 1988).

Studies carried out in our laboratory have shown that during the first few days following imbibition, arginase activity increases in cotyledons, stem, and roots (Fig. 4; Loyola-Vargas et al. 1988a), coinciding with the decrease in L-canavanine content observed by Rosenthal (1970). The levels of arginase are 15-fold higher in stem and roots than in the cotyledons, suggesting a major role for these organs in the degradation of L-canavanine. After translocation from cotyledons, if L-canavanine is to be used as a nitrogen source by other plant organs, increases in L-canavanine metabolizing pathways in those regions would be expected.

Urease activity increases in cotyledons during the early stages of development of the plant (Sehgal and Naylor 1966; Rosenthal 1974; Loyola-Vargas et al. 1988a) (Fig. 5), producing high levels of ammonia (Loyola-Vargas et al. 1988a). This ammonia will be then available for assimilation and incorporation into amino acids (Rosenthal and Rhodes 1984). Ammonia can be assimilated through three different pathways in higher plants. All three pathways are functional in *C.*

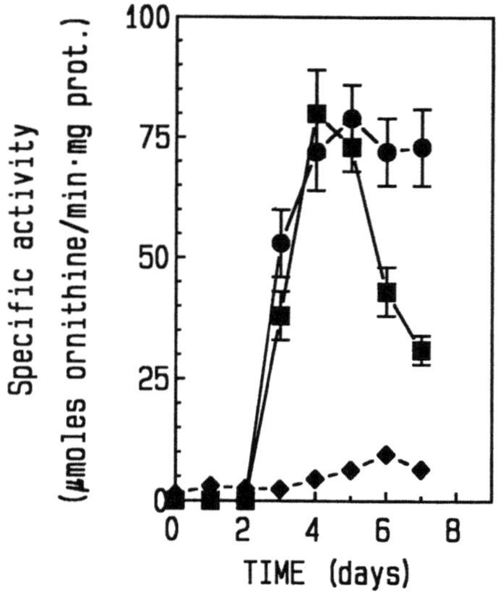

Fig. 4. Arginase activity in cotyledons (◆), stem (■), and roots (●). The data are the average of at least three separate experiments. *Bars* represent ±1 SE. (After Loyola-Vargas et al. 1988a)

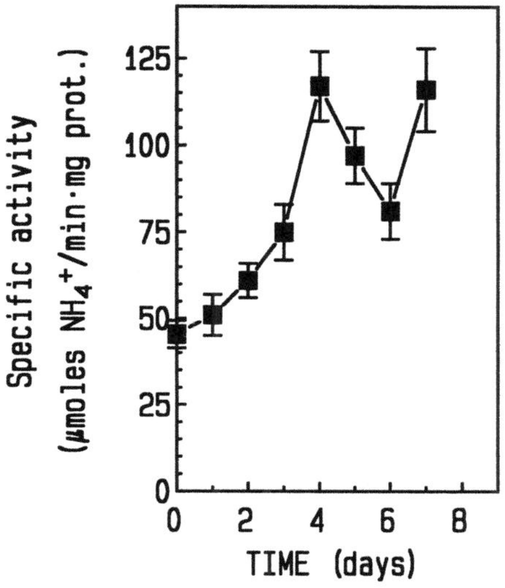

Fig. 5. Urease activity in cotyledons. The data are the average of at least three separate experiments. *Bars* represent ±1 SE. (After Loyola-Vargas et al. 1988a)

ensiformis. During the very early stages of development of the plant, aspargine synthetase activity is particularly high in the roots, where it would appear to be, at least initially, the main enzyme involved in nitrogen assimilation (Rosenthal, 1970; Loyola-Vargas et al. 1988b). After a few days, aspargine synthetase rapidly decreases and the enzymes of the glutamine synthetase/glutamate synthase pathway become active in the aboveground plant parts (Fig. 6), from day 4 onwards. At the same time, glutamate dehydrogenase activity increases in the roots (Fig. 7; Loyola-Vargas et al. 1988b). Therefore, in a given tissue, one or more alternative pathway(s) must exist for ammonia assimilation in *C. ensiformis*.

These results also support the proposed role of L-canavanine as a nitrogen reserve source (Rosenthal 1977b), during the early stages of heterotrophic growth

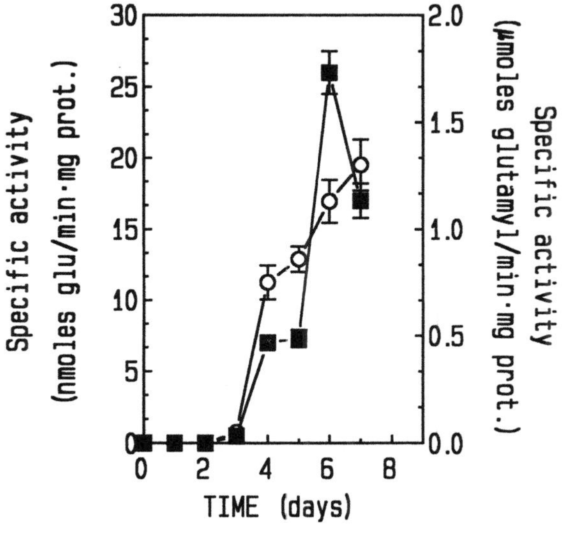

Fig. 6. Glutamine synthetase activity in leaves (○) and Fd-glutamate synthase activity in cotyledons (■). The data are the average of at least three separate experiments. *Bars* represent ±1 SE. (After Loyola-Vargas et al. 1988b)

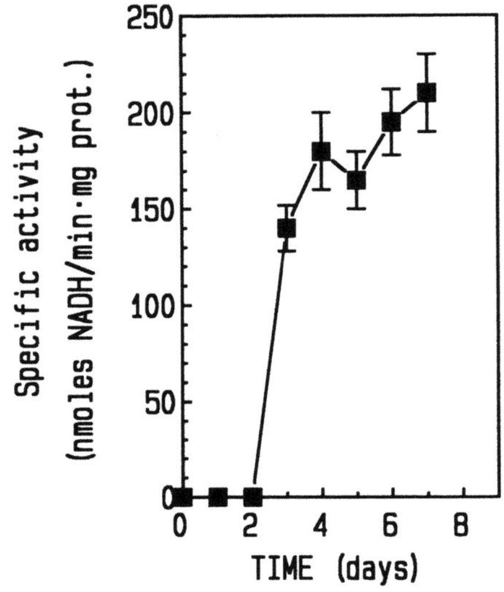

Fig. 7. NADH-glutamate dehydrogenase activity in roots. The data are the average of three separate experiments. *Bars* represent ±1 SE. (After Loyola-Vargas et al. 1988b)

when both the biosynthetic potential and energy charge of the cells are low, and the level of ammonia is high. Therefore, during this phase both aspargine synthetase and glutamate dehydrogenase may play a fundamental role in the reassimilation of ammonia in *C. ensiformis*. However, when the plant acquires full photosynthetic capacity and is capable of autotrophic growth, the glutamine synthetase/glutamate synthase cycle becomes the main pathway of reassimilation of ammonia in the aboveground plant parts.

3 Inactivation of Antimetabolites

Jackbean seed as an animal foodstuff has been used successfully for ruminants (Mora et al. 1981; Dixon et al. 1983; Pacheco and Rivera 1985). However, in monogastric animals, raw seed diets have been found to be toxic. Steady progress has been made in the development of processing treatments to reduce toxicity.

Heat processing is generally accepted as an effective means of inactivating heat-labile toxic constituents of legume grains. When diets including 500 g/kg of raw jackbean seed were fed to quails, 100% mortality was obtained; however, when the seed was autoclaved, there was no mortality (Jayne-Williams 1973). In similar studies using chicks (Belmar et al. 1985; Kessler et al. 1990), intestinal lesions, enlargement of pancreas and liver, a 35–55% reduction in growth rate, and over 50% mortality were obtained with a diet containing 30% raw seed. Autoclaving (0.5 or 1 h) of the seed completely reduced mortality and intestinal lesions. However, effects on pancreas, liver, and growth rates were not avoided, which is indicative of the persistence of heat-stable antimetabolites, in particular, L-canavanine (Kessler et al. 1990). Attempts have been carried out to eliminate L-canavanine. Belmar et al. (1985) reported the use of soaking plus autoclaving, and of boiling. Although they found a reduction of 25% in the L-canavanine content, growth rates did not significantly improve. Amino-acid supplementation of diets containing autoclaved jackbean seed have also been tested. In chicks, no beneficial effect was observed (D'Mello et al. 1989). When given to rats, performance was comparable to a casein-based diet (De la Vega and Sotelo 1982).

4 Conventional Practices for Its Propagation

Jackbean is propagated by seeds usually sown in rows 75 cm apart, with 45–60 cm between plants. Preparation before sowing may be limited to a light hoeing, followed by a dressing of a complete fertilizer that should be raked into the top layer of soil before seeds are sown. For green manure, 225 kg/ha superphosphate is recommended. When planted in rows, seed rate is 25–30 kg/ha. When planted as a green manure crop, seed rate is 40–60 kg/ha. Seeds should be covered to 2 cm (Duke 1981).

In spite of its potential, *C. ensiformis* is a plant which has been given little attention from an agronomic point of view. Although it is obvious that more research is required in order to use the seed as a constituent of diets for monogastric animals, it has been used successfully with ruminants. Therefore, convenient agronomic practices are necessary to obtain the best performance from the crop.

Attempts have been already made, as for example, those reported with excellent yields in Puerto Rico (Kay 1979) and Venezuela (Dixon et al. 1983). In a systematic study carried out in Yucatán from 1982–1986 by Kessler (1987), seed yield was found to vary according to the rainfall pattern of the season, with the highest seed yields occurring in the year of most prolonged rainfall. Despite variation between years, jackbean seed yields were consistently higher than the average yields of maize and beans obtained under similar conditions. The jackbean crops in common with local legumes appeared not to be nodular, however, seed yields were not affected by the use of fertilizers, including nitrogen and micronutrients. Large differences were found between weeded and unweeded treatments. No seed yield reduction was found when planting at high plant density (9 plants/m^2). On the other hand, seed yields were reduced with intercropping with maize, but jackbean yield was obtained in addition to that of the maize yields, which were not affected. Jackbean activated produced forage with a high nitrogen content, but regrowth after defoliation only occurred after a long growing season and seed yield was closely related to final crop biomass, therefore, crops intended for both seed and forage were not recommended unless a water supply can be guaranteed.

5 In Vitro Approaches

Although so far it has not been conclusively demonstrated that L-canavanine is the limiting factor in the utilization of the jackbean for animal food, a large body of evidence supports this theory. Attempts to eliminate it from the seed have only been partially successful. An alternative approach could be genetic improvement, which could also be useful to improve yield. This could be achieved by plant breeding. So far, there are no reports in the literature dealing with jackbean plant breeding. A shortcut may be taken through in vitro tissue culture techniques. This could be particularly convenient, since it has been claimed that there is very little variation within the species (Sastrapadja et al. 1976; NAS 1979).

Studies on tissue culture of jackbean have been scarce. Callus cultures were initially established to study the metabolism of 2,4-dichlorophenoxyacetic acid (2,4-D) (Feung et al. 1975) and the biosynthesis of the phytoalexin 3-hydroxy-9-methoxypterocarpan (medicarpin) (Gustine 1976; Gustine et al. 1978). In these studies, jackbean tissue cultures were used as a model system; the studies were not interested in jackbean *per se*. On the other hand, Collins et al. (1978) reported on the use of the synthetic auxin 4-amino-3,5,6-trichloropicolinic acid (picloram) to improve fresh-weight increase of the callus of different species including the jackbean. Picloram promoted a fourfold increase in the fresh weight of jackbean callus.

In this laboratory work, the tissue culture of jackbean initially dealt with callus and suspension cultures induced from ovules, ovaries, and anthers (Quiroz et al. 1987). Later, it was extended to cultures induced from hypocotyl, epicotyl, leaves, roots, and cotyledons (data unpublished).

5.1 Callus Induction

Reproductive plant material was obtained from flower buds harvested from field-grown jackbean plants. Anthers, at the uninucleate microspore stage, were excised from sterile 3–5-mm-long flower buds. Ovaries and ovules were excised from sterile flower buds, 5–9 mm long. Vegetative organ explants were obtained from plantlets germinated and grown under sterile conditions. Plant material was surface sterilized in 70% (v/v) ethanol for 15 min before transferring to 2% sodium hypochlorite for 15 min followed by thorough washing with several changes of sterile distilled water.

For callus induction from reproductive plant parts, four solid (agar, 11 g/l) media were tested: PC medium (Phillips and Collins 1979); MS medium (Murashige and Skoog 1962); B5 medium (Gamborg et al. 1968) in its original formulation and a modified form (B5+) containing inositol (320 mg/l); thiamine (25 mg/l); vitamin E (0.4 mg/l); nicotinic acid (1 mg/l); p-aminobenzoic acid (1 mg/l); biotin (0.2 mg/l); and calcium pantothenate (0.2 mg/l). In the case of callus induction from vegetative plant parts, only MS medium was used. Different concentrations (0, 2, and 5 mg/l) of α-naphthaleneacetic acid (NAA) and 6-benzylaminopurine (BA) were added to the media separately and in all possible combinations. Callus induction was assessed in terms of percentage of induction, weight gain during induction, and friability.

The best calli were induced from anther explants in B5+ medium (0.1 mg/l NAA and 5 mg/l BA); from ovule explant in B5+ (2 mg/l NAA and 2 mg/l BA); from ovary explants in PC medium (3 mg/l NAA and 3 mg/l BA). The best hypocotyl callus was obtained with PC medium (2 mg/l NAA and 5 mg/l BA). Calli from other plant parts were also induced, but without testing different medium formulations. Calli from apical meristems were induced in MS (0.18 mg/l NAA and 0.022 mg/l BA) and from hypocotyl, leaves, roots, and cotyledon explants in MS medium (0.9 mg/l NAA and 0.022 mg/l BA).

5.2 Callus Maintenance

The same medium (growth regulator) formulations tested for callus induction were assessed to determine the appropriate conditions for callus tissue growth. The best medium for anther, ovule, and ovary calli was PC (2 mg/l NAA and 5 mg/l BA). In all cases, callus tissue developed from every explant and an increase in fresh and dry weight of about tenfold was obtained after 15 days in culture (Fig. 8). Similar growth curves were obtained with calli grown in the dark. The color of the calli

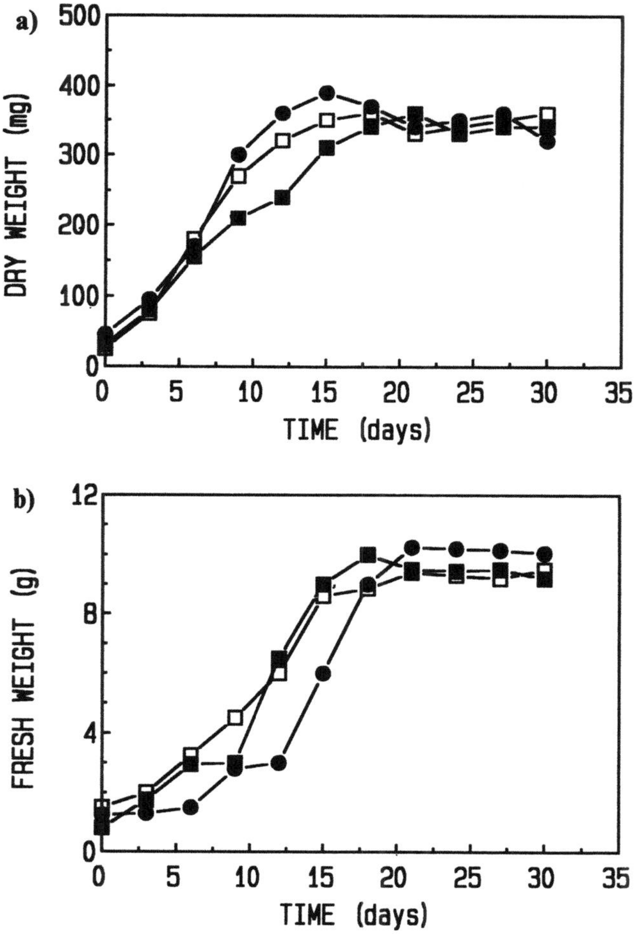

Fig. 8a, b. Increases in **a** dry and **b** fresh weight of continuously illuminated anther (●), ovary (□) and ovule (■) callus cultures at 25 °C on a PC medium containing 2 mg/l NAA and 5 mg/l BA. Data are the average of three separate experiments. (After Quiróz et al. 1987)

derived from anther, ovule, and ovary was found to be affected by the light regime under which the calli were cultured. Continuously illuminated calli were green and calli grown in the dark were white. When subcultured and cultivated in the dark, green calli developed into white calli in 3 weeks, and white calli readily developed into green calli within 3 weeks when cultured under continuous illumination.

In the case of calli derived from vegetative plant parts, different media formulations were not tested to determine appropriate growth conditions and the induction media was used instead. Weight gain was not recorded except for hypocotyl calli; fresh and dry weight increases were three- and twofold, respectively. No further improvement was attempted.

5.3 Cell Suspension Cultures

Anther, ovule, and ovary calli readily formed cell suspension cultures in a maintenance medium (PC, 2 mg/l and 5 mg/l BA) without agar. However, growth as measured by fresh weight increase was faster in LS (Linsmaier and Skoog 1965) medium than in other media tested [B5, B5+, MS, PC, and SH (Schenk and Hildebrandt 1972)]. The increase in fresh weight was tenfold after 30 days in liquid culture. Figure 9 shows a typical growth curve of anther suspension culture. Similar growth curves were obtained with ovule and ovary suspension cultures.

5.4 Differentiation

Initial attempts in our laboratory to obtain differentiation were based on changing the growth-regulator formulation in the media. Anther, ovule, and ovary calli were cultured in SH, LS, and PC medium. Different concentrations (ranging from

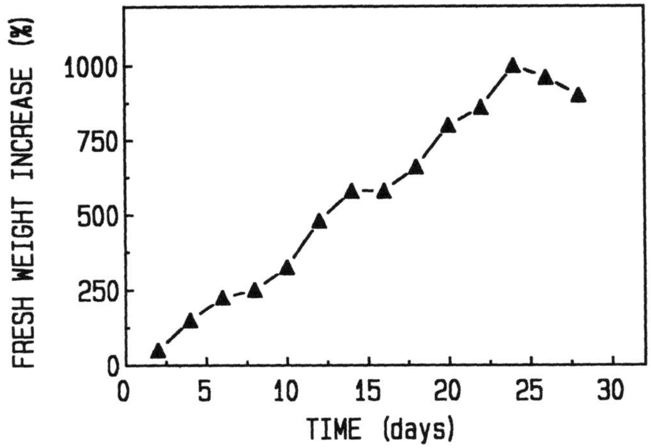

Fig. 9. Increase in fresh weight of continuously illuminated anther-cell suspension culture grown at 25 °C in an LS medium containing 2 mg/l NAA and 5 mg/l BA. Data are the average of three separate experiments. (After Quiróz et al. 1987)

0.1–10 mg/l) of indole-3-acetic acid (IAA), indole-3-butyric acid (IBA), 2,4-D, zeatin, and BA were added to the media singly and in different combinations (60 in total). The results were negative. Additionally, according to previous reports, the medium (PC) was also supplemented with different concentrations of sucrose (Kochba and Button 1974); adenine (Phillips and Collins 1981); amino acids (proline, arginine and ornithine) singly or in combination (Stuart and Strickland 1984); different nitrate/amonnia ratios (Walker and Sato 1981; Gleddie et al. 1983); complex organic supplements (peptone, casein hydrolysate, potato, carrot and yeast extract) (see Dodds and Roberts 1985). Again, no differentiation was observed in any case.

Plant regeneration has been quite difficult among the legumes. Grain legumes have generally been found recalcitrant to morphogenesis, and with a few exceptions, regeneration of plants has been achieved from undifferentiated tissues, e.g.: soybean (*Glycine max*; Barwale et al. 1986); winged bean (*Psophocarpus tetragonolobus*; Gregory et al. 1980; Wilson et al. 1985); pea (*Pisum sativum*; Mroginski and Kartha 1981); peanut (*Arachis hypogaea*; Bajaj et al. 1981); and mothbean (*Vigna aconitifolia*; Gill and Eapen 1986). Based on these reports we tried to obtain differentiation from jackbean explants. Since plant regeneration was reported for soybean (Ranch et al. 1985) from embryo-derived tissue, we cultured immature embryo explants in a modified LS medium (NH_4Cl instead of NH_4NO_3, and starch instead of agar) added with growth regulators in different combinations. Embryo-like structures developed from the explants in medium added with NAA/2,4-D/picloram/BA (2:1.25:0.05:4.5 mg/l) after 30 days in culture. Most of the structures

Fig. 10. Differentiation from apical meristem explants cultured in MS medium with different NAA/BA ratios (expressed as µmol/l). (Godoy, unpubl. data)

were globular, but heart-shaped forms were also observed (Quiróz et al. 1987). Unfortunately, there was no further development of these structures.

Finally, apical meristems were also cultured in MS medium with different concentrations added of NAA and BA (singly and in different combinations). Shoot formation was obtained (Fig. 10) when the auxin/cytokinin ratio in the medium was small, and root formation when the ratio was large. This is an approach with only limited potential for genetic improvement or propagation purposes, since the jackbean readily propagates from seed; however, it could be useful in developing a methodology for eliminating viral pathogens.

5.5 Canavanine Formation In Vitro

Studies of the biochemistry and physiology of L-canavanine have so far been restricted to the intact plant; however, a clear understanding of its importance in plants is yet to be achieved. Therefore *in vitro* cultures may provide a convenient model for further studies on the metabolism of L-canavanine under closely controlled conditions. A search for *in vitro* cultures with the ability to accumulate L-canavanine have been reported (Krikorian 1969: Avila and Garcia 1984). In our laboratory, the occurrance of L-canavanine in *C. ensiformis* tissue cultures was also studied (Ramirez et al. 1991). An initial attempt to test the presence of L-canavanine was carried out in calli derived from hypocotyl, ovule, and immature embryo. The calli were induced as mentioned in Sect. 5.1 and maintained in an LS medium (BA 4.5 mg/l; NAA 1.86 mg/l; picloram 0.05 mg/l; and 2,4-D 0.05 mg/l). No L-canavanine was detected in either of them. Then the effect of culture medium was tested in hypocotyl calli. L-canavanine was present in calli cultured in PC medium (BA 2 mg/l and NAA 5 mg/l). Analysis of L-canavanine was carried out with the colorimetric method reported by Natelson and Bratton (1984) and by high-performance liquid chromatography as reported by Oropeza et al. (1988). In addition, four transformed hypocotyl callus cultures were obtained (with four different strains of *Agrobacterium tumefaciens*). The media formulations used for induction and maintenance (PC) were the same used for untransformed hypocotyl

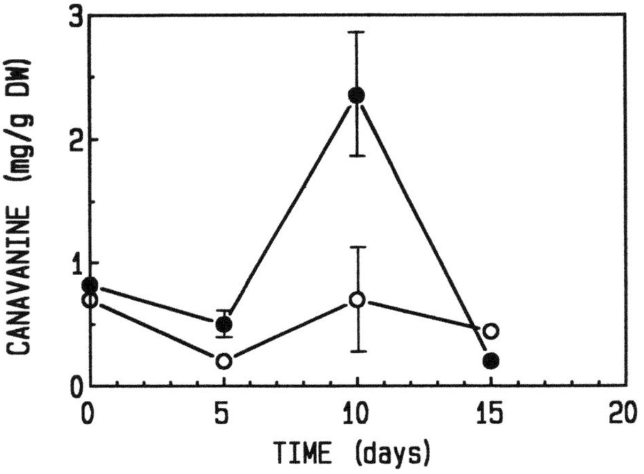

Fig. 11. L-Canavanine content of hypocotyl callus grown in medium containing 0% (●) and 4% PEG (○). Bars represent ± SD (n = 3). (After Ramirez et al. 1991)

calli. L-canavanine was also present in all four types of cultures, but (depending on the culture) accumulated from 30 to 90% times less L-canavanine than the untransformed calli. In order to modify the L-canavanine content of the cultures, untransformed hypocotyl calli were grown in a medium with and without polyethylene glycol (PEG, 40 g/l). The callus L-canavanine content decreased when PEG was added (Fig. 11).

6 Conclusions and Prospects

Jackbean is a legume with potential, but its usefulness as animal feed (and eventually for human nutrition) is so far restricted to ruminants. Jackbean toxicity, in particular the heat-labile fraction, is an obstacle which should be overcome in order to extend its use. So far this has not been possible by means of processing the seeds already formed. Although such efforts should continue, this task seems to be more feasible to be solved through genetic improvement. The evidence supports that L-canavanine is responsible for seed toxicity after heat processing, therefore, genetic improvement must be aimed at reducing the content of L-canavanine in the plant. However, it is important to demonstrate without any doubt that this amino acid is the cause of this residual toxicity, since other heat-stable constituents could contribute to the problem; for instance, the nitrate content of the seed is high (15 g/kg) (Yañez 1985). This problem could be approached by testing the effect of L-canavanine added to standard diets in economically important animals, such as chicks. We also need to extend our understanding of the importance of L-canavanine in plants.

References

Avila F, de Garcia E (1984) Effectos de la canavanina sobre el crecimiento de callos obtenidos de explantes de *Nicotiana tabacum* y *Canavalia ensiformis*. Acta Cient Venez 35: 265–269

Babar VS (1987) Biochemical investigations of Jackbean (*Canavalia ensiformis*): solubilization of proteins and effects of heat treatments and germination on trypsin inhibitor activity and polyphenols. M.Sc Thesis, Mahatma Phule Agric Univ, Rahuri, India

Babar VS, Kadam SS, Salunkhe DK (1989) Jack bean. In: Salunkhe DK, Kadam SS (eds) CRC handbook of world food legumes: nutritional chemistry, processing technology, and utilization. CRC Press Inc, Boca Raton, Florida, pp 107–113

Bajaj YPS, Kumar P, Labana KS, Singh MM (1981) Regeneration of plants from seedling explant and callus of *Arachis hypogaea* L. Indian J Exp Biol 19: 1026–1029

Barwale UB, Kerns HR, Widholm JM (1986) Plant regeneration from callus cultures of several soybean genotypes via embryogenesis and organogenesis. Planta 167: 473–481

Bell EA, Lackey JA, Polhill RM (1978) Systematic significance of canavanine in the Papilionoideae (Fabaideae). Biochem Syst Ecol 6: 201–212

Belmar R, Ellis N, Laviada E (1985) Utilización del grano de *Canavalia ensiformis* en dietas para aves. Memorias de la Primera Reunión sobre la Producción y Utilización del Grano de *Canavalia ensiformis* en Sistemas Pecuarios de Yucatán. Univ Autónoma de Yucatán. Fac Med Vet Zootec. Mérida Yucatán México, pp 45–56

Blair LC, Chastain CJ, Widholm JM (1988) Initiation and characterization of a cotton (*Gossypium hirsutum* L.) photoautotrophic cell suspension culture. Plant Cell Rep 7: 266–269
Carlini CR, Guimaraes JA (1981) Isolation and characterization of toxic protein from *Canavalia ensiformis* (jackbean) seeds distinct from concanavalin A. Toxicon (19)5: 667–676
Collins GB, Vian WF, Phillips GC (1978) Use of 4-amino-3,5,6-trichloropicolinic acid as a auxin source in plant tissue cultures. Crop Sci 18: 286–288
Damodaran M, Narayanan KGA (1940) A comparative study of arginase and canavanase. Biochem J 34: 1449–1459
De La Vega A, Sotelo A (1982) Valor nutritivo de la semilla de *Canavalia ensiformis* sola y suplementada con metionina y cereales. Rev Soc Quim Mex 26: 92–96
Dixon RM, Escobar A, Montilla J, Viera J, Mora M, Risso J, Parra R, Preston TR (1983) *Canavalia ensiformis*: a legume for the tropics. In: Farrel DJ, Vohra P (eds) Recent advances in animal nutrition in Australia. Univ of New England Publ Unit, Armidale, pp 129–140
D'Mello JPF, Acamovic T, Walker AG (1985) Nutritive value of jackbeans (*Canavalia ensiformis* L.) for young chicks. Trop Agric 62: 145–150
D'Mello JPF, Acamovic T, Walker AG (1989) Nutritive value of jackbeans (*Canavalia ensiformis* L. DC) for young chicks. Effects of amino acid supplementation. Trop Agric 66(3): 201–205
Dodds JH, Roberts LW (1985) Experiments in plant tissue culture, 2nd edn. Cambridge Univ Press, Cambridge, pp 35–53
Duke JA (1981) Handbook of legumes of world economic importance. Plenum Press, New York London, pp 39–41
Ellis N, Belmar R (1985) La composición química del grano de *Canavalia ensiformis*: Su valor nutritivo y sus factores tóxicos. Memorias de la Primera Reunión sobre la Producción y Utilización del Grano de *Canavalia ensiformis* en Sistemas Pecuarios de Yucatán. Univ Autónoma de Yucatán. Fac Med Vet Zootec. Mérida Yucatán México, pp 18–31
Fearon WR, Bell EA (1955) Canavanine: detection and occurrence in *Colutea arborescens*. Biochem J 59: 221–224
Feung CS, Hamilton RH, Mumma RO (1975) Metabolism of 2,4-dichlorophenoxyacetic acid. VII. Comparison of metabolites from five species of plant callus tissue cultures. J Agric Food Chem 23(3): 373–376
Gamborg OL, Miller RA, Ojima K (1968) Nutrient requirement of suspension cultures of soybean root cells. Exp Cell Res 50: 151–158
Gill R, Eapen S (1986) Plant regeneration from hypocotyl protoplast of mothbean (*Vigna aconitifolia*). Curr Sci 55: 100
Gleddie S, Keller W, Setterfield G (1983) Somatic embryogenesis and plant regeneration from leaf explant and cell suspensions of *Solanum melongena* (eggplant). Can J Bot 61: 656–666
Greenstein JF, Winitz M (1961) Chemistry of the amino acids, vol 3. John Wiley and Sons, New York, pp 2622–2628
Gregory HM, Haq N, Evans PK (1980) Regeneration of plantlets from leaf callus of the winged bean (*Psophocarpus tetragonolobus*) (L.) DC. Plant Sci Lett 18: 395–400
Gustine DL (1976) Biosynthesis of medicarpin in jackbean callus tissue cultures. Fed Proc 35: 1552
Gustine DL, Sherwood RT, Vance CP (1978) Regulation of phytoalexin synthesis in jackbean callus cultures: stimulation of phenylalanine ammonia-lyase and o-methyltransferase. Plant Physiol 61: 226–230
Hegdekar BM (1970) Amino acid analogues as inhibitors of insect reproduction. J Econ Entomol 63: 1950–1956
Herrera F (1983) Efecto de densidad de población sobre el rendimiento de semilla de *Canavalia ensiformis*. Prod Anim Trop 8: 166–169
Jayne-Williams DJ (1973) Influence of dietary jackbean (*Canavalia ensiformis*) and of concanavalin A on the growth of conventional and gnobiotic Japanese quail (*Coturnis coturnix* japonica). Nature, New Biol 243: 150–151
Kalyankar GD, Ikawa M, Snell EE (1958) The enzymatic cleavage of canavanine to homoserine and hydroxyguanidine. J Biol Chem 233: 1175–1178
Kay DE (1979) Jackbean. In: Crops and product digest No. 3. Food legumes. Trop Prod Inst, London, p 197
Kessler CDJ (1987) Agronomic studies of the tropical legume *Canavalia ensiformis* (L.) DC (jackbean) in Yucatan, Mexico. PhD Thesis, Univ Coll of North Wales, Bangor, Gwynedd, pp 1–128

Kessler M, Belmar R, Ellis N (1990) Effects of autoclaving on the nutritive value of the seeds of *Canavalia ensiformis* (jackbean) for chicks. Trop Agric (Trinidad) 67: 16–20
Kitagawa M, Tomiyama T (1929) A new amino-compound in the jackbean and a corresponding new ferment (I.). J Biochem (Tokyo) 11: 265–271
Kochba J, Button J (1974) The stimulation of embryogenesis and embryoid development in habituated ovular callus from the Shamouti orange (*Citrus sinensis*) as affected by tissue age and sucrose concentration. Z Pflanzenphysiol 73: 415–421
Krikorian AD (1969) The synthetic potentialities of cultured plant cell and tissues. PhD Thesis, Cornell Univ. In: Steward FC (ed) Plant physiol, Acad Press, New York London, pp 253–257
Liener IE (1980) Heat-labile antinutritional factors. In: Summerfield RJ, Bunting AH (eds) Advances in legume science. Univ of Reading, England, pp 157–170
Linsmaier EM, Skoog F (1965) Organic growth factor requirements of tobacco tissue cultures. Physiol Plant 18: 100–127
Loyola-Vargas VM, Roman ME, Quiroz J, Oropeza C, Robert ML, Scorer KN (1988a) Nitrogen metabolism in *Canavalia ensiformis* (L.) DC. I. Arginase and urease ontogeny. J Plant Physiol 132: 284–288
Loyola-Vargas VM, Yañez A, Caldera J, Oropeza C, Robert ML, Quiroz J, Scorer KN (1988b) Nitrogen metabolism in *Canavalia ensiformis* (L.) DC. II. Changing activities of nitrogen-assimilating enzymes during growth. J Plant Physiol 132: 289–293
Macallef BJ, Shelp BJ (1989) Arginine metabolism in developing soybean cotyledons. III. Utilization. Plant Physiol 91: 170–174
Molina MR, Bressani R, Elias LG (1977) Non-conventional legume grains as protein sources. Food Tech 22: 188
Mora M, Escobar A, Parra R (1981) Crecimiento de ovinos contres niveles de harina de granos de *Canavalia ensiformis* en raciones completas. Informe Final. IPA Fac Cent Agr Univ Centr Venezuela
Mroginski LA, Kartha KK (1981) Regeneration of pea (*Pisum sativum* L. cv. Century) plants by in vitro culture of immature leaflets. Plant Cell Rep 1: 64–66
Murashige T, Skoog F (1962) A revised medium for rapid growth and bioassays with tobacco tissue cultures. Physiol plant 15: 473–497
NAS (ed) (1979) Tropical legumes: resources for future. Washington DC, pp 54–58
Natelson S (1985) Canavanine to arginine ratio in alfalfa (*Medicago sativa*), clover (*Trifolium*) and jackbean (*Canavalia ensiformis*). J Agric Food Chem 33(3): 413–419
Natelson S, Bratton GR (1984) Canavanine assay of some alfalfa varieties (*Medicago sativa*) by fluorescence. Practical procedure for canavanine preparation. Microchem J 29: 26–43
Nutman PS (1976) IBP field experiments on nitrogen fixation by nodulated legumes. In: Nutman PS (ed) Symbiotic nitrogen fixation by modulated legumes. Cambridge Univ Press, Cambridge, pp 211–237
Oropeza C, Alpizar L, Loyola-Vargas VM, Quiroz J, Scorer KH (1988) Determination of L-canavanine and L-canaline in plant tissues by high-performance liquid chromatography. J Chrom 456: 405–409
Pacheco MA, Rivera J (1985) Utilización del grano de *Canavalia ensiformis* en dietas para rumiantes. Memorias de la Primera Reunión sobre la Producción y Utilización del Grano de *Canavalia ensiformis* en Sistemas Pecuarios de Yucatán. Univ Autónoma de Yucatán. Fac Med Vet Zootec. Mérida Yucatán México, pp 57–67
Phillips GC, Collins GB (1979) In vitro tissue culture of selected legumes and plant regeneration from callus cultures of red clover. Crop Sci 19: 59–64
Phillips GC, Collins GB (1981) Induction and development of somatic embryos from cell suspension cultures of soybean. Plant Cell Tissue Organ Cult 1: 123–129
Pines M, Rosenthal GA, Applebaum SW (1981) In vitro incorporation of L-canavanine into vitellogenin of the fat body of the migratory locust *Locusta migratoria* migratorioides. Proc Natl Acad Sci USA 78(9): 5480–5483
Piper CV (1925) The American species of *Canavalia* and *Wenderothia*. Contrib US Nat Herb 20: 555–585
Quiroz J, Godoy G, Gastro L, Oropeza C, Loyola-Vargas VM, Scorer KN (1987) *Canavalia ensiformis* (L.) DC tissue culture and embryo formation in vitro. In: Angarita Zerda A (ed) Int Congr Plant Tissue Cult Trop Species. Bogotá Colombia, p 23 (Abstr)
Rahiala EL, Kekomaki M, Janne J, Raina A, Raiha CR (1971) Inhibition of pyridoxal enzymes by L-canaline. Biochim Biophys Acta 227: 337–343

Ramírez M, Alpizar L, Quiróz J, Oropeza C (1991) Formation of L-canavanine in in vitro cultures of *Canavalia ensiformis* (L.) DC. Plant Cell Tissue Organ Cult (submitted)
Ranch JP, Oglesberry L, Zielinski AC (1985) Plant regeneration from embryo-derived tissue cultures of soybeans. In vitro Cell Dev Biol 21: 653–658
Rosenthal GA (1970) Investigations of canavanine biochemistry in the jackbean plant, *Canavalia ensiformis* (L.) DC. I. Canavanine utilization in the developing plant. Plant Physiol 46: 273–276
Rosenthal GA (1974) The interrelationship on canavanine and urease in seeds of the Lotoideae. J Exp Bot 25: 609–613
Rosenthal GA (1977a) Nitrogen allocation for L-canavanine synthesis and its relationships to chemical defense of the seed. Biochem Syst Ecol 5: 219–220
Rosenthal GA (1977b) The biological effects and mode of action of L-canavanine, a structural analogue of L-arginine. Q Rev Biol 52: 155–178
Rosenthal GA (1982) Plant nonprotein amino and imino acids. Biological, biochemical, and toxicological properties. Acad Press, New York, p 95
Rosenthal GA (1983) Biochemical adaptations of the bruchid beetle, *Caryedes brasiliensis* to L-canavanine, a higher plant allelochemical. J Chem Ecol 9(7): 803–815
Rosenthal GA (1988) The protective action of a higher plant toxic product. Bioscience 38(2): 104–109
Rosenthal GA, Rhodes D (1984) L-canavanine transport and utilization in developing jackbean, *Canavalia ensiformis* (L.) DC. (Leguminosea). Plant Physiol 76: 541–544
Rosenthal GA, Dahlman DL, Janzen DH (1976) A novel means for dealing with L-canavanine, a toxic metabolite. Science 192: 256–258
Rosenthal GA, Berge MA, Bleiler JA, Rudd TP (1987) Aberrant canavanyl protein formation and the ability to tolerate or utilize L-canavanine. Experientia 43(5): 558–561
Ruiter H de, Kollöffel C (1983) Arginine catabolism in the cotyledons of developing and germinating pea seeds. Plant Physiol 73: 525–528
Sastrapadja S, Lubis SHA, Lubis I, Sastrapadja D (1976) Studies in the javanese species of *Canavalia*. II Variation in mature and seedlings characters. Ann Bogor 6 : 97–110
Sauer J (1964) Revision of *Canavalia*. Brittonia 16: 106–181
Sauer J, Kaplan L (1969) *Canavalia* beans in American prehistory. Am Antiquity 34: 417–424
Schenk RU, Hildebrandt AC (1972) Medium and techniques for induction and growth of monocotyledonous and dicotyledonous plant cell cultures. Can J Bot 50: 199–204
Sehgal PP, Naylor AW (1966) Ontogenetic study of urease in jack beans, *Canavalia ensiformis* (L.) DC. Bot Gaz 127: 27–34
Shargool PD, Jain JC, McKay G (1988) Ornithine biosynthesis, and arginine biosynthesis and degradation in plant cells. Phytochemistry 27: 1571–1574
Smartt J (1985) Evolution of grain legumes II. Old and New World pulses of lesser economic importance. Exp Agric 21: 1–18
Stuart DA, Strickland SG (1984) Somatic embryogenesis from cell cultures of *Medicago sativa* L. I. The role of amino acid additions to the regeneration medium. Plant Sci Lett 34: 165–174
Thompson JF (1980) Arginine synthesis, proline synthesis, and related processes. In: Miflin BJ (ed) The biochemistry of plants. Acad Press, New York, pp 375–402
Vanderzant ES, Chremos JH (1971) Dietary requirements of the boll weevil for arginine and the effect of arginine analogues on growth and on the composition of the body amino acids. Ann Entomol Soc Am 64: 480–485
Walker KA, Sato SJ (1981) Morphogenesis in callus tissue of *Medicago sativa*: the role of ammonium ion in somatic embryogenesis. Plant Cell Tissue Organ Cult 1: 109–121
Warren RP, Hunt GE (1971) The biosynthesis of canavanine from $^{14}CO_2$ and its asymetric labeling in isolated pericarp tissue of *Canavalia ensiformis*. Planta (Berlin) 100: 258–261
Wilson VM, Haq N, Evans PK (1985) Protoplast isolation, culture and plant regeneration in the winged bean (*Psophocarpus tetragonolobus* L.). Plant Sci 41: 61–68
Wright JE, Srb AM (1950) Inhibition of growth in maize embryos by canavanine and its reversal. Bot Gaz 112: 53–57
Yañez IAA (1985) Comportamiento de las enzimas responsables en la asimilación del amonio en *Canavalia ensiformis*. Tésis de Licenciatura, Fac Química, Univ Nacional Autónoma de México, México, DF pp 71–72

IV *Capsicum* spp. (Peppers): In Vitro Production of Capsaicin

P.D. WILLIAMS[1]

1 General Account

Capsicum peppers are known by a variety of common names (chilli, paprika, pimiento, sweet, red, cayenne, and bird), which are more closely connected with their uses as foods and spices than with any taxanomic differences (Purseglove et al.

Fig. 1. *Capsicum annuum* cv. Ace plant and fruits

[1] Department of Chemical Engineering, University of Manchester Institute of Science and Technology, Manchester, UK

Biotechnology in Agriculture and Forestry, Vol. 21
Medicinal and Aromatic Plants IV (ed. by Y.P.S. Bajaj)

1981). However, the main division may be drawn between the sweet peppers, which are forms of *Capsicum annuum* and are eaten as a vegetable (Fig. 1), and the more pungent peppers, which can be *Capsicum annuum*, *C. frutescens*, *C. baccatum*, *C. pubescens* or *C. chinense*, and are generally used as spices (Fig. 2). The pungency of *Capsicum* peppers is derived mainly from the compound capsaicin (Fig. 3), one of the most pungent compounds known, which is detectable on the palate at dilutions of from 1 to 15 million (Todd 1959). It is with the pungent varieties of *Capsicum*, rather than the vegetable varieties, that the spice industry is concerned and the following figures are with reference to these varieties.

Fig. 2. *Capsicum frutescens* cv. Red Cherry plants and fruits

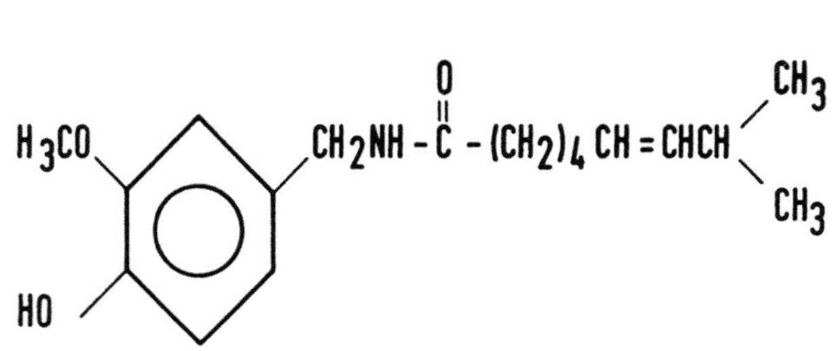

Fig. 3. The Capsaicin Molecule

The commercial cultivation of *Capsicum* spp. is widespread, with major producers including India, Pakistan, China, East Africa, the USA, and Mexico (Smith 1982). It has been estimated by the International Trade Centre (1982) that during the period 1976–1980, average imports of *Capsicum* ranged from 40 000 to 44 000 t/a, with a value of $67 million. They are, therefore, the second most important commodity in the international spice trade, after *Piper* pepper species. The main use of *Capsicum* fruits is for culinary purposes and seasoning, although extracted capsaicin does have a pharmaceutical role, internally as a powerful stimulant and externally as a counter-irritant in the treatment of such diseases as rheumatism. In western Europe and North America some two-thirds of *Capsicum* is processed by the food industry, going into products such as processed meats, spice mixes, curry powders, soups and a wide range of other prepared foods (Smith 1982).

Capsicum seeds are generally sown in a nursery bed or greenhouse, and plants are grown for 6–8 weeks before transplanting. Approximately 1 kg of seed should provide sufficient plants for 1 ha (Purseglove et al. 1981), with around 17 500 plants/ha (Boswell 1964), depending on cultivar. Harvesting takes place between 17–24 weeks after sowing (Gollifer 1973).

After harvesting, the dried spice is prepared from the fruit by reducing the water content from 65–80% down to approximately 10% (Purseglove et al. 1981). In all areas except the USA this is generally carried out by sun-drying, without any special preparation of the fruit. In the USA artificial drying takes place, with the fruit being exposed to a flow of air at 50–60 °C. Artificial drying gives a more consistent quality crop in a shorter time. The dried fruit may be stored whole or sliced, or as a ground spice powder.

An alternative to the ground spice, of increasing importance in industrial applications, is the oleoresin preparation. The bulk of oleoresin produced is used in the food industry and a small proportion (10–15% in the FRG, less in the USA and UK) is used by the pharmaceutical industry (Smith 1981). Oleoresins are prepared from highly pungent *Capsicum* species, dried down to 7–8% water content. These are ground and then extracted with a solvent such as acetone or ether. The solvent is finally removed from the crude oleoresin by distillation (Purseglove 1981).

2 In Vitro Approaches

2.1 Tissue Cultures for Secondary Metabolite Production

The production of the secondary metabolite capsaicin, the pungent flavour compound of peppers, by tissue cultures, has been investigated mainly in cultures of *Capsicum frutescens* Mill cv. annuum. In a series of publications Lindsey, Yeoman and co-workers (Yeoman et al. 1980; Lindsey et al. 1983; Lindsey and Yeoman 1984; Lindsey 1985, 1986) have demonstrated that the production of capsaicin by tissue cultures of *Capsicum* is inversely related to culture growth rate. They have further shown that secondary metabolism can be increased by manipulating culture conditions to limit primary metabolism.

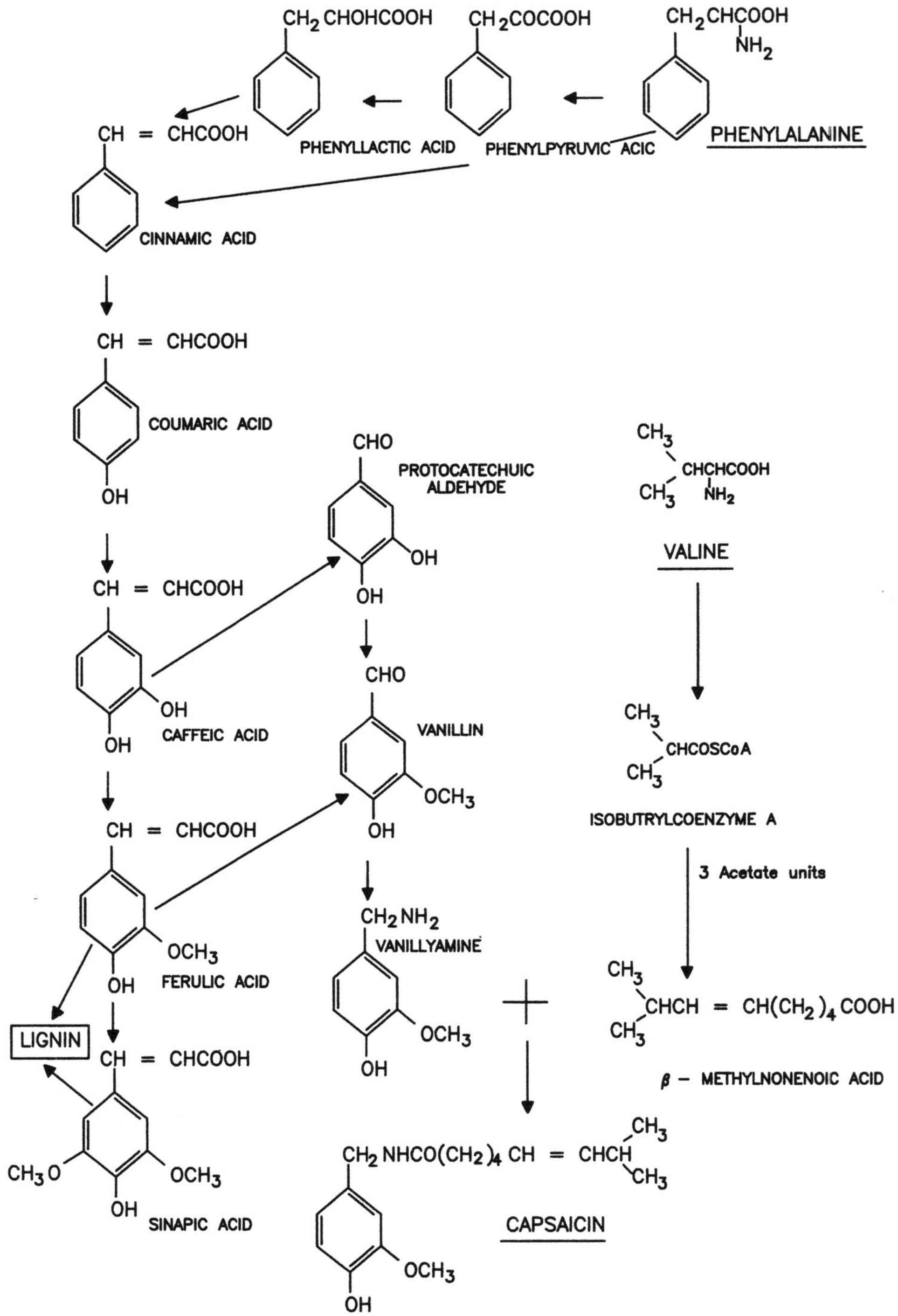

Fig. 4. Proposed biosynthetic pathway for capsaicin from phenylalanine and valine

A metabolic pathway, proposed by Yeoman et al. (1980), based on radiolabelling experiments and on information from the literature (McCalla and Neish 1959; Bennet and Kirby 1968; Leete and Loudon 1968), suggests that capsaicin is formed from phenylalanine and valine by two pathways which come together ultimately to form capsaicin from vanillylamine and 8-methyl nonenoic acid (Fig. 4). It can be seen from this pathway that there is a direct competition between primary metabolism, in the use of phenylalanine and valine for protein synthesis and for lignin synthesis, and secondary metabolism, in capsaicin synthesis. This was further confirmed by the inhibition of protein synthesis by colchicine, which gave an increased incorporation of phenylalanine and valine into capsaicin (Yeoman et al. 1980). Capsaicin was largely found in the culture medium, rather than in the cells of the culture. The reason for this preferential export of synthesized capsaicin is not fully understood, but it may be an active transport system, as in the transfer of capsaicin into secretory vesicles in the ripening fruit (Yeoman et al. 1980).

2.1.1 Media for Induction of Capsaicin Synthesis

Work on the capsaicin pathway suggested that the flow of precursors into primary metabolism (cell wall and protein synthesis) must be limited to allow their uptake into secondary metabolic pathways. Lindsey (1985) investigated the use of nutrient limitation in order to separate primary and secondary metabolism, using media with reduced concentrations of phosphate or of nitrate, or of both phosphate and nitrate. Culture of *Capsicum frutescens* cells on all of the depleted media resulted in decreased incorporation of labelled phenylalanine into protein and increased incorporation into capsaicin. This was also correlated with decreased growth rates. Best results, in terms of increased capsaicin production, were obtained in cultures in media with nitrate depletion only. It was therefore suggested that a two-phase system was required: firstly, growth in normal Schenk and Hildebrandt medium with 2,4-D, *p*-chlorophenoxyacetic acid and kinetin; secondly, capsaicin production in a medium with depleted nitrate and no growth regulators. Ravishankar et al. (1988) also found that a medium with eliminated nitrate gave increased production of capsaicin by cultures of *Capsicum annuum* immobilized in gel beads.

The elucidation of the pathway for capsaicin biosynthesis by Yeoman et al. (1980) allows the use of metabolic precursors to induce or increase capsaicin production. Mavituna et al. (1987a) used a nitrate-limited production medium, containing the precursors vanillylamine and valine, to produce capsaicin in 5l bioreactors from immobilized *Capsicum frutescens* (Sect. 2.1.4). The inclusion of sinapic acid in production media was also found to increase capsaicin levels (Mavituna et al. 1987b). The sinapic acid is thought to act as a sink for lignin synthesis, freeing precursors for secondary metabolic pathways (see Fig. 4).

2.1.2 Oxygen Stressing for Induction of Capsaicin Synthesis

The increase in capsaicin synthesis shown by cultures subjected to nutrient stressing suggested that other forms of applied stress might have similar effects. Wilkinson et al. (1988) used immobilized cultures of *C. frutescens* in bioreactors to examine the effects of oxygen limitation on cultures already undergoing nutrient limitation. They found that when aeration of the culture was discontinued and oxygen levels fell to 0–5% saturation, a significant increase in capsaicin production was achieved. The possibility that production was induced by the accumulation in the culture medium, during the period of non-aeration, of an essential volatile, was countered by an experiment in which the aeration was replaced by sparged nitrogen, giving a decrease in dissolved oxygen, without a cessation of gas-stripping. An increase in capsaicin production was again shown. It was thought that the effect of oxygen stressing was to limit primary metabolism, in the same manner as nutrient stressing, releasing precursors for capsaicin synthesis.

In all of the studies with oxygen stressing, it was found that restoration of oxygen levels to approximately 60% saturation led to the disappearance of capsaicin from the culture medium. This disappearance was shown to be dependent on the presence of viable cells in the culture on restoration of the oxygen supply, suggesting that capsaicin may be reabsorbed by the cells.

2.1.3 Immobilization

The use of separate growth and production media, the fact that the produced capsaicin is exported into the culture medium and is produced by slow growing or non-growing cultures, all suggest the possibility of using cell immobilization for capsaicin synthesis from tissue cultures. Immobilization has many advantages including ease of handling, a reduced shear environment and the possibility to use continuous or semi-continuous culture systems without loss of the cells. This last factor is important as it allows the long lead time of growing the *Capsicum* cultures to be offset against a long production time in which cells are used for continuous or repeated capsaicin synthesis. Lindsey et al. (1983) investigated the use of immobilized cultures of *C. frutescens*, using a novel immobilization technique previously developed for use with microorganisms (Atkinson et al. 1979). The method is dependent on the passive invasion of the pores of reticulate polyurethane foam matrices by the cells of a suspension culture. Lindsey and Yeoman (1984) found that this method of immobilization gave improved viability, with prolonged and enhanced ability to synthesize capsaicin. This was further illustrated by feeding labelled phenylalanine and labelled cinnamic acid to cultures of *C. frutescens* (Lindsey 1986), which showed increased incorporation of the label into capsaicin in immobilized cultures, compared to suspension cultures, particularly in nitrate-limited production medium. It was also shown that, in actively growing suspension cultures, labelled phenylalanine was largely incorporated into soluble protein and labelled cinnamic acid into a cell wall fraction. This again enforces the view that, under conditions favouring primary metabolism, precursors will be utilized in cell wall and protein synthesis at the expense of capsaicin synthesis.

Immobilization by passive entrapment is dependent on matching size distribution of the culture to pore size of the immobilization matrix. However, *C. frutescens* tends to grow in suspension culture as aggregates, with increasing aggregate size, and possibly agglomeration, during typical batch growth (Fig. 5; Mavituna et al. 1987a). Park and Mavituna (1984) interacted suspensions of *Capsicum* with foams of pore sizes of 2.5, 1.27 and 0.85 mm (10, 30, and 40 pores per inch) and showed

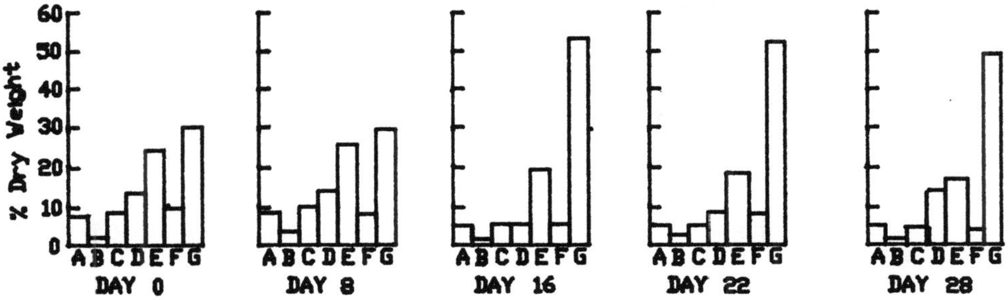

Key: A <0.25 mm, B 0.25–0.50 mm, C 0.50–1.00 mm, D 1.00–1.25 mm, E 1.25–2.75 mm, F 2.75–3.00 mm, G > 3.00 mm

Fig. 5. Changes in cell aggregate size distribution of *Capsicum frutescens* cultures during growth over 30 days

that only a limited portion of the spectrum of aggregate size in the culture was effectively entrapped, depending on the pore size of the foam matrix (Fig. 6).

Cell aggregates which are not entrapped will continue to grow in the bulk liquid of a culture and will compete for substrates, with negligible contribution to capsaicin production. For this reason, Williams et al. (1988) developed a method which allows some control of size distribution of plant cell aggregates. The aggregates were subject to high shear for a very short period of time, which gave a more homogeneous, finer culture (Fig. 7). Although the resulting cultures had initially low viabilities, this rapidly recovered (Fig. 8). The cultures resulting from this treatment are particularly efficient at immobilizing by the entrapment method (Fig. 9).

Cells immobilized by entrapment in foam matrices form compact masses of tissue and the rate of transfer of substrates through this mass could conceivably play an important part in the overall rate of the process of capsaicin production. Mavituna et al. (1987a) examined the rate of diffusion of labelled glucose through callus and through immobilized *Capsicum* tissue. Little difference was found

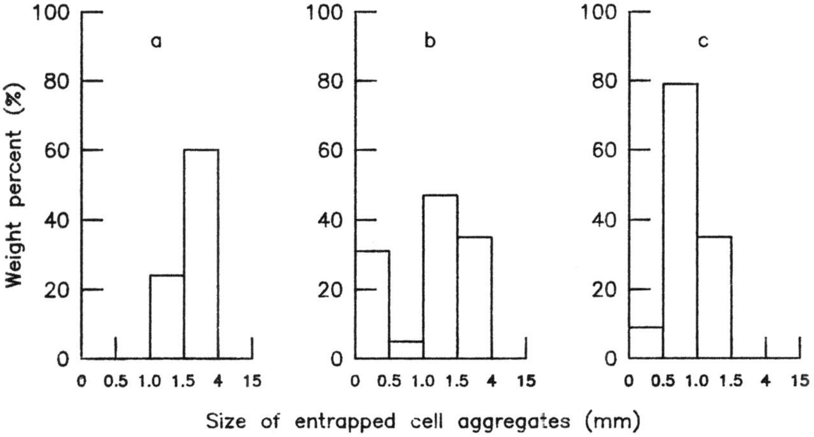

Fig. 6 a–c. Size distribution of cell aggregates of *Capsicum frutescens* entrapped from a suspension. Average pore size of foam: **a** 10 ppi; **b** 30 ppi; **c** 40 ppi

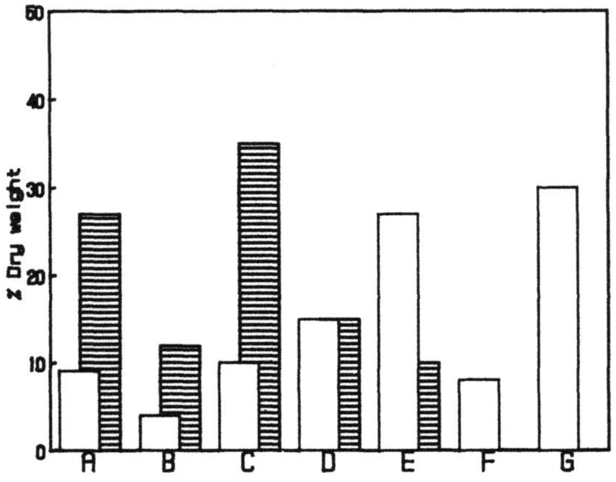

Key: A < 0.25 mm, B 0.25–0.50 mm, C 0.50–1.00 mm, D 1.00–1.25 mm, E 1.25–2.75 mm, F 2.75–3.00 mm, G > 3.00 mm

Fig. 7. Comparison of aggregate size distribution of normal suspension (*open*) and homogenized suspension (*hatched*) after 8 days growth

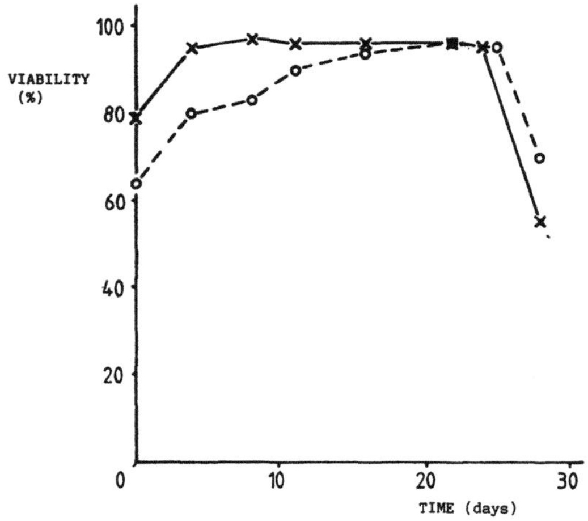

Fig. 8. Percentage viability of normal suspension culture and homogenized suspension culture. X——X Normal suspension culture. O– –O homogenized suspension culture

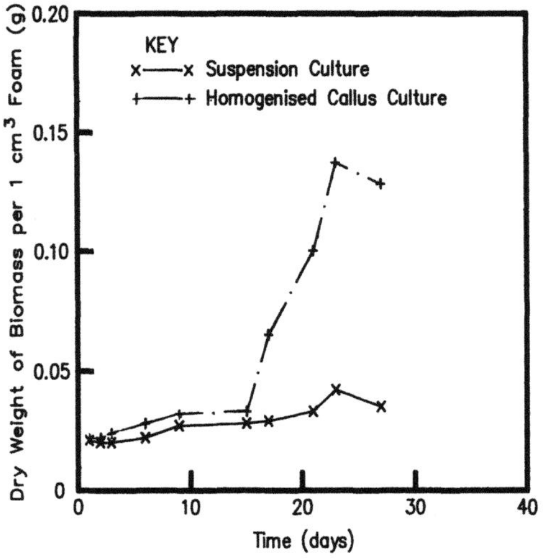

Fig. 9. Immobilization of normal suspension culture and homogenized porous foam matrices

between the two systems, with transfer of the labelled glucose in the immobilized cultures being, if anything, slightly more efficient than in the callus cultures. It was concluded that the rate of substrate diffusion was unlikely to be a problem in an immobilized system of this kind, and that entrapment in foam matrices may well compare favourably with immobilization in alginate beads on this point.

Cultures, in this case of *Capsicum annuum*, have also been immobilized in calcium alginate beads (Ravishankar et al. 1988). It was again found that cultures which showed no detectable capsaicin synthesis in suspension could be induced to produce capsaicin by immobilization.

2.1.4 Production of Capsaicin in Bioreactors

The use of two different forms of bioreactor for capsaicin production has been described by Mavituna et al. (1987b). The first of these was a circulating bed bioreactor, with 1000 l-cm^3 foam particles in 5-6 l of medium, circulated by an off-centre air sparge. In order to effect immobilization the foams were initially held as a packed bed by retractable grids, allowing the circulating culture to pass through, and be entrapped in the foams. Once the foams were full of immobilized culture the grids were withdrawn and the particles circulated. A second bioreactor design was an adaptation of a normal stirred tank, with foam sheets held as static vertical baffles around a central impeller. This gave efficient immobilization without the need for a separate mechanism for holding and releasing the immobilization matrix. It also had the advantage of being a simple adaptation of conventional equipment.

Figure 10 shows a production run with the circulating bed bioreactor, illustrating two methods of inducing capsaicin formation. An initial growth and immobilization period of 30 days was followed by a period of oxygen deprivation, during which capsaicin was detected at levels up to 0.45 mg/g dry cell weight. After restoration of the air supply, capsaicin disappeared from the medium and the growth medium was drained from the reactor and replaced by a production medium. This was nitrate-limited, contained no growth regulators and included vanillylamine, valine, and sinapic acid. The production of capsaicin at levels of up to 0.09 mg/g cell dry weight was promoted by this production medium, without the use of oxygen stressing.

A similar production run in the stirred bioreactor is shown in Fig. 11. The growth and immobilization period lasted for 37 days, after which time the growth medium was replaced by a production medium similar to that used in the previous example, but with ferulic acid as the only precursor in this case. Capsaicin levels reached 0.21 mg/g cell dry weight in the production medium.

2.1.5 Continuous Extraction of Capsaicin

In view of both the elusive nature of the product (Mavituna et al. 1987a), and the feedback inhibition which capsaicin production shows (Lindsey 1986), continuous extraction of capsaicin from the culture medium may be advantageous. Mavituna et al. (1987c) used food grade oil to extract capsaicin, finding that the solubility of capsaicin in oil was at least 500 times greater than in aqueous media. With

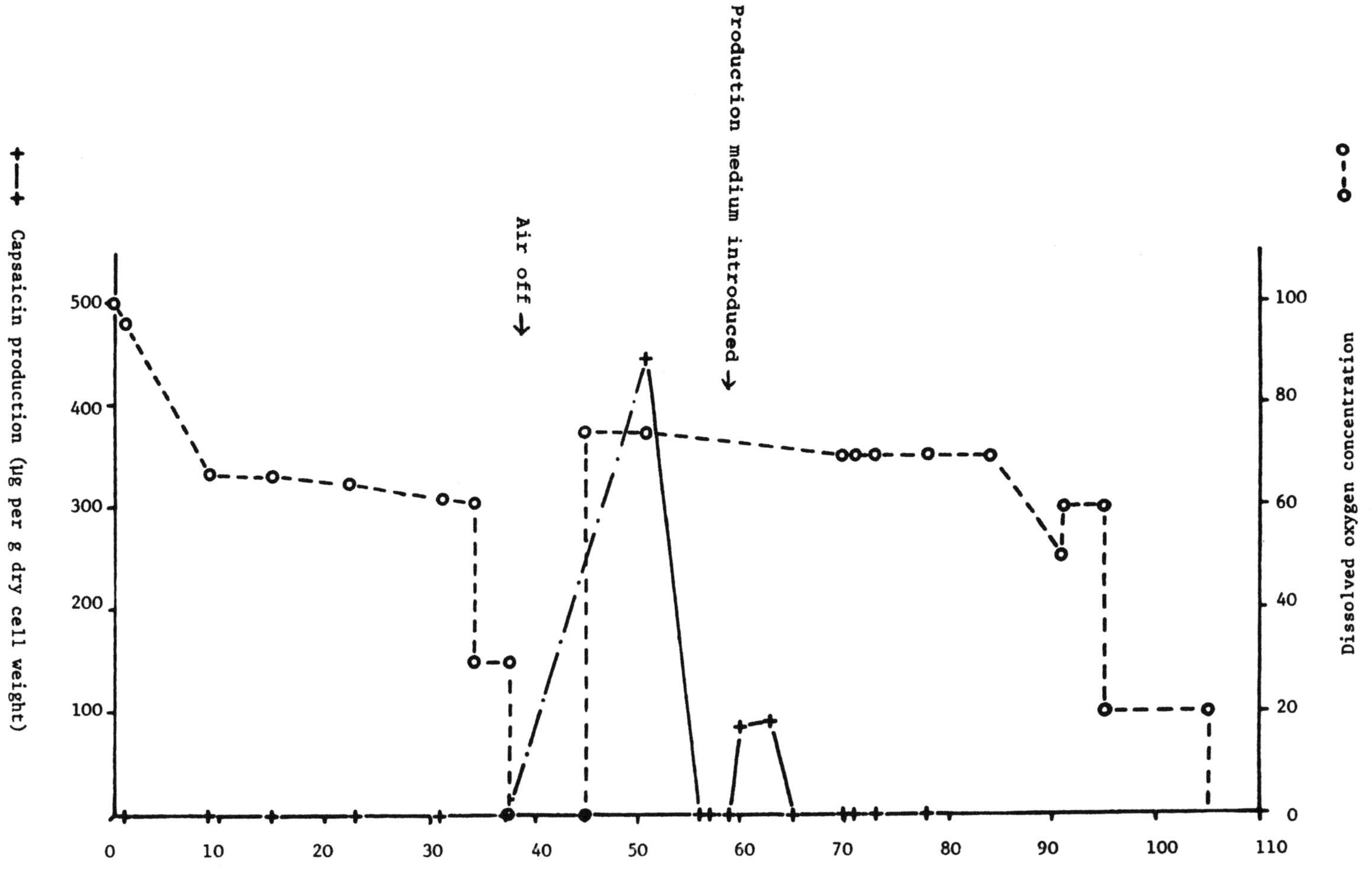

Fig. 10. Performance of a 5-l bioreactor with immobilized *C. frutescens* using S & H growth medium and production medium with vanillylamine, valine, and sinapic acid

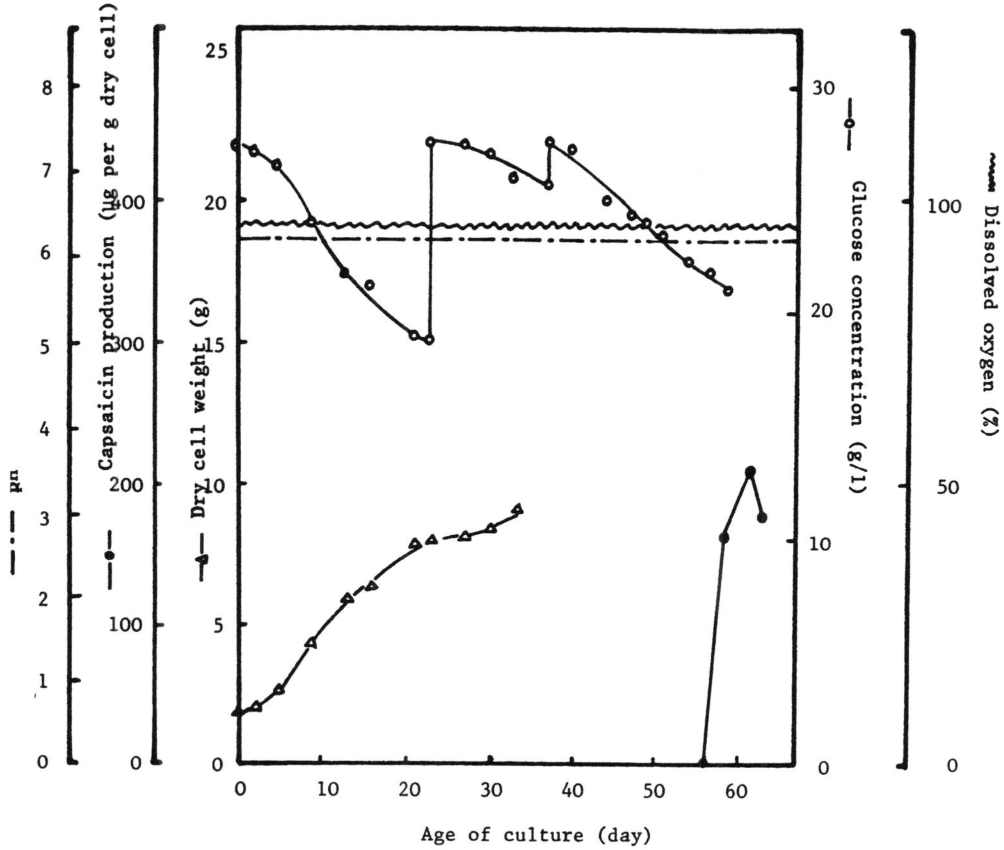

Fig. 11. Performance of a 5-l bioreactor with immobilized *C. frutescens* using modified S & H growth medium and production medium with ferulic acid

immobilized cells, when oil was present in the medium at a concentration of 17%, capsaicin production was significantly enhanced, up to yields of 1.5 mg/g cell dry weight. The feasibility of using oil extraction with bioreactors was examined with an oil column, through which the medium from a bioreactor of immobilized *Capsicum* was periodically passed. Capsaicin partitioned into the oil from the medium and capsaicin levels of up to 0.90 mg/g cell dry weight were produced.

3 Conclusions and Prospects

The use of immobilized *C. frutescens* cultures in production medium, in combination with continuous extraction, has given specific production rates of capsaicin, of 0.1 mg/g dry weight per day (Mavituna et al. 1987c), which compare well with those in ripening fruit, of 0.5 mg/g dry weight per day (Suzuki and Iwai 1984). However, many problems remain to be solved before the commercial production of capsaicin from tissue culture becomes feasible. Quoted levels of capsaicin are generally produced by selected high yielding cell lines, which are invariably unstable. The fluctuations of product levels in the culture media also requires further investigation, as do oxygen and nutrient stressing effects. Some initial work on a scale-up of the stirred immobilization bioreactor has been carried out at this laboratory, but further work is required on this, and on downstream processing and process integration.

4 Protocol

Callus cultures of *Capsicum* can be started from most types of explant, but leaf tissue is a convenient source. Schenk and Hildebrandt medium (1972), with 0.5 mg/l of 2,4-D, 2 mg/l of *p*-chlorophenoxyacetic acid and 0.1 mg/l of kinetin, is used for callus initiation and growth and also for suspension culture growth and maintenance. On this medium, at 25 °C, 50 ml cultures maintained in shaken 250-ml Erlenmeyer flasks will require transfer every 3–4 weeks. For initiating and increasing capsaicin yields, cultures are transferred to a production medium, which differs from normal Schenk and Hildebrandt medium in having potassium nitrate and ammonium dihydrogen phosphate replaced by potassium dihydrogen phosphate at 200 mg/l. The production medium also contains no growth regulators.

References

Atkinson B, Black GM, Lewis PJS, Pinches A (1979) Biological particles of given size, shape and density for use in biological reactors. Biotechnol Bioeng 21: 193–200

Bennett DJ, Kirby GW (1968) Constitution and biosynthesis of capsaicin. J Chem Soc C1: 442–446

Boswell VR (1964) Pepper production. US Dep Agric Inf Bull 276

Gollifer DE (1973) The introduction of spice crops into the British Solomon Islands. Proc Conf on Spices, Int Trade Centre, London

International Trade Centre UNCTAD/GATT (1982) Spices, a survey of the world market. Int Trade Centre, Geneva

Leete E, Loudon MCL (1968) Biosynthesis of capsaicin and dihydro-capsaicin in *Capsicum frutescens*. J Am Chem Soc 90: 6837–6841

Lindsey K (1985) Manipulation, by nutrient limitation, of the biosynthetic capacity of immobilised cells of *Capsicum frutescens* Mill. cv. annuum. Planta 165(1): 126–133

Lindsey K (1986) Incorporation of [^{14}C]phenylalanine and [^{14}C]cinnamic acid into capsaicin in cultured cells of *Capsicum frutescens*. Phytochemistry 25(12): 2793–2801

Lindsey K, Yeoman MM (1984) The viability and biosynthetic activity of cells of *Capsicum frutescens* Mill. cv. annuum immobilised in reticulate polyurethane. J Exp Bot 35: 1684–1696

Lindsey K, Yeoman MM, Black GM, Mavituna F (1983) A novel method for the immobilisation and culture of plant cells. FEBS Lett 155: 143–149

Mavituna F, Park JM, Williams PD, Wilkinson AK (1987a) Characteristics of immobilised plant cell reactors. In: Webb C, Mavituna F (eds) Process possibilities for plant and animal cell cultures. Ellis Horwood, Chichester, pp 92–115

Mavituna F, Wilkinson AK, Williams PD, Park JM (1987b) Production of secondary metabolites by immobilised plant cells in novel bioreactors. In: Moody GW, Baker PB (eds) Bioreactors and biotransformations. Elsevier, London, pp 26–38

Mavituna F, Wilkinson AK, Williams PD (1987c) Liquid-liquid extraction of a plant secondary metabolite as an integrated stage with bioreactor operation. In: Verrall MS, Hudson MJ (eds) Separations for biotechnology. Ellis-Horwood, Chichester, pp 333–340

McCalla DR, Neish AC (1959) Biosynthesis of phenolic cinnamic acids. Can J Biochem Physiol 37: 537–547

Park JM, Mavituna F (1986) Factors affecting the immobilisation of plant cells in biomass support particles. In: Webb C, Black GM, Atkinson B (eds) Process engineering aspects of immobilised cell systems. Pergamon Press, Oxford, pp 295–303

Purseglove JW, Brown EG, Green CL, Robbins SRJ (1981) Spices, vol 1, Longman, London

Ravishankar GA, Sarma KS, Venkataraman LV, Kadyan AK (1988) Effect of nutritional stress on capsaicin production in immobilised cell cultures of *Capsicum annuum*. Curr Sci 57: 381–383

Schenk RU, Hildebrandt AC (1972) Medium and techniques for induction and growth of monocotyledenous and dicotyledenous plant cell cultures. Can J Bot 50: 199–204

Smith A (1982) Selected markets for chillies and paprika. Trop Prod Inst, London

Suzuki T, Iwai K (1984) Constituents of red pepper species: chemistry, biochemistry, pharmacology and food science of the pungent principle of *Capsicum* species. In: Brossi A (ed) The alkaloids, vol 23. Academic Press, London, pp 227–299

Todd EC (1959) New extraction process improves paprika colour. Food Eng (Aug): 82–83

Wilkinson AK, Williams PD, Mavituna F (1988) The effect of oxygen stress on production of capsaicin by immobilised *Capsicum frutescens*. 8th Int Biotechnol Symp, Paris, Societe Francaise de Microbiologie, Abstr, p 285

Williams PD, Wilkinson AK, Lewis JA, Black GM, Mavituna F (1988) A method for the rapid production of fine plant cell suspension cultures. Plant Cell Rep 7: 459–462

Yeoman MM, Miedzybrodzka MB, Lindsey K, McLauchlan WR (1980) The synthetic potential of cultured plant cells. In: Sala F, Parisi B, Cella R, Ciferri O (eds) Plant cell cultures: results and perspectives. Elsevier, Amsterdam, pp 327–343

V *Cassia didymobotrya* (Wild Senna): In Vitro Culture, Biotransformation and the Production of Secondary Metabolites

B. Botta and G. Delle Monache[1]

1 Introduction

1.1 Distribution and Importance

The genus *Cassia* (tribe *Caesalpinoideae*, family *Leguminosae*) consists of more than 70 species that grow preferentially in the hottest areas of the world. Many species are used in folk medicine in the treatment of constipation, rheumatism, gonorrhea, diabetes, etc.

Cassia didymobotrya (Masai e senetoi, Kilongo mchora) Fres. (Fig. 1) is a plant widely used as a purgative and an antimalarial medicine; a decoction of the leaves is also used to stimulate hunger as well as to cure stomach diseases (Watt and

Fig. 1. *Cassia didymobotrya*

[1] Centro Chimica dei Recettori, Istituto di Chimica, Università Cattolica del S. Cuore, Largo F. Vito 1, 00168 Roma, Italy

Biotechnology in Agriculture and Forestry, Vol. 21
Medicinal and Aromatic Plants IV (ed. by Y.P.S. Bajaj)

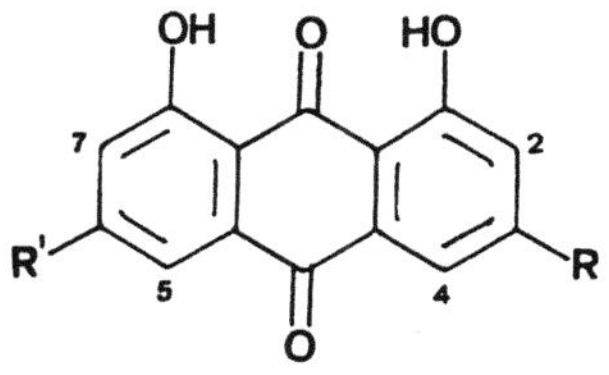

	R	R'
1	CH_3	H
2	CH_3	OCH_3
1a	CH_2OH	H
2a	CH_2OH	OCH_3
1b	COOH	H
2b	COOH	OCH_3

Fig. 2. Anthraquinones isolated from leaves of *Cassia didymobotrya*. (El-Sayyad and Ross 1983; Alemayehu 1989)

Breyer-Brandwijk 1962). Leaves and roots of *Cassia didymobotrya* are reported to contain the anthraquinones (Fig. 2), chrysophanol (*1*), aloe-emodin (*1a*) and rhein (*1b*), which account for the laxative effect, and flavonoid glycosides (El-Sayyad and Ross 1983). A trisaccharide, raffinose, and choline have been isolated from the seeds, while a small amount of an alkaloid, cassein, was obtained from the pod (Watt and Breyer-Brandwijk 1962). Recently, further anthraquinones (Fig. 2), i.e., physcion (*2*), fallacinol (*2a*), and parietinic acid (*2b*), as well as torosachrysone (see below), have been reported from the leaves (Alemayehu et al. 1989).

2 In Vitro Approaches

2.1 Establishment of Cell Suspension

Our work represents the first study of cell culture development with *C. didymobotrya*: for the purpose of establishing cell culture the orthogonal design method (Botta et al. 1989a) was applied after an appropriate growing callus was developed.

Seeds of *C. didymobotrya* were collected during the summer of 1985 at the University of Addis Ababa (Ethiopia). The seeds were washed, sterilized according to standard methods (Yeoman and MacLeod 1977), and aseptically transferred to sterile Petri dishes containing 10 ml of MS basal salt medium (Murashige and Skoog 1962) enriched with sucrose (3%) and solidified with 0.9% (w/v) agar. Explants of hypocotyl and stem, excised from 5–7-day-old seedlings, were inoculated aseptically, with two explants per 100-ml Erlenmeyer flask containing 20 ml of basal culture medium to which 2,4-D (1 ppm) and kinetin (0.1 ppm) were added. Induction of callus occurred after 3–5 weeks cultivation in the dark at 26 °C. The pH of the basal medium for both callus and cell suspensions was adjusted to 5.5 before autoclaving. Suspension cultures were initiated by transferring callus (0.5 g fresh weight) to 250-ml Erlenmeyer flasks containing 65 ml of the abovementioned basal medium containing 2,4-D and kinetin. The initial suspension cultures were grown in the dark at 26 °C in a rotary shaker (New Brunswick G 10) at 120 rpm. Subculturing was performed every 3 weeks to maintain the parent culture.

In order to obtain higher levels of biomass, the parent culture was filtered through a nylon mesh and then grown in different media containing the following four combinations of hormones: (1) 2,4-D (1 mg/l) and kinetin (0.1 mg/l); (2) 2,4-D (1 mg/l) and BAP (0.1 mg/l); (3) IAA (1 mg/l) and BAP (0.1 mg/l); (4) IAA (1 mg/1) and BAP (0.1 mgl).

Cell growth and metabolite production of the cultures obtained from the above studies were followed by determination of dry weight of the cells and weight of the EtOH extracts, respectively. Having established that the combination of 2,4-D and kinetin (condition 1, Fig. 3) was optimum, the culture was then grown according to the following method. To 65 ml of basal culture medium enriched with 2,4-D and kinetin, 1 g (fresh weight) of actively growing cell cultures was added. Cell growth and metabolite accumulation were monitored by systematic measurements (every 2 days) of the following parameters: dry weight of the cells, refractive index of the medium, pH and weight of EtOH extract expressed as percentage of cell dry weight (Fig. 4).

In order to optimize the growth of the suspension cultures as well as the secondary metabolite production, a series of experiments were carried out in which the hormonal concentration, the amount of sucrose, and the weight of inoculum were varied according to a fractional factorial design (Box et al. 1978), namely, the orthogonal design method. Similar multivariate designs have been applied for optimization of an organic synthesis (Wold et al. 1986) and for toxicological evaluation of chemical data (Tosato et al. 1988).

In brief, let us consider the information derived from a series of n^k experiments (where k represents the number of factors which can be changed in n different ways, i.e., at n different levels). Significant information can be obtained when a number of carefully selected experiments are completed. In Table 1, a typical example of orthogonal design is illustrated. In this instance, nine selected experiments are

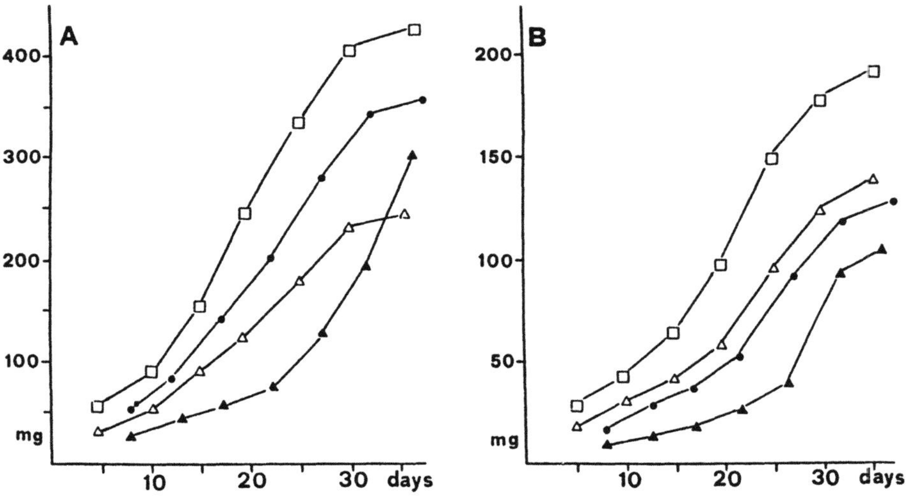

Fig. 3A, B. Selection of the pairs of hormones in cell culture of *C. didymobotrya*. Inoculum of 0.5 g (fresh wt.) in MS basal medium enriched with the following hormones: —□—□— 2,4-D (1 mg/1), kinetin (0.1 mg/1); —△—△— 2,4-D (1 mg/1), BAP (0.1 mg/l); —●—●— IAA (1 mg/l), kinetin (0.1 mg/l); —▲—▲— IAA (1 mg/1), BAP (0.1 mg/1). **A** Cell growth; **B** metabolite production. (Botta 1989a)

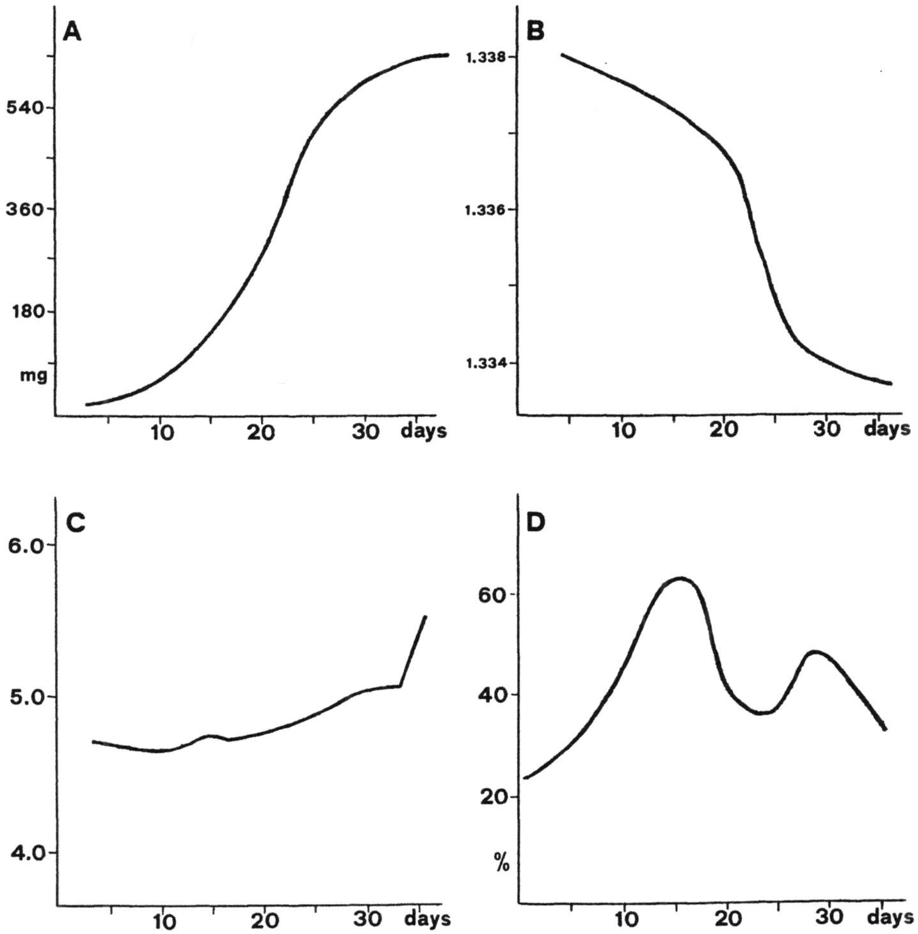

Fig. 4A–D. Growth control parameters of suspension cultures of *C. didymobotrya*. **A** Dry weight; **B** refractive index; **C** pH; **D** percentage of EtOH extract. (Botta 1989a)

Table 1. Example of orthogonal design. (Botta 1989a)

Test \ Row	A	B	C	D
I	1	2	3	1
II	2	1	1	1
III	3	3	2	1
IV	1	1	2	2
V	2	3	3	2
VI	3	2	1	2
VII	1	3	1	3
VIII	2	2	2	3
IX	3	1	3	3

performed by varying the four parameters A, B, C, and D at three (1, 2, 3) different levels. The general rules of an orthogonal design are expressed as follows:

1. Any level appears the same number of times in each row; for example, levels 1, 2, and 3 appear three times in row A.
2. In any two rows (for example, A and B) the horizontal combinations of any two numerals appear the same number of times. That is, the combination 1(2) appears only *once* in rows A and B when reading these columns in a vertical manner.

Similarly, the combinations 2(1), 3(3), 1(1), etc. also appear only once in rows A, B. Three experiments are then conducted for each level, while each of the other factors is changed at three different levels. Thus, in row A, three experiments (tests I, IV, and VII) involve level 1 at different levels 1, 2, and 3 in rows B, C, and D. We now wish to evaluate the influence of a given parameter at the different levels of the other parameters used in any given experiment. For example, if we consider parameter B, the influence of parameter B at level 1 is expressed as a summation, ΣRB(1), of the results derived from experiments II (R2), IV (R4), and IX (R9), i.e., ΣRB(1) = R2 + R4 + R9. In a similar fashion, ΣRB(2) = R1 + R6 + R8, while ΣRB(3) = R3 + R5 + R7 (see Table 1).

Thus, the algebraic difference (ΔΣRB) between the highest and lowest values obtained from ΣRB(1), ΣRB(2), and ΣRB(3), compared with the other parameter summations (ΔΣRA, ΔΣRC, and ΔΣRD), will account for the influence (major *or* minor) of parameter B in the experiment. For a practical example, in Table 2, we illustrate the effect of kinetin (parameter B) at the different concentrations

Table 2. Optimization experiments for n = 30 days. (Botta 1989a)

Test	2,4-D (mg/l)	Kinetin (mg/l)	Sucrose (g/l)	Inoculum[a] (mg)	P_n (mg)	P_n/P_0	E_n (mg)	E_n/E_0
I	1.0	0.1	30	36	546	15	191	24
II	0.5	0	10	36	145	4	24	3
III	1.5	0.2	50	36	503	14	128	16
IV	1.0	0	50	18	207	11.5	72	18
V	0.5	0.2	30	18	101	5.5	36	9
VI	1.5	0.1	10	18	91	5	52	13
VII	1.0	0.2	10	54	283	4	80	6.5
VIII	0.5	0.1	50	54	353	6.5	181	15
IX	1.5	0	30	54	295	5.5	107	9
Cell growth								
ΣR(1)	16	21	13	22				
ΣR(2)	30.5	26.5	26	33				
ΣR(3)	24.5	23.5	32	16				
ΔR	14.5	5.5	19	17				
Metabolite production								
ΣR(1)	27	30	22.5	40				
ΣR(2)	48.5	52	42	43				
ΣR(3)	38	31.5	49	30.5				
ΔR	21.5	22	26.5	12.5				

[a] The amount of the inoculum is expressed as dry weight: 18, 36, and 54 mg correspond to 0.5, 1, and 1.5 g fresh weight, respectively. The weights (E_0) of the extracts (22%) are 4, 8, and 12 mg, respectively.

(0 mg/l = level 1; 0.1 mg/l = level 2; and 0.2 mg/l = level 3) with the other parameters also at varying concentrations: parameter A, 2,4-D, at level 1 (1 mg/l), level 2 (0.5 mg/l), level 3 (1.5 mg/l); parameter C, sucrose, at level 1 (10 g/l), level 2 (50 g/l), level 3 (30 g/l): parameter D, inoculum concentration at level 1 (1 g, fresh weight in 65 ml), level 2 (0.5 g), level 3 (1.5 g). The results R are expressed as the total increment in cell growth = P_n/P_0 (P_n = dry weight of the cells after n days in culture; P_0 = dry weight of the inoculum) and total increment in metabolite production = E_n/E_0 (E_n = amount of the secondary metabolite production after n days in culture; E_0 = amount of the secondary metabolite production in the inoculum). In order to determine the influence of kinetin at the different levels for cell dry weight, Table 2 reveals the highest ΣR value of 26.5 and the lowest value of 21, thereby providing the value for $\Delta\Sigma RB$ of 5.5. A similar calculation for the EtOH extract provides a $\Delta\Sigma RB$ of 22. Calculations of the other parameters are shown in Table 2.

The conclusions of Table 2 are expressed in graphic form (tendency diagram) in Fig. 5. For example, Fig. 5A illustrates the influence of 2,4-D on cell dry weight (—) and EtOH extract (---), while Fig. 5B shows the effect of kinetin, etc.

Both cell growth and metabolite accumulation followed the same qualitative course when the different combinations of hormones were varied as shown in Fig. 3. The combination of 2,4-D and kinetin afforded the best cell growth (up to 18 times from day 0) as well as optimum metabolite production. As a result, this system was employed in the following experiments.

Cell growth and metabolite production were then monitored by systematic determination of dry weight (Fig. 4A), refractive index (Fig. 4B), pH (Fig. 4C), and weight of EtOH extract expressed as percentage of dry weight (Fig. 4D). Analysis of Fig. 4A, B reveals that biomass increase coincided with a corresponding decrease in refractive index. On the other hand, the change in pH (Fig. 4C) was insignificant over the 35-day growth period. Finally, metabolite production (mg of total extract versus mg of dry cells) varied with the age of the cultures, ranging from 22% in the initial inoculum to 63% at day 15 (Fig. 4D).

The use of the orthogonal design method was highly effective in determining the most important growth parameters for secondary metabolite production within a given cell culture system.

As Table 2 indicates, the kinetin concentration has a relatively minor importance for the cell growth, however, it is an influential parameter for the production of secondary metabolites. The optimum concentration is 0.1 mg/l (Fig. 5B). Conversely, the weight of inoculum (Table 2) is more important for the production of biomass and less important for the accumulation of metabolites. The optimum amount of inoculum is 1 g (fresh weight) of cells in 65 ml of medium (Fig. 5C). Noteworthy is that an amount of inoculated cells higher than 1 g (fresh weight) produces a negative effect.

The addition of 2,4-D has the same influence in both cases (Table 2) with the best results obtained at 1 mg/l (Fig. 5A). With regard to sucrose (Fig. 5D) a definite maximum is not observed, but the concentration of 30 g/l was selected for the maintenance of the suspension culture on comparison of the increases obtained, ranging from 10–30 g/l and from 30–50 g/l, respectively. Similar results (not shown) were obtained in the application of the orthogonal design method for 22 days.

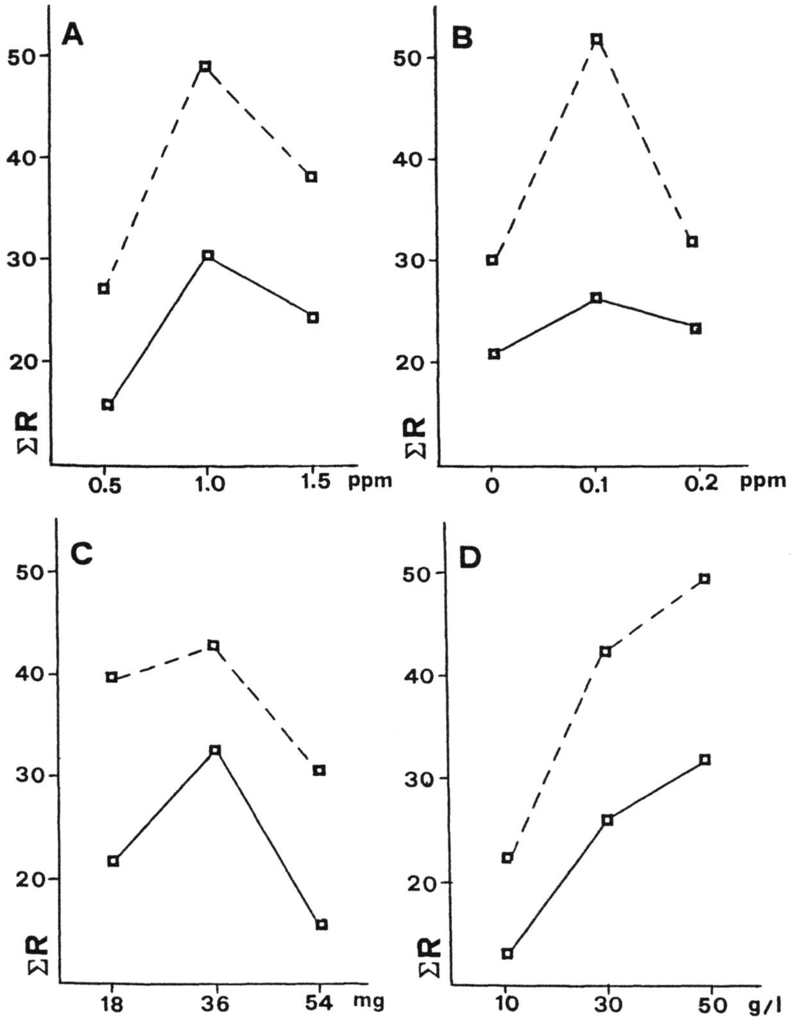

Fig. 5A–D. Tendency diagrams relative to cell growth (——) and metabolite production (---) of 30-day-old suspension cultures of *C. didymobotrya*. **A** 2,4-D; **B** kinetin; **C** inoculum; **D** sucrose. (Botta 1989a)

The use of nine experiments to establish these optimum conditions is indeed attractive since normally many more experiments are required in a conventional study.

2.2 Extraction and Identification of Cell-Produced Metabolites

2.2.1 Standard Extraction Procedure

The filtered cells were washed thoroughly with water and extracted with EtOH under reflux. The residue of three combined extracts was suspended in water and extracted initially with chloroform and then with ethyl acetate.

Column chromatography of the former extract afforded anthraquinones and stilbenes, whereas the latter extract provided mainly flavonoids. The main metabolites thus isolated were evaluated by HPLC, employing different solvent systems in accord with the polarity of the extracts.

2.2.2 Standardization of Metabolites

The residue of the partial extracts at different ages of the cells was diluted with MeOH to a concentration of 5 mg/ml and a portion of 20 μl was injected in a Perkin Elmer LC 85 spectrophotometer supplied with the following equipment: pump system, LC 10; recorder, LKB 2210; column, Spheri-5 RP 18 (100 × 4.6 mm); guard column, New Guard Cartridge.

The anthraquinones were monitored by performing HPLC profiles with isocratic elution (1 ml/min) in a solvent system of MeOH/water, 85 : 15. Detection was by absorbance at 430 nm. Conversely, the conditions for the flavonoids were as follows: solvent system MeOH/water, 55 : 45, containing 1% TFA; UV detector at 370 nm.

It is highly interesting to note that secondary metabolite accumulation showed two maxima, corresponding to the early exponential phase (15th day) and to the early stationary phase (28th day). Notably, the composition of the extracts was different in that anthraquinones were predominant in the former phase, whereas flavonoids were the major secondary metabolites in the latter. Retention times of the main anthraquinones and flavonoids are reported in Table 3. The five major anthraquinones, initially produced in the calli, are similarly present in the early stages of growth in suspension cultures (up to 15 days) but gradually disappear, in part at the expense of anthraquinone analogs. The total amount of 4% (weight versus dry weight of cells) of anthraquinones was obtained from 15-day-old cultures. On the other hand, flavonoids do not appear in the callus but start to accumulate in the cell suspension (after approx. 10 days or at the late lag phase) and reach the maximum concentration of 3.5% (weight versus dry weight of cells) at

Table 3. Retention time for anthraquinones and flavonoids isolated from cell cultures of *C. didymobotrya*. (Botta 1989a)[a]

Compound	*Retention time*
Anthraquinones[b]	
Questin	2 min 30 s
Emodin	3 min 6 s
Chrysophanol	3 min 36 s
Physcion	4 min 36 s
Nataloe emodin 8-methyl ether	5 min
Flavonoids[c]	
Quercetin	3 min 24 s
Quercetin 3-methyl ether	3 min 54 s
Kaempferol	4 min 48 s

[a] Flow rate 1 ml/min.
[b] Solvent system: MeOH/water, 85 : 15.
[c] MeOH-water-TFA (55 : 45 : 1).

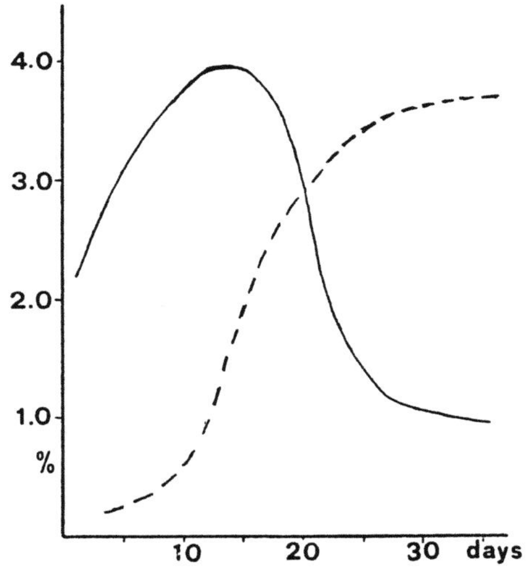

Fig. 6. Selective metabolite accumulation in suspension cultures of *C. didymobotrya* (wt. versus dry wt. of cells). —— Anthraquinones; ---- flavonoids. (Botta 1989a)

day 28. The above findings, which are summarized in Fig. 6, clearly reveal a different rate of biosynthesis for these two families of compounds.

2.3 Structure Elucidation

Several metabolites were isolated by partition of the total EtOH extract among hexane, $CHCl_3$, and EtOAc, followed by extended chromatography. Details of the structure determination of these compounds are reported in Delle Monache et al. (1991). The following classes of products were represented.

Anthraquinones (Fig. 7). Among the anthraquinones produced from the suspension cultures of *C. didymobotrya*, only the common chrysophanol (*1*) and physcion (*2*) were present in the parent plant. Emodin (*3*), quite profuse in the *Cassia* genus, was also produced.

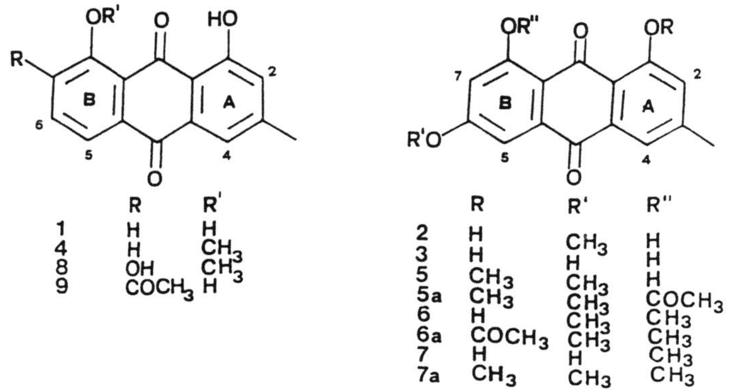

Fig. 7. Anthraquinones isolated from in vitro cultures of *C. didymobotrya*. (Delle Monache 1991)

A typical feature of most of the isolated metabolites was the presence of an 8-methoxy group, as in chrysophanol-8-monomethyl ether (*4*), which was identified by comparison of the ^{1}H-NMR spectrum with those of the methyl derivatives of chrysophanol (de Oliveira et al. 1977). The identification was confirmed by the ^{1}H-NMR spectrum of the acetyl derivative, which showed a low-field shift of the H-2 and H-4 proton signals. Chrysophanol-8-methyl ether has been reported from *C. speciosa* Schrad (de Oliveira et al. 1977) and from *C. obtusifolia* (Kitanaka and Takido 1986).

Emodin-1,6-dimethyl ether (*5*) and emodin-6,8-dimethyl ether (*6*) were distinguished on the basis of the low-field shifts induced in the ^{1}H-NMR spectrum by acetylation on the aromatic protons of rings B and A, respectively. Compound *6* has been isolated from several plants (Thomson 1987), but not from *Cassia*, whereas the presence of metabolite *5* in a fungus (Bodo et al. 1980) was not supported by spectroscopic details.

Questin (*7*), isomeric with physcion (*2*), resulted in compound *6* after treatment with CH_2N_2, and was thus assigned the structure of emodin-8-methyl ether. Since compound *7* was found in several cultures of microorganisms (Thomson 1987), the suspended cells were microscopically controlled before extraction and the presence of bacteria was excluded. On the other hand, questin has also been isolated from *Cassia obtusifolia* Linn (de Oliveira et al. 1977).

A fifth anthraquinone featured a C-8 methoxy group, whose resonance (δ 62) in the ^{13}C-NMR spectrum revealed, together with the proton signals of the ^{1}H-NMR spectrum, the substitution pattern of structure *8*. The compound, namely, nataloe-emodin-8-methyl ether was obtained by $FeCl_3$ oxidation of homonataloin, an anthrone C-glycoside present in *Aloe* spp. (Haynes et al. 1960). As a free quinone, it has been reported only from *Aloe speciosa* Bak (van Rheede van Oudtshoorn 1964).

The last anthraquinone isolated from the tissue culture of *C. didymobotrya* displayed in the ^{1}H-NMR spectrum the typical signals of the A ring of the above reported anthraquinones. Conversely, the B ring was characterized by the resonances for two *ortho*-coupled protons (at 7.86 and 8.16 ppm), a chelated hydroxy group (at 13.18 ppm), and an acetyl substituent (3H at 2.77 ppm). The presence of the latter was supported in the IR spectrum by an absorption band at $\nu = 1680\ cm^{-1}$ and in the mass spectrum by the prominent loss of a methyl group, typical of acetophenones (Budzikievicz et al. 1967). The low-field values of 8-OH and H-6 signals require the location of the acetyl group at C-7. The compound was thus assigned the new structure of 7-acetylchrysophanol (*9*).

Bianthrones (Fig. 8). Characteristic features of bianthrones are the longer wavelength UV absorption (360 nm) and the H-10, H-10′ proton resonances in the ^{1}H-NMR spectrum (at 4.3–4.5 ppm).

The metabolites *10* and *12* were identified with chrysophanol-10,10′-bianthrone and physcion-10,10′-bianthrone, respectively. The two compounds were, as expected (Delle Monache 1985), a mixture of ($\pm$) and *meso* forms. Notably, the ^{1}H-NMR spectrum of *11* may be seen as a result of the superimposition of those of compounds *10* and *12*. Analogously, in the mass spectrum of *11*, two fragments, corresponding to chrysophanol-anthrone (*m/z* 240) and to physcion-anthrone (*m/z* 270), were present. Therefore, the metabolite *11* was assigned the structure of

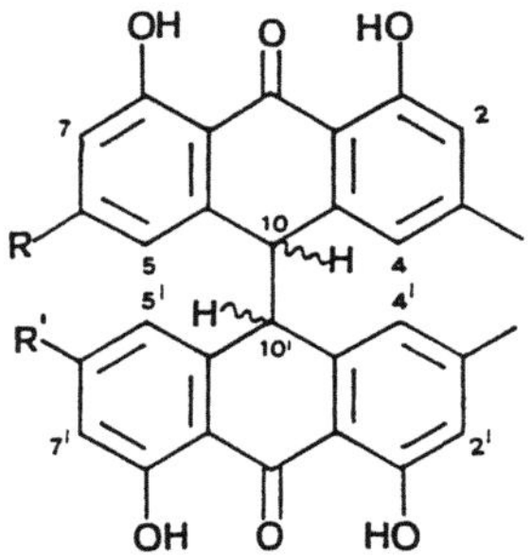

	R	R'
10	H	H
11	H	OCH_3
12	OCH_3	OCH_3

Fig. 8. Bianthrones isolated from in vitro cultures of *C. didymobotrya.* (Delle Monache 1991)

chrysophanol-physcion-10,10′-bianthrone. The formation of these compounds from tetrahydroanthracenes will be discussed below.

Stilbenes (Fig. 9). The stilbenes (UV λ_{max} at 313 nm) *13* and *14* were isomers, the only difference being the olefinic double bond, which was *trans* in the former (signals at δ 7.01 and 6.87 with J = 16 Hz) and *cis* in the latter (signals at δ 6.53 and 6.44 with J = 12 Hz). The substitution pattern of *13* and *14*, both possessing one hydroxy and two methoxy groups, was disclosed by the signal multiplicities in the aromatic region. The H-4′ proton showed *meta* couplings with H-2′ and H-6′ protons. These two were not equivalent and thus a 3′-hydroxy-5′-methoxy substitution was attributed to ring B. The coupling pattern in the ^{1}H-NMR spectrum of the aromatic protons and the resonance values in the ^{13}C-NMR spectrum (at δ 148–149) of the oxygenated carbons required a 3,4-dimethoxy substitution for ring A. In conclusion, the metabolites *13* and *14* were assigned the structures of (E)- and (Z)-3′-hydroxy-3,4,5′-trimethoxystilbene, respectively. It must be noted that compound *14* may have been formed from *13* by photoisomerization (Drewes and Fletcher 1974).

The ^{1}H- and ^{13}C-NMR spectra of the (E)-stilbene *15*, featuring two hydroxy and two methoxy groups, revealed the same substitution pattern of *13* and *14* for

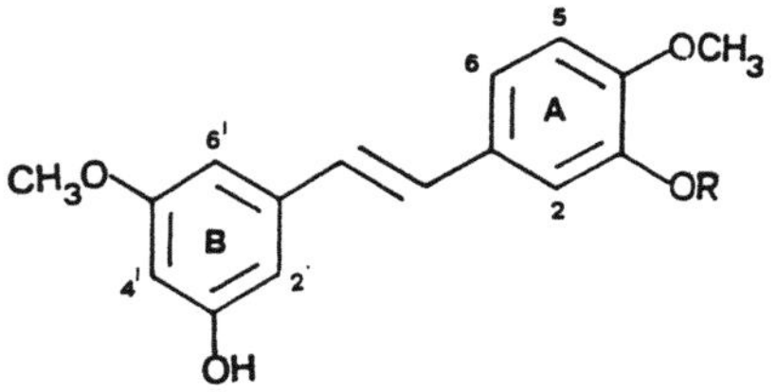

	R
13	CH_3
14	CH_3 (E)-isomer
15	H

Fig. 9. Stilbenes isolated from in vitro cultures of *C. didymobotrya.* (Delle Monache 1991)

the B ring, whereas the lower-field signal [δ 7.13 versus 7.06 in (*13*)] of the H-2 proton suggested a 3-hydroxy-4-methoxy substitution for the A ring. Therefore, metabolite *15* was assigned the structure of (E)-3,3′-dihydroxy-4,5′-dimethoxystilbene. A stilbene with the same oxygenation pattern of *13–15* (piceatannol) has been isolated from *C. marginata* (Rao and Rajadurai 1968) and from *C. dentata* Vogt (de Oliveira et al. 1977).

Tetrahydroanthracenes (Fig. 10). The yellow fluorescent pigments *16* and *17* showed the peculiar features of 2H,4H-3,8,9-trihydroxytetrahydroanthracenes:

1. In the ^{1}H-NMR spectrum, a weakly chelated OH signal (at δ 9.7–9.8) and a strongly chelated OH resonance (at δ 16.5–17);
2. In the IR spectrum, a CO absorption at 1630 cm^{-1};
3. In the UV spectrum, an absorbance band at 400 nm;
4. Two methylene signals at 2.8 and 3 ppm in the ^{1}H-NMR spectrum and at 38 and 46 ppm in the ^{13}C-NMR spectrum;
5. In the mass spectrum, a base peak corresponding to the loss of H_2O from the molecular ion.

By examination of the other spectral data, compounds *16* and *17* were identified with germichrysone and torosachrysone, respectively. Germichrysone has been previously isolated from seedlings (Takahashi et al. 1976), which is noteworthy, and by tissue culture (Noguchi and Sankawa 1982) of *C. torosa*, while torosachrysone was obtained from seeds of the plant (Takido et al. 1977). As mentioned above, the latter has been isolated from *C. didymobotrya* itself.

Since tetrahydroanthracenes easily lose water and isomerize to anthrones, which, in turn, are oxidized to bianthrones and/or anthraquinones, the presence of bianthrones *10–12* and anthraquinones *1* and *2* in the extract was expected; the anthrones *16a* (from *16*) and *17a* (from *17*) act as artifacts for the above compounds during the isolation process. By contrast, the accumulation of chrysophanol (*1*) and

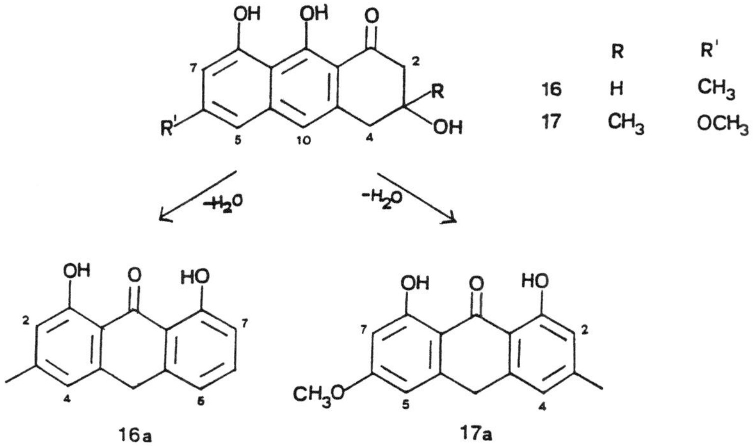

16a 17a

Fig. 10. Tetrahydroanthracenes isolated from in vitro cultures of *C. didymobotrya* and their dehydration and isomerization products. (Delle Monache 1991)

physcion (*2*) in the cell extract must be ascribed to a real enzymatic process, if we consider that *1* and *2* are the logical precursors of a series of biotransformation products (*4*, *5*, *6*, *8*, and *9*). The existence of a tetrahydranthracene precursor for emodin may also be postulated. The nonisolation of such a compound may be due either to its low concentration or to its high reactivity.

Flavonoids (Fig. 11). Apigenin (*18*), luteolin (*19*), kaempferol (*20*), quercetin 3-methyl ether (*21*), and quercetin (*22*) were identified by comparison of the spectral data to those of the literature (Mabry et al. 1970).

The blue fluorescent compounds *23* ($C_{17}H_{14}O_6$) and *24* ($C_{18}H_{16}O_6$) exhibited a similar oxygenation pattern (5, 7, 3′, 4′) in the ^{1}H-NMR spectra, the only difference being a third methoxy group signal (at δ 3.80) in *23* instead of a singlet (at δ 6.55). Accordingly, the UV spectrum of *23* showed a band II at a longer wavelength (342 nm versus 326 nm) than that of *24*. These data indicated that *24* was a flavone and *23* was the corresponding 3-methoxy derivative. As expected, both compounds in the mass spectra displayed the same fragments $a_1 + H^+$ and b_2^+ (Mabry and Markham 1975), whose values (*m/z* 167 and 151) demonstrated the presence of a hydroxy and a methoxy group in both rings A and B. The absence of a chelated OH (no signal in the ^{1}H-NMR spectrum and no bathochromic shift in the UV spectrum with the addition of $AlCl_3$) required the location of a methoxy group at C-5, while the down-field shift of the H-5′ proton in the ^{1}H-NMR spectrum in C_5D_5N (Alves De Lima et al. 1982) was in agreement with the presence of a 4′-hydroxy group. In conclusion, the two metabolites were assigned the structures of 7,4′-dihydroxy-5,3′-dimethoxyflavone (*23*) and 7,4′-dihydroxy-3,5,3′-trimethoxyflavone (*24*). Compound *23* has been previously isolated from *Medicago × varia* (Gehring and Geiger 1980), whereas *24* is a new compound.

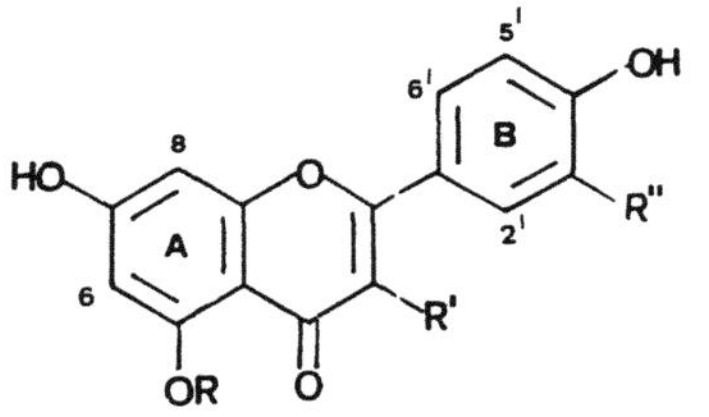

	R	R'	R''
18	H	H	H
19	H	H	OH
20	H	OH	H
21	H	OCH_3	OH
22	H	OH	OH
23	CH_3	H	OCH_3
24	CH_3	OCH_3	OCH_3

Fig. 11. Flavonoids isolated from in vitro cultures of *C. didymobotrya*. (Delle Monache 1991)

2.4 Biotransformation Studies

Flavonoids have potential importance in medicine as inhibitors of AMPc-phosphodiesterase (Nikaido et al. 1989): in agriculture, since isoflavonoids and rotenoids are considered phytoalexins (Brooks and Watson 1985); in the food industry because dihydrochalcones (Horowitz and Gentili 1963; Antus et al. 1978) and flavanonols (Kinghorn 1987), with the appropriate substitution in ring B, are considered nonnutritive edulcorants and therefore can be exploited in the hypocaloric diet and in cosmetics for the preparation of toothpaste without a cariogenic effect.

The above reasons prompted us to investigate the possibility of utilizing the above plant tissue cultures with the aim of obtaining higher amounts of potentially interesting "target flavonoids". In considering the use of appropriate precursors in such a study, it is highly desirable to consider the biosynthetic pathways leading to the end products, since such information aids in establishing the most likely chemical structures which, under enzymatic conditions, can be transformed to the target molecules.

The biosynthetic pathway of the flavonoid family, from acetyl CoA and phenylalanine, suggests that chalcones represent plausible "late stage" intermediates. It was therefore appropriate to initiate our biotransformation experiments with 2′,4′-dihydroxychalcone (*25*) as precursor. The first experiments involved the incubation of *25* with "suspended" cells of *C. didymobotrya.* The cells, grown in MS medium, were removed by filtration at different ages of growth (1, 15, 22, and 28 days, respectively), resuspended in a phosphate buffer (pH 7.2), and incubated with *25* for varying time periods. However, no biotransformation of *25* could be observed. One of the difficulties associated with such experiments concerns the permeability of the cell membranes in order for the precursors to enter the cell where the biotransformation is most likely to occur. It was possible that the negative result obtained was due to the cell membrane impermeability. For this reason, the next experiments involved crude enzyme systems (cell-free extracts) obtained by isolation from the cell culture.

2.4.1 Chalcone-Flavanone Isomerase

Preliminary experiments with cell-free extracts (CFE) of *C. didymobotrya* revealed the presence of a chalcone-flavanone isomerase (EC 5.5.1.6) (Shimokoriyama 1957; Begley et al. 1982), the enzyme responsible for the biotransformation $a \rightarrow b$, as shown in Fig. 12.

A series of experiments (Botta et al. 1989b) were performed to establish the variation of the enzymatic activity at different ages of the cultures. The cell suspensions, grown in 250-ml Erlenmayer flasks, were harvested at different times in the transition period between the early lag phase and late stationary phase, specifically at 1, 15, 22, and 28 days of growth. The cultures were homogenized in a buffer solution and, after centrifugation, the activity of the supernatant was tested with *25* as a substrate. As shown in Table 4, the highest activity was found in the crude extracts from 22-day-old cultures. These extracts afforded 75% of the corresponding flavanone after a 1.5-h incubation. It should also be noted (Table 4)

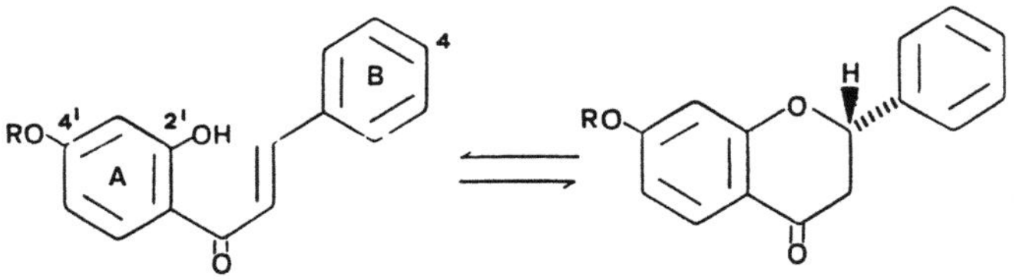

Fig. 12. Chalcone-flavanone interconversion. (Botta 1989b)

Table 4. Biotransformation of 2′,4′-dihydroxychalcone. (Botta 1989b)

Experiment[a]	Age of cells (days)	Overall substrate conversion (%)	Flavanone (%)	Reaction time (h)
A	0	10	10	8
B	15	40	40	4
C	22	75	75	1.5
D	28	50	10	8

[a] Optimum pH = 7.7 ~ 8; temperature = 37 °C.

that other enzymes occurring in older cells (28 days) are competitive with the chalcone-flavanone isomerase. For example, in experiment D, a 50% conversion of substrate was noted, but only 10% represented flavanone. CFE from younger cultures did not give any other unexpected product, that is, conversion of substrate resulted only in flavanone formation (see experiments A and B).

An interesting observation was noted in the comparison of the enzymatic activity versus the age of culture from which the enzyme system was isolated. It appears that in *younger* cells (5–9 days old) two competing enzyme systems are present, namely, an enzyme responsible for the conversion $a \rightarrow b$ (Fig. 12) and an enzyme involved in the reverse process ($b \rightarrow a$). At low substrate concentrations, the substrate *25* (structure *a*) is converted to the cyclic compound *b* (for example, 75% conversion in 1.5 h) but on further incubation *b* starts converting back to *a* (only 40% of *b* in 2 h) and finally after 5 h, *b* is converted almost completely back to *a* (only 5% of *b* present in the mixture). On the other hand, at a higher concentration of substrate, *a* converts to *b* (for example, 40% of *b* in 1.5 h) but the reverse process is inhibited (40% of *b* still present after 5 h). At very high concentrations, *no* conversion of *a* to *b* occurs, perhaps due to enzyme denaturation. With *older* cells (15–22 days) the reverse process is completely inhibited, that is, only the forward $a \rightarrow b$ conversion is noted and the enzyme system for the $b \rightarrow a$ conversion appears absent. It was thus decided to perform all subsequent experiments with 22-day-old cells and to isolate the enzymes from these cells.

The specificity of the chalcone-flavanone isomerase system was then studied, with chalcone substrates featuring different substitution patterns in the A ring (Fig. 13).

The results, as summarized in Table 5, reveal that a prenyl side chain at the 3′-position inhibits conversion of substrate to the corresponding flavanone (see

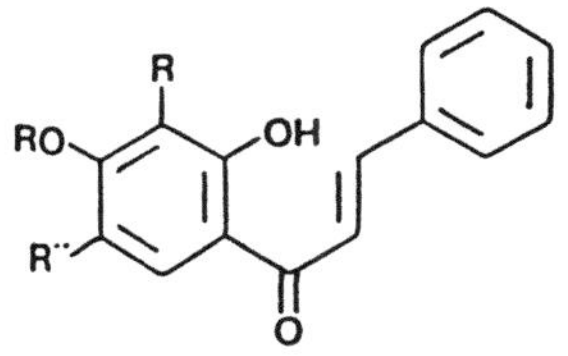

	R	R'	R''
25	H	H	H
26	Pr	H	H
27	H	Pr	H
29	Me	H	H
30	Me	H	Pr
31	Me	Pr	H
32	H	H	Pr

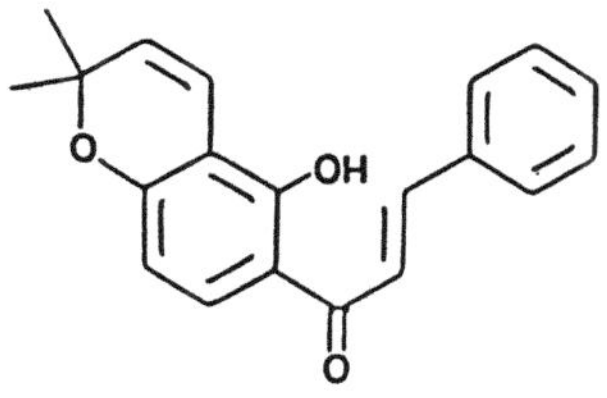

28

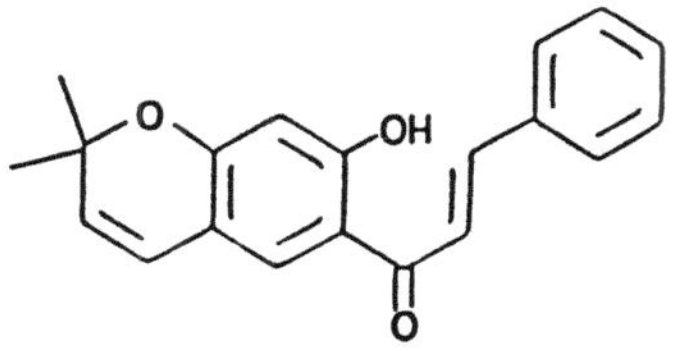

33

Pr ≡ 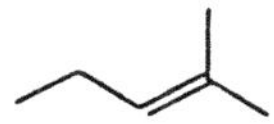

Fig. 13. A-ring-substituted chalcone substrates. (Botta 1989b)

Table 5. Chalcone substrate specificity (ring A). (Botta 1989b)[a]

Substrate	Reaction time (h)	Overall substrate conversion (%)	Flavanone (%)	$[\alpha]_D$	(Solvent)
25	1.5	75	75	−28	(MeOH)
26	1.5	50	50	−43	($CHCl_3$)
27	1.5	–	–	–	
28	1.5	48	48	−81	($CHCl_3$)
29	1.5	30	20	−4	($CHCl_3$)
30	1.5	75	75	−15	($CHCl_3$)
31	1.5	–	–	–	
32	1.5	45	45	−25	(MeOH)
33	0.5	45	45	−47	($CHCl_3$)

[a] The enzymes were extracted from 22-day-old cell cultures.

lack of conversion of substrates *27* and *31*). On the other hand, chalcones possessing a prenyl side chain at the 5′-position (see conversion of *30* and *32*) do not inhibit conversion. Perhaps the 3′-prenyl function prevents enzymatic conversion due to steric hindrance as a result of its close proximity to the chelated hydroxyl group. As noted in the studies with substrates *26* and *30*, various oxygenated side chains can be tolerated by the enzyme.

Although the cell cultures normally produce flavonoids having the phloroglucinol substitution pattern (2′, 4′, and 6′), the isolated enzyme system in this study is capable of utilizing other substitution patterns, as shown in substrates *25*, *26*, *28*,

29, *30*, *32*, and *33*, to form the corresponding flavanones. It thus appears that a rather broad substrate specificity exists within our enzyme systems.

The next goal was to study the enzyme specificity for precursors having different substitution patterns in ring B. It would be desirable to maintain the "normal" phloroglucinol-type pattern in ring A. However, in our study, it was shown in controlled experiments (no enzyme and buffer at pH 7.7 ~ 8) that such precursors rapidly converted to the corresponding flavanones; we therefore turned our attention to the more stable precursors with the 2′,4′-(resorcinol-type) substitution patterns (Fig. 14).

Chalcone *34* was incubated with the CFE obtained from the 22-day-old cell cultures. An interesting difference in product formation was observed. Although substrate *34* was utilized by the enzyme to an extent of 60% with a 3-h incubation time, only a 45% yield of the expected flavanone was obtained. The other products (see below) were obtained in the 15% yield.

As can be seen in Table 6, flavanone formation occurs to a significant extent with substrate *34*, bearing only a C-4 hydroxyl function, but is retarded when an additional C-3 methoxyl group is present (see substrate *40*). Other substituents in

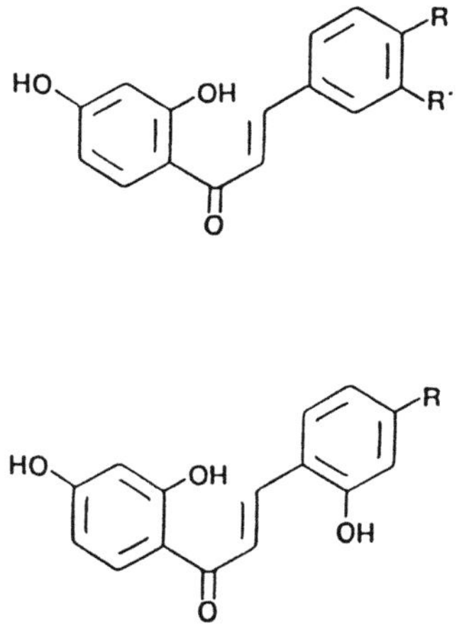

	R	R
34	OH	H
35	OMe	H
36	H	OH
38	OMe	OH
40	OH	OMe

	R
37	H
39	OMe

Fig. 14. B-ring-substituted chalcone substrates. (Botta 1989b)

Table 6. Chalcone substrate specificity (ring B). (Botta 1989b)[a]

Substrate	Reaction time (h)	Overall substrate conversion (%)	Flavanone (%)	$[\alpha]_D$	(Solvent)
34	3	60	45	−12	(MeOH)
35	3	–	–	–	
36	3	–	–	–	
37	3	–	–	–	
38	3	–	–	–	
39	3	–	–	–	
40	1	100	16	−9	($CHCl_3$)

[a] The enzymes were extracted from 22-day-old cell cultures.

ring B, for example, substrates *35–39*, are similarly detrimental to enzyme-catalyzed flavanone formation. Comparison of the data in Table 6 with those of Table 5, the latter involving substrates lacking any substituents in ring B, reveals *highest* yields of flavanone formation when *no* substitution is present in ring B. It thus appears that the "steric bulk" of ring B substituents prevents appropriate association of the substrates with the active sites of the enzyme.

2.4.2 Polyphenol Oxidase

As our studies with enzymes derived from older cells (22–28 days) continued, it was clear that in addition to the cyclization process, chalcone → flavanone, another enzymatic reaction was occurring and additional products were being obtained.

Detailed experiments with chalcones *34* and *40* were performed (Botta et al. 1989c). Since it is well known that "polyphenol oxidases" are widely distributed in the plant kingdom (Mayer 1987), it was thought that they could also be present in our cell suspension cultures. These enzymes are known to catalyze (Brown 1967) phenol oxidation and are usually divided into three main classes: (1) tyrosinases or o-diphenol oxidases; (2) laccase or p-diphenol oxidases; and (3) peroxidases. The last family is dependent on H_2O_2, while the first two do not appear to require this cofactor. In our initial studies, a series of experiments were performed in order to establish the variation of the polyphenol oxidases produced at different ages of the cultures. Guaiacol (2-methoxyphenol) was used as a substrate. Table 7 shows that such enzymatic activity was higher in older cells, specifically cells grown for 29 days. Since it was previously shown that maximum accumulation of the chalcone-flavanone isomerase (CHFI) enzymes occurred in 22-day-old cultures, it was now appropriate to compare the competition of these two types of enzyme systems for the substrate chalcone at the various time periods (0, 15, 22, and 29 days). In order to distinguish the two possible types of enzymatic conversions, cyanide was added, since it is known that this ion inhibits polyphenol oxidase (PPO) activity (Rathmel and Bendal 1972).

Table 7. Variation of polyphenol-oxidase activity in cell cultures of *C. didymobotrya*. (Botta 1989c)

Age of cells (days)	0		15		22		29	
Experiment[a]	%[b]	Time[c] (min)	%	Time (min)	%	Time (min)	%	Time (min)
A[d]	4	60	5	10	15	60	59	30
B	5	60	22	60	42	60	68	60
C	7	60	17	10	33	10	41	10
D	7	60	10	60	35	60	40	60

[a] pH = 6.5 ~ 7; temperature = 37 °C.
[b] % Transformed guaiacol.
[c] Optimum time of conversion.
[d] A: Cell-free extract (CFE); B: CFE + H_2O_2; C: CFE + horseradish peroxidase (HRP) D: CFE + H_2O_2 + HRP.

Therefore, the relative extent of CHFI conversion versus that for PPO activity could be readily demonstrated. For this study, 2,4,4'-trihydroxychalcone (*34*) was utilized as substrate. Table 8 summarizes the data in the bioconversion of *34* with an enzyme system derived from 29-day-old cells. It is noted in experiment A, in which only the cell-produced enzyme system is employed, that *34* is utilized to the extent of 60% with 3-h incubation and, as Fig. 15 indicates, products arising from

Table 8. Biotransformation of 2',4,4'-trihydroxychalcone (*34*) by polyphenol oxidase. (Botta 1989c)

Experiment	Time of reaction (h)	Chalcone *34* converted[b] (%)	Products formed (see Fig. 15) (%)			
			41	*42*	*43*	*45*
			(R = H)			
A[a]	3	60	5	5	15	30
B	3	79	8	9	15	30
C	1	100	8	9	15	30
D	1	100	8	9	15	30

[a] A, B, C, D: see Table 7.
[b] Conversion of *34* was determined by HPLC employing *34* as internal standard for purposes of calibration.

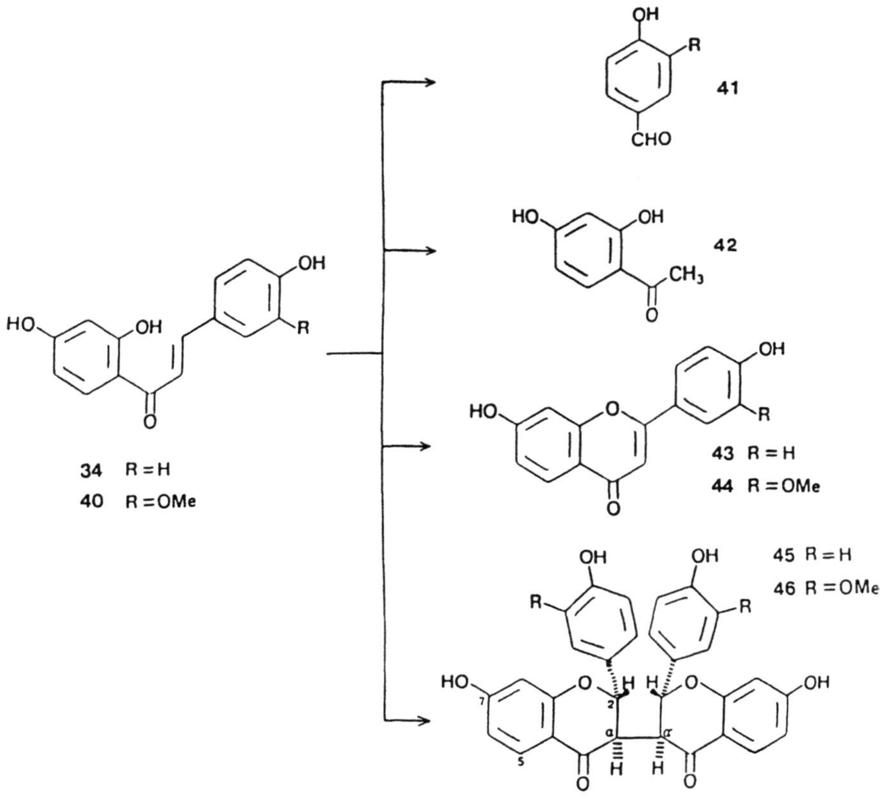

Fig. 15. Biotransformation of chalcones by polyphenol oxidase. (Botta 1989c)

the PPO type of activity are produced. We could detect no evidence of CHFI activity, that is, no flavanone products. In experiment B, where H_2O_2 is added as cofactor, there is an increase in the yield of compounds *41* (R = H) and *42* as well as other unaccounted products, but *no change* in yields of *43* and *45*. In a similar fashion, the addition of commercial horseradish peroxidase (HRP, experiment C) increases the yield of *41* (R = H) and *42* and, in particular, a substantial increase in yields of other unaccounted products. Experiment D confirms the data concerning the overall effects of H_2O_2 and HRP in the bioconversion of *34*. In summary, flavone *43* and the novel compound *45* are products truly resulting from cell-produced enzymes reacting with substrate *34*.

The above results have revealed that the chalcone substrate *34* does not provide flavanone products with the cell-free extract which exhibits a PPO type of activity. Furthermore, when liquiritigenin (4′,7-dihydroxyflavanone) was incubated with our cell-free extract derived from the 29-day-old cultures, the biflavanone product *45* was not formed. This result thus excludes a flavanone intermediate as a precursor to *45*.

When flavone *43* was similarly incubated with the cell-free extract, *45* was not formed. It is therefore clear that chalcone *34* proceeds via two independent pathways to *43* and *45*. Therefore, the enzyme-catalyzed process to *45* is not in accord with the mechanistic pathway proposed by Pelter in his chemical studies (Pelter et al. 1971). On the other hand, the enzymatic conversion of *34* to *43* could well correspond to the Pelter mechanism for flavone formation (Pelter et al. 1971).

In order to provide further data on the possible generality of the above process, we investigated the enzyme-catalyzed conversion of substrate *40*, which possessed an additional methoxyl group in ring B. Table 9 summarizes the results.

In this instance, substrate *40* is more rapidly consumed (1 h versus 3 h for *34*) and the yields of flavone *44* and biflavanone *46* are higher. Again, H_2O_2 and HRP play no role in affording conversions of *40* to *44* and *46*, but simply alter the yields of *41* (R = OCH_3) and *42*, presumably via a retro-aldol-like process. It is also clear that an account for products [*41* (R = OCH_3), *42*, *43*, and *46*] comprises 85–95% of the final product mixture, so that "other" products (for example, 45% in experiment A, Table 8) are minimized in the bioconversion of *40* when compared to

Table 9. Biotransformation of 2′,4,4′-trihydroxy-3-methoxychalcone (*40*) by polyphenol oxidase. (Botta 1989c)

Experiment	Time of reaction (h)	Chalcone *40* converted[b] (%)	Products formed (see Fig. 15) (%)			
			41 (R = OCH_3)	*42*	*44*	*46*
A[a]	1	100	10	10	20	45
B	1	100	15	15	20	45
C	1	100	15	15	20	45
D	1	100	15	15	20	45

[a] A, B, C, D: see Table 7.

[b] Conversion of *40* was determined by HPLC employing *40* as internal standard for purposes of calibration.

34. In summary, the cell-free extract isolated from 29-day-old cultures does not appear to be of the "peroxidase" type, since H_2O_2 is not involved in the bioconversions. Further studies on the nature of these enzymes are underway.

3 Summary and Conclusion

The optimization of both cell growth and the accumulation of secondary metabolites in cell suspension cultures of *Cassia didymobotrya* was achieved for the first time by applying an orthogonal design method. This method allowed the use of a limited number of experiments, compared with the much higher number of experiments normally required in a conventional study.

The screening of the main metabolites, anthraquinones and flavonoids, at varying ages of the culture, revealed a different rate of biosynthesis for the two families of compounds.

A chalcone-flavanone isomerase was also isolated from younger cells; from the studies of its activity on several chalcones it was established that one of the most important criteria in the chalcone-flavanone interconversion is the overall "bulk" of the substrate and its proclivity to fit into the active site of the enzyme.

Conversely, a mixture of polyphenol oxidases was isolated from older cells; since this cell-free extract was able to biotransform chalcones bearing a free hydroxy group in position 4 (ring B) without addition of H_2O_2, it was demonstrated that "peroxidases" were not involved in the bioconversion.

4 Protocol

The explants of hypocotyl and stem, excised from 5-7-day-old seedlings were shown to be the best. The best basal salt medium was MS (Murashige and Skoog 1962) enriched with kinetin, 2,4-D and sucrose at the following concentrations:

1. A kinetin concentration (0.1 mg/l) has relatively minor importance for the cell growth; however, it is an influential parameter for the production of secondary metabolites;

2. The addition of 2,4-D (1 mg/l) has the same influence in both cases;

3. With regard to sucrose, a definitive maximum is not observed, but the concentration of 30 g/l was selected for the maintenance of the suspension culture;

4. The optimum amount of the inoculum is 1 g (fresh weight) of cell in 65 ml medium.

Acknowledgements. This work was supported by C.N.R. Target Project "Biotechnology and Bioinstrumentation", and by Ministero della Pubblica Istruzione.

Contributions from the following researchers are gratefully acknowledged: A. Carbonetti, G. Dall' Olio, M.C. De Rosa, F. Ferrari, P. Iacomacci, D. Misiti, B. Monacelli, G. Pasqua, R. Scurria, and V. Vinciguerra.

The authors also wish to thank Prof. James P. Kutney (University of British Columbia, Vancouver, Canada) for very helpful discussions of the manuscript.

References

Alemayehu G, Abegaz B, Snatzke G, Duddeck H (1989) Quinones of *Senna didymobotrya.* Bull Chem Soc Ethiop 3: 37–40

Alves De Lima R, Delle Monache G, Botta B (1982) Pyridine-d_5 induced shift in PMR spectra as a base for structural elucidation of natural polyphenols. Rev Latinoam Quim 13: 61–64

Antus S, Farkas L, Gottsegen A, Nogradi M, Pfliegel T (1978) Dihydrochalcone type sweetening agents. I. Structure-taste relationship. Acta Chim Acad Sci Hung 98: 225–230

Begley MJ, Crombie L, London M, Savin J, Whiting D (1982) Chalaurenol: a novel product from enzymatic oxidation of 2,4,4′-trihydroxychalcone. J Chem Soc Chem Commun: 1319–1321

Bodo B, Molho L, Davonst D, Molho L (1980) Isolation and structure elucidation of achaetolide and achaetolidone, two new macrolides from a fungus. IUPAC 12th Int Symp Chemistry of Natural Products, Tenerife (Canary Islands), Abstr B98

Botta B, Dall'Olio G, Ferrari F, Monacelli B, Pasqua G, Scurria R, Delle Monache G (1989a) Cell suspension cultures of *Cassia didymobotrya*: optimization of growth and secondary metabolite production by application of the orthogonal design method. J Plant Physiol 135: 290–294

Botta B, De Rosa MC, Vinciguerra V, Scurria R, Iacomacci P, Ferrari F, Delle Monache G, Misiti D (1989b) Studies in cell suspension of *Cassia didymobotrya.* Part II. The biotransformation of chalcones to flavanones. Heterocycles 29: 2167–2174

Botta B, Vinciguerra V, De Rosa MC, Scurria R, Carbonetti A, Ferrari F, Delle Monache G, Misiti D (1989c) Studies in cell suspension of *Cassia didymobotrya.* Part III. The biotransformation of chalcones to flavones and biflavanones. Heterocycles 29: 2175–2183

Box GEP, Hunter WG, Hunter JS (1978) Statistics for experimenters. Wiley, New York, pp 374–418

Brooks CJW, Watson DG (1985) Phytoalexins. Nat Prod Rep 2: 441–451

Brown BR (1967) Biochemical aspects of oxidative coupling of phenols. In: Taylor WI, Battersby AR (eds) Oxidative coupling of phenols. Marcel Dekker, New York, pp 167–201

Budzikievicz H, Djerassi C, Williams DH (1967) Mass spectrometry of organic compounds. Holden Day, San Francisco, pp 162–173

Delle Monache F (1985) Chemistry and biological activity of the secondary metabolites of Vismieae. Rev Latinoam Quim 16: 5–15

Delle Monache G, De Rosa MC, Scurria R, Monacelli B, Pasqua G, Dall'Olio G, Botta B (1991) Metabolites from in vitro cultures of *Cassia didymobotrya.* Phytochemistry 30: 1849–1854

De Oliveira A, Fernandes M, Shaat V, De Vasconcelos I, Gottlieb OR (1977) Constituents of *Cassia* species. Rev Latinoam Quim 8: 82–85

Drewes SE, Fletcher IP (1974) Polyhydroxystilbenes from the heartwood of *Schotia brachypetala.* J Chem Soc Perk Trans I: 961–962

El-Sayyad SM, Ross SA (1983) A phytochemical study of some *Cassia* species cultivated in Egypt. J Nat Prod 46: 431–432

Gehring E, Geiger H (1980) Die Flavonoids der Samen von *Medicago* x *varia* Martyn c.v. Cardinal (Fabaceae). Z Nat 35c: 380–383

Haynes LJ, Henderson JI, Tyler JM (1960) The structure of homonataloin and the synthesis of nataloe-emodin. J Chem Soc: 4879–4885

Horowitz CM, Gentili B (1963) Dihydrochalcone derivatives and their use as sweetening agents. US Patent 3.083.821 (1963); C.A. 5, P116500

Kinghorn AD (1987) Biologically active compounds from plants with reputed medicinal and sweetening properties. J Nat Prod 50: 1009–1024

Kitanaka S, Takido M (1986) Studies on the constituents in the roots of *Cassia obtusifolia* L. and the antimicrobial activities of constituents of the roots and the seeds. Yakugaku Zasshi 106: 302–306

Mabry TJ, Markham KR (1975) Mass spectrometry of flavonoids. In: Harborne JB, Mabry TJ, Mabry H (eds) The flavonoids. Chapman & Hall, London, pp 78–126

Mabry TJ, Markham KR, Thomas MB (1970) The systematic identification of flavonoids, ch 5. Springer, Berlin Heidelberg New York

Mayer AM (1987) Polyphenol oxidase in plants – recent progress. Phytochemistry 26: 11–20

Murashige T, Skoog F (1962) A revised medium for rapid growth and bioassays with tobacco tissue culture. Physiol Plant 15: 473–497

Nikaido T, Ohmoto T, Kinoshita T, Sankawa U, Delle Monache F, Botta B, Tamimori T, Miyaichi Y,

Shirataki Y, Yokoe I, Komatsu M (1989) Inhibition of adenosine 3′,5′-cyclic monophosphate phosphodiesterase by flavonoids. Chem Pharm Bull 37: 1392–1395
Noguchi H, Sankawa U (1982) Formation of germichrysone by tissue cultures of *Cassia torosa*. Phytochemistry 21: 319–323
Pelter A, Bradshaw J, Warren RF (1971) Oxidation experiments with flavonoids. Phytochemistry 10: 835–850
Rao VSS, Rajadurai S (1968) Isolation of 3,4,3′,5′-tetrahydroxystilbene (piceatannol) from *Cassia marginata* heartwood. Aust J Chem 21: 1921–1922
Rathmel WG, Bendal BS (1972) The peroxidase-catalyzed oxidation of a chalcone and its possible physiological significance. Biochem J 127: 125–132
Shimokoriyama M (1957) Interconversion of chalcones and flavanones of a phloroglucinol-type structure. J Am Chem Soc 79: 4199–4202
Takahashi S, Takido M, Sankawa U, Shibata S (1976) Germichrysone, a hydroanthracene derivative from seedlings of *Cassia torosa*. Phytochemistry 15: 1295–1296
Takido M, Takahashi S, Masuda K, Yasukawa K (1977) Torosachrysone, a new tetrahydroanthracene derivative from the seeds of *Cassia torosa*. J Nat Prod 40: 191–194
Thomson RH (1987) Naturally occurring quinones: recent advances, 3rd edn. Academic Press, London
Tosato ML, Cesareo D, Galassi S, Vigano I, Cruciani G, Clementi S, Skakerberg B (1988) Multivariate design in toxicological evaluation of chemical data. Chimicaoggi 6(3): 41–45
Van Rheede van Outdshoorn MCB (1964) Chemotaxonomic investigation in Asphodelae and Aloineae (Liliaceae). Phytochemistry 3: 383–390
Watt JM, Breyer-Brandwijk MG (1962) Medicinal and poisonous plants of southern and eastern Africa, 2nd edn. E and S Livingstone, Edinburgh
Wold S, Sjostrom M, Carlson R, Lundstedt T, Hellberg S, Skakerberg B, Wikstrom C, Ohman J (1986) Multivariate design. Anal Chim Acta 191: 17–32
Yeoman NM, Macleod AJ (1977) Tissue (callus) cultures – Techniques. In: Street HE (ed) Plant tissue and cell culture. Botanical monographs, vol 11. Univ California Press, Berkeley, pp 31–59

VI *Cephaelis ipecacuanha* A. Richard (Brazilian Ipecac): Micropropagation and the Production of Emetine and Cephaeline

K. YOSHIMATSU and K. SHIMOMURA[1]

1 Introduction

The genus *Cephaelis* (family Rubiaceae) is comprised of about 100 species and is native to the tropics of the Americas, Africa, and Asia. Its species are mostly evergreen shrubs or small trees with opposite, undivided leaves and small flowers in terminal heads with an involucre of bracts. The plant generally starts flowering in the second year from germination, and the flowers have a five-toothed calyx, a five-lobed corolla, five stamens, and a two-branched style (Fig. 1). The fruits are small, usually two-seeded, berry-like drupes and their production becomes maximum three to four years after germination (Everett 1981; Chatterjee et al. 1982).

Fig. 1. Flowering *Cephaelis ipecacuanha* plant, propagated by layering, which has been grown in the greenhouse for 2 years

[1] Tsukuba Medicinal Plant Research Station, National Institute of Hygienic Sciences, Ministry of Health and Welfare, 1 Hachimandai, Tsukuba, Ibaraki 305, Japan

Biotechnology in Agriculture and Forestry, Vol. 21
Medicinal and Aromatic Plants IV (ed. by Y.P.S. Bajaj)

There are two pharmaceutically important species, *Cephaelis ipecacuanha* A. Richard (also known as *Psychotria ipecacuanha* Stokes and *Uragoga ipecacuanha* Baill), and *C. acuminata* Karsten. The drug ipecac (ipecacuanha root) is the dried root or rhizome of these plants, which was used in Amazonian folk medicine and was introduced into Europe in 1672 to treat amoebic dysentery (Trease and Evans 1989).

1.1 Distribution and Importance of the Ipecac

C. ipecacuanha, known in commerce as Rio or Brazilian ipecac, is found over a large area in Brazil, particularly in the moist and shady forests of Matto Grosso and Minas Geraes. Plantations have been established in the Matto Grosso area. It is cultivated to some extent in Malaysia, Burma, and the Darjeeling Hills of West Bengal. *C. acuminata*, known in commerce as Cartagena, Nicaragua, or Panama ipecac, is exported from Colombia, Nicaragua, and Costa Rica. The main differences between the Rio and Cartagena drugs are listed in Table 1 (see also Fig. 2, Trease and Evans 1989).

Ipecac is commonly used as an expectorant in the treatment of bronchitis, croup, asthma and whooping cough, as an emetic in cases of poisoning, and an amoebacide in amoebic dysentery. It is rich in the isoquinoline alkaloids emetine,

Table 1. The main differences between the Rio (*C. ipecacuanha*) and Cartagena (*C. acuminata*) drugs. (After Trease and Evans 1989)

	Rio ipecacuanha	Cartagena ipecacuanha
Usual diameter	1–4 mm	4–6.5 mm
Color	Brick-red to brown	Grayish-brown
Annulations	Very crowded	Less crowded and less projecting
Starch	Individual grains up to 15 μm	Individual grains up to 20 μm
Alkaloid ratio	Emetine > cephaeline	Emetine < cephaeline

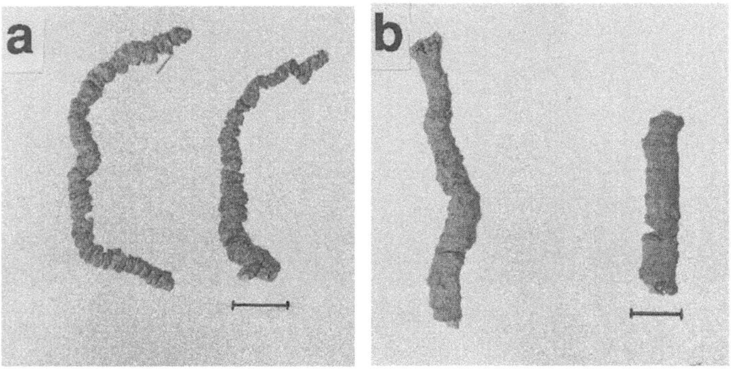

Fig. 2. a The Rio (*C. ipecacuanha*) and **b** the Cartagena (*C. acuminata*) ipecac. *Bars* = 1 cm. The Rio root is smaller and more annulated than the Cartagena root. These samples were kindly supplied by Prof. Dr. H. Becker, Universität des Saarlandes, FRG

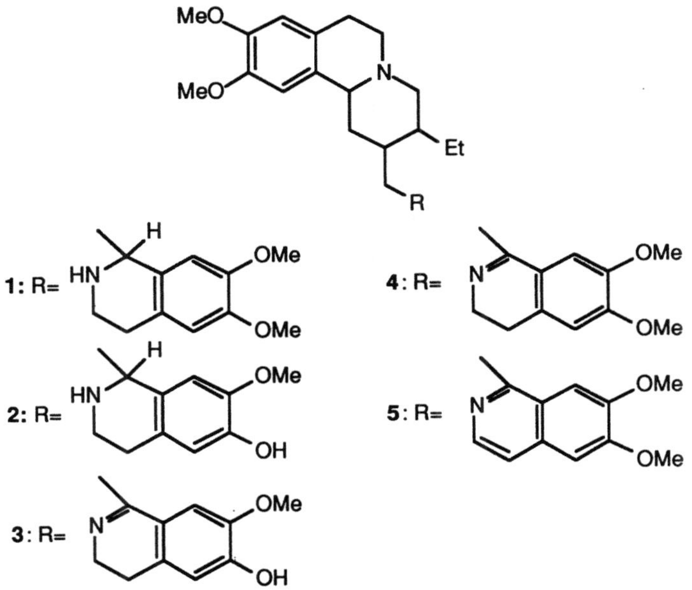

Fig. 3. The structures of emetic alkaloids. *1* Emetine; *2* cephaeline; *3* psychotrine; *4* O-methylpsychotrine; *5* emetamine

cephaeline, psychotrine, *O*-methylpsychotrine, and emetamine (Wiegrebe et al. 1984), emetine and cephaeline being the principal active ingredients. Because emetine has a more expectorant and less emetic action than cephaeline (Radomski et al. 1952), and emetine hydrochloride is used in the treatment of amoebic dysentery by injection, it is the alkaloid usually required in medicine.

1.2 Cultivation of Ipecac

Various varieties of ipecac contain different proportions of the principal alkaloids. The Rio variant (*C. ipecacuanha*) contains 2–2.4% alkaloids (dry wt.) of which 60–75% is emetine. The varieties derived from *C. acuminata* yield 2–3.5% alkaloids (dry wt.), of which emetine may constitute 30–50% (Trease and Evans 1989). Thus *C. ipecacuanha* is regarded as the best source of ipecac.

The supply of ipecac has fluctuated for many years because it is collected mainly from wild plants. In Matto Grosso, the collector partially digs the plant up from the ground and, having removed most of the roots, replaces it in the ground to produce further plants. The roots are dried in the sun or by fire and transported downriver to ports such as Rio de Janeiro, Bahia, and Pernambuco, to be exported in bales. The Rio variant is now becoming difficult to obtain commercially, and the high price favors cultivation, but only modest success has resulted from the efforts to grow the drug in Malaya and India.

C. ipecacuanha can grow from seeds as well as from leaf and stem cuttings. The plants prefer artificial shade and soil rich in humus and acidic in nature. Chatterjee et al. (1982) studied the field cultivation of this plant to provide information for the economically feasible propagation of ipecac in West Bengal, and faced the following problems:

1. Seeds *lost* viability after storage, although treatment with limewater or hydrogen peroxide could improve germination.
2. The yield of roots was maximum after 4 years and the alkaloid content was maximum during the third and fourth year. Taking all aspects into consideration such as cost of cultivation, yield, and alkaloid contents, maximum benefit is obtained by harvesting the roots between the third and fourth years.
3. Covering the plants increased the yields of roots about threefold, but this might become a major share of the total expenditure when developing the plantation. Also, selection of a cover for ipecac plants to grow is one of the very important considerations.

2 In Vitro Approaches

C. ipecacuanha has appeared in the Japanese Pharmacopoeia (1986) as ipecac or powdered ipecac. In Japan, field cultivation of this plant is very difficult, therefore ipecac is imported from other countries. Teshima et al. (1984) quantitatively analyzed emetic alkaloid levels in ipecac roots being purchased from South American countries and powdered ipecac in Japanese markets. They revealed that both the value of alkaloids and their ratio (emetine/cephaeline) varied widely, and concluded that imported ipecac drugs were not qualitatively identical.

Some tissue culture studies have been done to propagate the homogeneous plants efficiently (Ikeda et al. 1988; Yoshimatsu and Shimomura 1991a) and produce its alkaloids in vitro (Jha et al. 1988; Teshima et al. 1988; Yoshimatsu and Shimomura 1991a, b)

Tissue culture studies of *C. ipecacuanha* are summarized in Tables 3 and 4.

Table 2. Contents of emetic alkaloids in commercial powdered ipecac (sample Nos. 1–4) and commercial Crude Ipecacuanha (sample Nos. 5–11). (After Teshima et al. 1984)

Sample No.	Emetine (% dry wt.)	Cephaeline (% dry wt.)	Emetine/cephaeline
1	0.75 ± 0.01	1.52 ± 0.01	0.49
2	1.18 ± 0.01	1.34 ± 0.01	0.88
3	0.75 ± 0.01	1.65 ± 0.03	0.45
4	0.75 ± 0.01	1.51 ± 0.01	0.49
5	0.91 ± 0.02	0.72 ± 0.02	1.26
6	0.60 ± 0.01	3.33 ± 0.03	0.18
7	1.43 ± 0.01	2.67 ± 0.01	0.54
8	1.38 ± 0.01	2.41 ± 0.03	0.57
9	1.42 ± 0	2.79 ± 0.05	0.51
10	1.09 ± 0.01	2.13 ± 0.06	0.51
11	0.84 ± 0	2.30 ± 0	0.37

Table 3. In vitro propagation of *C. ipecacuanha*[a]

Method	Shoot multiplication				Rooting		References
	Explant source	Tissue examined	Basal medium[b]	Phytohormone (mg/l)	Basal medium[b]	Phytohormone (mg/l)	
I	Shoot tip	Node	B5	NAA + BA (0.01)(1–3)	B5	IAA (3)	Ikeda et al. (1988)
II	Shoot tip	Internode	WP	BA or Kin (0.01–0.1)(0.1)	WP or B5	HF	Yoshimatsu and Shimomura (1991a)

[a] The tissue was cultured under 16 h light at 25 °C for 8 weeks.
[b] The node segments were placed vertically on solid medium (method I) and the internode segments were placed horizontally on solid or in liquid medium (method II).

Table 4. Production of emetic alkaloids in vitro

Explant source	Tissue examined	Basal medium[a]	Phytohormone[b] (mg/l)	Alkaloid production	References
In vitro leaf	Callus	MS	2,4-D	Trace	Teshima et al. (1988)
	Root culture	MS	IAA (1–3)	+	
			NAA (1)	+	
			NAA (1) + Kin (0.1)	+	
In vitro leaf	Root culture	MS	4-Cl-IAA (0.1–1)	+	Yoshimatsu and Shimomura (1991b)
			5,6-Cl_2-IAA (0.01–0.1)	+	
			NAA (5)	+	
Hypocotyl	Callus	MS	IBA (8) + IAA (4) + NAA (4)	+	Jha et al. (1988)
		SH + MS	IBA (8) + IAA (4) + NAA (4)	+	
In vitro internode	Shoot	WP	BA (0–0.1)	+	Yoshimatsu and Shimomura (1991a)

[a] Media: B5 (Gamborg et al. 1968); WP, woody plant (Lloyd and McCown 1980); MS, Murashige and Skoog (1962); SH, Schenk and Hildebrandt (1972); SH + MS, SH macrosalts + all other components as in MS.
[b] Phytohormone: NAA, 1-naphthaleneacetic acid, 2,4-D, 2,4-dichlorophenoxyacetic acid, IBA, indole-3-butyric acid. IAA, indole-3-acetic acid, 4-Cl-IAA, 4-chloro-indole-3-acetic acid. 5,6-Cl_2-IAA, 5,6-dichloro-indole-3-acetic acid. BA, 6-benzylaminopurine. Kin, kinetin. HF, phytohormone-free.

2.1 Propagation of Plants In Vitro and Evaluation of Regenerates

Two kinds of successful propagation in vitro have so far been described; one through shoot meristem culture (method I), and the other through adventitious shoot formation (method II). Both methods include two different cultural steps: shoot multiplication and rooting.

Method I. Node segments were chosen for routine shoot multiplication, and five shoots per segment were formed on a B5 medium containing 0.01 mg/l NAA + 3–5 mg/l BA after a 2-month culture (Fig. 4). It is reported that the use of

Fig. 4. Multiple shoots formed on a nodal segment cultured on B5 medium containing 0.01 mg/l NAA and 3 mg/l BA for 8 weeks. *Bar* = 1 cm. (Ikeda et al. 1988)

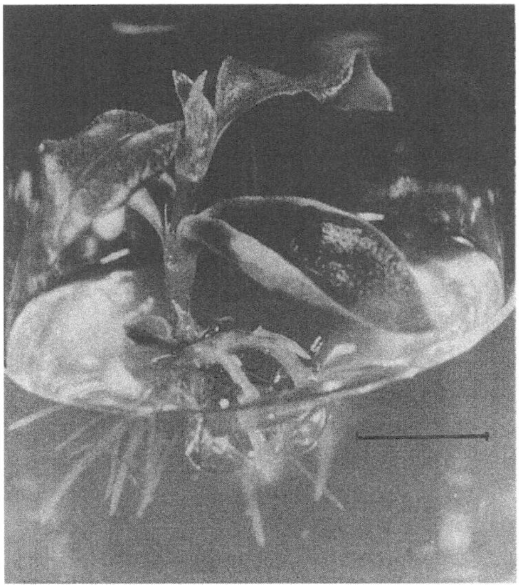

Fig. 5. Rooted plantlet cultured on B5 medium containing 3 mg/l IAA for 2 months. *Bar* = 1 cm. (Ikeda et al. 1988)

cytokinin in combination with a low concentration of auxin was more effective for shoot formation than a single cytokinin addition. The multiple shoots were separated and rooted on B5 medium containing 3 mg/l IAA after 2 months (Fig. 5). NAA also stimulated root formation; however, it induced callus at the cut end of the shoots, resulting in a low establishment rate to soil. The regenerated plants grew well (Fig. 6) and accumulate alkaloids with the same pattern and distribution

Fig. 6. The regenerated plants through shoot meristem culture, grown in a greenhouse. From *left* to *right*: 1; 2; and 7-month-old plants after transplantation to soil, respectively. *Bar* = 5 cm

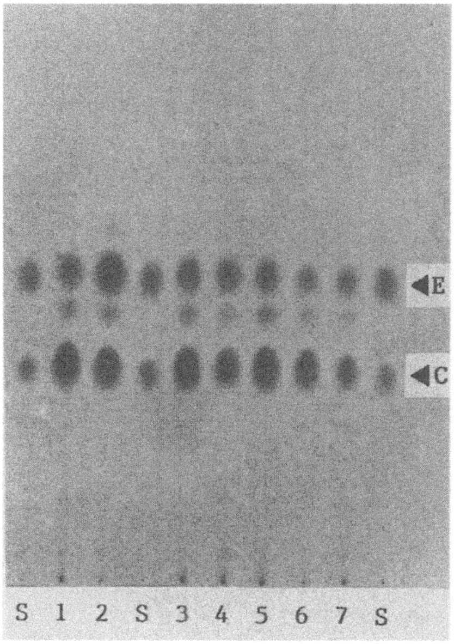

Fig. 7. TLC pattern of the alkaloid in the regenerates. *E* Emetine; *C* cephaeline; *S* authentic compounds. *1* and *2* leaves and roots of the parental plant; *3–5* leaves of the 1; 2; and 7-month-old regenerated plants; *6* roots of the 2-month-old regenerated plants; *7* leaves of shoot culture. (Ikeda et al. 1988)

Table 5. Alkaloid content (% dry wt.) in the regenerated plants and the parental plants. (After Ikeda et al. 1988)

Material	Regenerated plants[a]						Parental plants	
	1 month		2 months		7 months			
	Emetine	Cephaeline	Emetine	Cephaeline	Emetine	Cephaeline	Emetine	Cephaeline
L-1	0.705	2.060	0.567	1.595	0.647	3.571	1.085	2.990
L-2	0.496	2.657	0.239	1.309	0.243	1.135	0.380	1.058
L-3					0.225	1.145	0.196	1.005
L-4					0.167	0.853	0.195	0.955
L-5							0.179	1.040
L-6							0.179	1.281
L-7							0.105	1.060
Stem							0.072	1.013
Roots	0.251	2.185	0.513	2.357	0.673	1.828	1.025	1.662
Rootlets					1.291	1.435	1.621	1.873

[a] The regenerated plants were cultivated in a phytotron for 1 month under the following conditions; 16 h light/8 h dark, 25 °C, relative humidity 80%, and then in a greenhouse. The parental plants propagated by layerings were cultivated in the greenhouse for 2 years. L-1 represents the youngest expanded leaf, while L-2 to L-7 represent those leaves which came below L-1.

as the parental plants (Fig. 7, Table 5). The roots of 4-year-old regenerated plants (Fig. 8) are annulated and very similar to crude ipecac (Fig. 2a).

Method II. Internodal segments without axillary buds selected for shoot formation. Numerous adventitious shoots (20–40 in number) were formed on the epidermis of internodal segment cultured horizontally on a WP solid or liquid medium containing low concentrations of cytokinin (0.01–0.1 mg/l BA or 0.1 mg/l Kin; Fig. 9a), and they grew enough to transfer to the rooting medium after

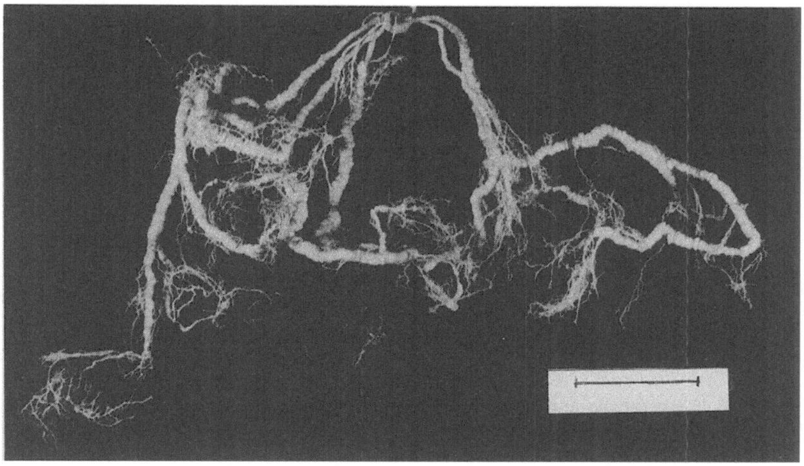

Fig. 8. The roots of 4-year-old regenerated plants. *Bar* = 5 cm

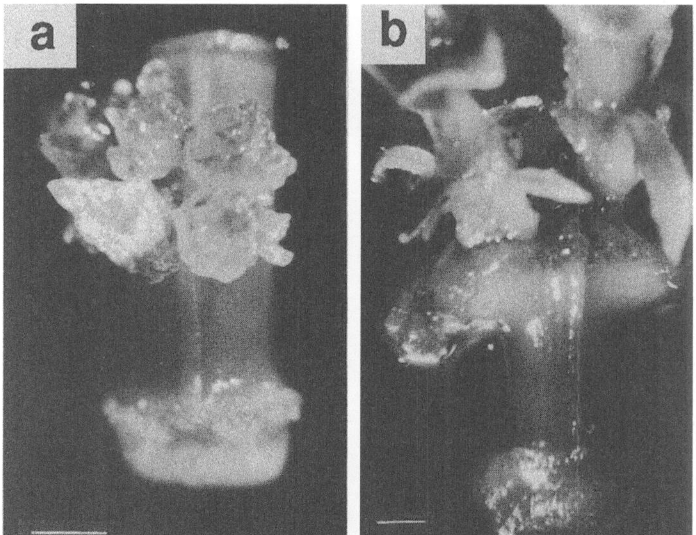

Fig. 9. Comparison of shoot formation on **a** an internodal and **b** a nodal segment after 4-week culture. *Bars* = 1 mm. (Yoshimatsu and Shimomura 1991a)

2-month culture. Whereas no adventitious shoots were formed on the epidermis of nodal segment under these conditions (Fig. 9b), it is noteworthy that the lack of shoot meristem apparently permitted adventitious shoot formation on the epidermis and decreased the cytokinin requirement. Histological observations of shoot initiation area (Fig. 10) indicate that adventitious shoots were directly formed on

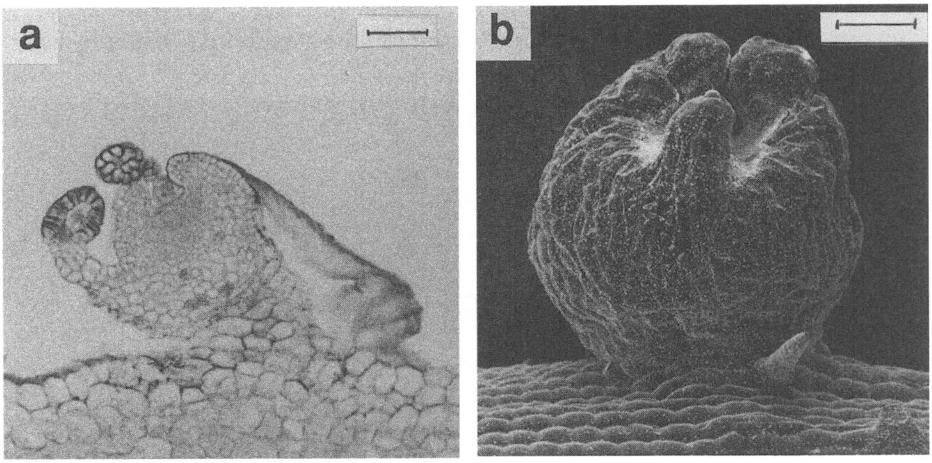

Fig. 10. a Cross-section and **b** scanning electron microscopy of initial shoots (after 2–3 weeks culture). *Bars* = 0.1 mm. (Yoshimatsu and Shimomura 1991a)

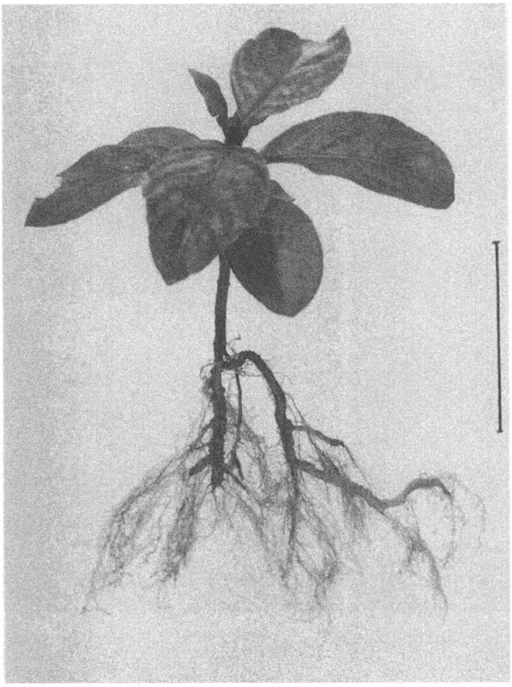

Fig. 11. One-year-old regenerated plant from an internodal segment. *Bar* = 5 cm. (Yoshimatsu and Shimomura 1991a)

Table 6. Alkaloid content (% dry wt.) in the regenerated plants from internodal segments. (After Yoshimatsu and Shimomura 1991a)[a]

Part	Emetine	Cephaeline
Leaves	0.084 ± 0.004	0.661 ± 0.033
Roots	0.820 ± 0.185	2.156 ± 0.416
Rootlets	1.340 ± 0.047	1.383 ± 0.166

[a] The regenerated plants were cultivated in a greenhouse for 1 year. Alkaloid content is shown as the means of two plants.

Table 7. Theoretical numbers of ipecac plants which can be propagated in vitro. (Ikeda et al. 1988; Yoshimatsu and Shimomura 1991a)

Method	Material	Period (year)	No. of plants
I	Axenic shoot	1	600
II	Axenic shoot	0.5	>1000

the epidermis of internodal segment without callusing. It is preferable that shoot formation does not occur through the intermediary of callus, when obtaining genetically identical plants. The shoots thus obtained rooted after individual transplantation to B5 or WP solid medium without auxin. The plants regenerated from the internodal segments could be cultivated in a greenhouse, exhibited quite normal features (Fig. 11), and accumulated alkaloids (Table 6) in a similar content to parental plant (Table 5).

The theoretical number of ipecac plants which can be propagated in vitro are summarized in Table 7. Method II is theoretically more feasible than the method I, although both enable practical propagation; method II requires a shorter culturing period and less phytohormones (of both cytokinin and auxin).

2.2 Emetic Alkaloid Production by Tissue Culture

Root (Teshima et al. 1988; Yoshimatsu and Shimomura 1991b), callus (Jha et al. 1988; Teshima et al. 1988) and adventitious shoot culture (Yoshimatsu and Shimomura 1991a) of *C. ipecacuanha* have produced emetic alkaloids.

Teshima et al. (1988) established callus and root culture from leaves of an axenic shoot (Fig. 12), and examined their productivity of emetic alkaloids. The callus grew satisfactory in the presence of 1 mg/l 2,4-D, but it lost the capability of alkaloid production after the third culture passage. However, root cultures, especially in liquid medium, demonstrated stable levels of alkaloids (0.164–0.514% dry wt. emetine and 0.625–1.319% dry wt. cephaeline) even after the third culture passage. The culture conditions of root culture in liquid medium were investigated in further experiments (Yoshimatsu and Shimomura 1991b), then the roots cultured in MS medium (50 ml per 100-ml flask) containing 0.01 mg/l 5,6-Cl_2-IAA for

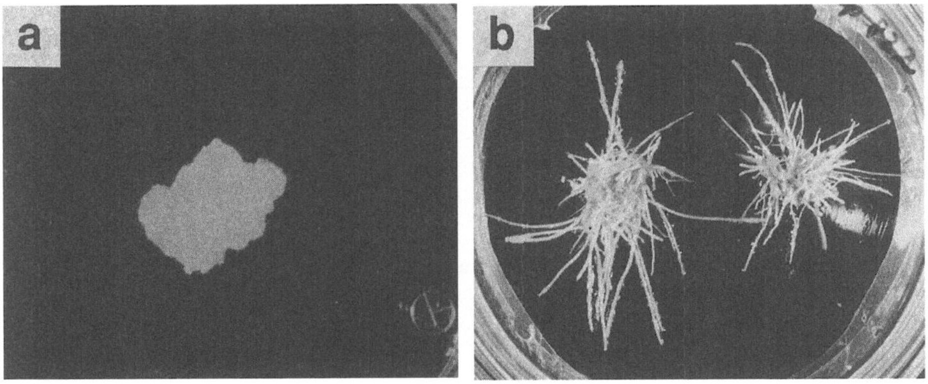

Fig. 12a, b. Callus (**a** MS, 1 mg/l 2,4-D) and root (**b** MS, 1 mg/l NAA) culture of *C. ipecacuanha* after 2-month culture

7 weeks (Figs. 13, 14) yielded 0.6 mg emetine and 2.4 mg cephaeline. This yield is comparable to that in roots of 1-year-old regenerated plants through tissue culture (ca. 0.2 g total dry wt.).

Jha et al. (1988) induced callus from the hypocotyl of a seeding in vitro on MS medium containing 4 mg/l 2,4-D + 4 mg/l NAA, and examined the effects of macrosalt formulation and the combination of auxins for alkaloid accumulation. In this case, callus which was cultured on the medium, of which the macrosalt was SH and the other components were MS containing 8 mg/l IBA + 4 mg/l IAA + 4 mg/l NAA, accumulated 0.346% dry wt. emetine and 0.93% dry wt. cephaeline after 3–4 weeks, though the growth index was relatively low (harvest dry wt./inoculum dry wt. = 3.5). These results seem to be in conflict with Teshima et al. (1988); however, different initial materials were used, which might influence the subsequent characteristics in culture.

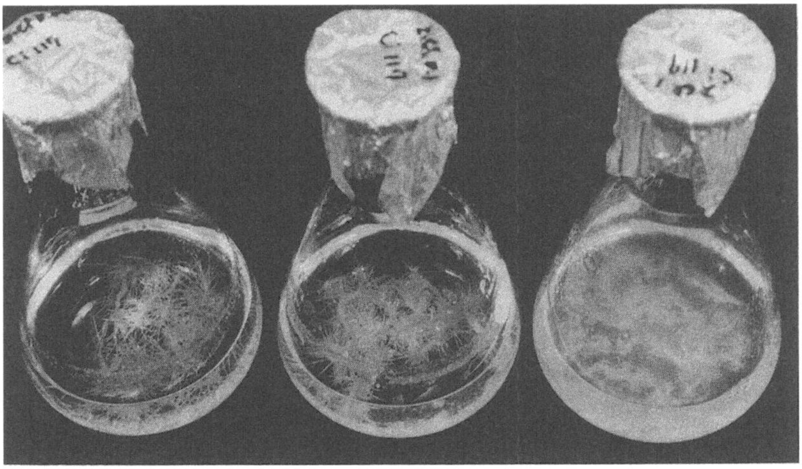

Fig. 13. The root cultures of *C. ipecacuanha* in MS liquid medium (after 7-week culture). From *left* to *right*: 0.01, 0.1 and 1 mg/l 5,6-Cl_2-IAA

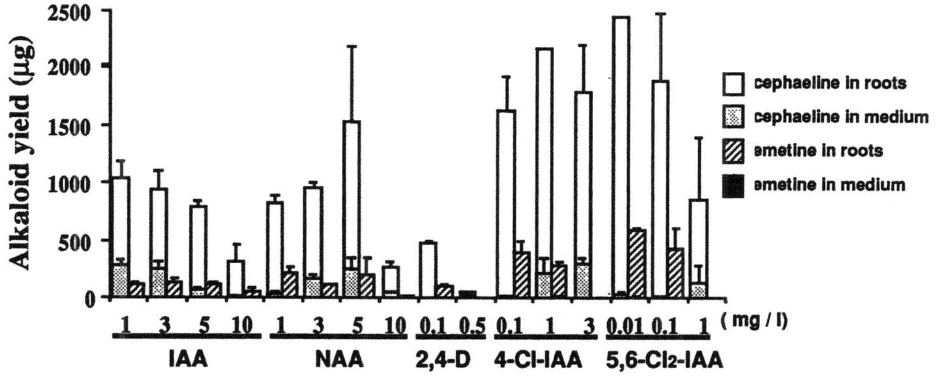

Fig. 14. Effects of auxins on alkaloid yields. The roots (*ca.* 10–30 mg) were cultured in MS liquid medium (50 ml/100 ml Erlenmeyer flask) in the dark for 7–9 weeks on a rotary shaker (100 rpm) at 25 °C. Bars represent standard deviation of the means, n = 2. (After Yoshimatsu and Shimomura 1991b)

Adventitious shoot cultures induced on internodal segments with WP medium containing 0.01–0.1 mg/l BA produced 0.04–0.07% dry wt. emetine and 0.4–0.5% dry wt. cephaeline (Yoshimatsu and Shimomura 1991a; Fig. 15). It is of interest that adventitious shoot cultures without roots could synthesize alkaloids as well as root cultures, and both of them accumulated much more cephaeline than emetine, although intact plant contained almost the same levels of emetine and cephaeline in its roots (Table 5). It might indicate the different capability of alkaloid biosynthesis between the in vitro culture and the parental plant.

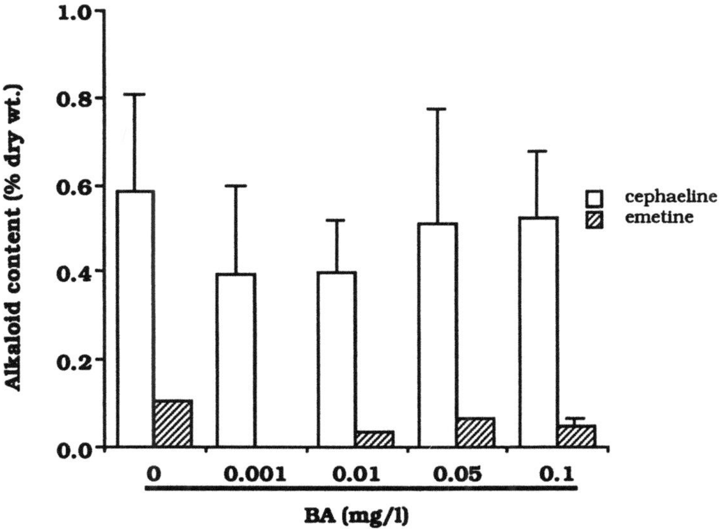

Fig. 15. Alkaloid content in adventitious shoots formed on internodal segments, grown in liquid medium. The segments (3 segments/Petri dish) were cultured in WP liquid medium under 16 h light at 25 °C on a rotary shaker (30 rpm) for 8 weeks. Bars represent standard deviation of the means, n = 6. (Yoshimatsu and Shimomura 1991a)

3 Conclusions and Prospects

The ipecac plants are commercially cultivated in the lower hills of Darjeeling in India, and they are usually propagated by stem or root cuttings. Micropropagation in vitro (method II), which enables over 1000 plants to be propagated from a single axenic culture within six months (Table 7), is much faster and more feasible than the conventional methods. However, micropropagation in vitro should not be accounted economically feasible when considering it from a commercial standpoint, because it requires special facilities, reagents, and many instruments. It depends on the relation between the cost and profit as to whether commercial cultivation of ipecac combined with micropropagation in vitro is possible or not.

Organ, root, and shoot cultures of *C. ipecacuanha* are very attractive for the production and study of the biosynthesis of emetic alkaloids. The roots, which were cultured for 7 weeks in 100-ml flasks under optimized conditions, yielded the comparable amount of alkaloids to the roots of 1-year-old regenerated plant (Fig. 14), although the ratio of emetine to cephaeline was much lower than the roots of parental plant (Fig. 14, Table 5). In solanaceous species, it is commonly accepted that the root is the organ of alkaloid biosynthesis. In *C. ipecacuanha*, however, shoot cultures (without roots) as well as root cultures produced alkaloids (Fig. 15). This implies that biosynthesis of emetic alkaloids occurs over a whole plant. It is noteworthy that both root and shoot cultures in vitro accumulated much more cephaeline than emetine, while the roots of the parental plant contained these two alkaloids at almost the same level. The ratio of emetine to cephaeline in regenerated plants seems to change as the plants become mature.

4 Protocols for Micropropagation and Production of Alkaloids

A systematic protocol for the tissue culture of *C. ipecacuanha* is summarized in Fig. 16.

4.1 Explant Source

Shoot tips or seeds are generally chosen as initial plant materials. Axenic shoot culture of *C. ipecacuanha*, which is one of the best experimental sources for tissue culture study, was established by using the shoot tip as an initial explant. Although the establishment of tissue culture from seeds might be possible, precautions are needed; because Chatterjee et al. (1982) reported a decrease of seed viability during storage, Jha et al. (1988) treated seeds with 20 mg/l GA_3 after surface-sterilization to induce germination.

4.2 Establishment of Axenic Shoot Culture From Shoot Tips

1. Dip the young shoots (3-cm length, with leaves removed) of 2–3-year old plants in 75% EtOH for 10 seconds, and rinse in sterilized H_2O.

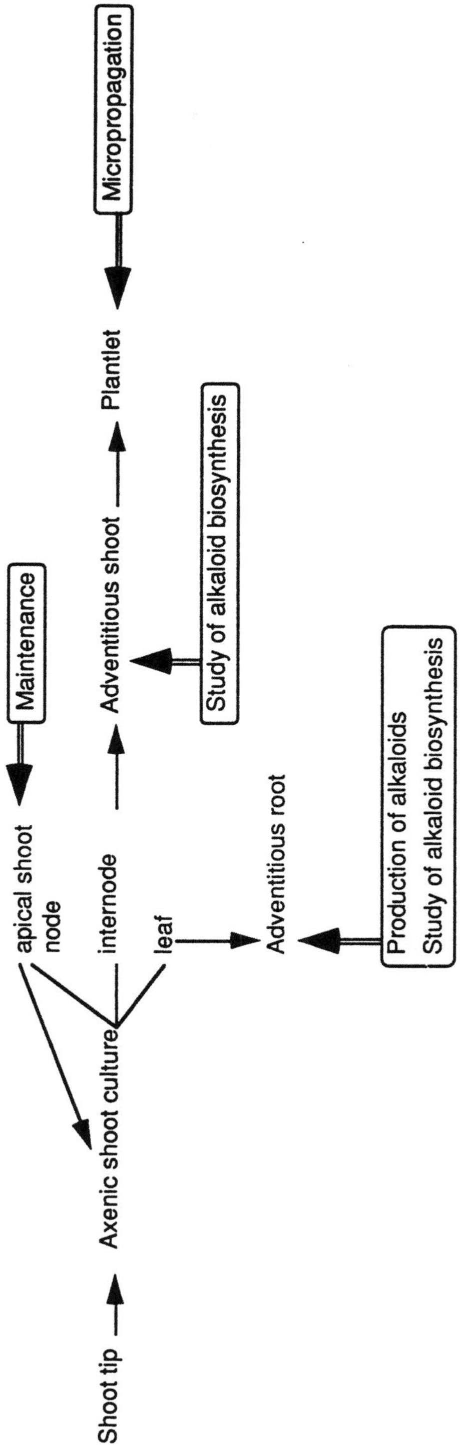

Fig. 16. Systematic protocol for tissue culture of *C. ipecacuanha.*

2. Surface-sterilize in 3% NaOCl with Tween 20 (1 drop/40 ml) for 10 min, and wash with sterilized H_2O three times.

3. Excise shoot tips (1 mm height) under a binocular microscope, and inoculate them on MS solid medium containing 0.01 mg/l NAA + 0.1 mg/l Kin.

4. Culture shoot tips at 25 °C in 16 h light (80 μE $m^{-2} S^{-1}$) per day. Initial shoot tips grow very slowly. Approximately 3 months later, they can be separated into node and apical shoot segments.

5. Cut shoot and node segments and transfer to fresh medium.

6. Subculture them every 2–3 months under the same culture conditions.

4.3 Micropropagation

1. Excise internode segments (without axillary meristem) from axenic shoot culture and place them horizontally on WP solid medium supplemented with 0.01–0.1 mg/l BA or 0.1 mg/l Kin.

2. Culture them at 25 °C in the light (80 μE $m^{-2} S^{-1}$, 16 h). After 8 weeks, numerous adventitious shoots (20–40) will develop from epidermis of internodal segments.

3. Transfer the shoots separated individually to phytohormone-free, WP or B5 solid medium for rooting, and culture under the same conditions.

4. Transplant the rooted plantlets to pots (soil: sand: leaf mold = 5:1:1) and cultivate for 1 month in a phytotron under 25 °C, 16 h light (200 μE $m^{-2} S^{-1}$) 8 h dark, 80% relative humidity for adaptation to outside.

5. Cultivate established plants in a greenhouse.

4.4 Establishment of Root Culture and Production of Alkaloids

1. Excise leaf segments (5 × 5 mm) from axenic shoot cultures, and place on MS solid medium supplemented with 0.5–3 mg/l NAA.

2. Culture them at 25 °C in the dark. After 4 to 6 weeks, adventitious roots will appear.

3. Subculture the roots in liquid, transferring the root segments (1–2 cm) to fresh medium (50 ml per 100-ml flask) every 2–3 months, and culturing at 25 °C in the dark on a rotary shaker (100 rpm).

4. For the production of alkaloids, culture the roots in MS liquid medium containing NAA (5 mg/l), 4-Cl-IAA (0.1–1 mg/l) or 5,6-Cl_2-IAA (0.01–0.1 mg/l) for 7–9 weeks under the same conditions.

Acknowledgment. The authors are grateful to Prof. Dr. Hans Becker, Universität des Saarlandes, FRG, for providing the dried root samples of *C. ipecacuanha* and *C. acuminata*.

References

Chatterjee SK, Nandi RP, Ghosh NC (1982) Cultivation and utilization of ipecac in West Bengal. In: Atal CK, Kapur BM (eds) Cultivation and utilization of medicinal plants. Regional Research Laboratory, CSIR, Jammu-Tawi, India, pp 295–301

Everett TH (1981) *Cephaelis*. In: The New York Botanical Garden Illustrated Encyclopedia of Horticulture, vol. 2. Garland, New York, pp 679–680

Gamborg OL, Miller RA, Ojima K (1968) Nutrient requirements of suspension culture of soybean root cells. Exp Cell Res 50: 151–158

Ikeda K, Teshima D, Aoyama T, Satake M, Shimomura K (1988) Clonal propagation of *Cephaelis ipecacuanha*. Plant Cell Rep 7: 288–291

Jha S, Sahu NP, Mahato SB (1988) Production of the alkaloids emetine and cephaeline in callus cultures of *Cephaelis ipecacuanha*. Planta Med: 504–506

Lloyd G, McCown B (1980) Commercially-feasible micropropagation of mountain laurel, *Kalmia latifolia*, by use of shoot-tip culture, vol. 30. In: Int. Plant Propag Soc Combd Proc for 1980, pp 421–427

Murashige T, Skoog F (1962) A revised medium for rapid growth and bioassays with tabacco tissue cultures. Physiol Plant. 15: 473–497

Radomski JL, Hagan EC, Fuyat HN, Nelson AA (1952) The pharmacology of ipecac. J Pharmacol Exp Ther 104: 421–426

Schenk RU, Hildebrandt AC (1972) Medium and techniques for induction and growth of monocotyledonous and dicotyledonous plant cell cultures. Can J Bot 50: 199–204

Teshima D, Tsuchiya T, Aoyama T, Horioka M (1984) Quantitative determination of main emetic components (emetine and cephaeline) in ipecac. IYAKUHIN KENKYU 15: 63–71

Teshima D, Ikeda K, Satake M, Aoyama T, Shimomura K (1988) Production of emetic alkaloid by in vitro culture of *Cephaelis ipecacuanha* A. Richard. Plant Cell Rep 7: 278–280

The Pharmacopoeia of Japan, 11th edn (1986) English version. The Ministry of Health and Welfare, pp 1273–1274

Trease GE, Evans WC (1989) Ipecacuanha. In: Pharmacognosy 13th edn. Bailliére Tindal, London, pp 595–599

Wiegrebe W, Kramer WJ, Shamma M (1984) The emetic alkaloids. J Nat Prod 47: 397–408

Yoshimatsu K, Shimomura K (1991a) Efficient shoot formation on internodal segments and alkaloid formation in the regenerates of *Cephaelis ipecacuanha* A. Richard. Plant Cell Rep 9: 567–570

Yoshimatsu K, Shimomura K (1991b) Emetic alkaloid formation in root culture of *Cephaelis ipecacuanha*. Phytochemistry 30: 505–507

VII *Cornus officinalis*: In Vitro Culture and the Production of Gallotannins

K. YAZAKI and T. OKUDA[1]

1 Introduction

1.1 Distribution and Importance of the Plant

Cornus officinalis Sieb. et Zucc. (Japanese: *San-shu-yu*; Cornaceae) is a deciduous medium-sized tree, brought to Japan as a medicinal plant in the eighteenth century from China via Korea, and is cultivated as a garden tree because of its yellow flower (Fig. 1) and red fruit (Fig. 2). The fruit (common name: cornus fruit) has also been used as an astringent, a tonic, and a hemostatic in these countries, particularly

Fig. 1. Flowers of *C. officinalis*. The photograph was taken in March 1988

[1] Faculty of Pharmaceutical Sciences, Okayama University, Tsushima, Naka 1-1, Okayama 700, Japan

Biotechnology in Agriculture and Forestry, Vol. 21
Medicinal and Aromatic Plants IV (ed. by Y.P.S. Bajaj)

Fig. 2. Fruits of *C. officinalis*: The tree is ca. 20 years old. The photograph was taken in November 1987 in Japan

often as one of the ingredients in traditional prescriptions such as *Hachimi-gan* and *Rokumi-gan* in Japanese (the Chinese terms are *ba-wei-wan* and *liu-wei-wan*) which are popular tonics for preventing and improving symptoms of aging, including pollakiuria and cataract.

1.2 Medicinal Components and Conventional Practices

Several noteworthy biological activities of tannins have been recently reported (Okuda et al. 1989a), and the fruit of *C. officinalis* is one of the rich sources of hydrolyzable tannins (Okuda et al. 1981, 1984; Hatano et al. 1989a, b). Generally, the majority of tannins are classified into two large groups of polyphenolic compounds: hydrolyzable tannins, which are metabolites of gallic acid; and condensed tannins, which are also called proanthocyanidins (Haslam 1989). The former group of compounds is again classified into several groups: gallotannins, which are galloyl esters of sugar (mostly glucose) or polyalcohol; ellagitannins, which are hexahydroxydiphenoyl esters; and oxidation products from ellagitannins, such as dehydroellagitannins with dehydrohexahydroxydiphenoyl (DHHDP) groups; and others.

"Tannic acid", among the commercially available tannins, is generally a gallotannin mixture of variable composition and purity, produced from the insect gall of

Rhus, *Quercus*, and other species of plants. Tannic acid is listed in the pharmacopoeia of many countries, based on its medicinal applicability. Berberine tannate, arbumin tannate, diphenhydramine tannate, etc. are examples of medicines prepared from tannic acid, and there are various industrial applications of tannic acid.

There are also a variety of drugs of plant origin, in which tannins are regarded as the active principles. Myrobalans (the fruit of *Terminalia chebula*) in Ayurveda medicine, and *Geranium thunbergii* in Japanese pharmacopoeia, which are both rich in ellagitannins, are examples of popular medicinal plants of this kind. *C. officinalis* also can be considered as belonging to this group.

The propagation of *C. officinalis* in Japanese horticulture is conventionally done by layering, although the most efficient way of the propagation may be sowing. Whatever the method, it takes a long time to raise adult trees which are capable of producing fruits for medicinal use, and the fruit production is reduced as the tree ages.

2 In Vitro Approaches

There have been some reports concerning tannin production in plant cell cultures (Lees 1986; Krajci and Gross 1987; Zaprometov 1988). However, few of them reported the isolation and quantitation of each compound of tannin. The cell culture systems producing gallotannins (Yazaki and Okuda 1989), ellagitannins (Yazaki and Okuda 1990a), and condensed tannins (Yazaki and Okuda 1990b) have been established at our laboratory to study the biosynthesis of these polyphenolic compounds with large molecular weights. The focus of these studies was concentrated on these isolated compounds of defined chemical structures with comparatively large molecular weights. The color reactions often used for detection of tannins also detect the phenolic compounds of low molecular weight such as mono- and digalloylglucose. However, these compounds show almost no binding activity, as does tannin (Okuda et al. 1985). Therefore, in our work described here, each galloylglucose larger than a trigalloylglucose was designated as a gallotannin. As for *C. officinalis*, the present investigation is the first study of in vitro cell culture of this plant leading to the production of large amounts of gallotannins.

2.1 Establishment of Cell Culture

For the induction of callus, the surface-sterilized yellow buds of *C. officinalis* were used. These buds were collected in March at the botanical garden of the Faculty of Pharmaceutical Sciences, Okayama University. The buds were placed on LS agar medium (Linsmaier and Skoog 1965) containing 3% sucrose, 1 μM 2,4-dichlorophenoxyacetic acid (2,4-D) and 10 μM 6-benzyladenine (BA), and were kept at 25 °C in the dark. The callus tissues were obtained from part of the peduncle, and were further subcultured on LS agar medium containing the same plant growth regulators as above at 25 °C in the dark at intervals of 1 month for 1 year. The calli appear pale yellow or light brown, and tend to aggregate to form clusters (Fig. 3).

Fig. 3. Callus tissues of *C. officinalis* cultured on LS medium containing 1 μM 2,4-D and 10 μM BA, which were subcultured for 1 year (1988)

The cell suspension cultures were initiated by transferring callus (7 g fresh wt.) from the fourth generation of the callus cultures to LS liquid medium (90 ml) containing 1 μM 2,4-D and 10 μM BA in 300-ml Erlenmeyer flasks. The flasks were agitated on a reciprocal shaker at a speed of 100 strokes/min at 25 °C in the dark, and were subcultured every 3 weeks.

In both callus and suspension cultures, a large amount of gallotannin mixture was accumulated and detected on a TLC plate sprayed with $FeCl_3$. This mixture was found by HPLC to consist of four galloylglucoses, among which 1,2,3,6-tetra-*O*-galloyl-*β*-D-glucose was the main component (Table 1).

In order to optimize the cell growth and gallotannin production, the effects of plant-growth regulators were examined with various combinations of auxins and BA, and the cell cultures were collected and extracted with a five fold volume of 70% aqueous acetone. After filtration, the extracts were evaporated to dryness in a nitrogen stream, and the residue was dissolved in methanol and analyzed by reversed-phase HPLC after centrifugation.

The reversed-phase HPLC (Okuda et al. 1989b) was conducted on a LiChrospher RP-18 column (4 × 250 mm, Merck) in an oven at 40 °C, with a solvent system consisting of 0.01 M H_3PO_4–0.01 M KH_2PO_4-ethanol-ethyl acetate (85:85:20:10), at a flow rate of 1.2 ml/min, monitoring with UV absorption at 280 nm. The quantitation was based on the peak areas recorded by Chromatopac C-R6A (Shimadzu).

Table 1. Isolated galloylglucoses from callus cultures of *C. officinalis*

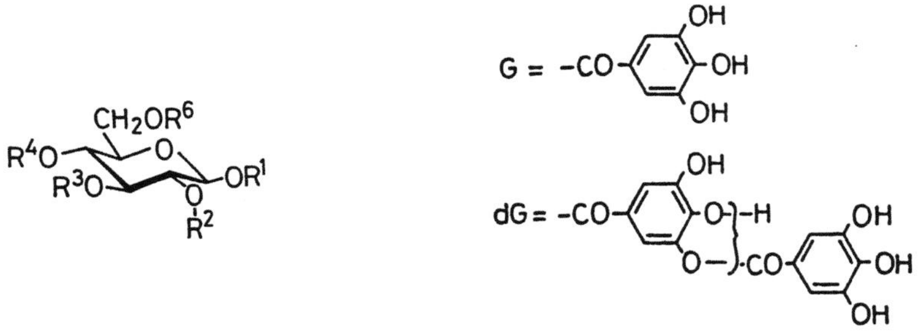

Compound	Abbreviation	R[1]	R[2]	R[3]	R[4]	R[6]
1,2,6-Tri-*O*-galloyl-β-D-glucose	3GG	G	G	H	H	G
1,2,3,6-Tetra-*O*-galloyl-β-D-glucose	4GG	G	G	G	H	G
1,2,3,4,6-Penta-*O*-galloyl-β-D-glucose	5GG	G	G	G	G	G
6-*O*-Digalloyl-1,2,3-tri-*O*-galloyl-β-D-glucose	6dG	G	G	G	H	dG

The effects of 2,4-D and 1-naphthaleneacetic acid (NAA) as auxins are shown in Figs. 4 and 5. The tannin content in these figures and also in Fig. 9 is the sum of the amount of each galloylglucose shown in Table 1, because of the absence of the other galloylglucoses, as mentioned below.

The effects of 3-indoleacetic acid (IAA) and kinetin were very similar to those of 2,4-D and BA, respectively (data not shown). As observed in Figs. 4 and 5, BA is required for efficient tannin production, and the best concentration is at about 10 μM. As the auxin, 2,4-D is preferable to NAA for the tannin production, because cell growth was better in the presence of 0.1–1 μM 2,4-D than in NAA at the 10 μM BA level.

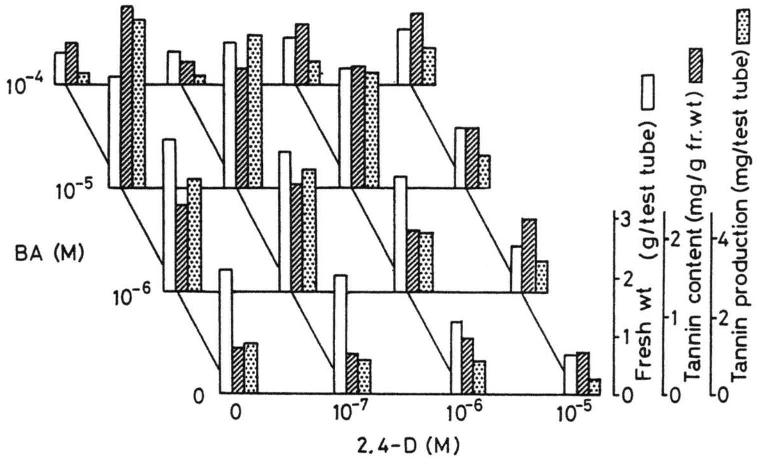

Fig. 4. Effects of 2,4-D and BA on cell growth and gallotannin production in callus cultures of *C. officinalis*

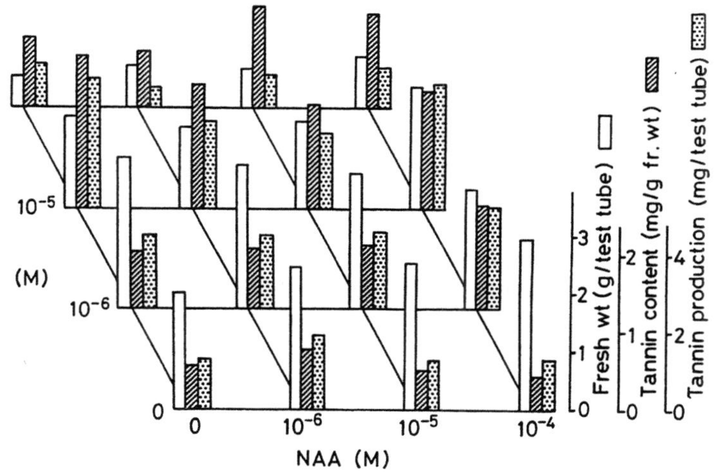

Fig. 5. Effects of NAA and BA on cell growth and gallotannin production in callus cultures of *C. officinalis*

The tannin content was the highest (2.29 mg/fresh wt. = 46.5 mg/dry wt.; Fig. 4) when 10 μM of BA was used as the sole growth regulator, and was the lowest in the medium containing 10 μM NAA alone; however, cell growth was rapid in the latter (NAA-containing medium). The tannin content was even lower (3.84 mg/dry wt.) when the cells were subcultured in the medium with NAA alone.

These cell lines cultured in the media of two different types, which produced either a high or low yield of tannins, will be useful for biochemical studies, such as those concerning enzymes for tannin biosynthesis.

The amounts of each galloylglucose, produced under various combinations of 2,4-D and BA, are shown in Fig. 6. Although the differences in the content ratios of

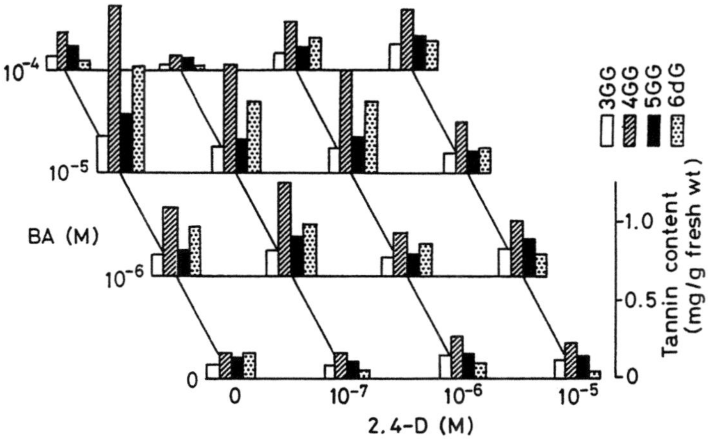

Fig. 6. Effects of 2,4-D and BA on the contents of individual galloylglucoses in callus cultures of *C. officinalis*

the four tannins are generally not large, the content of 5GG among them is comparatively constant under these various conditions, whereas the contents of 4GG and 6GG are relatively large.

The time course of individual galloylglucose contents per fresh weight is depicted in Fig. 7, in which the callus cells were cultured on agar medium under the same conditions as those of the subcultures. The data in Figs. 7 and 8 are expressed as means ± standard errors. The content of 3GG increased during 14 days after inoculation, and then decreased slowly, as observed in Fig. 7. The production of 4GG lagged behind that of 3GG, and reached its maximum after 28 days of culture, whereas the changes in the 5GG and 6GG content were small, if any.

The effects of light on the galloylglucose production was also examined and the results are shown in Fig. 8. The irradiation experiments were carried out as follows:

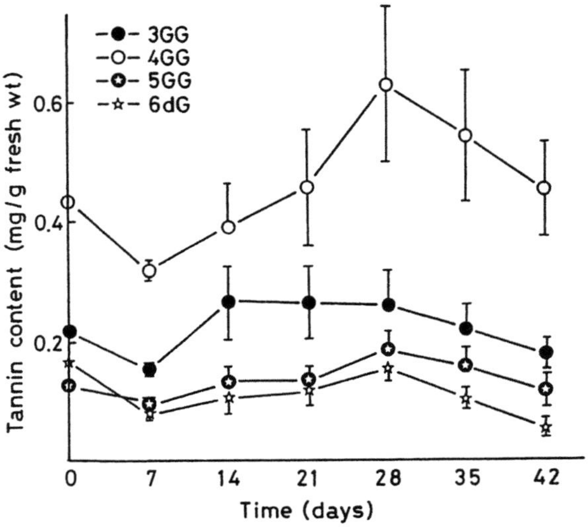

Fig. 7. Time course of the contents of individual galloylglucoses in callus cultures of *C. officinalis* grown in the dark. The results are expressed as means ± SE of four test tubes

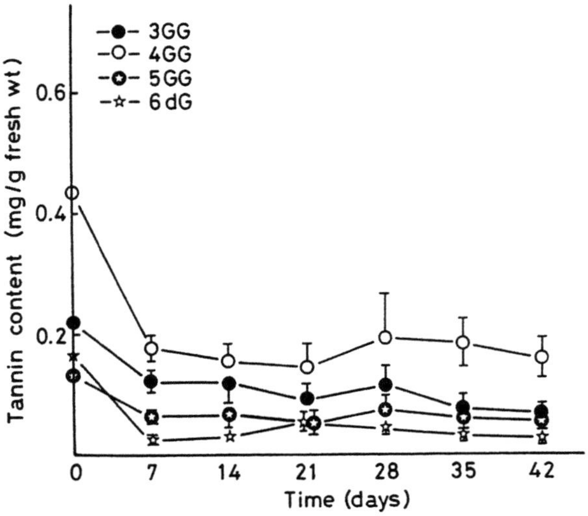

Fig. 8. Time course of the contents of individual galloylglucoses in callus cultures of *C. officinalis* grown under illumination (6000 lx) from fluorescent lamps. The results are expressed as means ± SE of four test tubes

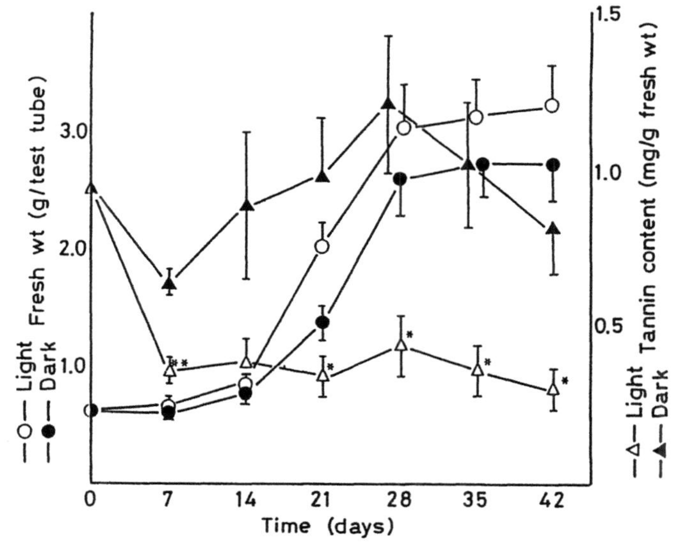

Fig. 9. Time course of cell growth and gallotannin content in callus cultures of *C. officinalis* grown in the dark, and under illumination (6000 lx) from fluorescent lamps. The results are expressed as means ± SE of four test tubes: $^{*}P < 0.05$, $^{**}P < 0.01$

the callus cultured in the dark was inoculated on LS agar medium containing the same growth regulators as above, and illuminated with fluorescent lamps of 6000 lx for 12 h/day during the whole culture period.

The production of each galloylglucose, particularly of 4GG, the main component, was inhibited by this illumination. The time course of cell growth and the total content of gallotannins in the callus cultures, in the dark and under the illumination, are shown in Fig. 9. The bars show standard errors of means. The statistical analyses were made with Student's t-test.

In the dark, the total amount of gallotannins increased with cell growth, up to the 28th day after inoculation, and then decreased. The illumination inhibited tannin formation in this cell cultures, and suppression was about 64% at the early stationary phase. However, cell growth was somewhat improved by illumination, to show that the inhibitory effect of light is limited to the tannin production.

2.2 Extraction and Structures of Gallotannins

The tannin-rich crude extract was prepared by homogenizing fresh cells (188 g) cultured on agar medium in 70% acetone (450 ml × 3 times), and then concentrating the filtrate to ca. 50 ml under reduced pressure at 40 °C. This concentrate was extracted with ethyl acetate (50 ml × 10 times), and the combined ethyl acetate layer was evaporated to dryness under reduced pressure at 40 °C to yield a pale yellow residue (407 mg).

This residue was chromatographed on a column (1.1 × 23 cm) of Toyopearl HW-40 fine (Tosoh) with 40 and 70% aqueous ethanol in a stepwise gradient mode, collecting fractions of each of 300 drops. Galloylglucoses in the fractions were detected by TLC developed with 7% acetic acid and visualized by spraying with $FeCl_3$ solution.

Combined fractions 13–16 were evaporated to afford 1,2,6-tri-*O*-galloyl-β-D-glucose (20 mg) as an off-white amorphous powder. Fractions 21–35 gave 1,2,3,6-tetra-*O*-galloyl-β-D-glucose (72 mg). Fractions 41–54 yielded 1,2,3,4,6-penta-*O*-galloyl-β-D-glucose (30 mg), and fractions 57–66 gave 6-*O*-digalloyl-1,2,3-tri-*O*-galloyl-β-D-glucose. Each galloylglucose was identified with the authentic specimen (Okuda et al. 1984; Yoshida et al. 1988) by direct comparison (TLC, HPLC, ^{1}H-NMR, etc.).

The galloylglucoses isolated were tri-pentagalloylglucoses (Table 1). Gallic acid and mono- and digalloylglucoses, of almost negligible binding activity, were detected by HPLC. However, their quantities were too small to be isolated. Hexagalloylglucose and larger galloylglucoses were not even detected by HPLC.

The galloylglucose contents in callus tissues and in fruits and leaves of the intact plants are shown in Table 2. The fresh fruits and leaves were collected in October and November at the above-mentioned botanical garden, and were extracted and treated in a way similar to that applied to the cultured cells. The data are expressed as milligrams per tissue dry wt., since about 90% of the weight of fresh cultured cells is water. The tannin contents in the cell suspension cultures obtained from these callus cells, in the same medium without agar, are also given in Table 2.

The amounts of the individual gallotannins produced in the callus tissues are almost the same as those in the cell suspension cultures, but they are much higher than those in the intact fruits and leaves, as exemplified by the 4GG and 5GG contents, as follows: (1) the 4GG content in dried cells of the suspension cultures is 36 and 12 times that in the intact fruits and leaves, respectively; and (2) the 5GG content in the cultured cells is 103 and 8 times that in the dried fruits and leaves, respectively. No 6dG was detected in the intact plant in the present study.

Table 2. Contents of gallotannins in callus cultures, cell suspension cultures, intact fruits, and intact leaves of *C. officinalis*. (Yazaki and Okuda 1989)

Compound	Tannin content (mg/g dry wt.)			
	Callus	Suspension cultures[a]	Fruits	Leaves
3GG	2.44	2.83	–[b]	–[b]
4GG	12.98	14.63	0.41	1.23
5GG	8.77	9.30	0.09	1.21
6dG	8.57	10.48	n.d.[c]	n.d.
Total[d]	32.76	37.24	–[e]	–[e]

[a] Fresh weight of the cells cultured in 90 ml medium was 31 g, 3 weeks after inoculation.
[b] Quantitative analysis was impossible because the peak of 3GG was hidden by that of ellagitannins.
[c] Not detected.
[d] The sum of the contents of the four compounds shown in this table.
[e] Total gallotannin content cannot be calculated because the 3GG content is unknown.

3 Conclusions and Prospects

Gallotannin mixtures composed of tri-, tetra-, and pentagalloylglucose were produced by callus and suspension cultures of *Cornus officinalis* Sieb. et Zucc. The amount of 1,2,3,6-tetragalloylglucose, the main component thus produced in the cell suspension cultures, was 36 times its content in the intact fruits. 1,2,6-Trigalloylglucose, 1,2,3,4,6-pentagalloylglucose, and 6-digalloyl-1,2,3-trigalloylglucose were also produced in these cultures. The ratios of the amount among these tannins did not vary much during the culture period, and were not much affected by differences in the combination of plant growth regulators in the medium. The tannin production was stimulated by BA, whereas cell growth required 2,4-D or NAA.

It is notable that almost no ellagitannin or its metabolites, which are the main constituents in the intact plant cells, were produced in the cell cultures. Biogenetically, the HHDP group in ellagitannins is regarded as being produced by the oxidative coupling between two galloyl groups in a molecule of gallotannin (Hatano et al. 1986). The absence of ellagitannin, in spite of the production of a large amount of gallotannins in the cultures, is therefore attributed to the lack of enzymatic activity catalyzing this oxidative coupling. These results may support the view that the galloylation of glucose is a biosynthetic step preceding the formation of the HHDP group in ellagitannin biosynthesis.

The substitution patterns in the isolated galloylglucoses are also of interest, because most of them lack the galloyl group on O-4 of glucose. The galloyllation at O-4, therefore, may be a step occurring at a later stage of gallotannin formation. This view supports the proposal concerning gallotannin biosynthesis in oak leaves by Cammann et al. (1989), based on an enzymatic study. The cell cultures established by the present work are expected to offer an enzyme source useful in the study of the biosynthetic pathways of gallotannins and ellagitannins. The *Cornus* cell cultures may also be useful in the practical production of gallotannins because of their high yield.

References

Cammann J, Denzel K, Schilling G, Gross GG (1989) Biosynthesis of gallotannin: β-glucogallin-dependent formation of 1,2,3,4,6-pentagalloylglucose by enzymatic galloylation of 1,2,3,6-tetragalloylglucose. Arch Biochem Biophys 273: 58–63

Haslam E (1989) Plant polyphenols, vegetable tannins revisited. Cambridge Univ Press, Cambridge

Hatano T, Kira R, Yoshizaki M, Okuda T (1986) Seasonal changes in the tannins of *Liquidamber formosana* reflecting their biogenesis. Phytochemistry 25: 2787–2789

Hatano T, Ogawa N, Kira R, Yasuhara T, Okuda T (1989a) Tannins of cornaceous plants I. Cornusiins A, B and C, dimeric, monomeric and trimeric hydrolyzable tannins from *Cornus officinalis*, and orientation of valoneoyl group in related tannins. Chem Pharm Bull 37: 2083–2090

Hatano T, Yasuhara T, Okuda T (1989b) Tannins of cornaceous plants II. Cornusiins D, E and F, new dimeric and trimeric hydrolyzable tannins from *Cornus officinalis*. Chem Pharm Bull 37: 2665–2669

Krajci I, Gross GG (1987) Formation of gallotannins in callus cultures from oak (*Quercus robur*). Phytochemistry 26: 141–143

Lees GL (1986) Condensed tannins in the tissue culture of sainfoin (*Onobrychis viciifolia* Scop.) and birdsfoot trefoil (*Lotus corniculatus* L.). Plant Cell Rep 5: 247–251
Linsmaier EM, Skoog F (1965) Organic growth factor requirements of tobacco tissue cultures. Physiol Plant 18: 100–127
Okuda T, Hatano T, Yasui T (1981) Revised structure of terchebin, isolated from *Cornus officinalis*. Heterocycles 16: 1321–1324
Okuda T, Hatano T, Ogawa N, Kira R, Matsuda M (1984) Cornusiin A, a dimeric ellagitannin forming four tautomers, and accompanying tannins in *Cornus officinalis*. Chem Pharm Bull 32: 4662–4665
Okuda T, Mori K, Hatano T (1985) Relationship of the structures of tannins to the binding activities with hemoglobin and methylene blue. Chem Pharm Bull 33: 1424–1433
Okuda T, Yoshida T, Hatano T (1989a) Ellagitannins as active constituents of medicinal plants. Planta Med 55: 117–122
Okuda T, Yoshida T, Hatano T (1989b) New methods of analyzing tannins. J Nat Prod 52: 1–31
Yazaki K, Okuda T (1989) Gallotannin production in cell cultures of *Cornus officinalis* Sieb. et Zucc. Plant Cell Rep 8: 346–349
Yazaki K, Okuda T (1990a) Ellagitannin formation in callus cultures of *Heterocentron roseum*. Phytochemistry 29: 1127–1130
Yazaki K, Okuda T (1990b) Condensed tannin production in callus and suspension cultures of *Cinnamomum cassia*. Phytochemistry 29: 1559–1562
Yoshida T, Chen L, Shingu T, Okuda T (1988) Tannins and related polyphenols of euphorbiaceous plants IV. Euphorbins A and B, novel dimeric dehydroellagitannins from *Euphorbia hirta* L. Chem Pharm Bull 36: 2940–2949
Zaprometov MN (1988) Tannins, lignans, and lignins. In: Constabel F, Vasil IK (eds) Cell culture and somatic cell genetics of plants, vol V. Academic Press, San Diego, pp 89–97

VIII *Cucurbita pepo* (Pumpkin): In Vitro Production of Ascorbate Oxidase

M. Esaka[1]

1 Introduction

Pumpkins (*Cucurbita* spp.), which belong to the family Cucurbitaceae, were originally found in Central and South America, and have long been cultivated throughout the world. The fruits, which contain a large amount of starch, proteins, free amino acids, and vitamins A, B, and C, have popularly been used as a vegetable, for pies, and for livestock feed. In Japan, the custom prevails of eating pumpkin fruits in the winter solstice, as they are thought to prevent a cold or palsy. The young vines and leafstalks have also been eaten as a vegetable. Dried ripe seeds (Cucurbitae semen) of cultivated varieties of *Cucurbita pepo* contain fixed oil, myosin, vitellin, and sugar, and have been used as food. Furthermore, the dried seeds have been used as a teniacide, since they contain resin (peporesin).

Ascorbate oxidase (EC 1.10.3.3) is a copper-containing enzyme that occurs in cucurbitaceous plants such as pumpkin (Lee and Dawson 1973) and cucumber (Nakamura et al. 1968) and plays a role in secondary metabolism (Stark and Dawson 1962; Weis 1975). The precise biological function of ascorbate oxidase has not been clarified, although it has been reported that the enzyme may participate in redox system involving ascorbic acid (Weis 1975). However, the enzyme has been recently used as a valuable reagent for clinical and food analyses of L-ascorbic acid (Matsumoto et al. 1981; Esaka et al. 1985).

Recently, much attention has been focused on the induction mechanisms of enzymes responsible for secondary metabolism in cultured plant cells, with a view to higher productivity of useful metabolites (Ayabe et al. 1980; Leube and Grisebach 1983; Zenk et al. 1985; Kaulen et al. 1986).

In this chapter, our work on the production and secretion of ascorbate oxidase in cultured pumpkin cells is summarized.

2 In Vitro Culture

2.1 Establishment of Pumpkin Callus

Pumpkin fruits (*Cucurbita pepo*, Ebisu Nankin; Fig. 1) were purchased from a local market and stored at about 15 °C until use. Callus was induced from the sarcocarp

[1] Faculty of Applied Biological Sciences, Hiroshima University, Kagamiyama, Higashi-Hiroshima 724, Japan

Biotechnology in Agriculture and Forestry, Vol. 21
Medicinal and Aromatic Plants IV (ed. by Y.P.S. Bajaj)

Fig. 1. Pumpkin fruit

Fig. 2. Pumpkin callus induced from fruit tissues. Callus (1 g), which had been subcultured at about 4-week intervals for more than 3 years, was inoculated into a 100 ml Erlenmeyer flask containing 50 ml of Murashige and Skoog's basal medium with 1 mg/l 2,4-D, 0.1 mg/l kinetin, 3% sucrose and 0.8% agar, and cultured at 25 °C in the dark for 3 weeks

tissue (Esaka et al. 1988a). Disks (8 mm in diameter, 2 mm thick), prepared using a cork borer and a surgical blade under aseptic conditions, were cultured in a 100-ml Erlenmeyer flask (four disks/flask) containing 50 ml of Murashige and Skoog's (MS) basal medium with 1 mg/l 2,4-D, 0.1 mg/l kinetin, 3% (w/v) sucrose and 0.8% (w/v) agarose. The material was grown at 25 °C in the dark, and the cultures were maintained by transferring about 1 g of the callus every 4 weeks. The yellow, undifferentiated pumpkin callus is shown in Fig. 2.

2.2 Preparation of Crude Extract from Pumpkin Callus and Assay for Ascorbate Oxidase Activity

Pumpkin callus was homogenized in 0.05 M Tris-HCl buffer, pH 7, with a glass homogenizer (Esaka et al. 1988a). The homogenates thus prepared were squeezed through nylon gauze and filtered through a Toyo No. 2 filter. The filtrates were used as crude extracts and were assayed for ascorbate oxidase activity. Ascorbate oxidase activity was assayed at 25 °C by following the decrease in A_{265} of the reaction mixture containing 0.05 M potassium phosphate buffer (pH 7), 0.5 mM EDTA, 0.002% metaphosphoric acid, 0.15 mM L-ascorbic acid and enzyme solution in a final volume of 3 ml, according to the method of Oberbacher and Vines (1963). One unit of enzyme activity was defined as the amount of enzyme which oxidizes 1 µmol of L-ascorbic acid per minute.

2.3 Formation of Ascorbate Oxidase in Pumpkin Callus

The changes in ascorbate oxidase activity during callus formation from pumpkin fruit tissue in the dark or under continuous light were investigated (Esaka et al.

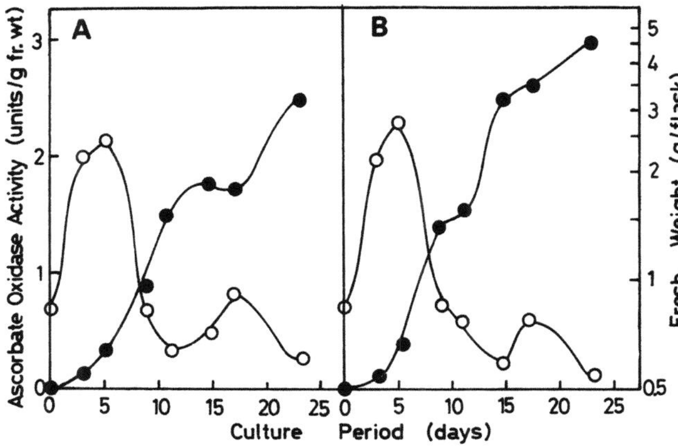

Fig. 3A, B. Changes in ascorbate oxidase activity during callus formation from pumpkin fruit tissue **A** in the dark and **B** under continuous light. Pumpkin fruit tissue slices were inoculated into each flask containing MS agar medium with 1 mg/l 2,4-D and 0.1 mg/l kinetin, and cultured at 25 °C. The cultured tissues were harvested at the indicated periods, weighed and homogenized for the enzyme assay as described (see text). ○, Ascorbate oxidase activity; ●, fresh weight. (Esaka et al. 1988a)

1988a). In the dark, ascorbate oxidase activity increased after transfer of fruit tissue to the callus-forming medium containing 1 mg/l 2,4-D and 0.1 mg/l kinetin, reached a maximum after 5 days, and subsequently declined (Fig. 3A). Later, another smaller peak of activity appeared at 17 days. Under continuous light, the changes in ascorbate oxidase activity were similar to those in the dark, although the yellow callus turned green (Fig. 3B). Thus, light had little effect on the appearance of ascorbate oxidase activity during callus formation.

Furthermore, ascorbate oxidase activity during growth of callus which had been subcultured at about 4-week intervals for more than 1 year, was determined. The activity increased after transfer of the callus to fresh medium and reached a maximum in the early logarithmic phase of growth. It then declined so that in the stationary phase, it was less than one-tenth of the maximum level. Continuous light, which induced greening of callus, had little effect on the changes in ascorbate oxidase activity during callus growth. We have also shown that ascorbate oxidase activity increases during germination of seeds (Esaka et al. 1989a). There is a possibility that ascorbate oxidase participates in respiration.

2.4 Marked Increase in Ascorbate Oxidase in Pumpkin Callus by Adding Copper

Since ascorbate oxidase is a copper-containing enzyme, the effect of the copper concentration in a culture medium on ascorbate oxidase activity in callus was investigated (Fig. 4). Ascorbate oxidase activity in the callus grown in the medium containing 10 μM $CuSO_4$ (Murashige and Skoog's basal medium contains 0.1 μM $CuSO_4$) was about eight times that in the basal medium. However, a large excess of $CuSO_4$ (100 μM $CuSO_4$) inhibited callus growth with a consequent decrease in the enzyme activity. When $CuSO_4$ was omitted from the medium, the activity was approximately one-third of that found with callus grown in the basal medium. Thus, ascorbate oxidase activity in pumpkin callus clearly depends on the $CuSO_4$ concentration in the culture medium. Changes in ascorbate oxidase activity during growth of pumpkin callus in the medium containing 10 μM $CuSO_4$ were compared

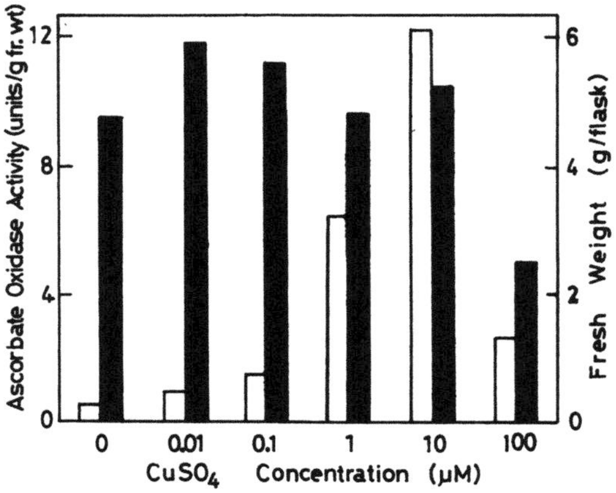

Fig. 4. Effect of $CuSO_4$ concentration in the medium on ascorbate oxidase activity and fresh weight of pumpkin callus cultured for 15 days. ○, Ascorbate oxidase activity; ●, fresh weight. (Esaka et al. 1988a)

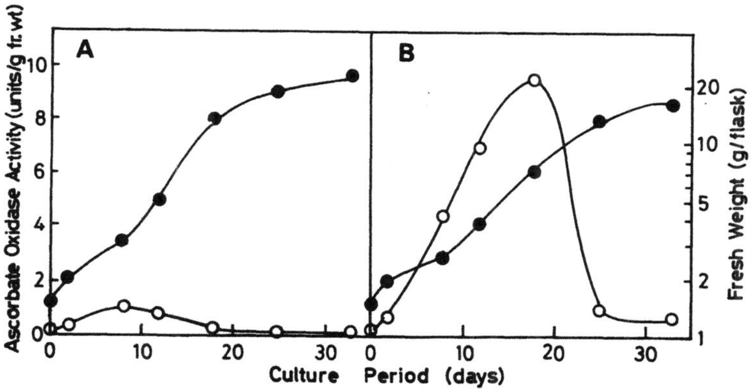

Fig. 5A, B. Changes in ascorbate oxidase activity during growth of pumpkin callus. Calluses were inoculated into each flask containing MS agar medium with 0.1 (A) or 10 μM $CuSO_4$ (B), and cultured at 25 °C. The calluses were harvested at the indicated periods, weighed and homogenized for the enzyme assay as described (see text). ○, Ascorbate oxidase activity; ● fresh weight. (Esaka et al. 1988a)

with those of callus grown in the basal medium containing 0.1 μM $CuSO_4$ (Fig. 5). Ascorbate oxidase activity at the peak, at 17 days after transfer to the fresh medium containing 10 μM $CuSO_4$ (Fig. 5B), was about ten times the maximal activity attained by the culture containing 0.1 μM $CuSO_4$ (Fig. 5A).

There have been only a few reports that the enzyme in cultured plant cells is induced by its prosthetic metal. Polacco (1977) has reported that urease activity in soybean tissue cultures is increased by adding nickel, which is a component of urease. Ikeda et al. (1982) have also reported that Cyt aa_3 content in cultured tobacco cells increases twofold after subculturing seven times in a high copper concentration medium. Delhaize et al. (1986) have reported that the amount of diamine oxidase protein in clover leaves is controlled by the copper concentrations of the leaves. Jones and Carbonell (1984) have reported that the amount of α-amylase in aleulone layers of barley is controlled by calcium, the metal cofactor of the enzyme. Furthermore, in *Chlamydomonas reinhardtii*, the amount of plastocyanin apoprotein has been shown to be regulated by the copper concentration of the medium (Wood 1978). We investigated whether or not the increase in ascorbate oxidase activity during the growth of pumpkin callus correlated with an increase in the enzyme protein (Esaka et al. 1988b).

To produce a specific antiserum against ascorbate oxidase, we purified ascorbate oxidase from a commercially available source (Boehringer Mannheim Co.). After fractionation with saturated $(NH_4)_2SO_4$ solution (40 to 80% saturation), the enzyme was applied to a Sepharose 6B column. Figure 6 gives the elution profiles of ascorbate oxidase activity and protein (A_{280}) from the Sepharose 6B column. The activity-peak fractions (Nos. 80 to 90) all showed almost the same specific activity, suggesting that the fractions were in a highly purified state. We used only the most active fraction (No. 85) for further work. When this peak fraction was subjected to SDS-PAGE, a single polypeptide band was observed (Fig. 6B). Despite many reports on ascorbate oxidase, the molecular properties of this enzyme have not been fully elucidated. Avigliano et al. (1983) have reported that ascorbate oxidase from pumpkin is composed of two noncovalently linked

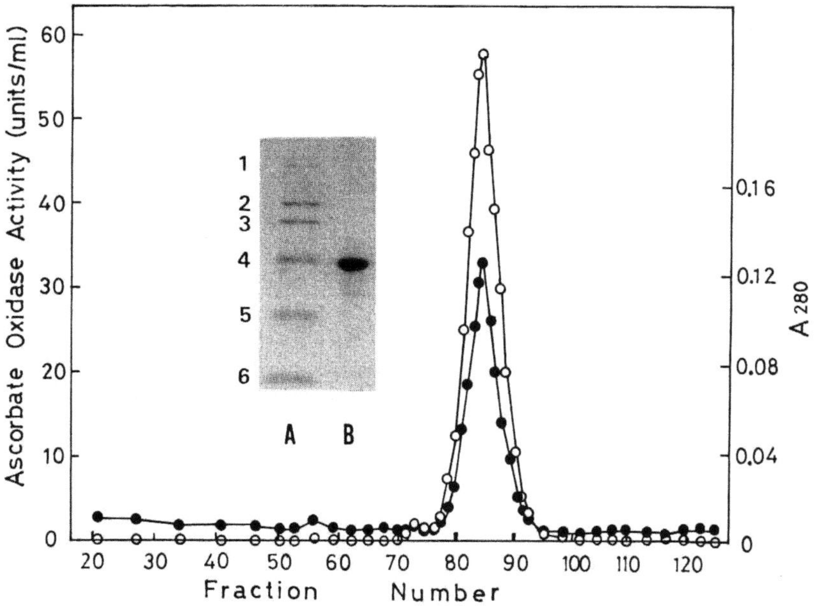

Fig. 6. Sepharose 6B column chromatography of ascorbate oxidase preparation, and SDS-PAGE of the peak fraction. The partially purified preparation after a $(NH_4)_2SO_4$ fractionation was applied on a Sepharose 6B column as described in text and fractions (0.6 ml each) were assayed for ascorbate oxidase activity (○) and protein (A_{280}, ●). The peak fraction and mol wt standards were subjected to a 10% (w/v) polyacrylamide gel containing 0.1% (w/v) SDS. **A** Mol wt standards; *1* myosin (205 000); *2* β-galactosidase (116 000); *3* phosphorylase (97 400); *4* albumin bovine (66 000); *5* albumin egg (45 000); and *6* carbonic anhydrase (29 000). **B** purified ascorbate oxidase preparation. (Esaka et al. 1988b)

subunits of slightly different molecular weights (75 000 and 72 000, respectively). We showed that the enzyme was composed of polypeptides with molecular weight 64 000 (Fig. 6). Antiserum against the purified ascorbate oxidase preparation was produced in a rabbit by injecting twice at an interval of about 20 days with a mixture of the purified preparation (about 100 μg protein) and complete Freund's adjuvant, followed by an injection of the mixture (about 20 μg protein) after 1 week. When the specificity of the antiserum was examined with Ouchterlony double immunodiffusion tests (Ouchterlony and Nilsson 1973), the antiserum formed a single precipitin line with the antigen (Fig. 7A). Furthermore, the antiserum also formed a single precipitin line with the crude extract from pumpkin fruit tissue or pumpkin callus (Fig. 7B). The precipitin lines formed with the purified preparation and each crude extract fused to another (Fig. 7B), suggesting that they were identical in their immunological properties. The amount of ascorbate oxidase protein could be determined by the single radial immunodiffusion method described by Mancini et al. (1965). There was a linear correlation between the square of the diameter (D^2) of the immunoprecipitating circle and the amount of purified ascorbate oxidase added to the well (Fig. 8). Analysis with the immunological method showed that the amount of ascorbate oxidase protein increased during the growth of pumpkin callus (Fig. 9A), suggesting that the increase in ascorbate

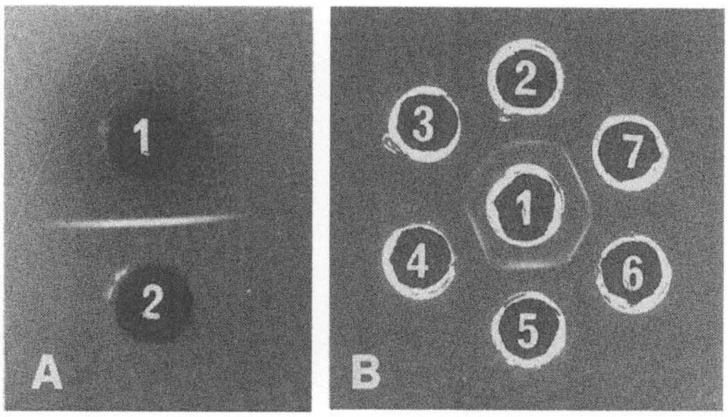

Fig. 7A, B. Double immunodiffusion tests. **A** Well *1* anti-ascorbate oxidase immunoglobulin G fraction (about 10 μg protein); well *2* purified ascorbate oxidase preparation (about 1 μg protein). **B** *Center* (well *1*), anti-ascorbate oxidase immunoglobulin G fraction (about 5 μg protein); wells *2* and *5* purified ascorbate oxidase preparation (about 0.5 μg protein each); wells *3* and *6* crude extract from pumpkin fruit tissue; wells *4* and *7* crude extract from pumpkin callus. (Esaka et al. 1988b)

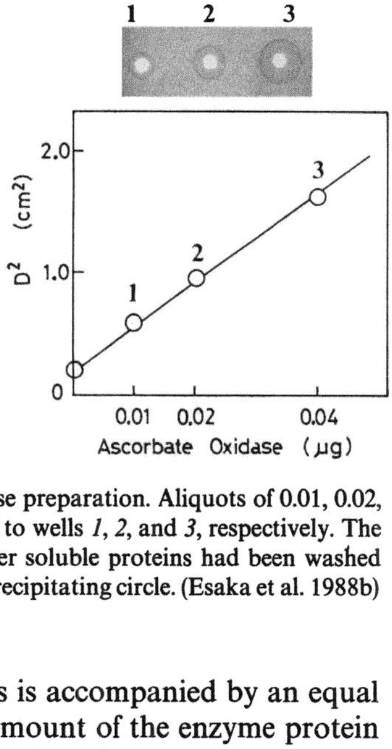

Fig. 8. Single radial immunodiffusion of purified ascorbate oxidase preparation. Aliquots of 0.01, 0.02, and 0.04 μg of purified ascorbate oxidase preparation were added to wells *1*, *2*, and *3*, respectively. The precipitin lines were stained with Coomassie brilliant blue R, after soluble proteins had been washed away. The *ordinate* shows the square of diameter (D^2) of immunoprecipitating circle. (Esaka et al. 1988b)

oxidase activity during the growth of pumpkin callus is accompanied by an equal increase in the enzyme protein. The increase in the amount of the enzyme protein was almost the same as that of the activity (Fig. 9A). Furthermore, the analysis showed that the amounts of ascorbate oxidase protein at 1 and 2 weeks after transfer to the fresh medium containing 10 μM $CuSO_4$ were about five and eight times those in a culture containing 0.1 μM $CuSO_4$, although the increase in the ascorbate oxidase protein was a little less than that of the activity (Fig. 9). This shows that the marked increase in ascorbate oxidase activity by adding copper is also accompanied by an increase in the enzyme protein. However, it is not yet known whether copper enhances the de novo synthesis of the enzyme or prevents the proteolytic breakdown of the enzyme. Keyhani and Keyhani (1975) have reported that Cyt c oxidase activity in a *Candida utilis* yeast cell increases in

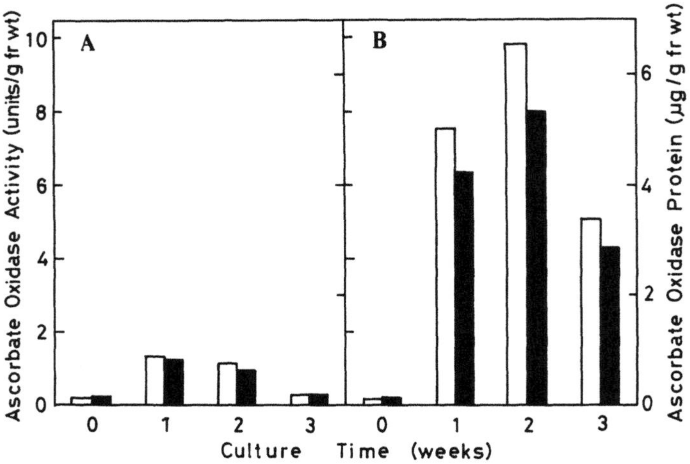

Fig. 9A, B. Changes in the activity of ascorbate oxidase and the amount of immunoreactive protein during growth of pumpkin callus. Callus was inoculated into each flask containing MS agar medium with 0.1 (**A**) or 10 µM $CuSO_4$ (**B**), and cultured at 25 °C. The callus was harvested at the indicated times and homogenized. The crude extracts were assayed for the enzyme activity and the amount of immunoreactive protein. *White bar* ascorbate oxidase activity expressed as units per g fresh weight; *black bar* the amount of immunoreactive protein determined by single radial immunodiffusion method and expressed as µg per g fresh weight. (Esaka et al. 1988b)

response to increased copper concentration in the medium, and Shatzman and Kosman (1978) have also reported that galactose oxidase activity in the fungus *Dactylium dendroides* depends on copper concentration. However, the amounts of apoproteins of these enzymes were not controlled by the copper concentration. They have suggested that copper may control the holoenzyme assembly. On the other hand, Delhaize et al. (1986) have reported that the synthesis of apoprotein of diamine oxidase in clover leaves is regulated by the copper of the leaves so that no apoprotein is synthesized in copper-deficient leaves. Recently, Merchant and Bogorad (1987) indicated that the copper-dependent expression of Cyt c-552 synthesis in *Chlamydomonas reinhardtii* was affected directly by the concentration of copper. We also showed that the amount of ascorbate oxidase protein in pumpkin callus was controlled by copper, a prosthetic metal of the enzyme. The mechanism of metal mediated regulation of protein biosynthesis is of general interest. Deikman and Jones (1985) have suggested that the amount of mRNA coding for B-group α-amylases in aleurone layers of barley was not affected by calcium, although the amount of the enzyme protein is controlled by the metal cofactor. They have proposed that calcium exerted its control at translation or at a site after translation. Further study will be required to determine the site where copper controls the amount of ascorbate oxidase protein in pumpkin callus.

2.5 Secretion of Ascorbate Oxidase by Suspension-Cultured Pumpkin Cells

It is convenient that the useful metabolites and enzymes are secreted by cultured plant cells. In regard to enzymes, peroxidase (Huystee and Lobarzewski 1982;

Chibbar et al. 1984), phosphatase (Ueki and Sato 1971; Ciarrochi et al. 1981), glucosidase (Yamazaki and Konno 1985), laccase (Bligny and Douce 1983) and so on have been suggested to be secreted by cultured plant cells. Since ascorbate oxidase is a valuable enzyme which is used for the determination of L-ascorbic acid, the secretion of ascorbate oxidase by cultured pumpkin cells is also expected.

Suspension-cultures were established by inoculating callus cells, which had been subcultured at about 4-week intervals for more than 1 year, into a 200-ml Erlenmeyer flask containing 50 ml of the MS liquid medium with 1 mg/l 2,4-D, 0.1 mg/l kinetin and 3% (w/v) sucrose. These cultures were agitated on a rotary shaker at 100 rpm at 25 °C, and maintained by transferring about 1 g (fresh weight) of the cells to fresh liquid medium at about 4-weeks intervals (Esaka et al. 1989b).

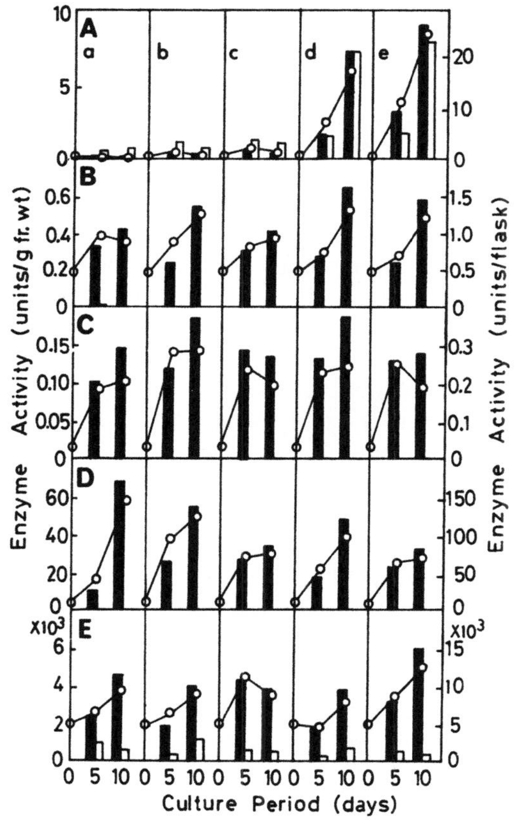

Fig. 10A–E. Ascorbate oxidase, alcohol dehydrogenase, glucose-6-phosphate dehydrogenase, catalase and peroxidase activities in cells and in the medium after pumpkin cell suspension cultures in the presence of $CuSO_4$ of various concentrations. Pumpkin cells were inoculated in MS liquid medium containing 0 (*a*), 0.01 (*b*), 0.1 (*c*), 1 (*d*) or 10 µM $CuSO_4$ (*e*) and cultured at 25 °C. The cultures were harvested at 5 or 10 days, and the cells were separated from the cultures and homogenized as described in experimental. The cell homogenate and the culture medium were assayed for ascorbate oxidase (**A**), alcohol dehydrogenase (**B**), glucose-6-phosphate dehydrogenase (**C**), catalase (**D**) and peroxidase activities (**E**), *o* Enzyme activity expressed as units per g fresh weight in the cells; *black bar* enzyme activity expressed as units per flask in the cells; *white bar* enzyme activity expressed as units per flask in the medium. (Esaka et al. 1989b)

In the cell suspension cultures grown in the medium containing 0, 0.01, 0.1, 1, or 10 μM $CuSO_4$ for 5 or 10 days, ascorbate oxidase activity in the medium was compared with that in the cells (Fig. 10). Ascorbate oxidase activity in the cells clearly depended on the $CuSO_4$ concentration in the culture medium. In the same manner, ascorbate oxidase activity in the medium was also increased by adding copper to the fresh culture medium. In the cultures grown in the medium containing 0, 0.01, and 0.1 μM $CuSO_4$, ascorbate oxidase activity in the medium was above twice that in the cells. In the case of 1 μM $CuSO_4$, the activity in the medium was almost the same that in the cells. Ascorbate oxidase activity in the medium, in the presence of 10 μM $CuSO_4$, was a little lower than that in the cells. The decrease in the rate of the activity in the medium to that in the cells along with an increase in $CuSO_4$ concentration may be due to a limit in ability of the enzyme secretion. At any rate, a large amount of ascorbate oxidase was found to be released into the culture medium. On the other hand, the activities of alcohol dehydrogenase and glucose-6-phosphate dehydrogenase, cytosolic enzymes which must be easily released from damaged cells, was not detected in the culture medium. Furthermore, the activity of catalase, a microbody enzyme which is also detected in cytoplasmic fraction (Esaka and Asahi 1982), was also undetected in the medium. The activity of peroxidase, which is suggested to be a secretory enzyme, was slightly detected in the medium. Thus, the possibility that ascorbate oxidase activity in the medium is from damaged cells can be denied. Finally, ascorbate oxidase is suggested to be secreted by cultured pumpkin cells. The subcellular localization of ascorbate oxidase has not yet been established, although there have been some reports that the enzyme is localized in the cytoplasm (Yamauchi et al. 1984), microsomes, vacuole (Vines and Oberbacher 1963), or cell wall (Honda 1955). We found that ascorbate oxidase was an extracellular enzyme.

Most of secretory proteins have been suggested to be glycoproteins (Kornfeld and Kornfeld 1985). For instance, Bligny and Douce (1983) have reported that a laccase-type polyphenol oxidase secreted by sycamore cells contains 45% carbohydrate. We investigated whether or not ascorbate oxidase was a glycoprotein. The purified ascorbate oxidase preparation on the gel was stained by periodic acid-Schiff method (data not shown), suggesting that ascorbate oxidase is a glycoprotein (Esaka et al. 1989b). Takahashi et al. (1986) have discussed a structural study of oligosaccharides of laccase from sycamore cells. A structural study of oligosaccharides of ascorbate oxidase will also be required.

2.6 Stimulation of Ascorbate Oxidase Secretion by Divalent Cations

The roles of divalent cations in cell function and regulation have been widely studied (Cheung 1980). In particular, calcium is known to play an important role in the secretion of proteins in various animal cells such as the pancreatic cells (Jamieson and Palade 1977; Gardner et al. 1980), and the requirement of the intracellular elevation of calcium for exocytosis has been firmly established. Recently, much attention has been focused on elucidating the mechanism of the cell regulation by Ca^{2+} at the molecular level. Calcium has also been implicated in the secretion of proteins by plant cells. For example, Chrispeels and Varner (1967), and

Jacobsen et al. (1970) reported that, in barley aleurone layers, the secretion of several enzymes including α-amylase is markedly influenced by Ca^{2+}. Mitsui et al. (1984) reported that the secretion of α-amylase by rice-seed scutellar epithelium is regulated by the cytoplasmic Ca^{2+} concentration. Furthermore, Jones and Jacobsen (1983) reported that Ca^{2+} increases the amount of one particular group of α-amylase isozymes secreted into the incubation medium, and Deikman and Jones (1985) suggested that the presence of Ca^{2+} is necessary either for the translation of α-amylase mRNA in vivo or for the posttranslational processing and transport of α-amylase isozymes in barley aleurone cells. We investigated whether or not $CaCl_2$ and $MgCl_2$ stimulated the secretion of ascorbate oxidase in cultured pumpkin cells (Esaka et al. 1989c).

Figure 11 shows the changes in the activities of ascorbate oxidase and peroxidase, another secreted enzyme, in the culture medium during the growth of pumpkin cells cultured in the presence of 0, 5, 10, 20, 40, and 80 mM $CaCl_2$. On adding $CaCl_2$ to the medium, ascorbate oxidase activity in the culture medium markedly increased. The activities in the culture medium after growing in the presence of 40 mM $CaCl_2$ for 10 and 20 days were about 10 and 50 times those in the absence of $CaCl_2$ respectively (Fig. 11A). On the other hand, peroxidase activities in the culture medium in the presence of 40 mM $CaCl_2$ at 10 and 20 days were only 1.7 and 4 times those in the absence of $CaCl_2$ (Fig. 11B). The amount of total protein in the culture medium also increased by adding $CaCl_2$ (Table 1). However, the specific activity of ascorbate oxidase in the culture medium was increased by adding $CaCl_2$, since the increase in ascorbate oxidase activity in the culture medium was much more than that of total protein. The specific activities of ascorbate oxidase in the culture medium in the presence of 20 and 40 mM $CaCl_2$ were about 25 and 22 times those in the absence of $CaCl_2$ (Table 1). On the other

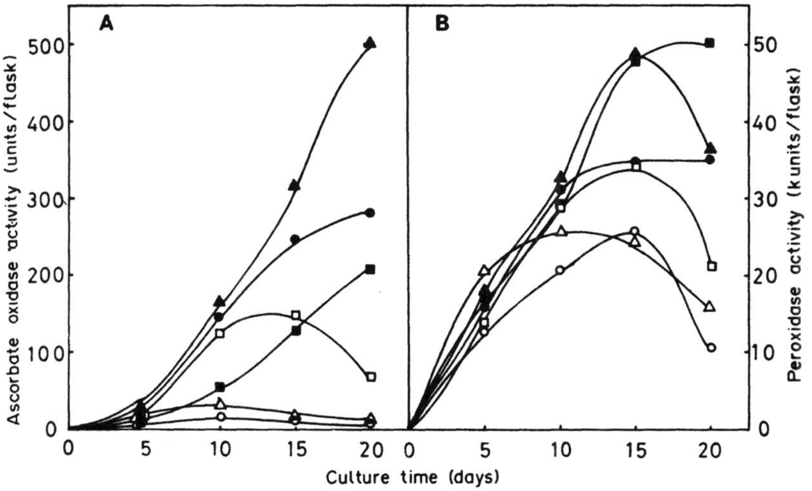

Fig. 11A, B. Changes in the activities of **A** ascorbate oxidase and **B** peroxidase in the culture medium during growth of pumpkin cell suspension cultures in the presence of various concentrations of $CaCl_2$. Cells were inoculated into MS liquid medium containing 10 µM $CuSO_4$ with 0 (○), 5 (△). 10 (□), 20 (●), 40 (▲) or 80 mM $CaCl_2$ (■), and cultured at 25 °C. (Esaka et al. 1989c)

Table 1. Effect of the $CaCl_2$ concentration of the culture medium on the specific activities of ascorbate oxidase and peroxidase in the culture medium of pumpkin cell suspension cultures. (Esaka et al. 1989c)[a]

$CaCl_2$ (mM)	Protein (mg)	Ascorbate oxidase (units/mg)	Peroxidase (units/mg)
0	3.83	1.87	2500
5	5.70	2.94	2740
10	6.27	10.7	3480
20	5.99	47.7	6210
40	13.2	41.1	2950
80	9.58	24.4	5570

[a] Cultures grown for 20 days.

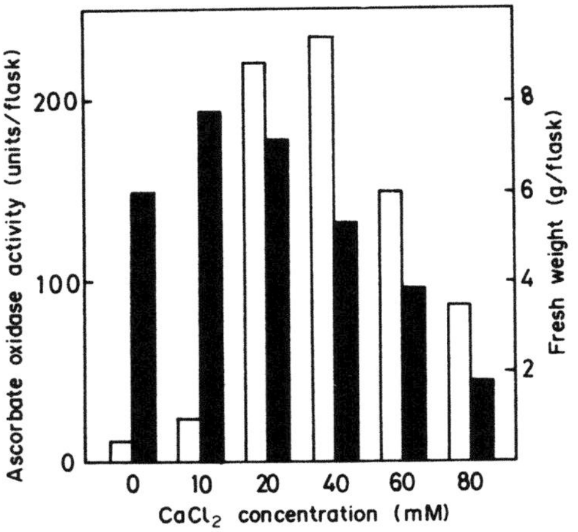

Fig. 12. Effect of $CaCl_2$ concentration in the medium on ascorbate oxidase activity in the medium and fresh weight of pumpkin cells after suspension culture for 15 days. □, ascorbate oxidase activity in the medium; ■, fresh weight. (Esaka et al. 1989c)

hand, the specific activity of peroxidase in the culture medium was little increased by adding $CaCl_2$ (Table 1). These results suggest that the marked increase in ascorbate oxidase activity in the culture medium by adding $CaCl_2$ is not due to the increase in protein washed from cell wall surface by the increase in ionic strength, but may be due to the stimulation of ascorbate oxidase secretion. The effect of $CaCl_2$ on the growth of the pumpkin cells was also investigated (Fig. 12). In the presence of 80 mM $CaCl_2$, cell growth was significantly inhibited, so that ascorbate oxidase activity in the culture medium was relatively low. In the other cases, there was no relationship between cell growth and ascorbate oxidase activity in the culture medium, suggesting that the stimulation of ascorbate oxidase secretion by $CaCl_2$ was not the result of the effect of calcium on cell growth.

Ascorbate oxidase acitivity in the culture medium was compared with that in the cells (Table 2). Ascorbate oxidase activity in the cells was hardly affected by $CaCl_2$ concentration, whereas that in the culture medium was markedly increased by adding $CaCl_2$. Consequently, total ascorbate oxidase activity in a flask (the sum of the ascorbate oxidase activities in the cells and in the culture medium), was

Table 2. Effect of the $CaCl_2$ concentration of the culture medium on ascorbate oxidase activities in the cells and in the culture medium of pumpkin cell suspension cultures. (Esaka et al. 1989c)[a]

$CaCl_2$ (mM)	Cells (units/flask)	Medium (units/flask)	Total (units/flask)
0	11.9	33.1	45.0
10	13.2	38.0	51.2
20	5.01	152	157
30	11.1	229	240
40	5.98	192	198
50	4.47	166	170
60	12.0	154	166
80	6.54	102	109

[a] Cultures grown for 15 days.

Table 3. Effects of the $MgCl_2$ and KCl concentrations of the culture medium on the specific activities of ascorbate oxidase and peroxidase in the culture medium of pumpkin cell suspension cultures. (Esaka et al. 1989c)[a]

$MgCl_2$ (mM)	KCl (mM)	Growth rate (−)	Protein (mg)	Ascorbate oxidase (units/mg)	Peroxidase (units/mg)
–	–	3.7	8.21	1.68	2790
10	–	3.5	8.40	4.79	2130
20	–	3.7	12.9	17.1	793
40	–	3.3	16.3	20.0	541
80	–	2.1	10.5	14.2	343
–	10	3.8	7.91	1.63	2250
–	20	4.2	9.72	1.83	1860
–	40	4.0	10.9	1.21	1860
–	80	3.4	12.9	3.27	2560

[a] Cultures grown for 15 days.

increased by adding $CaCl_2$ (Table 2). Therefore, we infer that the synthesis of ascorbate oxidase may be also stimulated by adding $CaCl_2$.

Table 3 shows the effects of $MgCl_2$ and KCl on the ascorbate oxidase secretion. Ascorbate oxidase activity in the culture medium increased by adding $MgCl_2$, while the peroxidase activity in the culture medium had a tendency to decrease by adding $MgCl_2$. When the effect of KCl was investigated, ascorbate oxidase activities in the culture medium in the presence of 10, 20, and 40 mM KCl were almost the same as that in the absence of KCl. In the presence of 80 mM KCl, ascorbate oxidase in the medium was about three times that of in the absence of KCl. However, the increase may not be specific for ascorbate oxidase, since peroxidase activity in the medium was about three times that of in the absence of KCl. However, the increase may not be specific for ascorbate oxidase, since peroxidase activity in the medium was also increased by adding 80 mM KCl. Finally, we showed that ascorbate oxidase secretion in cultured pumpkin cells was stimulated by adding $CaCl_2$ or $MgCl_2$, but not by adding KCl. It cannot be denied that the stimulation effect is nonspecific on membrane permeability, since the effect required the high concentration of $CaCl_2$ or $MgCl_2$. However, we infer that the stimulation effect is specific for ascorbate oxidase secretion, considering that the specific activity of ascorbate oxidase in the culture medium markedly increased by adding $CaCl_2$ or $MgCl_2$ (Tables 1 and 3) and that peroxidase activity in the culture medium increased little by adding these cations (Fig. 11, Table 3). Furthermore, there is a possibility that the stimulation effect is due to the release of ascorbate oxidase from cells which had been damaged by adding high concentrations of $CaCl_2$ or $MgCl_2$. However, the results that cell growth was not inhibited by adding 10 or 20 mM $CaCl_2$ (or $MgCl_2$), which was effective in stimulating ascorbate oxidase secretion, and that total ascorbate oxidase activity in a flask (the sum of the ascorbate oxidase activities in the cells and in the culture medium) increased by adding these cations (Table 2) suggest that the stimulation effect is not due to the release from the damaged cells. In addition, there is no denying the possibility that the stress which is caused by adding a high concentration of $CaCl_2$ or $MgCl_2$ stimulates ascorbate oxidase secretion. Recently, we showed that Eosine Yellowish and salicylate, which are known to be inducers of pathogenesis-related proteins, markedly stimulated the accumulation of ascorbate oxidase in the culture medium (Esaka et al. 1990a). Thus, ascorbate oxidase may be regarded as a stress protein such as pathogenesis-related proteins. Further studies will be required to investigate why these divalent cations stimulate ascorbate oxidase secretion.

3 Conclusions and Prospects

Ascorbate oxidase in cultured pumpkin cells was studied. Its activity rapidly increased during callus formation from fruit tissue. The activity reached a maximum at 5 days after transfer and then declined. In callus which had been subcultured at about 4-week intervals for more than 1 year, the activity also increased after transfer of fresh medium and reached a maximum in the early logarithmic phase of growth. Light had little effect on the appearance of ascorbate oxidase activity in callus. In the callus grown in the presence of 10 μM $CuSO_4$, the activity was about 10 times that in the presence of 0.1 μM $CuSO_4$, suggesting that the formation of ascorbate oxidase in callus is stimulated by copper, a prosthetic metal of the enzyme. The study with a specific antibody against purified ascorbate oxidase showed that the increase in ascorbate oxidase activity during the growth of callus was accompanied by an equal increase in enzyme protein. Furthermore, the marked increase in the enzyme activity by adding copper to the culture medium was also accompanied by an increase in enzyme protein. It remains to be established whether copper enhances the de novo biosynthesis of the enzyme or prevents its degradation.

Ascorbate oxidase was found to be released into the medium in pumpkin-cell suspension cultures. Ascorbate oxidase activity in the medium, as well as that in the cells, increased soon after transfer to fresh medium, and reached a maximum at about 5 days. Alcohol dehydrogenase and glucose-6-phosphate dehydrogenase (cytosolic enzyme) and catalase (microbody enzyme) were not detected in the medium, whereas peroxidase, which is known to be a secretory protein, was slightly detectable. Purified ascorbate oxidase on acrylamide gel was stained by the periodic acid-Schiff method. Thus, ascorbate oxidase is a secretory glycoprotein. Perhaps ascorbate oxidase is actively secreted during the growth of cultured pumpkin cells. Ca^{2+} markedly stimulated the accumulation of ascorbate oxidase in the culture medium. Furthermore, the specific activity of the enzyme in the culture medium was markedly increased by adding $CaCl_2$. On the other hand, Ca^{2+} had little effect on the accumulation of peroxidase in the medium. Mg^{2+}, but not K^+ was shown to be effective in the stimulation of ascorbate oxidase secretion, suggesting that Mg^{2+} can be substituted for Ca^{2+} in stimulating the secretion of ascorbate oxidase. Further studies will be required to elucidate the process of extracellular secretion of ascorbate oxidase by cultured pumpkin cells and to investigate why divalent cations stimulate ascorbate oxidase secretion. Recently, we succeeded in cloning ascorbate oxidase cDNA in pumpkin callus (Esaka et al. 1990b).

4 Protocol for Ascorbate Oxidase Production by Cultured Pumpkin Cells

Pumpkin callus was induced from the sarcocarp tissue of fruits. Disks (8 mm in diameter, 2 mm thick), prepared using a cork borer and a surgical blade under aseptic condition, were cultured in a 100-ml Erlenmeyer flask (four disks/flask) containing 50 ml of Murashige and Skoog's basal medium with 1 mg/l 2,4-D, 0.1 mg/l kinetin, 3% (w/v) sucrose and 0.8% (w/v) agarose. The material was grown at 25 °C in the dark, and the cultures were maintained by transferring about 1 g of the callus every 4 weeks.

The best yield of ascorbate oxidase was obtained when pumpkin callus cells were cultured on MS basal medium containing 1 mg/l 2,4-D, 0.1 mg/l kinetin, 3% (w/v) sucrose, 10 μM $CuSO_4$ and 40 mM $CaCl_2$ at 25 °C in the dark.

References

Avigliano L, Vecchini P, Sirianni P, Marcozzi G, Marchesini A, Mandovi B (1983) A reinvestigation on the quaternary structure of ascorbate oxidase from *Cucurbita pepo medullosa.* Mol Biochem 56: 107–112

Ayabe S, Yoshikawa T, Kobayashi M, Furuya T (1980) Biosynthesis of a retrochalcone echinatin: involvement of O-methyltransferase to licodione. Phytochemistry 19: 2331–2336

Bligny R, Douce R (1983) Excretion of laccase by sycamore (*Acer pseudoplatanus* L.) cells. Biochem J 209: 489–496

Cheung WY (1980) Calmodulin plays a pivotal role in cellular regulation. Science 207: 19–27

Chibbar RN, Cella R, Albani D, Huystee RBV (1984) The growth of and peroxidase synthesis by two carrot cell lines. J Exp Bot 35: 1846–1852

Chrispeels MJ, Varner JE (1967) Gibberellic acid-enhanced synthesis and release of α-amylase and ribonuclease by isolated barley aleurone layers. Plant Physiol 42: 398–406

Ciarrochi G, Cella R, Nielsen E (1981) Release of nucleotide cleaving and phosphatase from carrot cells grown in suspension culture. Physiol Plant 53: 357–377

Deikman J, Jones RL (1985) Control of α-amylase mRNA accumulation by gibberellic acid and calcium in barley aleurone layers. Plant Physiol 78: 192–198

Delhaize E, Dilworth MJ, Webb J (1986) The effects of copper nutrition and developmental state on the biosynthesis of diamine oxidase in clover leaves. Plant Physiol 82: 1126–1131
Esaka M, Asahi T (1982) Purification and properties of catalase from sweet potato root microbodies. Plant Cell Physiol 23: 315–322
Esaka M, Suzuki K, Kubota K (1985) Determination method for L-ascorbic acid in foods with immobilized ascorbate oxidase. Agric Biol Chem 49: 2955–2960
Esaka M, Imagi J, Suzuki K, Kubota K (1988a) Formation of ascorbate oxidase in cultured pumpkin cells. Plant Cell Physiol 29: 231–235
Esaka M, Uchida M, Fukui H, Kubota K, Suzuki K (1988b) Marked increase in ascorbate oxidase protein in pumpkin callus by adding copper. Plant Physiol 88: 656–660
Esaka M, Uchida M, Suzuki K, Kubota K (1989a) Increases in the activity and amount of ascorbate oxidase during the germination of pumpkin seeds. Agric Biol Chem 53: 1181–1182
Esaka M, Fukui H, Suzuki K, Kubota K (1989b) Secretion of ascorbate oxidase by suspension-cultured pumpkin cells. Phytochemistry 28: 117–119
Esaka M, Nishitani I, Fukui H, Suzuki K, Kubota K (1989c) Stimulation of ascorbate oxidase secretion from cultured pumpkin cells by divalent cations. Phytochemistry 28: 2655–2658
Esaka M, Suzuki K, Kubota K (1990a) Stimulation of ascorbate oxidase secretion from cultured pumpkin cells by Eosine Yellowish and potassium salicylate. Phytochemistry 29: 1547–1549
Esaka M, Hattori T, Fujisawa K, Sakajo S, Asahi T (1990b) Molecular cloning and nucleotide sequence of full-length cDNA for ascorbate oxidase from cultured pumpkin cells. Eur J Biochem 191: 537–541
Gardiner JD, Walker MD, Rottman AJ (1980) Effect of A23187 on amylase release from dispersed acini prepared from guinea pig pancreas. Am J Physiol 238: 9457–9466
Honda SI (1955) Ascorbic acid oxidase in barley roots. Plant Physiol 30: 174–181
Huystee RBV, Lobarzewski J (1982) An immunological study of peroxidase release by cultured peanut cells. Plant Sci Lett 27: 59–67
Ikeda T, Matsumoto T, Obi Y (1982) Influences of copper concentration on cytochrome aa_3 formation and growth in cultured tobacco cells. Agric Biol Chem 46: 565–566
Jacobsen JV, Scandalios JG, Varner JE (1970) Multiple forms of amylase induced by gibberellic acid in isolated barley aleurone layers. Plant Physiol 45: 367–371
Jamieson JD, Palade JE (1977) Production of secretory proteins in animal cells. In: Brinkley BR, Porter KR (eds) International cell biology 1976–1977. Rockefeller Univ Press, New York, pp 308–317
Jones RL, Carbonell J (1984) Regulation of the synthesis of barley aleurone α-amylase by gibberellic acid and calcium ions. Plant Physiol 76: 213–218
Jones RL, Jacobsen JV (1983) Calcium regulation of the secretion of α-amylase isozymes and other proteins from barley aleurone layers. Planta 158: 1–9
Kaulen H, Schell J, Kreuzaler F (1986) Light-induced expression of the chimeric chalcone synthase-NPTII gene in tobacco cells. EMBO J 5: 1–8
Keyhani E, Keyhani J (1975) Cytochrome c oxidase biosynthesis and assembly in *Candida utilis* yeast cells. Arch Biochem Biophys 167: 596–602
Kornfeld R, Kornfeld S (1985) Assembly of asparagine-linked oligosacchrides. Annu Rev Biochem 54: 631–664
Lee MH, Dawson CR (1973) Ascorbate oxidase: further studies on the purification of the enzyme. J Biol Chem 248: 6596–6602
Leube J, Grisebach H (1983) Further studies on induction of enzymes of phytoalexin synthesis in soybean and cultured soybean cells. Z Naturforsch 38: 730–735
Mancini G, Carbonara AO, Heremans GF (1965) Immunochemical quantitation of antigens by single radial immunodiffusion. Immunochemistry 2: 235–254
Matsumoto K, Yamada K, Osajima Y (1981) Ascorbate electrode for determination of L-ascorbic acid in food. An Chem 53: 1974–1979
Merchant S, Bogorad L (1987) Metal ion regulated gene expression: use of a plastocyanin-less mutant of *Chlamydomonas reinhardtii* to study the Cu(II)-dependent expression of cytochrome c-552. EMBO J 6: 2531–2535
Mitsui T, Christeller JT, Hara-Nishimura I, Akazawa T (1984) Possible roles of calcium and calmodulin in the biosynthesis and secretion of α-amylase in rice seed scutellar epithelium. Plant Physiol 75: 21–25
Nakamura T, Makino N, Ogura Y (1968) Purification and properties of ascorbate oxidase from cucumber. J Biochem 64: 189–195

Oberbacher MF, Vines HM (1963) Spectrophotometric assay of ascorbic acid oxidase. Nature 197: 1203–1204

Ouchterlony O, Nilsson LA (1973) Immunodiffusion and immunoelectrophoresis. In: Weir DM (ed) Handbook of experimental immunology. Blackwell, Oxford, 19, pp 1–39

Polacco JC (1977) Nitrogen metabolism in soybean tissue culture: II. Urea utilization and urease synthesis require Ni^{2+}. Plant Physiol 59: 827–830

Shatzman AR, Kosman DJ (1978) The utilization of copper and its role in the biosynthesis of copper-containing proteins in the fungus *Dactylium dendroides*. Biochim Biophys Acta 544: 163–179

Stark GR, Dawson CR (1962) On the accessibility of sulfhydryl groups in ascorbic acid oxidase. J Biol Chem 237: 712–716

Takahashi N, Hotta T, Ishihara H, Mori M, Tejima S, Bligny R, Akazawa T, Endo S, Arata Y (1986) Xylose-containing common structural unit in N-linked oligosaccharides of laccase from sycamore cells. Biochemistry 25: 388–395

Ueki K, Sato S (1971) Effect of inorganic phosphate on the extracellular acid phosphatase activity of tobacco cells cultured in vitro. Physiol Plant 24: 506–511

Vines HM, Oberbacher MF (1963) Citrus fruit enzyme: 1. Ascorbic acid oxidase in oranges. Plant Physiol 38: 333–337

Weis W (1975) Ascorbic acid and electron transport. Ann NY Acad Sci 258: 190–200

Wood PM (1978) Interchangeable copper and iron proteins in algal photosynthesis. Studies on plastocyanin and cytochrome c-552 in chlamydomonas. Eur J Biochem 87: 9–19

Yamauchi H, Yamawaki K, Ueda Y, Chachin K (1984) Subcellular localization of redox enzymes involving ascorbic acid in cucumber fruits. J Jpn Soc Hortic Sci 53: 347–353

Yamazaki Y, Konno H (1985) Three forms of α-glucosidase from suspension-cultured rice cells. Agric Biol Chem 49: 3383–3390

Zenk MH, Rueffer M, Amann M, Neumann BD, Nagakura N (1985) Benzylisoquinoline biosynthesis by cultivated plant cells and isolated enzymes. J Nat Prod 48: 725–738

IX *Elettaria cardamomum* Maton (Cardamom): Aromatic Compounds, In Vitro Culture Studies, and Clonal Propagation

Y.P.S. BAJAJ[1], B.R. REGHUNATH[2], and P.K. GOPALAKRISHNAN[2]

1 General Account

Cardamom is a polymorphic species of the monotypic genus *Elettaria*. True cardamom or lesser cardamom is a monocot belonging to the family Zingiberaceae under the natural order Scitaminae. The varietal status of true cardamom has been designated as *Elettaria cardamomum* var. *cardamomum* (syn. var. *minor* Watt; var. *minuscula* Burkhill, Purseglove 1975). The seeds, contained in the dried fruits (capsules) and possessing a characteristic pleasant aroma, are the cardamom of commerce. Rosengarter (1969) ranked cardamom as the third costliest spice in the world. In India it is the second most important spice next to black pepper (Pruthi 1979).

The plant is a tall perennial shrub (Fig. 1), the aerial part of which consists of 10–20 erect, leafy shoots (pseudo-stem), 2–5.5 m tall and made of leaf sheaths. The shoots and the panicle emerge from a horizontal subterranean woody rhizome. Each panicle bears numerous small, white or pale-green flowers characterized by a conspicuous labellum with violet streaks radiating from the center. The flowers are hermaphrodites. The ripe fruit (Fig. 2) is an ovoid trilocular capsule containing 15–20 aromatic seeds.

Cardamom cultivation is mainly concentrated in the southern states of India, i.e., Kerala, Karnataka, and Tamil Nadu. Guatemala, Tanzania, and Sri Lanka are the other major cardamom producing countries. New Guinea, Costa Rica, El Salvador, and Honduras have also emerged in recent years as producers and suppliers of cardamom in the world market. The average annual world production of cardamom is presently estimated around 10 000 t, shared by India (40%), Guatemala (45%), and other countries (15%) (Cherian 1987).

The dried fruits (capsules) of *E. cardamomum* var. *cardamomum* are the true cardamoms. The spice obtained by drying the fruits of various *Amomum* and *Aframomum* species and also that of wild *Elettaria* varieties are often referred as "false cardamoms". They are regarded as being much inferior in aroma and flavor (Purseglove et al. 1981). Based on the size of the fruit, a botanical variety of true cardamom other than var. *cardamomum* has been recognized, namely, var. *major*

[1] Former Professor of Tissue Culture, Punjab Agricultural University, Ludhiana, India. (Present address: A-137 New Friends Colony, New Delhi 110065, India)
[2] Department of Plantation Crops and Spices, College of Horticulture, Kerala Agricultural University, Trichur 680654, India

Biotechnology in Agriculture and Forestry, Vol. 21
Medicinal and Aromatic Plants IV (ed. by Y.P.S. Bajaj)

Fig. 1. A 5-year-old plant of *Elettaria cardamomum* growing in virgin forest land

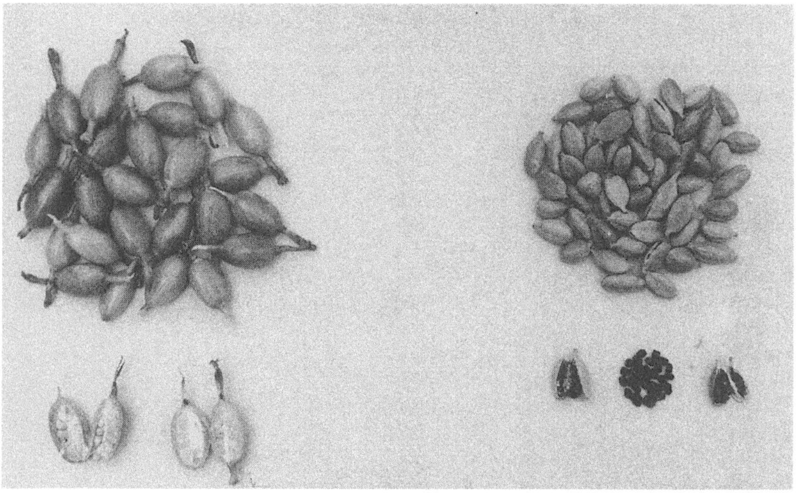

Fig. 2. Capsules (fruits) and seeds: green (*left*) and dried (*right*)

Thwaites. The fruit of the var. *major* is larger and longer (2.5–5 cm), with numerous larger and less aromatic seeds than that of the var. *cardamomum*. It is known as Long Wild Cardamom and is less common in trade compared to var. *cardamomum*.

2 Medicinal Importance and Secondary Metabolites

Cardamom is valued for its dried capsules, containing highly aromatic seeds used directly as a flavoring material in three forms: whole decorticated seeds and ground. It is from the decorticated seeds that the essential oil (steam distillation) and oleoresin (solvent extraction) are extracted. In the whole or ground form, cardamom is used mainly for domestic culinary purposes. In Asia, it is a masticatory as well. It is used as a flavoring in a variety of spiced rice, meat, and vegetable preparations, puddings, sweets, and certain beverages. In the Middle East, cardamom has been traditionally used to flavor coffee. In European countries, the spice is used in curry powders, sausages, soups, canned fish, and also to flavor tobacco. A range of baked goods including cakes, buns, pastries, and bread are flavored with cardamom.

Cardamom oil finds its main application in the flavoring of processed foods, cordials, bitters, and liqueurs and occasionally in perfumery. The oleoresin also has similar applications but is less extensively used.

Cardamom has distinctive medicinal properties as well, and is official in most pharmacopoeias. In India, the tincture of cardamom seed is used in the indigenous systems of medicine to strengthen the digestive, respiratory, and urogenital systems

Table 1. Approximate composition of ground seeds of *E. cardamomum*. (Nayar 1989)

Constituent	Unit	Quantity (per 100 g edible portion)	Quantity (per teaspoon)
Water	(g)	8.4	0.150
Food energy	(cal.)	360	6
Protein	(g)	9.8	0.176
Fat	(g)	2.9	0.052
Total carbohydrate	(g)	74.2	1.336
Fiber carbohydrate	(g)	9.9	0.178
Ash	(g)	4.7	0.085
Calcium	(g)	0.3	0.005
Phosphorus	(g)	0.21	0.004
Sodium	(g)	0.01	0.0002
Potassium	(g)	1.20	0.022
Iron	(mg)	11.60	0.209
Thiamine	(mg)	0.18	0.003
Riboflavin	(mg)	0.23	0.004
Niacin	(mg)	2.30	0.041
Ascorbic acid	(mg)	12.00	0.216
Vitamin A	(IU)	175	3

(Pruthi 1979; Warrier 1989). Kirtikar and Basu (1935) reported tonic, stimulant, stomachic, and emmenagogricue properties of the spice. According to them, it is administered internally in diseases of the liver and uterus, and externally to tumors of the uterus. Cardamom is reported to have curing properties in cases of rheumatism, bronchitis, poisoning, itching, and furuncles (Warrier 1989). The aqueous extract of the rhizome of the cardamom plant is a good tonic and is claimed to have laxative properties. Cardamom oil has an antibacterial activity (Sreenivasamurthy and Krishnamurthy 1959). Nayak and Dutta (1961) reported the bactericidal property of cardamom oil against *Vibrio comma* in vitro at dilutions 1:1000–1:15000, ineffective against cholera in vivo.

Cardamom seed contains volatile oil, fixed (fatty) oil, protein, cellulase, pentosans, sugars, starch, silica, calcium oxalate and minerals (Shankaracharya and

Table 2. Volatile constituents in the essential oil of *E. cardamomum*. (Purseglove et al. 1981)

Group	Constituent
Monoterpene hydrocarbons	p-cymene
	camphene
	d-limonene
	myrcene
	α-phellandrene
	α- and β-pinene
	sabinene
	α- and γ-terpinene
	α-thujene
Monoterpene epoxides and carbonyls	1,4- and 1,8-cineole
	ascaridole
	dl-camphor
	cis- and *trans*-citral
	citronellal
Monoterpene alcohols	dl-borneol
	citronellol
	geraniol
	linalool
	cis- and *trans*-p.2-menthen-1-olnerol
	trans- sabinene hydrate
	α- and β-terpinenol
	4-terpinenol
Monoterpene esters	geranyl acetate
	neryl acetate
	linalyl acetate
	α-terpinyl acetate
	4-terpinenyl acetate
Sesquiterpenes	bisabolene
	farnesol
	cis- and *trans*-nerolidol
Miscellaneous	methyl heptanone
	heptacosane
	2-undecane
	3-tridecane

Natarajan 1971). The major constituent of the seed is starch (45–50%) and that of the capsule (husk) is crude fiber (20–30%) (Pruthi 1979). The composition of the spice may vary to some extent between individual sample lots, owing to intrinsic differences between cultivars, environmental variations in the growing tract, maturity at harvesting, curing or drying and packing procedures, and duration of storage (Purseglove et al. 1981). Several reports are available on the extent of variations in the composition of nonvolatile as well as volatile constituents of spice entering international trade (see Purseglove et al. 1981). The approximate composition of nonvolatile constituents in the ground spice and major volatile constituents in cardamom oil are provided in Tables 1 and 2, respectively.

E. cardamomum seed on solvent extraction gives a greenish oleoresin containing about 70% volatile oil (Shankaranarayana and Krishnamurthy 1987). Nearly 40 constituents have been isolated from the cardamom oil obtained by steam distillation (Table 2) and many of them are present only in traces. Nevertheless, they contribute towards the characteristic aroma of cardamom. The fruity sweet aroma of cardamom is contributed by the alcohol linalool and the esters linalyl acetate and alpha-terpinyl acetate (Fig. 3). The camphorous odor is imparted by the cineoles, particularly 1,8-cineole (Purseglove et al. 1981; Shankaranarayana and Krishnamurthy 1987; Fig. 3). The extent of variation in the concentration of

Table 3. Concentration of 1,8-cineole and alpha-terpinyl acetate in the essential oil of some commercial types of *E. cardamomum*. (Pillai et al. 1984)[a]

Oil	Percentage	
	1,8-cineole	alpha-terpinyl acetate
Alleppey	34.2	37.7
Alleppey I	38.8	33.3
Alleppey Green	26.5	34.5
Ceylon type	36.0	30.0
Coorg Green	41.0	30.0
Guatemala I	36.4	31.8
Guatemala II	38.0	38.4
Guatemalayan I	39.1	40.3
Guatemalayan II	35.4	41.0
Malabar I	28.0	45.5
Malabar II	43.5	45.1
Malabar-type Ceylon[b]	31.0	52.5
Malabar-type Guatemala	23.4	50.7
Mangalore I	56.1	23.2
Mangalore II	51.2	35.6
Mysore	41.0	30.0
Mysore I	49.5	30.6
Mysore II	41.7	45.9
Mysore-type Ceylon[b]	44.0	37.0
Papua New Guinea	63.0	29.1

[a] All the oil samples were isolated by steam distillation unless otherwise mentioned.
[b] Cold-pressed.

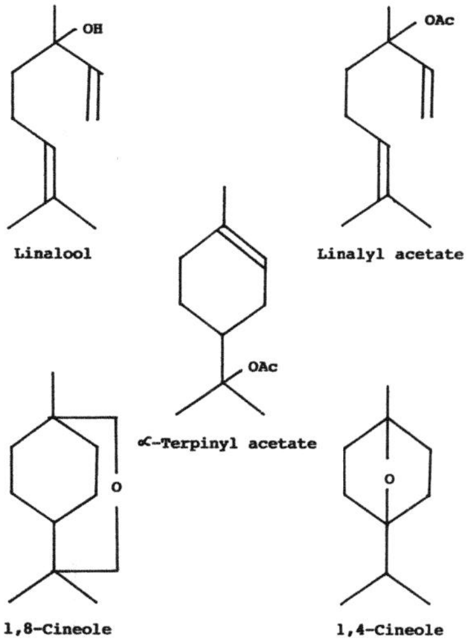

Fig. 3. Major monoterpene alcohol, esters, and oxygenated constituents of cardamom oil

1,8-cineole and alpha-terpinyl acetate in certain commercial cardamom oil samples is given in Table 3.

3 In Vitro Culture Studies

3.1 Review

Various tissue culture studies conducted on cardamom are summarized in Table 4.

3.2 Callus Cultures

Shoot primordium tips (3 mm) excised from both greenhouse-grown plants and plants raised in vitro (through axillary branching) of the cultivar Mysore were utilized for raising callus cultures. The cultures were maintained at $23 \pm 2\,^{\circ}C$ temperature under white fluorescent light of 1000 lx and a 16-h photoperiod.

Habituation of the explant to in vitro conditions and the least chances of microbial contamination of the explants resulted in callusing in 100% of the cultures compared to the cultures from the greenhouse-grown plants. Callusing was more or less similar in the two media used, i.e., MS medium (Murashige and Skoog 1962) and SH medium (Schenk and Hildebrandt 1972); however, the dry weight of callus obtained from MS medium was higher than that on SH medium (Table 5).

The cultures raised by supplementing the culture medium with NAA (4 mg/l) and BAP (1 mg/l) resulted in maximum callus production (Table 5). However, 2,4-D produced profuse, white and friable callus which did not result in shoot organogenesis; however, rhizogenesis was observed in certain cases.

Table 4. Summary of in vitro studies conducted on *E. cardamomum*

Explant	Medium (mg/l)[a]	Growth response	Reference
Seed sprout	MS + 2,4-D (2) or IBA (2) or NAA (2) + BAP (2) + CM (18)%	Callus shoots	Srinivasa Rao et al. (1982)
Shoot bud	MS + BAP (0.5) + Kinetin (0.5) + IAA (2) + CM (5%) + Vitamin	Shoots, plants	Nadgauda et al. (1983)
Inflorescence bud	MS + BAP (1) + Kinetin (0.5) + NAA (0.5) + CM (10%) + Vitamin	Shoots, plants	Kumar et al. (1985)
Shoot bud	MS + NAA (4) + BAP (1) + CM (5%)	Callus, shoots	Reghunath (1989), Reghunath and Bajaj (1992)
Shoot bud	SH + BAP (3) + Kinetin (1.5) + NAA (1) + CM (15%)	Shoots, plants	Reghunath (1989), Reghunath and Bajaj (1992)

[a] MS = Murashige-Skoog (1962); SH = Schenk-Hildebrandt (1972); BAP = 6-benzylaminopurine; CM = coconut milk; 2,4-D = 2,4-Dichlorophenoxyacetic acid; IAA = indole-3-acetic acid; IBA = indolebutyric acid; NAA = α-naphthaleneacetic acid.

Table 5. Callus production in various media from shoot tips of *E. cardamomum*[a]

BAP (mg/l) / Auxins (mg/l)	Murashige-Skoog medium: Callus fresh weight: dry weight (mg)[b]		Schenk-Hildebrandt medium: Callus fresh weight: dry weight (mg)[b]	
	0.5	1	0.5	1
NAA 0.5 + 2,4-D 0.1	308:6.17	453:13.83	406 : 4.83	420 : 9.33
NAA 2 + IAA 1	348:8.67	483:9.17	393:7.67	453:9.50
2,4-D 0.5 + IAA 1	320:6.33	527:15.50	350:4.17	428:10.67
2,4-D 1 + NAA 4	403:6.67	411:10.50	291:5.67	410:10.17
	495:9.67	615:16.33	375:8.00	446:10.83

[a] Reghunath (1989).
[b] Mean of three observations; culture period 28 days.

The calli obtained from the treatment combination involving NAA and BAP were compact and semihard in nature (Fig. 4); and when subcultured on an auxin-free medium containing 3 mg/l of BAP and 0.5 mg/l of kinetin or 2 mg/l of BAP and 1 mg/l of kinetin, resulted in shoot meristemoid production (Fig. 4B).

Each culture produced six to nine meristemoids which upon subculture to same medium produced shoots (9 ± 2 mm; Fig. 4c) within 28 days (Table 6). The different stages of development of a shoot tip explant (3 mm) to a whole plant through callusing is depicted in in Fig. 4D.

The addition of coconut milk to the culture medium favored caulogenesis. The mean number of green meristemoids produced in the callus mass was enhanced to 12 by adding 5% coconut milk.

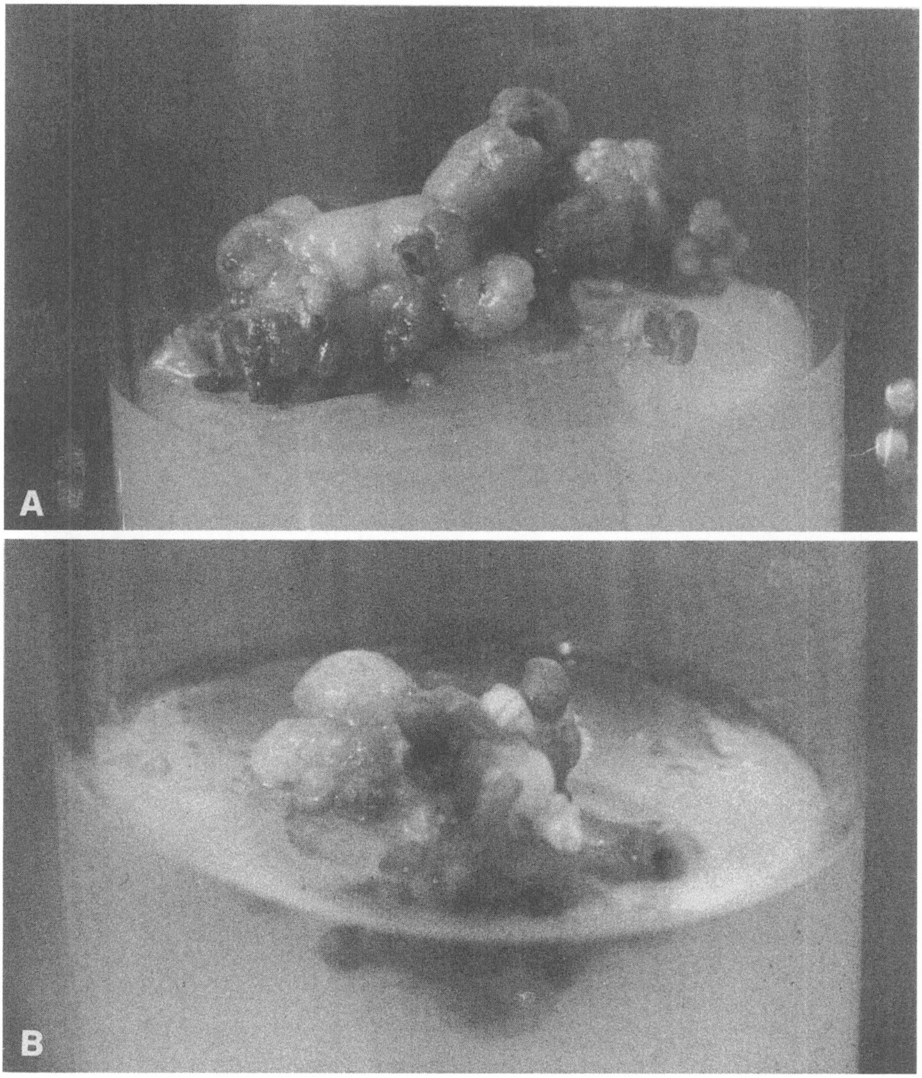

Fig. 4A–D. Establishment of callus cultures and regeneration of plants. **A** Compact callus induced from precocious shoot tip cultured for 28 days in modified MS medium containing NAA and BAP. **B** Induction of shoot meristemoids in auxin-free cytokinin-rich MS medium (14 days after subculture). **C** Regeneration of shoots (21–28 days after subculture) in the same medium. **D** Whole-plant regeneration from a shoot tip explant via callus mediated process; different stages

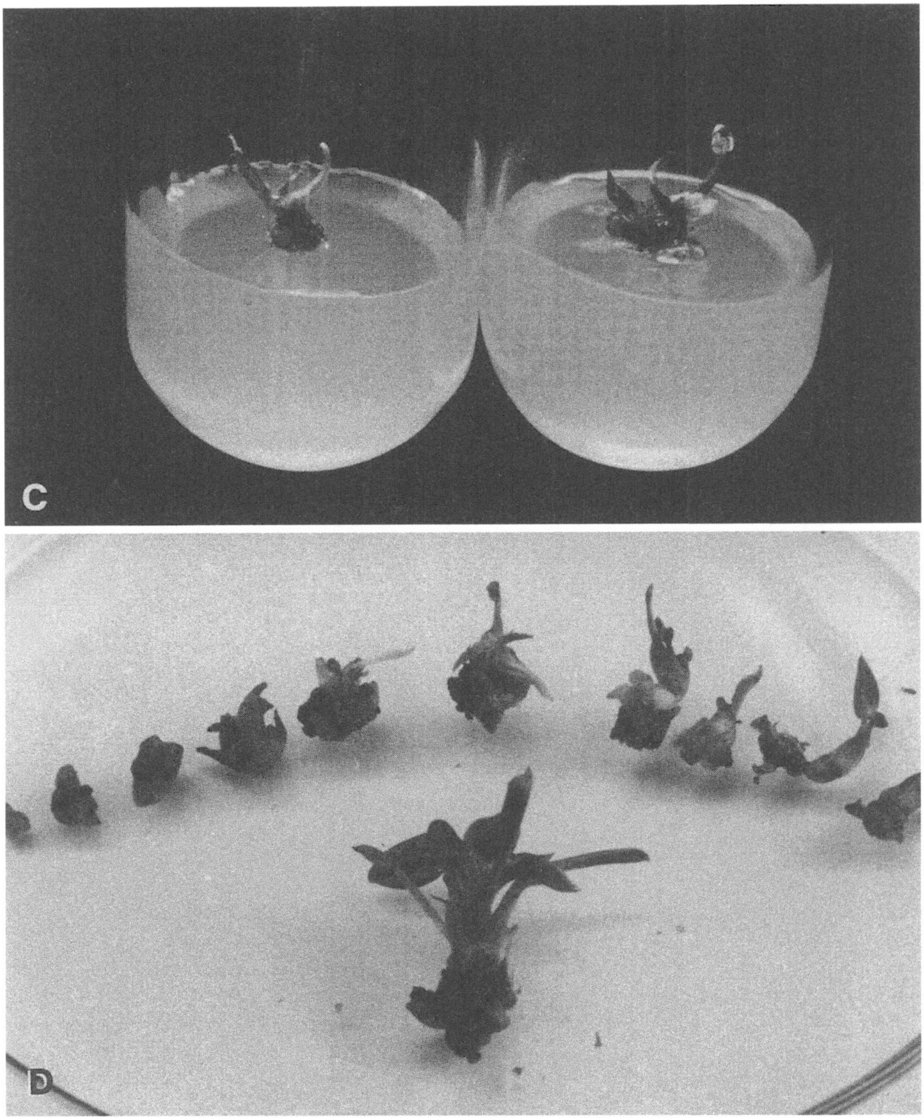

Fig. 4C and D

3.3 Clonal Propagation

Shoot primordia (20–30 mm) excised from green-house-grown plants of the cultivar Mysore were initially utilized, however, later precocious shoots (20–30 mm) obtained from initial-shoot primordium cultures were made use of. In MS medium, the regenerated shoots expressed leaf-tip drying and chlorosis symptoms. Reduction of inorganic salts in MS to one-half, or the use of SH medium, produced normal shoots. SH medium was superior to half-MS medium since it gave a 31% higher shoot dry weight (Table 7).

Table 6. Caulogenesis in callus cultures of *E. cardamomum*[a]

Kinetin (mg/l) \ BAP (mg/l)	Murashige-Skoog medium				
	Caulogenesis				
	0	0.5	1	2	3
0	$+^{0}$	$+^{0}$	$+^{2}$	$+^{2}$	$+^{3}$
0.5	$+^{0}$	$+^{1}$	$+^{2}$	$+^{3}$	$+^{4}$
1	$+^{1}$	$+^{2}$	$+^{2}$	$+^{4}$	$+^{3}$
2	$+^{1}$	$+^{2}$	$+^{2}$	$+^{3}$	$+^{0}$
3	$+^{2}$	$+^{3}$	$+^{2}$	$+^{0}$	$+^{0}$

[a] Reghunath (1989).
[b] Based on three visual observations (similarity 83.2 ± 16.6); culture period 28 days. $+^{0}$ Not tested. $+^{1}$ Callus turns to compact pale yellow masses without any further development. $+^{2}$ Callus turns to compact masses and certain white tubular protuberances appear from beneath. $+^{3}$ As in $+^{2}$ above; produces one or two greenish meristemoids from above. + As in $+^{3}$ above; produces six to nine greenish meristemoids which upon subculture to same medium produces shoots (9 ± 2 mm) within 28 days.

Table 7. Axillary branching in precocious shoots of *E. cardamomum* in various culture media

Observations[c]	Murashige-Skoog medium[b]			Schenk-Hildebrandt medium[b]		
	Inorganic salt strength (%)			Inorganic salt strength (%)		
	25	50	100	25	50	100
Axillary branches/shoot (No.)	3.25	5.50	5.00	3.00	4.75	5.75
Fresh weight of a branch (mg)	1370	1430	1500	1310	1390	1460
Dry weight of a branch (mg)	72.60	100.10	109.50	65.50	82.00	131.40

[a] Unpublished data.
[b] Medium contain 4 mg /l 6-benzylaminopurine and 0.5 mg/l α-naphthaleneacetic acid.
[c] Mean of four observations; culture period 35 days.

Table 8. Axillary branching in precocious shoots of *E. cardamomum* as influenced by growth regulators and coconut milk[a]

Treatments	Murashige-Skoog medium
	Axillary branches/shoot explant (No.)[b]
4 mg/l BAP + 0.5 mg/l NAA	5.00
1.5 mg/l Kn + 3 mg/l BAP + 0.5 mg/l NAA	5.00
12% CM + 4 mg/l BAP + 0.5 mg/l NAA	7.75

[a] Unpublished data.
[b] Mean of four observations; culture period 35 days.

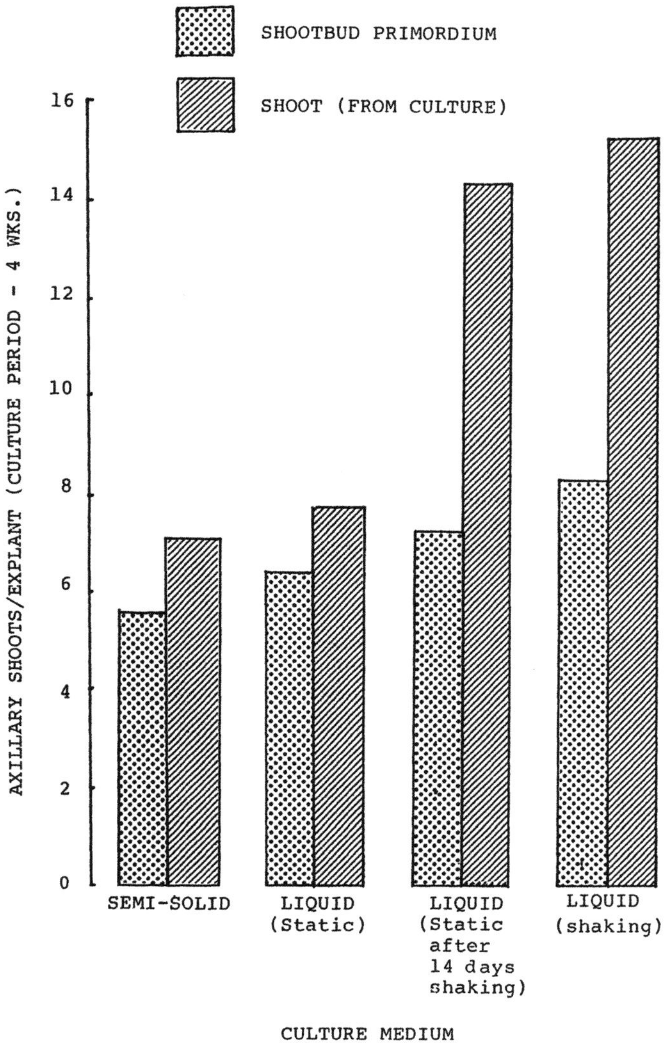

Fig. 5. Effect of gyratory shaking of liquid culture medium on the rate of axillary branching in in vitro cultures of cardamom (medium SH: culture period 28 days)

Precocious shoots planted in liquid medium in vials under gyratory shaking produced 111.57% more axillary branches than those in agar medium (Fig. 5). The maximum number of axillary branching was observed when the medium was supplemented with BAP (4 mg/l) + NAA (0.5 mg/l). At higher levels of these growth regulators, although a greater number of branches were produced, a large proportion of them were highly compressed and small. The addition of coconut milk (12%)to the medium significantly increased axillary branch induction (Table 8).

3.4 In Vitro Rooting and Transfer to Soil

Rooting of excised axillary shoots (30–40 mm) was achieved by culturing them initially in an agar (0.6%) medium containing half-strength MS salts and 0.5% activated charcoal for 1 week. They were then transferred on to a liquid MS/2 medium containing IBA (1.5 mg/l) for another week under high light intensity (3500 lx). Transfer to the IBA-containing medium helped to obtain a short, thin and fibrous root system, whereas prolonged culture in charcoal medium resulted in long, thick and inactive roots (Fig. 6). Rooted shoots were then transferred to a non-sterile MS/2 liquid medium containing only mineral salts and were then shifted to a greenhouse where partial sunlight was available. After 1–2 weeks, planting was done in polythene bags filled with vermiculite and fine sand 1:1 v/v) and the plants were kept under mist (90–100% RH) for 2–3 weeks (Fig. 7A). With this method, 100% rooting occurred, and the survival rate was 83.3% (Fig. 7B).

A field evaluation of the micropropagated plants showed that in spite of their initial setback, the micropropagated plants were at par or even better than the seedlings. At the 12th week after field planting, maximum plant height (97.4 cm), a greater number of leaves (18.8), and higher rate of tillering (50% or more) were recorded in micropropagated plants as compared to seedlings (see Reghunath and Bajaj 1992). This indicates that an enhanced axillary branching technique can be effectively utilized for the rapid multiplication of elite clones, especially those [which give high yield and show drought tolerance. For mass multiplication of virus-free clones, the in vitro method of shoot tip culture seems to be most suitable. The various modes of micropropagation in cardamom are presented in Fig. 8.

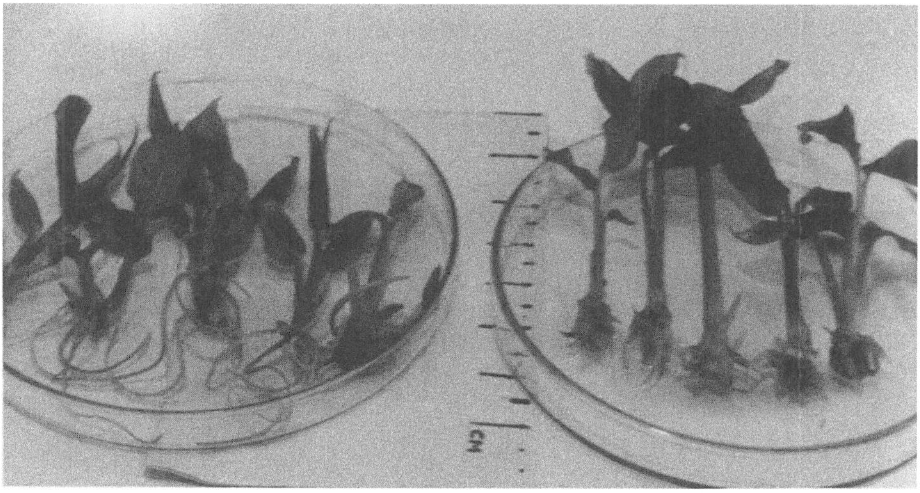

Fig. 6. Rooted plantlets of cardamom with long, thick, inactive roots raised in agar MS/2 medium containing activated charcoal (*left*); short, fibrous, active roots in MS/2 liquid medium containing IBA (*right*) 28 days after culturing

Fig. 7A, B. Transfer of in vitro plantlets from test tube to soil in polythene bags. A Rooted plantlets in greenhouse under partial sunlight and mist (28 days after planting out). B Micropropagated plants transferred to soil in polythene bags (16–20 weeks after planting out)

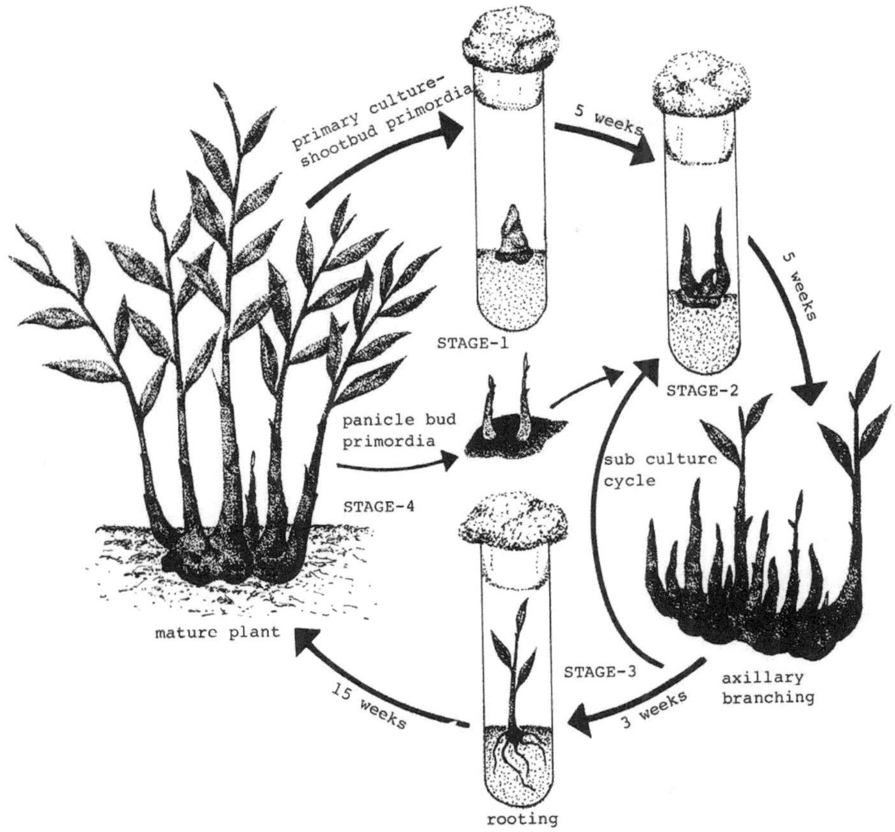

Fig. 8. Diagrammatic representation showing various modes of micropropagation of cardamom

4 Conclusions

Cardamom is valued as a spice for its dried capsules containing highly aromatic seeds. It is from the decorticated seeds from which the essential oil and oleoresin are extracted. Cardamom oil finds its main application in the flavoring of processed foods, cordials, bitters, and liqueurs and occasionally in perfumery. It has distinctive medicinal properties as well. It is reported to have curing properties in rheumatism, bronchitis, poisoning and itching, and has antibacterial and laxative properties. Selected plants have been clonally multiplied in vitro through shoot primordium culture and were field tested. Studies on the in vitro production of secondary metabolites, though lacking, would be highly rewarding for the selection of high-yielding clones through cell culture. (see Bajaj 1988).

5 Protocol for Callusing and Caulogenesis

Best Explant. Shoot meristem (3–4 mm long) from in vitro shoot proliferation cultures, precultured in MS basal medium containing 1% activated charcoal for 7 days.

Callus Induction Medium. MS basal medium + 4 mg/l NAA, 1 mg/l BA, 5% coconut milk. The medium is solidified with 5 g/l agar.

Shoot Regeneration Medium. MS basal medium + 0.5–1 mg/l kinetin, 2–3 mg/l BA, 5% coconut milk. his medium is solidified with 5 g/l agar.

Optimal Conditions. 21–25 °C, 16-h photoperiod (500–1000 lx).

Rooting Medium. MS salts (half-strength) gelled with agar and 0.5–1% activated charcoal.

Acclimation. 3–4 weeks at high humidity (90–100% RH) at 20 °C in vermiculite-sand (1:1 v/v) mixture.

References

Bajaj YPS (ed) (1988) Biotechnology in agriculture and forestry, vol 4. Medicinal and aromatic plants I. Springer, Berlin Heidelberg New York Tokyo

Cherian A (1987) Recent trends on increasing productivity and reducing cost of production of cardamom (Small) in India. Cardamom 20: 5–11.

Kirtikar KR, Basu ED (1935) *Elettaria cardamomum*. In: Indian medicinal plants, vol. 4, 1975 reprint edn. Periodical excerpts, Delhi, pp 2442–2444

Kumar KB, Prakashkumar P, Balachandran SM, Iyer RD (1985) Development of clonal plantlets from immature panicles of cardamom. J Plant Crops 13: 31–34

Murashige T, Skoog F (1962) A revised medium for rapid growth and bioassays with tobacco tissue cultures. Physiol Plant 15: 473–497

Nadgauda RS, Mascarenhas AF, Madhusoodanan KJ (1983) Clonal multiplication of cardamom (*Elettaria cardamomum* Maton) by tissue culture. J. Plant Crops 11: 60–64

Nayak KP, Dutta NK (1961) Role of essential oils and allied drugs in experimental cholera of the rabbit. Indian J Res 49: 51–54

Nayar KG (ed) (1989) The nutritional composition of spices. In: Spice Fair 1989 Cochin, India, April 9–11, 1989 Commemorative volume. Spices Board, Cochin, India pp 192–193

Pillai OGN, Mathulla T, George KM, Balakrishnan KV, Verghese J (1984) Studies in cardamom II; an appraisal of the excellence of Indian cardamom (*Elettaria cardamomum* Maton). Indian Spices 21: 17–25

Pruthi JS (1979) Spices and condiments. Natl Book Trust, New Delhi, India, pp: 53–54

Purseglove JW (1975) Tropical crops-monocotyledons, vols 1, 2. ELBS and Longman, London

Purseglove JW, Brown EG, Green CL, Robbins SRJ (1981) Spices, vol 2. Longman, London, pp 581–643

Reghunath BR (1989) In vitro studies on the propagation of cardamom (*Elettaria cardamomum* Maton). PhD Thesis, Kerala Agricultural Univ, Trichur, India

Reghunath BR, Bajaj YPS (1992) Micropropagation of cardamom (*Elettaria cardamomum* Maton). In: Bajaj YPS (ed) Biotechnology in agriculture and forestry, vol 19. High-Tech and micropropagation III. Springer, Berlin Heidelberg New York, pp 175–198

Rosengarter F (1969) The book of spices. Livinston, Philadelphia

Schenk RU, Hildebrandt AC (1972) Medium and techniques for induction and growth of monocotyledonous and dicotyledonous plant cell cultures. Can J Bot 50: 199–204

Shankaranarayana ML, Krishnamurthy N (1987) Research and development in spices, spice products and their end uses. Indian Spices 24: 5–13

Shankaracharya NB, Natarajan CP (1971) Cardamom – chemistry, technology and uses. Indian Food Packer 25: 28–36

Sreenivasamurthy V, Krishnamurthy K (1959) Place of spices and aromatics in Indian dietary. Food Sci 8: 284–288

Srinivasa Rao NK, Narayanaswamy S, Chacko EK, Doreswamy R (1982) Regeneration of plantlets from callus of *Elettaria cardamomum* Maton. Proc Indian Acad Sci 91: 37–41

Warrier PK (1989) Spices and Ayurveda – the combination that heals. In: Nayar KG (ed) Spice Fair 1989 Cochin, India, April 9–11, 1989 Commemorative vol. Spices Board, Cochin, India, p 105

X *Eupatorium cannabinum* L. (Hemp Agrimony): In Vitro Culture and the Production of Benzofurans

P. Proksch[1] and R. Siebertz[2]

1 Introduction

The genus *Eupatorium* L. (Asteraceae, tribe Eupatorieae) is comprised of some 38 perennial species that occur mainly in eastern North America and eastern South America (Robinson and King 1977). From this center of distribution, however, species of *Eupatorium* have reached parts of Europe and Asia as well. *E. cannabinum* L. (commonly known as hemp agrimony) is frequently found in moist woods and meadows in northern and central Europe where it often grows along streams (Fig. 1). It is a perennial which reaches a height of ca. 0.5–1.5 m. The flowering heads are pink in color.

E. cannabinum was used as a medicinal plant by the Greeks and Romans, as well as in medieval times to aid digestion and prevent constipation (Bianchini and Corbetta 1977). The roots of *E. cannabinum* are especially known to possess cholagogic properties and are also used in the treatment of liver diseases (Hoppe 1958). No information on the bioactive constituents from *E. cannabinum*, however, has been available so far, since phytochemical studies on *E. cannabinum* have been sparse and were mainly limited to the study of toxic pyrrolizidine alkaloids from aerial parts of this species (Hendriks et al. 1983). In the course of our continued interest in biologically active chromenes and benzofurans from the Asteraceae we have recently isolated for the first time a series of benzofurans from roots of *E. cannabinum* that are absent in the aerial parts of this species (Siebertz et al. 1989). Benzofurans are known to exhibit interesting biological activities, e.g., they are bacteriostatic and piscicidal, photoxic against several bacteria and yeasts, and cytostatic (Bonner and De Graw 1962; Zalkow et al. 1962; Wu et al. 1973; Proksch et al. 1983). In spite of these documented bioactivities, very little has been known so far of the biogenetic origin of this group of natural products in the Asteraceae. We have now for the first time established a root culture of *E. cannabinum* and have found it to be a suitable and easy accessible in vitro system for studies on the accumulation and biogenesis of benzofuran derivatives.

[1] Institut für Botanik und Pharmazeutische Biologie, Universität Würzburg, Mittlerer Dallenbergweg 64, W-8700 Würzburg, FRG
[2] Present address: Institut für Pharmakologie der Universität Düsseldorf, Moorenstr. 5, W-4000 Düsseldorf, FRG

Biotechnology in Agriculture and Forestry, Vol. 21
Medicinal and Aromatic Plants IV (ed. by Y.P.S. Bajaj)

Fig. 1. *Eupatorium cannabinum* (mature plant, approx. 3 years old, growing in the vicinity of Braunschweig, FRG)

2 In Vitro Approaches

2.1 Establishment and Growth of the Root Culture

Prior to this study, no information on in vitro culture regarding the production of secondary metabolites in *E. cannabinum* was available. The root culture of *E. cannabinum* (Fig. 2) employed in this study was established from sterile seedlings grown from achenes. Excised roots (1–2 cm) from 4–5 seedlings were transferred to 10 ml MS medium (Murashige and Skoog 1962), pH 5.8, containing 4% sucrose and no phytohormones. The flasks with the excised roots were kept on a gyratory shaker (120 rpm at 25 °C) in the dark. As soon as the root culture was established it was subcultured in 100-ml flasks. Young growing roots from the cultures were transferred every 3–4 weeks (for growth of root cultures, see Fig. 5). When the present study was initiated, the root culture had been in continuous cultivation for more than 2 years.

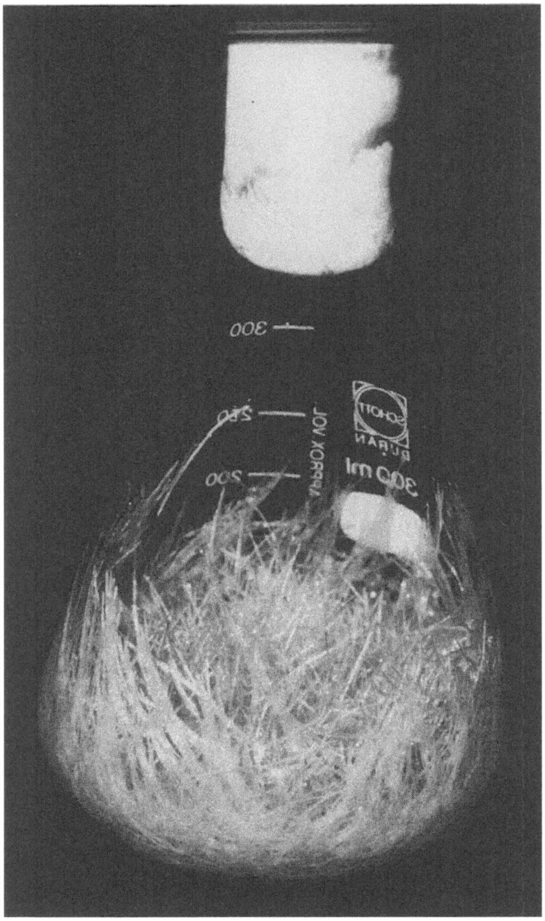

Fig. 2. Root culture of *E. cannabinum* (2 weeks following inoculation into fresh medium; for medium, see protocol)

2.2 Phytochemical Analysis of Root Cultures

The root culture of *E. cannabinum* accumulated a complex pattern of benzofurans (*1–9*, Fig. 3) that contributed to the dominating phenolic constituents as revealed by HPLC analysis (Fig. 4). The benzofurans were identical to those found in roots of flowering plants. Aerial parts of *E. cannabinum* are devoid of these compounds (Siebertz et al. 1989). The major benzofurans (except the unknown "x" which proved too labile for isolation) were isolated and identified based on their UV, MS, and NMR spectra. Only compounds *1–3* were previously known from *E. cannabinum* (Zdero and Bohlmann 1987). Benzofurans *4–7* as well as *9* have previously been isolated from other plant sources (Proksch and Rodriguez 1983), whereas benzofuran *8* was a new natural product. Dimeric benzofurans such as *9* are especially rare and have previously only been described from the roots of *Ageratum houstonianum* (Asteraceae) (Breuer et al. 1987).

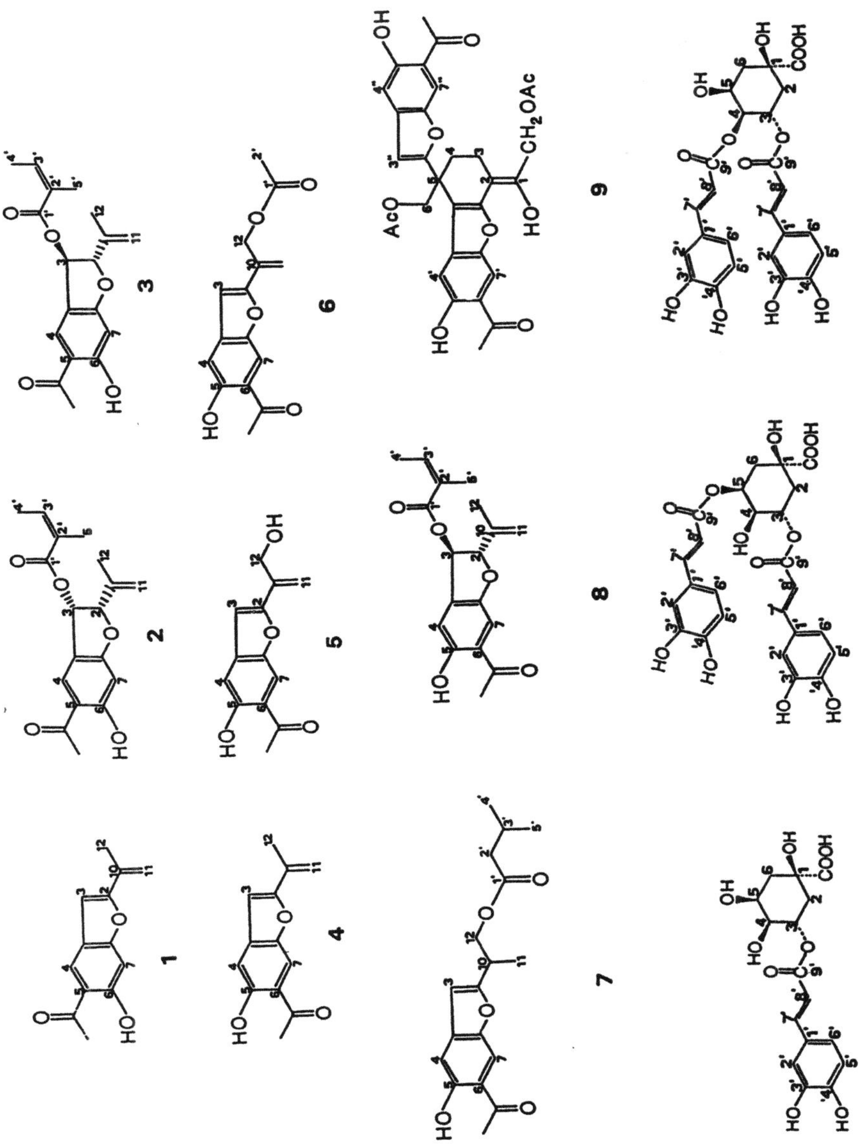

Fig. 3. Benzofurans (*1–9*) and caffeic acid derivatives (*10–12*) from roots and root cultures of *E. cannabinum*

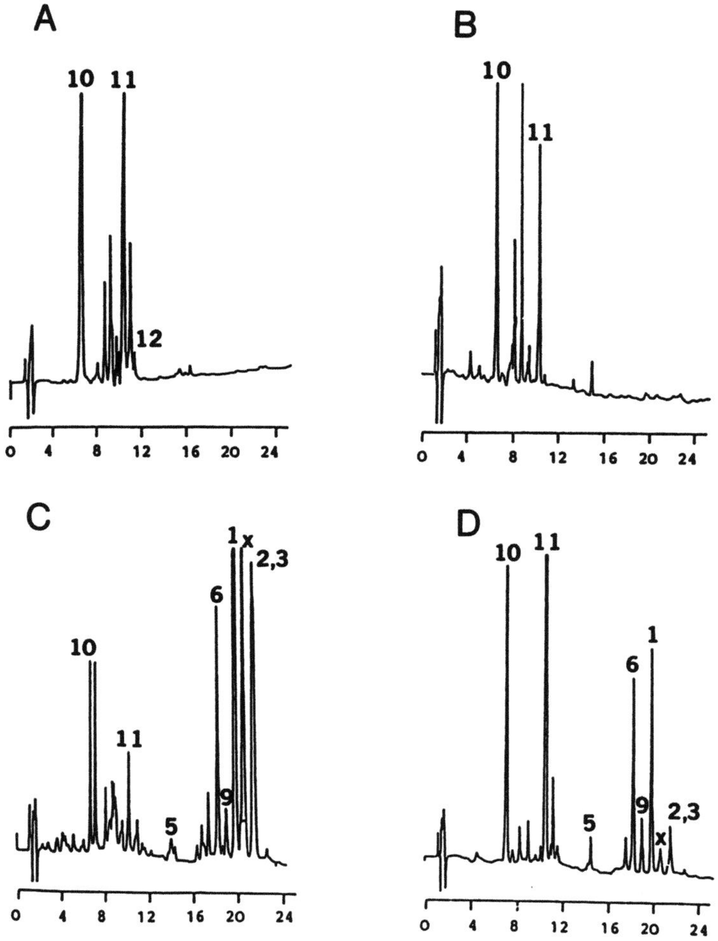

Fig. 4A–D. HPLC analysis of crude extracts from **A** flowering heads; **B** stems and leaves; **C** roots (all from flowering plants); **D** root cultures of *E. cannabinum*. The extracts were separated on an RP 8 column (5 μm, 125 × 4 mm) employing a linear gradient from **A** (10% CH_3CN, 90% H_2O, 1% H_3PO_4) to **B** (80% CH_3CN, 20% H_2O, 1% H_3PO_4) in 25 min, flow 1 ml/min. Detection was at 300 nm

The benzofurans from *E. cannabinum* belong to two major structural types. Compounds *1–3* are characterized by the acetyl substituent at C-5 (para to the oxygen atom in the heterocycle whereas benzofurans *4–9* exhibit the acetyl substituent at C-6 (meta to the heterocyclic oxygen). Benzofurans of the former structural type are common in the Asteraceae, whereas those belonging to the latter type are very rare (Proksch and Rodriguez 1983). Benzofurans *1–3* can thus be considered as derivatives of 2,4-OH-acetophenone, whereas benzofurans *4–9* may be regarded as derivatives of 2,5-OH-acetophenone. Both types of acetophenone derivatives can be distinguished by their UV spectra (Table 1). The UV spectra of 2,3-dihydrobenzofurans basically resemble those of their acetophenone moieties and can be used as a first hint of the linkage point of the acetyl substituent.

Further structural information is obtained from ^{1}H NMR spectra (Table 2). ^{1}H nOe experiments proved especially helpful in elucidating the position of the various substituents in the benzofuran molecule, including the C-acetyl group. The chemical shifts of H-4 and H-7 for example could be distinguished by irradiation at H-3,

Table 1. UV absorbance of 2,4-OH- and 2,5-OH-acetophenones and structurally related 2,3-dihydrobenzofurans from root cultures of *E. cannabinum*. (Siebertz et al. 1989)

Compound	UV λ_{max} (in MeOH)
2,4-OH-acetophenone	276, 315
Benzofurans *2* and *3*	274, 318
2,5-OH-acetophenone	258, 372
Benzofuran *8*	256, 366

Table 2. ^{1}H NMR data of benzofurans *1–9* from root cultures of *E. cannabinum*[a]

		1		*2*		*3*
H-2		–	dd	5.23	dd	5.20
H-3a	s	6.54	d	6.36	d	6.11
H-3b		–		–	d	–
H-4	s	7.90	s	7.97	s	8.01
H-7	s	6.97	s	6.37	s	6.40
H-11a	br s	5.75	dd	5.22	dd	5.08
H-11b	br s	5.18	dd	5.10	dd	4.98
H-12a	br s	2.10	br s	1.82	br s	1.79
H-2′		–		–		–
H-3′[a]		–	qq	6.12	qq	6.17
H-4′[a]		–	dq	1.94	dq	1.98
H-5′		- -	dq	1.81	dq	1.88
CH_3CO	s	2.68	s	2.55	s	2.55
OH	s	12.52	s	13.00	s	13.00
J(3a-2)		–		7.3		2.3
J(3b-2)		–		–		–
J(3a-3b)		–		–		–
J(12a-12b)		–		–		–
J(3′-4′)		–		7.3		7.3
J(3′-5′)		–		1.5		1.5
J(4′-5′)		–		1.5		1.5

		4		*5*		*6*
H-3	s	6.56	s	6.69	s	6.64
H-4	s	7.01	s	7.03	s	7.03
H-7	s	7.78	s	7.79	s	7.79
H-11a	br s	5.84	br s	5.99	br s	6.07
H-11b	br s	5.28	br s	5.58	br s	5.60
H-12	br s	2.11	br s	4.52	br s	4.93
H-2′		–		–	s	2.10
CH_3CO	s	2.67	s	2.67	s	2.67
OH	s	12.13	s	12.25	s	12.10

Table 2. Continued.

		7		*8*
H-2		–	d	5.09
H-3	br s	6.42	d	6.21
H-4	s	7.00	s	7.04
H-7	br s	7.77	s	7.42
H-10		3.32		–
H-11a	d	1.40	br s	5.13
H-11b		–	br s	5.00
H-12a	dd	4.37	s	1.82
H-12b	dd	4.28		–
H-2′	d	2.17		–
H-3′	m	2.06	dd	6.22
H-4′	d	0.90	dd	1.99
H-5′	d	0.90	dd	1.91
CH_3CO	s	2.68	s	2.67
OH	s	12.15	s	12.20
^{J}(3-2)		–		3.1
^{J}(12a-12b)		10.0		–
^{J}(12a-10)		6.7		–
^{J}(12b-10)		6.1		–
^{J}(10-11)		7.0		
^{J}(2′-2′)		7.0		–
^{J}(4′-3′)		6.6		7.3
^{J}(5′-3′)		6.6		1.5
^{J}(4′-5′)		–		1.5

	9	
H-1a	d	4.59
H-1b[b]	d	4.30
H-3a[b]	m	2.05
H-3b[a]	m	2.05
H-4a[b]	m	2.35
H-4b	m	2.35
H-6a	d	4.72
H-6b	d	4.70
6-$OCOCH_3$	s	1.97
H-3′	br s	6.28
H-4′	a	6.93
H-4″	s	7.09
H-7′	s	7.75
H, -7″	br s	7.59
6,, -CH_3CO	s	2.65
6-CH_3CO	s	2.68
OH-5,,	s	12.10
OH-5	s	12.03

[a] Abbreviations: s: singlet; d: doublet; br: broad; dd: double doublet; qq: quartet quartet; dq: double quartet.
[b] Assignments may be reversed. Spectra of compounds *1, 4, 5, 6, 7* were recorded in $CDCl_3$, *2, 3* and *8* in CD_3OD; and *9* in DMSO-d_6.

which gave a nOe only for the signal of H-4 but not for H-7. Irradiation at the signal of the methyl protons H-14 results in an increased intensity only of the signal of the proton which is ortho to the acetyl group. If the acetyl substituent is located at C-5, an increase in intensity is observed for the signal of H-4; if the acetyl group is present at C-6, an increase in intensity results for the signal of H-7. Distinction of 5-acetyl- versus 6-acetylbenzofurans can further be achieved by increment calculation. The signals of protons adjacent to a OH-group are shifted upfield, compared to a downfield shift for the signals of protons adjacent to an acetyl group.

The use of GC–MS in the structure elucidation of the benzofurans from *E. cannabinum* was limited to certain compounds only (*1*, *4–8*). Upon injection of the labile esters *2* and *3* into a hot injection port, only euparin (*1*) was detected. Formation of this artefact was avoided by on-column injection. The dimer *9* proved likewise unsuitable for GC–MS analysis, due to its polarity and molecular weight. A satisfactory mass spectrum of the latter compound could only be obtained by FAB–MS.

In addition to the benzofurans, the well-known caffeic acid derivatives chlorogenic acid (3-caffeoylquinic acid) (*10*), 3,4-dicaffeoylquinic acid (*11*) as well as 3,5-dicaffeoylquinic acid (*12*) were isolated from the root cultures and identified by spectroscopic means. In comparison to the benzofurans, the caffeic acid derivatives are present in all organs of flowering plants from *E. cannabinum* (Fig. 4).

2.3 Accumulation of Benzofurans in Root Cultures

The presence of benzofurans in root cultures of *E. cannabinum* was limited to the intact roots. Benzofurans were not detected in the liquid growth media except with old roots that already were subject to autolysis. This demonstrates unequivocally that roots of *E. cannabinum* are capable of benzofuran biosynthesis. The total concentrations of benzofurans (in µmol/g fresh weight), however, found in root cultures of *E. cannabinum* were only one-half of those present in roots of flowering plants (Table 3). Whereas an average of 12 µmol of benzofurans per gram fresh weight was detected in roots from flowering plants, this concentration dropped to approximately 6 µmol/g fresh weight in root cultures. The opposite trend was observed for the caffeic acid derivatives (*10–12*). The total concentration of caffeic acid derivatives in root cultures amounted to 11.7 µmol/g fresh weight compared to only 0.9 µmol/g fresh weight in roots from flowering plants. For this comparison,

Table 3. Abundance of major benzofurans (*1–3*, *6*) and caffeic acid derivatives (*10–12*) in root cultures and roots from flowering plants of *E. cannabinum*. (After Siebertz et al. 1989)

Abundance of compounds	Concentrations (µmol/g fresh wt.)					
	1	*2, 3*	*6*	*10*	*11*	*12*
Root cultures	2.5	1.2	1.8	4.3	7.2	0.2
Roots in vivo	6.1	4.2	1.2	0.5	0.4	+[a]

[a] 0.1 µmol/g fresh wt.

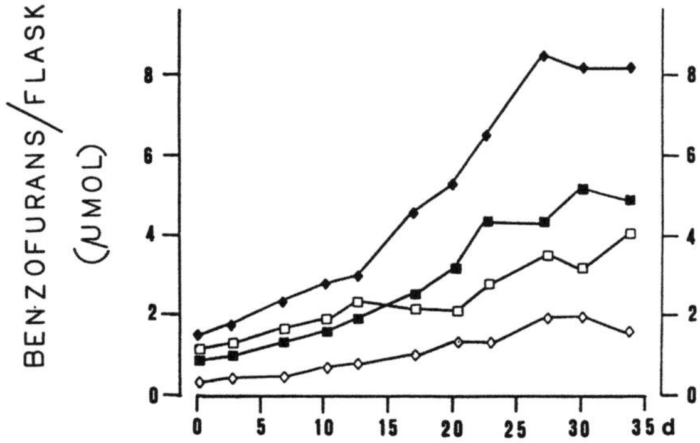

Fig. 5. Growth of root cultures and accumulation of benzofurans. Growing roots (ca. 500 mg) were inoculated at day 0. At fixed time intervals, roots were harvested, weighed, and extracted for benzofuran analysis by HPLC. All analyses were carried out in duplicate. ■, Fresh weight of roots; ◆, *1*; ◇, *2*, *3*; □, *6*

however, it must be kept in mind that it is not known at present if the concentrations of benzofurans and caffeic acid derivatives reported here for roots of flowering plants are stable, or subject to seasonal or ontogenetic variation.

Following inoculation of the roots into fresh growth medium, the biomass of roots increased from approximately 500 mg/flask at the time of inoculation to approximately 8 g/flask within 25–30 days of the culture period (Fig. 5). Thereafter the stationary phase was reached, followed by gradual autolysis of the roots. Accumulation of the benzofurans was limited to growing roots only, and stopped as soon as the roots reached the stationary phase, as shown for example for the major benzofuran euparin (*1*) that increased from ca. 1.5 µmol/flask to ca. 8 µmol/flask during the growth phase.

Further studies on root cultures obtained from different asteraceous species indicated that these root cultures accumulated the same pattern of benzofurans as found in roots of differentiated plants of the respective species. Examples include root cultures from *Ageratum houstonianum*, *Eupatorium purpureum*, and *E. pauciflorum* (all of tribe Eupatorieae) (Siebertz and Proksch, unpublished results) as well as *Tagetes patula* (tribe Tageteae) (Parodi *et al.* 1988). The capacity for the biosynthesis and accumulation of benzofurans in roots and root cultures is therefore not limited to *E. cannabinum*, but seems to be of broader occurrence within the Asteraceae.

2.4 Biogenesis of Benzofurans in Root Cultures

Growing root cultures of *E. cannabinum* offered a suitable and easy accessible experimental system for studies on the biogenetic origin of the acetophenone moiety of the benzofurans. A plethora of ^{14}C-labeled precursors, including L-phenylalanine, L-tyrosine, trans-cinnamic acid, benzoic acid, acetophenone, and acetic acid was fed to the cultured roots via the liquid growth media (Table 4). Uptake of all precursors by the cultured roots except acetophenone (uptake only 40–50%) was between 90–98% of the applied dose as calculated from the radioactivity remaining in the growth media 48 h after feeding. All precursors were metabolized by the root cultures, giving rise to numerous labeled products in the respective root extracts. Only ^{14}C-phenylalanine and ^{14}C-cinnamic acid, however,

Table 4. Incorporation of ^{14}C-labeled precursors into benzofurans *1–3*, *5*, and *6* by root cultures of *E. cannabinum*. (Siebertz et al. 1989)[a]

^{14}C-Tracers (dose: 1 μCi = 37k Bq each)	Assay conc. (mM)	Uptake by root cultures (% of applied dose)	Incorporation into benzofurans (% of applied dose)				
			1	*2*	*3*	*5*	*6*
L-(U-^{14}C) Phenylalanine	0.5	94–97	2.1	1.5	2.3	0.7	4.5
trans-(side chain-3-^{14}C) cinnamic acid	0.1	93–97	2.0	1.3	1.9	0.5	4.5
Sodium (U-^{14}C) acetate	0.5	97–99	n.d.[b]	n.d.	n.d.	n.d.	1.9
L-(U-^{14}C) Tyrosine-hydrochloride	0.1, 0.5, or 1.0	93–97	n.d.	n.d.	n.d.	n.d.	n.d.
(Ring-U-^{14}C) Benzoic acid	0.1, 0.5, or 1.0	94–97	n.d.	n.d.	n.d.	n.d.	n.d.
(Ring-U-^{14}C) Acetophenone	0.1, 0.5, or 1.0	40–54	n.d.	n.d.	n.d.	n.d.	n.d.

[a] Incubation conditions: growing roots (0.5 g fresh wt.) in 5 ml culture medium plus tracer, 48 h at 25 °C, 120 rpm.
[b] No measurable incorporation detected.

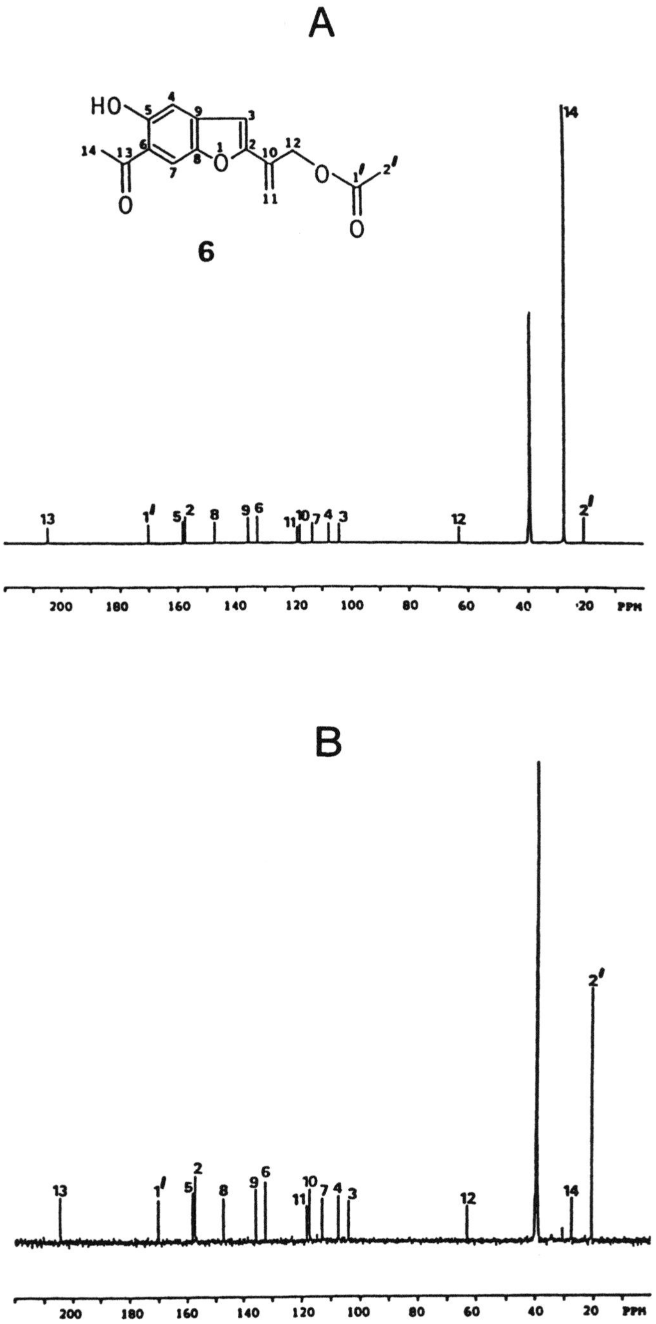

Fig. 6A, B. ^{13}C NMR spectra of benzofuran *6* following feeding of **A** L-(2-^{13}C)-phenylalanine and **B** (2-^{13}C)-acetate to root cultures of *E. cannabinum*. The spectra were recorded in DMSO-d_6 at 100 MHz were incorporated into all of the benzofurans studied. The incorporation of the latter two precursors into the benzofurans was comparable, as shown for the predominating compounds *5* (incorporation of 0.5–0.7% relative to the applied dose); *6* (4.5%); *1* (2%); *2* (1.3–1.5%); and *3* (1.9–2.3%) (Table 4). Feeding of ^{14}C-acetate in comparison resulted only in an incorporation into benzofuran *6* (1.9%), whereas no radioactivity was found in any of the other benzofurans (Table 4). Upon alkaline hydrolysis of *6*, the radioactivity following feeding of ^{14}C-acetate was lost, suggesting that ^{14}C-acetate had only been incorporated as an esterifying acid.

The experiments with ^{14}C-labeled tracers indicated the biogenetic origin of the acetophenone moiety (in the C_6–C_2 part of the molecules) of the benzofurans from phenylalanine and cinnamic acid. To obtain unequivocal proof for the incorporation of phenylalanine into the benzofurans L-(2-^{13}C)-phenylalanine was fed to root cultures of *E. cannabinum*. The major benzofurans *1* and *6* were subsequently isolated and subjected to ^{13}C NMR analysis. The ^{13}C spectra of both compounds were assigned from considerations of the DEPT spectra and from comparison of the two sets of data. By using substituent chemical shift increments for the acetyl and hydroxyl groups, the benzofuran ring protons could clearly be distinguished and assigned in the respective spectra.

Following feeding of L-(2-^{13}C)-phenylalanine, the spectra of *1* and *6* showed significantly increased intensities of the C-14 signals when compared to the signal intensities of the other C-atoms. The ^{13}C NMR spectrum of *6* following feeding of ^{13}C-phenylalanine is shown in Fig. 6. This experiment proved that the C-acetylsubstituent of the benzofurans *1* and *6* originated from the phenylpropanoid side chain

Table 5. Characteristic signals in the mass spectra of benzofurans *1* and *6* following feeding of L-(ring-$^2H_5$) phenylalanine to root cultures of *E. cannabinum*. (Siebertz et al. 1989)

Ions (m/z)	Relative intensity (%) following feeding L-(ring-$^2H_5$) phenylalanine[a]	Relative intensity (%) in controls
Benzofuran *1*:		
218 $(M+2)^+$	4.16	1.16
217 $(M+1)^+$	14.23	13.72
216 $(M)^+$	100	100
203 $[(M+2)-15]^+$	4.10	1.10
202 $[(M+1)-15]^+$	12.70	12.40
201 $(M-15)^+$	100	100
Benzofuran *6*:		
276 $(M+2)^+$	6.80	1.51
275 $(M+1)^+$	15.60	15.10
274 $(M)^+$	100	100
261 $[(M+2)-15]^+$	7.14	1.67
260 $[(M+1)-15]^+$	14.90	14.49
259 $(M-15)^+$	100	100

[a] The relative intensities of the $(M)^+$ and $(M-15)^+$ signals were set to 100%.

of phenylalanine through the loss of the carboxylic group. Since ^{14}C-cinnamic acid (^{13}C-cinnamic acid was not commercially available) was also effectively incorporated (Table 4), it is assumed that the biogenesis of benzofurans proceeds from phenylalanine via cinnamic acid rather than via phenylethylamine. When (2–^{13}C)-acetate was fed to the root cultures, an increased signal intensity was only observed for C-2 of compound *6* but not for any of the remaining carbons of this compound (Fig. 6). In addition, no increase in intensity was observed for any of the carbons in the ^{13}C NMR spectrum of *1*. This experiment unequivocally proved that acetate was not a precursor of the benzofuran skeleton, but was incorporated into *6* only as an esterifying acid, as suspected from the feeding experiments with ^{14}C-labeled tracers.

In addition to the ^{13}C-labeled precursors, L-(ring-$^{2}H_5$)phenylalanine was fed to the root cultures of *E. cannabinum*. The incorporation of this stable isotope into benzofurans *1* and *6* was determined by GC–MS and comparison with mass spectra of the nondeuterated compounds. The mass spectra of the deuterated benzofurans clearly showed higher $(M + 2)^+$ signal intensities, as well as $(M + 2 - 15)^+$ signal intensities when compared to the mass spectra of the respective controls (Table 5). These data were compatible with the expected incorporation of two deuterium atoms into the benzofurans, and furthermore indicated that no NIH-shift had occurred during hydroxylation at C-6 (*1*) or C-5 (*6*).

3 Conclusions

The biosynthesis of benzofurans from *E. cannabinum* in regard to its C_6C_2-part as elucidated in this study shows similarities to that of pungenine (3,4-dihydroxy-acetophenone), the major acetophenone derivative present in needles of *Picea pungens* (Pinaceae). Following feeding of ^{14}C-labeled tracers to needles of *P. pungens* L-^{14}C-phenylalanine, ^{14}C-cinnamic acid and ^{14}C-caffeic acid were effectively incorporated into pungenine, whereas ^{14}C-acetate was not (Neish 1959). In a previous study on the biogenetic origin of the acetophenone moiety of benzofurans in leaves of *Eupatorium rugosum*, the authors reported acetate as being the precursor of the C_6–C_2 part of the benzofuran dehydrotremetone (6-desoxyeuparin; euparin = *1*) (Lin et al. 1974). The incorporation of ^{14}C-acetate into dehydrotremetone, however, was only 0.06% of the applied dose, which is two orders of magnitude smaller than the incorporation of the ^{14}C-phenylalanine obtained in this study using root cultures of the related species *E. cannabinum*. The close systematic relationship of the two *Eupatorium* species studied, as well as the close similarity of the benzofurans euparin (*1*) and dehydrotremetone, calls for a reinvestigation of the biosynthesis of dehydrotremetone.

4 Protocol

For cultivation of the root cultures of *E. cannabinum*, as well as for the experiments conducted, the medium was composed of the following:

Substance	*mg/l*	*Substance*	*mg/l*
NH_4NO_3	1650	H_3BO_4	6.2
KNO_3	1900	$MNSO_4$ $4H_2O$	22.3
$CaCl_2$ $2H_2O$	440	$ZNSO_4$ $7H_2O$	8.6
$MgSO_4$ $7H_2O$	370	Na_2MO_4 $2H_2O$	0.25
KH_2PO_4	170	$CuSO_4$ $5H_2O$	0.025
Sucrose	40000	$CoCl_2$ $6H_2O$	0.025
NaFeEDTA	40	Nicotinic acid	1
myo-Inositol	100	Thiamin HCl	10
KJ	1	Pyridoxin HCl	1

The pH of the medium was adjusted to 5.8.

Ackinowledgments. Financial support of this project by a grant of the DFG to P.P. is gratefully acknowledged. We would further like to thank Dr. V. Wray (GBF, Braunschweig) for NMR measurements and Dr. L. Witte (Inst. für Pharmazeutische Biologie, Braunschweig) for MS analysis.

References

Bianchini F, Corbetta F (1977) Health plants of the world. Newsweek Books, New York

Bonner WA, De Graw JE (1962) Ketones from white snakeroot *Eupatorium uticaefolium.* Tetrahedron 18: 1295–1309

Breuer M, Budzikiewicz H, Siebertz R, Proksch P (1987) Benzofuran derivatives from Ageratum houstonianum. Phytochemistry 26: 3055–3057

Hendriks H, Malingre Th M, Elema ET (1983) Pyrrolizidine alkaloids, flavonoids and volatile compounds in the genus *Eupatorium.* Pharm Weekblad 5: 281–286

Hoppe HH (1958) Drogenkunde Cram, De Gruyter, Hamburg, p 374

Lin TJ, Ramstad E, Heinstein P (1974) In vivo biosynthesis of isopentenylacetophenones in *Eupatorium rugosum.* Phytochemistry 13: 1809–1815

Murashige T, Skoog F (1962) A revised medium for rapid growth and bioassays with tobacco tissue culture. Physiol Plant 15: 473–497

Neish AC (1959) Biosynthesis of pungenine from ^{14}C-labelled compounds by Colorado spruce. Can J Bot 37: 1085–1100

Parodi FJ, Fischer NH, Flores HE (1988): Benzofuran and bithiophenes from root cultures of *Tagetes patula.* J Nat Prod 51: 594–595

Proksch P, Rodriguez E. (1983) Chromenes and benzofurans of the Asteraceae, their chemistry and biological significance. Phytochemistry 22: 2335–2348

Proksch P, Proksch M, Towers GHN, Rodriguez E (1983) Phototoxic and insecticidal activities of chromenes and benzofurans from *Encelia.* J Nat Prod 46: 331–334

Robinson H, King RM (1977) Eupatorieae – systematic review. In: Heywood VH, Harborne JB, Turner BL (eds) The biology and chemistry of the Compositae. Academic Press, London, pp 437–485

Siebertz R, Proksch P, Wray V, Witte L (1989) Accumulation and biosynthesis of benzofurans in root cultures of *Eupatorium cannabinum.* Phytochemistry 28: 789–793

Wu WC, Lampe KF, Mende TJ (1973) Metabolic change induced in chickens by administration of tremetol. Biochem Pharmacol 22: 2835–2841

Zalkow LH, Burke N, Cabat G, Grula EA (1962) Toxic constituents of rayless goldenrod. J Med Chem 5: 1342–1351

Zalkow LH, Ekpo BA, Gelbaum LT, Harris RN, Keinan E, Novak JR, Ramming CT, Van Derveer D (1979) The benzofurans of *Isocoma wrightii.* Structure and stereochemistry. J Nat Prod 42: 203–219

Zdero C, Bohlmann F (1987) Eupatoriopicrin, 19-O-linolenate and other constituents from *Eupatorium cannabinum.* Planta Med 53: 169–174

XI *Genipa americana* L. (Marmalade Box): In Vitro Production of Iridoid Glucosides

S. UEDA[1]

1 General Account

1.1 Distribution and Importance of the Plant

Rubiaceous plants of the genus *Genipa* grow throughout tropical America, and *Genipa americana* L. (synonym *G. caruto* H.B. & K.) (Fig. 1A) is native to the wet or moist areas of Cuba, Hispaniola, Puerto Rico, the Virgin Islands, and from Guadeloupe of Trinidad; also from southern Mexico to Panama, and from Colombia and Venezuela to Peru, Brazil, Bolivia, Paraguay, and Argentina. The fruit (Fig. 1B, C), which is edible and popular as a source for beverages, has many colloquial names: marmalade box in the former British West Indies, *genipa*, *jagua* or *caruto* in Puerto Rico and several other Spanish-speaking countries; *genipapo*, *jenipapo* or *jenipa* in parts of Columbia and Brazil; *chipara*, *chibara* or *guanapay* among Colombian Indians; *carucarutoto*, *caruto rebalsero*, or *guaricha* in Venezuela, *tapoeripa* in Surinam, *lana* in Guyana; *bi*, *bicito* or *totumillo* in Bolivia; *huitoc*, *vito*, *vitu* or *palo colorado* in Peru; *maluco* in Mexico; *crayo*, *irayol de montana*, *guaitil* or *tapaculo* in Costa Rica; *irayol*, *tambor* or *tine-duentes* in El Salvador; and *guayatil colorado* or *jagua blanca* in Panama (Morton 1987; Silva et al. 1977). This tree is strictly tropical and grows well in a humid atmosphere and deep loamy soil. The erect trunk of the tree with spreading branches is over 30 m in height. The leaves, deciduous, especially at the rapid lowering of ambient temperatures, are opposite but mostly clustered at the tip of the branch, oblong, 10–30 cm in length, and have a prominent midrib. The tubular five-petaled flowers are yellowish, and ca. 4 cm in width. The elliptic fruit, ca. 10 cm in length, and bearing a short hollow tube at the apex, contains white flesh, which turns yellow to bluish-purple and finally to jet-black on exposure to the air (Morton 1987). It was reported that the natives bathed their limbs and sometimes the whole body, when tired, in a clear juice obtained from the fruits, which turned everything it touched as black as fine and polished jet (Oviedo 18th century).

The tree first fruited in the Philippines in 1913 and is occasionally planted there (Morton 1983). It was introduced into Java in 1913 (Heyne 1950; Kasahara 1986).

Fig 1. A *Genipa americana* trees (Rio de Janeiro, 1990). **B** *Genipa americana* leaves and fruit (Rio de Janeiro, 1990). **C–F** *Genipa americana* fruits (Rio de Janeiro 1990). *Genipa americana* 2 years, 4 years (in summer), and 4 years (in winter) after redifferentiation from callus tissues induced from seedlings of *Genipa americana* collected at Bogor Botanic Garden, 1983, respectively.

[1] Faculty of Pharmaceutical Sciences, Kyoto University, Sakyo-ku, Kyoto 606-01, Japan

Biotechnology in Agriculture and Forestry, Vol. 21
Medicinal and Aromatic Plants IV (ed. by Y.P.S. Bajaj)

A
B
C
D
E
F

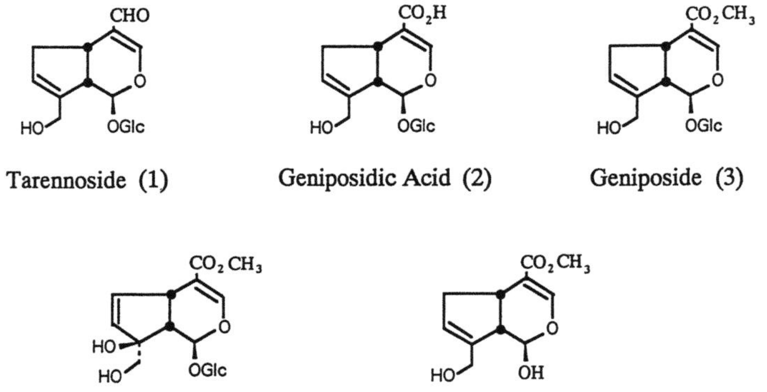

Tarennoside (1) Geniposidic Acid (2) Geniposide (3)

(a) Gardenoside (4) Genipin (5)

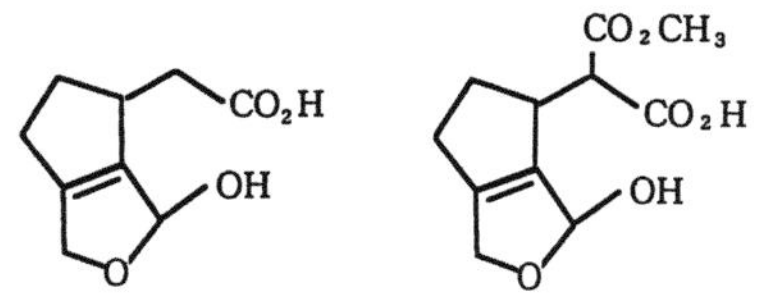

(b) Genipic Acid (6) Genipinic Acid (7)

Fig 2. a Iridoids of *Genipa americana* plant and cell cultures. **b** Cyclopentanoid monoterpenes from Puertorican *Genipa americana* plant

Although the fruit has long been used to make drinks (Nadal et al. 1965; Guedes and Oria 1978) and dyestuffs (Mell 1922, 1928), the discovery of various medicinal properties have made the *Genipa* plant an important medicinal resource. Genipin (5) was isolated from the ripe fruit (Djerassi et al. 1960) and the structure was elucidated (Djerassi et al. 1961; Fig. 2a). Two cyclopentanoid monoterpenes, genipic acid (6) and genipinic acid (7), were isolated from Puerto Rican *Genipa americana* fruit (Fig. 2b). Genipin (5), however, was not obtained from that source (Tallent 1964). Later, geniposidic acid (2) was isolated as a powder from the leaves of *G. americana* collected in Darien, Panama (Guarnaccia et al. 1972). *Gardenia jasminoides* Ellis, a related rubiaceous plant growing throughout subtropical to temperate East Asia from Vietnam to the southern part of Japan (Ohwi 1965) also produces geniposidic acid (2), geniposide (3), a glucoside of genipin (5), and genipin gentiobioside along with gardenoside (4) (Fig. 2a; Endo and Taguchi 1973; Inouye et al. 1974). Later, tarennoside (1) (Takeda et al. 1976), the main iridoid glucoside of Okinawan *Tarenna gracilipes* (Hay.) Ohwi (Hatsushima 1975) together with 2, 3, and 4 were obtained from the cultured cells of *Gardenia jasminoides* (Ueda et al. 1981). Through administration of the isotope-labeled putative precursors to the *Gardenia jasminoides* cell cultures, the biosynthetic sequence 1 → 2 → 3 → 4, started from mevalonic acid, has been established (Kobayashi et al. 1985; Uesato et al. 1986). Genipin (5) has purgative (Yamauchi et al. 1976) and choleretic activities (Miyagoshi et al. 1988). This compound, as well as geniposide (3), inhibited the germination of *Gardenia jasminoides* seeds and watercress (*Nasturtium officinale* R. Br.) seeds and the growth of Chinese cabbage (*Brassica rapa* L. var. *pervidis* Bailey) Shimomura et al. 1983). When genipin (5) was reacted with glycine, genipocyanin

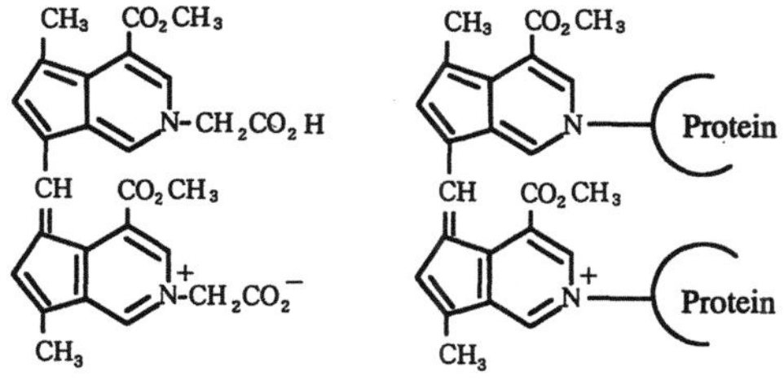

Fig. 3. Genipocyanin (8), an azaphilone iridoid dimer of a 1:1 adduct between glycine and genipin (5)

Table 1. Inhibitory effects of *Genipa americana* iridoids against TPA – induced Epstein-Barr virus early antigen activation

	Concentration (molar ratio/TPA)			
	1000	500	100	10
TPA (32 pmol) 100 = positive control	% to control (% viability)			
Tarennoside (1)	30.5 (80)	72.8	88.1	100
Geniposidic acid (2)	34.8 (80)	80.5	91.8	100
Geniposide (3)	62.4 (80)	91	100	100
Gardenoside (4)	92.6 (80)	100	100	100
Genipin (5)	11.5 (80)	44.2	64.2	100
1 + 2 + 3 + 4 + 5 (Molar ratio 1:1:1:1:1)	26.7 (70)		70	100
Retinoic acid	30 (10)		67 (40)	

G_1 (8), an azaphilone iridoid dimer of a 1:1 adduct between glycine and genipin (5) was formed (Fujikawa et al. 1987a). Reaction of genipin (5) with cytochrome C (horse heart) gave higher-molecular weight oligomers formed by the intermolecular cross-linkage of proteins mediated by the iridoid (Fujikawa et al. 1988; Fig. 3). This finding may open way for the development of a new type of cross-linking reagent. Recently, genipin (5) was found to have a remarkable tumor-promotion inhibitory activity on Epstein-Barr virus activation. Among the above described four iridoid glucosides, tarennoside (1) is the most active, but the activity decreases with the progress of the metabolism. Although the activity of geniposide (3) is lower than that of geniposidic acid (2), its aglycone genipin (5) is remarkably active (Ueda et al. 1991; Table 1). This finding is compatible with the general observations that glycosylation participates in reducing the biological activity (Umetani et al. 1982).

1.2 Conventional Practices for Its Propagation and the Production of Iridoids

The propagation of *Genipa americana* plants has been attempted at several sites in tropical areas with a view to the exploitation of the fruit for liquor manufacture. Despite some successful cases, the trials of the introduction of this strictly tropical tree into other areas have often failed. Sometimes even well-growing trees have never bloomed. Stricter conditions are required for fruiting. The leguminous *Erythrina glauca* tree has an advantage in increasing the soil-nitrogen content in addition to the role as a shade tree in a cacao plantation, while the rubiaceous *Genipa* tree does not contribute to the direct increase in soil nitrogen (Santana and Cabala-Rosand 1982).

Genipin (5), an iridoid, was reported as a constituent of *G. americana* fruit (Djerassi et al. 1960). This compound, however, was not detected in the methanol extracts of *G. americana* fruits, collected at Bogor, Indonesia in October 1983 and at Rio de Janeiro, Brazil in July 1990. The fresh fruits were extracted with hot methanol immediately after being cut into slices. A conventional work-up of both methanol extracts gave geniposide (3) and geniposidic acid (2), respectively, as colorless needles (Ueda et al. 1991). Dried fruit slices, colored bluish-purple, gave genipin (5), as reported previously. A crystalline component isolated from the wood of *G. americana* was found to be D-mannitol (Filho et al. 1965).

Genipin (5) itself has a potentially high demand in the world market, especially for choleretic and tumor inhibitory drugs. This compound (5), also obtainable by the hydrolysis of *Gardenia* geniposide (3) (Endo and Taguchi 1973; Inouye et al. 1974), could be a starting material for the synthesis of biologically active cyclopentanoid substances such as prostaglandins (Naruto et al. 1978; Berkowitz et al. 1982; Bonini et al. 1985) and triquinane sesquiterpenes (Weinges et al. 1989). Demand for the genipin-derived food dyestuffs could also be expected (Djerassi et al. 1960; Inouye et al. 1983; Fujikawa et al. 1987b).

2 In Vitro Approaches

Genipa americana callus tissues (Ueda 1986) were induced according to the method applied to the *Gardenia jasminoides* (Ueda et al. 1981). The callus tissues and the cell suspension cultures produce iridoid glucosides such as tarennoside (1), geniposidic acid (2), and gardenoside (4) in high yields like those of the *Gardenia jasminoides* cell cultures. However, geniposide (3), the main constituent of the *Genipa* and *Gardenia* fruits, was not detected in the *Genipa* cell cultures. Torpedo-shaped embryos formed from the *G. americana* callus tissues grew to seedlings on hormone-free MS medium (Murashige and Skoog 1962). They have grown to a height of 1.5 m in a greenhouse over the course of 7 years. The leaves of the redifferentiated plant contain geniposidic acid (2) as sole iridoid glucoside, as in the case of the parent plant (Ueda and Iwahashi 1991; Ueda et al. 1991). The seedlings were also obtained through the culture of apical meristems.

2.1 Establishment of Tissue Culture/Cell Suspensions

Seeds of *Genipa americana* were sterilized by soaking successively in 70% ethanol for 30 s and in sodium hypochlorite solution (1% active chlorine) for 30 min. The seeds, washed five times with sterile water, were placed on 2% agar plates and incubated at 26 °C in the dark. Within a period of 40 days, most of the seeds had

Table 2. Iridoid productivity (P, +) and callus viability (V, %) in the course of repeated long-term (100 days) cultures

		Cultivation temperature (°C)				
		15	20	25	30	35
Genipa americana	P	+	+ + +	+ + + + +	+ + + +	+ +
	V	3	85	98	69	14
Gardenia jasminoides	P	+ +	+ + + +	+ + + + +	+ + +	+
	V	45	98	86	57	13

germinated. The seedlings of about 5 mm in length were transferred to LS basal agar medium (Linsmaier and Skoog 1965) supplemented with 10^{-5} mol 2,4-dichlorophenoxyacetic acid (2,4-D) and incubated at 25 °C in the dark. Callus tissues induced from the seedlings during 1–3 weeks of incubation were transferred to Murashige-Skoog (MS) basal agar medium supplemented with 2×10^{-5} mol indoleacetic acid (IAA) and 10^{-5} mol kinetin, pH 5.8 before autoclaving, at 25 °C in the dark and subcultured every 4–5 weeks. Iridoid glucoside production in the callus tissues was recognized after three to five transfers. The iridoid productivity was monitored by treating the cells with mineral acids, such as hydrochloric or sulfuric acid. Iridoid-producing calli indicated blue to black coloration, depending on the iridoid content. The calli turned green under white light of approximately 6000 lx. The callus tissues were then transferred to a 500-ml flask containing 200 ml of the above-described MS medium from which agar was deleted. The liquid culture was stirred at 25 °C in the dark on a reciprocating shaker, 85 strokes/min, 8 cm in amplitude. The cells were subcultured every 4–5 weeks. *Genipa americana* cell suspension cultures were also incubated under the illumination of white light of approximately 2500 lx, resulting in the formation of green cell suspensions. The *Genipa americana* callus tissues grew well at around 20–30 °C, while they deteriorated to light brown cells at 15 °C, especially in the course of long-term cultivation up to 100 days on a limited amount of the medum. Within the range between 20–30 °C, high iridoid-producing cells survived long-term cultivation with a limited amount of water and nutrient. The growth rate slowed down at this temperature, as well as at 35 °C. High iridoid productivity has been maintained between 20–30 °C. These charcteristics are important factors for the selection of cell lines of high iridoid productivity (Table 2; Ueda and Iwahashi 1991; Ueda et al. 1991).

2.2 Extraction and Structure of Iridoids

2.2.1 Isolation of Iridoid Glucosides from Callus Tissues

Genipa americana callus tissues were extracted four times with methanol under reflux for 30 min. The combined methanol extracts were concentrated in vacuo to give a syrupy residue which was dissolved in water, and the insoluble materials were filtered off. The filtrate was washed with chloroform. The aqueous layer was concentrated, applied to a charcoal column, and eluted with water and aqueous

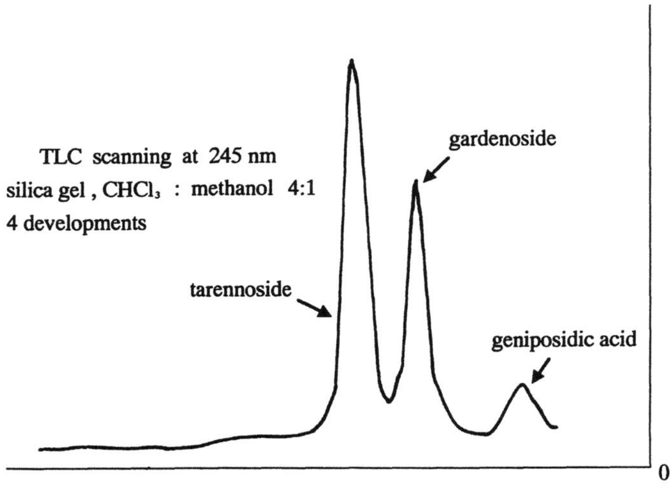

Fig. 4. Thin-layer chromatogram of *Genipa americana* callus methanol extract

methanol with the increasing methanol content. Fractions indicating the positive mineral acid test (Trim and Hill 1952) were combined and concentrated in vacuo to give a pale yellow solid residue, which was subjected to thin-layer chromatography (TLC) on silica gel 60 GF_{254} with a mixture of chloroform and methanol (4:1 by vol) as the solvent system and developed repeatedly. TLC chromatogram scanning at 245 nm clearly demonstrated three bands (Fig. 4). Preparative TLC (PLC; silica gel 60 PF_{254}) of the three major bands of the chromatogram afforded, through extraction with a mixture of chloroform and methanol (9:1) followed by concentration in vacuo, tarennoside (1), gardenoside (4), and geniposidic acid (2), in the order of the mobility on PLC. The yields were approximately 0.30, 0.15, and 0.04%, respectively, on a dry weight basis. The suspension-cultured cells also produced the three glucosides at almost the same levels. Recrystallization of geniposidic acid (2) from methanol gave colorless needles, mp 132–135 °C. Tarennoside (1) and gardenoside (4) were uncrystallizable. Identification of these iridoid glucosides with authentic samples were carried out through the comparisons of UV, IR, ^{1}H-NMR spectra, and optical rotations.

2.2.2 Isolation of Iridoid Glucosides from the Fruits

Fresh *Genipa americana* fruits collected at Bogor, Indonesia in October 1983 or at Rio de Janeiro, Brazil in July, 1990 were cut into slices and immediately extracted with boiling methanol. After extraction under reflux for 30 min, the fruit slices were filtered off and the filtrate was concentrated in vacuo. The greenish syrupy residue was dissolved in water and the insoluble materials were removed through filtration. The filtrate was concentrated in vacuo and the residue was subjected to column chromatography on charcoal, eluting with water and aqueous methanol with the increasing methanol content. Fractions positive to the mineral acid test for iridoid were combined and concentrated in vacuo. The resulting powdery residue was applied to silica-gel PLC plates and developed repeatedly with a mixture of chloroform and methanol (4:1). A work-up of the two major bands on the chromatogram afforded geniposide (3) and geniposidic acid (2). Recrystallization of geniposide (3) from ethanol and geniposidic acid (2) from methanol gave 3, mp 165 °C, and 2, mp 133–135 °C, as colorless needles. Both substances were identified with authentic samples by the comparisons of UV, IR and ^{1}H-NMR spectra, optical rotations and mixed melting point. The contents of 3 and 2 on a dry weight basis were about 0.2 and 0.01%, respectively.

2.2.3 Isolation of Iridoid Glucoside from the Leaves

Fresh leaves of *Genipa americana* plants collected at Honolulu in June 1976, at Bogor in October 1983, or at Rio de Janeiro in July 1990, were cut into pieces, and extracted immediately with boiling methanol. A conventional work-up of the extracts gave geniposidic acid (2) as the sole iridoid glucoside.

In the course of the long-term incubation of *Genipa americana* callus tissues under the illumination of white light of about 6000 lx over a period up to 20 weeks, torpedo-shaped embryos appeared. They grew to greenish shoots, which were transferred to the hormone-free MS agar medium to promote root formation. After growing the roots of the shoots to about 1 cm, the shoots were transferred to pots containing vermiculite and sand. Those plants bearing at least two intact healthy leaves were transferred to soil mixed with peat moss and sand. The plants have grown well to a height of 1.5 m in a greenhouse over the course of 7 years (Fig. 1D). The leaves were also extracted with methanol under reflux. The methanol extract, after concentration and chromatography, gave geniposidic acid (2) in the form of colorless needles.

3 Conclusions and Prospects for Commercial Use

Genipin (5), the aglycone of geniposide (3) has been found to be an important material for the production of food dyestuffs, beverages, and medicines (Okuyama et al. 1977). *Genipa americana* fruit was found to contain a considerable amount of geniposide (3), which is also one of the main constituents of *Gardenia jasminoides*, commonly distributed in subtropical and temperate East Asia. Although the present supply is considered as meeting the current demand, the finding of a highly inhibitory activity in the fruit against tumor promotion will raise the demand with the advance of the development of advantageous biological activities. The success of raising redifferentiated *Genipa americana* plants will contribute to the protection of tropical rain forests. Contrary to the conventional observations of reduced secondary metabolite productivity in the cell cultures, a combination of selection procedures has led to the establishment of high iridoid-producing *Genipa* cell lines. The selection procedures suggest a general way to improve the secondary metabolite production.

4 Protocol

Sprouts of about 5 mm in length are suitable for the induction of callus tissues on Linsmaier and Skoog basal agar medium supplemented with 10^{-5} mol 2,4-D at 25 °C in the dark. Murashige and Skoog medium supplemented with 2×10^{-5} mol IAA and 10^{-5} mol kinetin is effective for the subcultures of the callus tissues producing iridoid glucosides. The medium contains 3% sucrose. Selection by staining the cells with mineral acid to reveal the iridoid-producing cells, together with long-term cultures at 25–30 °C in the dark or under white light of approximately 6000 lx over 100 days on a limited amount of the medium to pick up the surviving cells of high iridoid productivity, is essential to acquire callus tissues of high iridoid productivity.

References

Berkowitz WF, Choudhry SC, Hrabie JA (1982) Conversion of asperuloside to optically active prostaglandin intermediates. J Org Chem 47: 824–829

Bonini C, Iavarone C, Trogolo C, DiFabio R (1985) One-pot conversion of 6-hydroxy-Δ^7-iridoid glucosides into cis-2-oxabicyclo[3,3,0]oct-7-enes and transformation into Corey's lactone analogue. J Org Chem 50: 958–961

Djerassi C, Gray JD, Kincl FA (1960) Naturally occurring oxygen heterocycles IX. Isolation and characterization of genipin. J Org Chem 25: 2174–2177

Djerassi C, Nakano T, James AN, Zalkow LH, Eisenbraun EJ, Shoolery JN (1961) Terpenoids XLVII. The structure of genipin. J Org Chem 26: 1192–1206

Endo T, Taguchi H (1973) The constituents of *Gardenia jasminoides*, geniposide and genipin gentiobioside. Chem Pharm Bull 21: 2684–2688

Filho LGP, Bobbio PA, Falanghe H (1965) Constituent from *Genipa americana* (Rubiaceae): mannitol. Monatsh Chem 96: 2017–2018

Fujikawa S, Fukui Y, Koga K, Iwashita T, Komura H, Nomoto K (1987a) Structure of genipocyanin G_1, a spontaneous reaction product between genipin and glycine. Tetrahedron Lett 28: 4699–4700

Fujikawa S, Fukui Y, Koga K, Kumada J (1987b) Brilliant skyblue pigment formation from *Gardenia* fruits. J Ferment Technol 65: 419–424

Fujikawa S, Nakamura S, Koga K (1988) Genipin, a new type of protein crosslinking reagent from *Gardenia* fruits. Agric Biol Chem 52: 869–870

Guarnaccia R, Madyastha KM, Tegtmeyer E, Coscia CJ (1972) Geniposidic acid, an iridoid glucoside from *Genipa americana*. Tetrahedron Lett 5125–5127

Guedes ZBL, Oria HF (1978) Nutritive value of edible fruits from Ceara, Brazil. Rev Bras Farm 59: 91–97

Hatsushima S (1975) Flora of Ryukyus, added and corrected. Okinawan Assoc Biol Education, Naha

Heyne K (1950) De Nuttige Planten Van Indonesië, 3e Druk. N.V. Uitgeverij W. van Hoeve-'s-Gravenhage/Bandung, p 1399

Inouye H, Takeda Y, Saito S, Nishimura H, Sakuragi R (1974) Studies on monoterpene glucosides and related natural products XXV. On the iridoid glucosides of *Gardenia jasminoides* Ellis forma *grandiflora* [Lour.] Makino (1). Yakugaku Zasshi 94: 577–586

Inouye H, Takeda Y, Inoue K, Kawamura I, Yatsuzuka M, Touyama R, Ikumoto T, Shingu T, Yokoi T (1983) Structure of blue pseudoazulene-skeleton pigment derived from genipin and amino acids. In: Abstr 26th Symp on the Chemistry of natural products, Kyoto, pp 577–584

Kasahara YS (1986) Medicinal herb index in Indonesia. PT Eisai Indonesia, Jakarta

Kobayashi K, Uesato S, Ueda S, Inouye H (1985) Studies on monoterpene glucosides and related natural products LV. Iridane skeleton formation from acyclic monoterpenes in the biosynthesis of iridoid glucosides in *Gardenia jasminoides* f. *grandiflora* cell suspension cultures. Chem Pharm Bull 33: 4228–4234

Linsmaier EM, Skoog F (1965) Organic growth factor requirements of tobacco tissue cultures. Physiol Plant 18: 100–127

Mell CD (1922) Interesting dyestuff plants. Text Color 44: 107–111
Mell CD (1928) Interesting source of natural dyestuffs. Text Color 50: 602–604
Miyagoshi M, Amagaya S, Ogihara Y (1988) Choleretic actions of iridoid compounds. J Pharmacobio-Dyn 11: 186–190
Morton JF (1987) Fruits of warm climates. Media, Incorporated, Greensboro
Murashige T, Skoog F (1962) A revised medium for rapid growth and bioassays in tobacco tissue cultures. Physiol Plant 15: 473–497
Nadal NGM, Rodriguez LV, Hernandes T (1965) Marmalade box genip, Am Perfumer Cosmet 80: 37–38
Naruto M, Ohno K, Naruse N (1978) The synthesis of useful chiral prostanoid intermediates and naturally occurring prostaglandins from aucubin. Chem Lett 1419–1422
Ohwi J (1965) Flora of Japan. In: Meyer FG, Walker EH (eds) Flora of Japan. Smithsonian Institution, Washington, DC, p 825
Okuyama H, Toyama R, Sawada Y (1977) Food coloring agent. J Patent Kokai 77 53932
Oviedo BV (18th century) Cualidades y Riquezas del nuevo Reino de Granada, 18th century manuscript published in 1930 by Cuervo LA. Biblioteca de Historia Nacional, Bogota
Santana MBM, Cabala-Rosand P (1982) Dynamics of nitrogen in a shaded cacao plantation. Plant Soil 67: 271–281
Shimomura H, Sashida Y, Nakata H, Yamamoto A, Kawakubo Y, Kawasaki J (1983) Germination and growth inhibitors in fruits of *Gardenia jasminoides*. Plant Cell Physiol 24: 123–126
Silva MF, Lisbõa PLB, Lisbõa RCL (1977) Names vulgares de plantas amazônieas. Conselho Nacional de Desenvolvimento científico e Tecnológico, Inst Nac Resquisas Amazonia, Manaus
Takeda Y, Nishimura H, Inouye H (1976) Studies on monoterpene glucosides and related natural products XXXII. Iridoid glucosides of *Tarenna kotoensis* var. *gyokushinka*. Chem Pharm Bull 24: 1216–1218
Tallent WH (1964) Two new antibiotic cyclopentanoid monoterpenes of plant origin. Tetrahedron 20: 1781–1787
Trim AR, Hill R (1952) The preparation and properties of aucubin, asperuloside and some related glycosides. Biochem J 50: 310–319
Ueda S (1986) Production of iridoid-type monoterpene glucosides by cell cultures of rubiaceous plants. In: Somers DA, Gengenbach BG, Biesboer DD, Hackett WP, Green CE (eds) Abstr 6th Int Congr Plant cell culture, Univ Minnesota, Minneapolis, p 139
Ueda S, Iwahashi Y (1991a) Habitat climate-dependent Gardenieae iridoid productivity revealed through callus induction. Naturwissenschaften 78: 171–172
Ueda S, Kobayashi K, Muramatsu T, Inouye H (1981) Studies on monoterpene glucosides and related natural products XL. Iridoid glucosides of cultured cells of *Gardenia jasminoides* f. *grandiflora*. Planta Med 41: 186–191
Ueda S, Iwahashi Y, Tokuda H (1991b): Production of anti-tumor-promoting iridoid glucosides in *Genipa americana* and its cell cultures. J Natural Products 54: 1677–1680
Uesato S, Ueda S, Kobayashi K, Miyauchi M, Itoh H, Inouye H (1986) Intermediacy of 8-epiiridodial in the biosynthesis of iridoid glucosides by *Gardenia jasminoides* cell cultures. Phytochemistry 25: 2309–2314
Umetani Y, Tanaka S, Tabata M (1982) Glucosylation of extrinsic compounds by various plant cell cultures. In: Fujiwara A (ed) Plant tissue culture 1982, Maruzen, Tokyo, pp 383–384
Weinges K, Iatridou H, Stammler H-G, Weiss J (1989) Chiral building blocks for the synthesis of triquinane sesquiterpenes: derivatives of 2-methylbicyclo[3,3,0]octan-3-ol from catalpol. Angew Chem Int Ed Engl 28: 447–448
Yamauchi K, Fujimoto N, Kawano S, Inouye H (1976) The mechanism of purgative action of geniposide, an iridoid glucoside of the fruit of *Gardenia*, in mice. Planta Med 30: 39–47

XII *Gentiana* Species: In Vitro Culture, Regeneration, and Production of Secoiridoid Glucosides

L. Skrzypczak, M. Wesołowska, and E. Skrzypczak[1]

1 Introduction

1.1 Distribution and Importance

Gentiana species belong to the family Gentianaceae, order Gentianales, superorder Gentiananae, subclass Asteridae, class Magnoliopsida (Takhtajan 1980). The species are divided into several sections according to the morphology of the aboveground organs (Sziszkin and Bobrow 1952; Hegi 1966; Tutin et al. 1972). The subgenera *Eugentiana* Kusnezow and *Gentianella* Kusnezow (Hegi 1966) are entered in Flora Europaea as separate genera: *Gentiana* L. and *Gentianella* Moench (Tutin et al. 1972).

The genus *Gentiana* comprises about 400 species distributed chiefly in mountain regions, especially in the Alps, the Carpathians, the Central Asia mountains, and the Andes in South America. Due to their impressive and colorful flowers, gentians decorate mountain meadows. Some species are also found in the monsoon zone of India, in New Zealand, and in southern Australia. More rarely, gentians are found in the temperate zone lowlands of the northern hemisphere (Hegi 1966; Index Kewensis). The yellow gentian root was already mentioned as a remedium stomachicum by Galen and Dioscorides (Dragendorff 1898). Apart from *Gentiana lutea* L., there are other medicinal species included in many pharmacopoeias and plant registers of the world (Penso 1983). According to most European pharmacopoeias, the official drug may also contain material from *G. pannonica Scop.*, *G. punctata L.*, and *G. purpurea L.*

The active constituents of the drug are bitter tasting secoiridoid glucosides which are applied in the treatment of gastrointestinal tract diseases (Maiwald 1987; Weiss 1988). A beneficial effect of these compounds on the immunological system of the entrails has also been observed (Zimmermann et al. 1986). Gentian roots are applied as a decoction, an extract, and a tincture. According to the Rote Liste (1986), Radix Gentianae is a component of about 100 drugs belonging to the groups amara, tonica, cholagoga, and laxantia. Once the gentian root was a component of Pilulae stomachicae cum Fello (Weiss 1980). Because of their bitter taste, extracts are used in the production of wines and liquors (Wenzel 1972).

[1] Department of Pharmaceutical Botany, K. Marcinkowski Medical Academy, Wieniawskiego 1, 61-712 Poznań, Poland

Biotechnology in Agriculture and Forestry, Vol. 21
Medicinal and Aromatic Plants IV (ed. by Y.P.S. Bajaj)

1.2 Conventional Methods for Propagation and Demand in the World Market

The difficulties encountered in the harvest of valuable gentian roots from wild populations prompted distributors to attempt the cultivation of these plants. So far, only *G. lutea* has been an object of interest (Franz and Fritz 1975, 1978; Paun and Mihalea 1977; Barralis et al. 1978; Schultze and Franz 1980; Procyk 1982; Gawłowska 1986). *G. lutea* cultivation was carried out in the presence of symbiotic fungi as well (Heymons et al. 1986). However, the results of cultivation are not very encouraging and primarily they are uneconomical. Plants require good soil conditions and laborious cultivation practices. Moreover, roots suitable for harvest are developed during the flowering period which takes place after some years of growth (Kozłowski 1983). In many countries gentians do not occur or are under protection. For example, in Poland, the entire yellow gentian root supply is imported. The yearly demand for the dry drug for medicinal use only is about 3000 kg. The drug is exported from Europe to Japan, where cultivation is also attempted (Hayaski and Yamagishi 1988). As reported earlier by Franz and Fritz (1975), a yearly production amounting to 1000 t of the fresh gentian root did not satisfy the demand for the drug.

1.3 Secondary Metabolites in *Gentiana* Species

Secoiridoid glucosides are the main compounds with medicinal properties in roots of *Gentiana* species. Acyl derivatives of glucosides – amarogentin and amaroswerin (Table 1, Fig. 1) – were found to be the most bitter compounds among natural products. Two other compounds, amaropanin and gentiopicroside, are also of biological importance. Gentiopicroside has the lowest bitter coefficient but its content in the drug is higher than that of acyl derivatives (Table 1). During enzymatic hydrolysis, gentiopicroside liberates a metabolite with fungistatic properties (Van der Sluis and Labadie 1981; Van der Nat et al. 1982) called gentiopicral (Ishiguro et al. 1983), previously gentiogenal (Van der Sluis et al. 1983). El-Sedawy

Table 1. Secoiridoid glycosides in the roots of *Gentiana* species. (Wagner and Vasirian 1974; Wagner and Münzing-Vasirian 1975; Wichtl 1978)

Compound	Bitter coefficient	Average contents (%)	Species	Presence of compounds, bitter coeff. of roots
Gentiopicroside – G	12000	2.0–4.0	*G. lutea* L.	G Ag 15000
Amarogentin – Ag	58000000	0.05–0.3	*G. pannonica* Scop.	G Ag As AP 91000
Amaroswerin – As	58000000	0.03–0.1	*G. punctata* L.	G Ag As Ap 135000
Amaropanin – Ap	20000000	0.05–0.2	*G. purpurea* L.	G Ag As Ap 175000

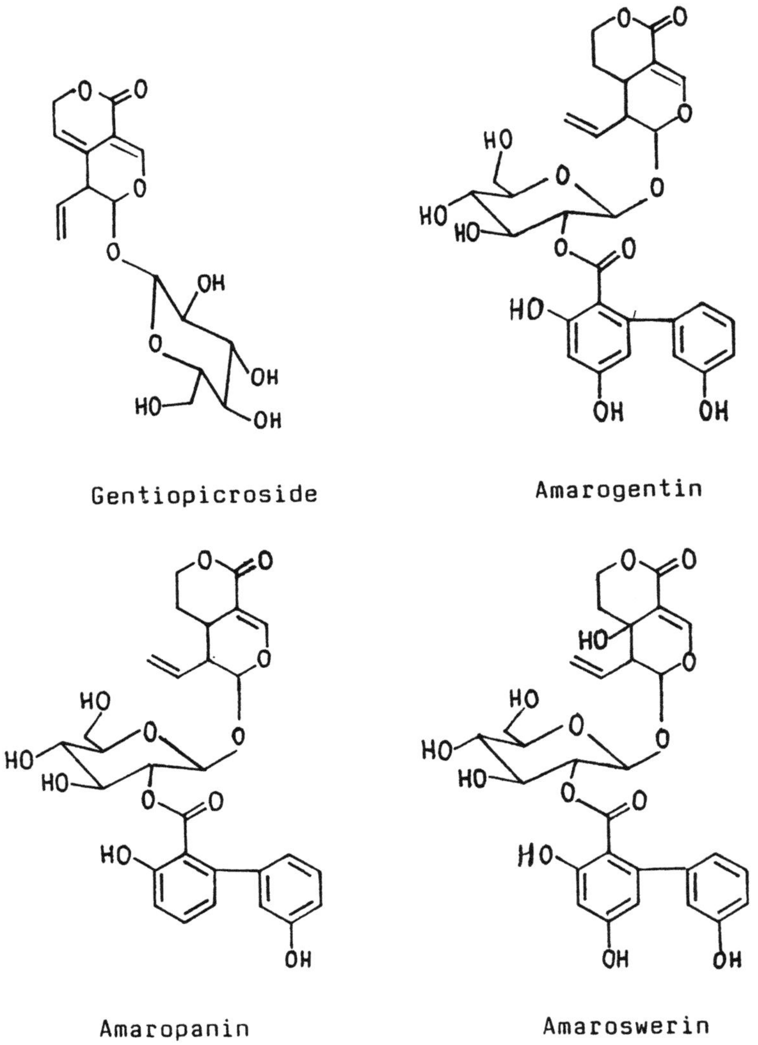

Fig. 1. Structures of the main secoiridoid glycosides of *Gentiana* species

et al. (1989) characterized the activity of gentiopicroside metabolites on several bacterial tribes and Isiguro et al. (1986) reported on their anticancer properties.

For a long time the bitter coefficient of gentian roots and other drugs are based on the sensorial method. After the structure of biologically active secoiridoids had been established (Canonica et al. 1961; Inouye et al. 1968, 1970; Inouye and Nakamura 1971), spectrophotometric methods were introduced for quantitative analyses (Wagner and Münzing-Vasirian 1975; Bakuridze et al. 1987a). For these purposes HPLC, FD-MS, DCI-MS, and TLC densitometry methods are equally used (Sticher and Meier 1978, 1980; Koshioka et al. 1980; Quercia et al. 1980; Sakamoto et al. 1983; Vanhaelen and Vanhaelen-Fastre 1983; Schaufelberger et al. 1984; Schaufelberger and Hostettmann 1985; Verotta 1985; Luo and Lou 1986;

Krupińska and Skrzypczak 1989; Krupińska et al. 1991). Bakuridze et al. (1987b) made a qualitative analysis of secoiridoids in 17 species (including also pharmacopoeial ones) of the genera *Gentiana* and *Gentianella* and found that all of them contained gentiopicroside, while amarogentin was present in only three species. A secoiridoid glucoside-trifloroside, found for the first time by Inouye et al. (1974) in *G. triflora* Pall. *var. japonica* (Kusnez.) Hara, was also detected in *G. scabra* Bge. and *G. pneumonanthe* L.

Yellow gentian roots also contain pharmacologically active xanthones (Verney and Debelmas 1973; Nikolaeva et al. 1983; Hayashi and Yamagishi 1988). The occurrence of some known compounds in *G. asclepiadea* L. and *G. cruciata* L. was investigated and compared with that of *G. lutea* and *G. punctata* (Menkovic et al. 1986). Flavonoids, xanthones, and L-(+)-bornesitol were found in *G. lutea* leaves (Bellmann and Jacot-Guillarmod 1973a, b; Hostettmann et al. 1973). The above-mentioned publications initiated a series of papers entitled *Contribution a la phytochimie du genera Gentiana*, dealing with studies on secondary metabolites, especially flavonoids and xanthones, and their meaning for the chemotaxonomy of the family Gentianaceae (e.g., Hostettmann and Wagner 1977). Similarly, Ghosal et al. (1983) published the series *Chemical constituents of Gentianaceae*. The isolation by these authors of 6-aryl-2-pyrones with adaptogenic and antistress properties from *G. pedicellata* Wall. was the first report on the occurrence of these compounds in the family Gentianaceae (Ghosal et al. 1983). In the same species an iridoid glucoside loganin, and its aryl derivative were also found for the first time (Garcia and Chulia 1986). New iridoid glucosides were found in *G. verne* L. and *G. campestris* L. (Mpondo and Garcia 1990; Mpondo et al. 1990a) and a new secoiridoid in *G. alpina* (Mpondo et al. 1990b). Moreover, some *Gentiana* species contain triterpenoids, phenolic acids, and oils (Hegnauer 1966; Chialva et al. 1986; Dombrowicz and Świątek 1987; Świątek et al. 1988).

2 In Vitro Approaches

2.1 Literature Review (Table 2)

According to literature data (Bajaj et al. 1988), there have been few medicinal species of the family Gentianaceae studied in in vitro cultures. In 1982 tissue cultures of species of the genus *Gentiana* were initiated in the Department of Pharmaceutical Botany of the Medical Academy in Poznań. After 3 years reproducible results were obtained (Wesołowska et al. 1985). As Table 2 shows, there are also in vitro cultures of other plants from Gentianaceae being grown. Lamproye et al. (1987) have published the results of their studies on the in vitro vegetative propagation of the species. *G. lutea* and *G. pneumonanthe* L. There has also been another short report on the positive results of in vitro cultures of *G. lutea* (Viola and Franz 1989). The genus *Gentiana* is mentioned by Pierik (1989) among plants subjected to in vitro cloning for commercial purposes in Holland. He also reports a twofold increase in plants produced in 1986 as compared with the previous year.

Table 2. In vitro culture studies on various species Gentianaceae

Species	Occurrence	Explant	Results	Reference
Blackstonia perfoliata L. Hudson	Europe-Mediterranean region	Parts of seedlings	Callus buds, shoots, plants	Skrzypczak et al. (1988b)
		Leaf segments		Skrzypczak and Wesołowska (1989)
Centaurium erythraea Rafn	Europe West Asia North Africa	Parts of seedlings Leaf segments	Callus, buds, shoots, plants	Our unpublished results
		Leaf segment or callus	Shoots, roots, callus	Barešová (1988)
		Callus, suspension	Callus, suspension embryogenesis	Barešová (1988)
Eustoma grandiflorum Shinn.	North America	Leaf segments	Buds, shoots, plants	Zenkteler and Zenkteler (1987)
		Shoot tips, internodal stem, leaf segments	Callus, shoots, plantlets	Semeniuk and Griesbach (1987)
		Parts of seedlings	Callus, buds, shoots, plants	Skrzypczak et al. (1988a,b)
		Leaf segments		Skrzypczak and Wesołowska (1989)
Gentiana cruciata L.	South and central Europe Asia Mnior West Siberia	Seedlings Leaf segments	Callus, buds Shoots, plants	Wesołowska et al. (1985)
Gentiana lutea L.	Mountains of central and southern Europe	Embryos, seedlings	Shoots, plantlets	Our unpublished results
		Roots	Callus	Lamproye et al. (1987)
		Parts of seedlings	Buds, shoots, plantlets	
		Meristem tips, axillary buds, shoot tips		Viola and Franz (1989)
Gentiana pneumonanthe L.	Europe	Parts of seedlings	Buds, shoots, plantlets	Lamproye et al. (1987)
Gentiana punctata L.	Mountains of central Europe	Embryos, seedlings	Shoots, plantlets	Our unpublished results
Swertia japonica Makino	Mountains of Japan, Korea and China	Seedling root	Callus shoots	Miura (1991)
Swertia pseudochinensis Hara	Japan, Korea and the temperate regions of Southeast Asia	Seedling hypocotyls and roots	Callus	Miura (1991)
		Embryogenic calli	Embryoids, plants	
		Seedling root tips	Excised root culture	

2.2 Establishment of Tissue Cultures and Plant Regeneration

Tissue cultures were derived from seeds obtained from various European botanical gardens and the Department of Botany of the Plant and Herbal Products Institute in Poznań. The seeds were placed in a refrigerator at a temperature of 0 to 4 °C for at least 1 week, and then transferred to Petri dishes, submerged in distilled water, and left for 24–48 h at about 25 °C. The seeds were then successively sterilized with 70% ethanol (30 s), a 0.1% water solution of mercuric chloride (6 min), and rinsed with sterile water several times. All these activities were carried out in a sterile room. The cultures were grown in a culture room at 25 °C (± 1 °C), a relative air humidity of 60–70%, and artificial light of 2000 lx. On the media were placed undamaged seeds with partly removed seed coats, and isolated embryos. During preliminary examinations, about 40 media enriched with growth regulators were tested. The basic media used were those prepared after Murashige and Skoog (MS, 1962), Nitsch and Nitsch (NN, 1969), Gamborg et al. (B_5, 1968), and modified MS (Zenkteler 1984).

Screening studies brought positive results for *G. cruciata* L. (Wesołowska et al. 1985). The callus tissue still shows vitality, while plants regenerated from it and placed in the ground flowered in the second vegetation period. Our further research focused on *G. lutea* L. and *G. punctata* L. The available seeds differed in their germination capacity. After the initial testing of several batches, those seeds showing high vitality were chosen for further experiments.

B_5 medium with various growth regulators was used. The best results were obtained using the variant with GA_3 (0.5 mg/l) and kinetin (0.25 mg/l). This

Fig. 2. Seedlings of *G. lutea* on B_5 medium without any growth regulators after 3 weeks of culture

Fig. 3. Shoots of *G. lutea* developed from adventitious buds on B_5 medium with 1 mg/l BAP and 0.1 mg/l IAA after 10 weeks of culture

medium was used to grow isolated embryos of *G. lutea* (about 1500) and *G. punctata* (about 100). In 3–5 days the embryos started to turn green and developed into seedlings (in *G. lutea* about 50%, and in *G. punctata* about 80%; Fig. 2). In order to obtain proper material for micropropagation, the seedlings were transferred to a B_5 medium without growth regulators. After 2 weeks the seedlings grew and provided adequate material for further studies.

To propagate the culture, seedlings without roots (with hypocotyls) or their leaves were transferred. The best results were obtained on a B_5 medium with BAP (1 mg/l) and IAA (0.1 mg/l).

After about 1 month, in 10–30% of *G. lutea* seedlings and about 80% of *G. punctata* seedlings adventitious buds appeared near the shoot tip which developed into shoots (Fig. 3). Attempts to induce bud formation on leaves were unsuccessful in these two species. For the induction of roots the shoots obtained by micropropagation were then transferred on to an MS medium with the addition of IBA or IAA (0.5 or 1 mg/l medium) for an appropriate period of time. Shoots (50%) of *G. punctata* and *G. lutea* started to root after 20–25 days only when they were transferred to an MS medium without growth regulators after 24-h contact with a liquid medium containing IAA (Fig. 4).

A callus culture was also started from fragments of roots obtained from *G. lutea* seedlings. When transferred to B_5 media with the addition of 2,4-D (Table 3), they started to develop a callus after 7–10 days (Fig. 5). The cultures underwent several passages. Under the conditions cited slow increments of the callus tissue could be observed.

Fig. 4. Rooted shoot of *G. lutea* on MS medium without any growth regulators after 8 weeks of culture

Fig. 5. Callus forming from roots of *G. lutea* on B_5 medium with 0.5 mg/l 2,4-D after 15 weeks of culture

Table 3. In vitro culture of species *Gentiana* L.

Species	Explant	B_5 medium with (mg/l)					Results
		GA_3	KIN	BAP	2,4-D	IAA	
G. lutea L.	Embryos	–	–	–	–	–	Seedlings
	"	0.5	0.25				"
	Seedling parts			1		0.1	Shoots
	" "			1			"
	Roots				0.25		Callus
	"				0.5		"
	"		1.0		0.5		"
G. punctata L.	Embryos	–	–	–	–	–	Seedlings
	"	0.5	0.25				"
	"	1.0	0.5				"
	Seedling parts			1		0.1	Shoots, plants
	" "			1			" "

2.3 Secondary Products in Tissue Culture (Fig. 6)

The occurrence of secondary metabolites in tissue cultures of species from Gentianaceae has been investigated with regard to *Gentiana cruciata* L. (Wesołowska et al. 1985) and *Centaurium erythraea* Rafn (Meravy 1987; Barešová 1988).

Thin layer chromatography (TLC) showed the presence of secoiridoid glucosides in the callus tissue of *G. cruciata* and plants developed from it. Meravy (1987) reported the occurrence of xanthones, phenolic acids, and flavonoids in the callus and suspension culture of *Centaurium erythraea.* The author indicated differences in the contents of those substances in the cultures investigated. On the basis of organoleptic tests, Barešová (1988) concluded that the bitter substances appeared only in minor quantities in embryogenic suspension cultures and in shoots formed in the light.

In our investigation the differentiated callus tissue and shoots without roots of *G. lutea* and *G. punctata* were screened for the production of secoiridoid glucosides. For comparison, these compounds were also analyzed in leaves of some parent plants of *Gentiana* species and in the pharmacopoeial drug, Radix Gentianae.

The materials were dried at 60 °C and after crumbling each 2-g sample was extracted with methanol (3 x 50 ml) in a water bath at 40–50 °C (each extraction for 1 h). The methanol was evaporated from the combined extracts under reduced pressure over a water bath below 40 °C. The residue was dissolved in 4 ml methanol for further analysis. A qualitative analysis was carried out on HPTLC plates (Merck) covered with silica gel G 60 F_{254}. The following mobile phases (v/v) were employed (Inouye et al. 1974; Wagner et al. 1984): chloroform-methanol (4 : 1) and ethyl acetate-methanol-water (77 : 15 : 8). Besides strictly determined amounts of extracts, standard compounds, gentiopicroside and amarogentin, were chromatographed. Chromatograms were analyzed under UV light (λ = 254 nm).

A quantitative analysis of the two compounds was performed by TLC densitometry (Krupińska et al. 1991). Drug samples (1–3 g) were extracted with

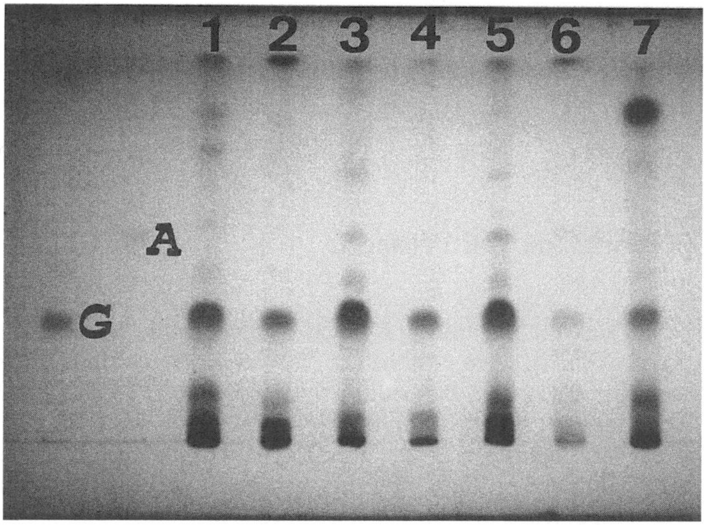

Fig. 6. Chromatogram in UV_{254} of methanolic fractions of *Gentiana* sp. HPTLC plate, silica gel G 60 F_{254}, mobile phase $CHCl_3$–MeOH (4:1). *G* Gentiopicroside; *A* amarogentin: *1,3,6* shoots from tissue culture; *2,5* leaves of parent plant; *4* differentiated callus with buds; *7* Radix Gentianae; *1,2 G. lutea*; *3,4,5 G. tibetica*; *6 G. punctata*

methanol as in the qualitative method. The extracts were then dissolved in 10 ml methanol. Extracts and standard compounds in appropriate amounts were chromatographed in a mixture of chloroform and methanol (4:1) as a mobile phase. Prior to the analysis of gentiopicroside, the extracts were diluted 5–20 times with methanol. The spots of compounds were marked under UV_{254} and determined densitometrically by measuring fluorescence quenching (Vitatron TLC 100, Holland). The scanning conditions were as follows: a mercury lamp with filters $\lambda = 254$ nm, $\lambda = 525$ nm on a photomultiplier, slit 0.5 mm, zigzag mode, plate and chart speed 1 cm/min. Spot areas are integrated automatically.

Plants from in vitro cultures contained 0.08 amarogentin and 3.7% gentiopicroside.

3 Protocol

Tissue Culture

1. Culture embryos on B_5 medium with GA_3 (0.5 mg/l) and kinetin (0.25 mg/l) to obtain seedlings.
2. Use B_5 medium with BAP (1 mg/l) and IAA (0.1 mg/l) to obtain shoots from parts of seedlings.
3. To induce rooting explants are transferred to MS medium without growth regulators after 24-h contact with a liquid MS medium containing IAA (0.5 mg/l).
4. Use B_5 medium with kinetin (1 mg/l) and 2,4-D (0.5 mg/l) to obtain callus from roots of seedlings.

Chemical Analysis. Amarogentin and gentiopicroside contents were determined using the TLC method. The extracts and standard compounds were spotted on HPTLC silica gel 60 F_{254} plates (Merck). The chromatogram was developed in a mixture of chloroform-methanol (4 + 1 v/v).

TLC densitometry was used for the quantitative determination of amarogentin and gentiopicroside.

4 Conclusion and Prospects

Conditions for the micropropagation and callus culture of some species of the genus *Gentiana* were established. Seedlings were obtained from seeds or isolated embryos placed on a B_5 medium with the addition of GA_3 and kinetin. It was also found that the stratification of seeds for a few months was not necessary, contrary to earlier suggestions (Buc and Krys 1975). Seeds collected in the given year developed into seedlings after they had been placed on the medium for several days at a temperature of 0–4 °C. The rooting of shoots was most effective when they were put in a liquid medium with IAA for 24 h, and then placed on an MS medium without growth regulators. Rooted plantlets were transferred to pots and then to the ground.

In vitro cultures of *G. lutea* and *G. pneumonanthe* were reported by Lamproye et al. (1987), who initiated them also from seeds by placing them on an MS medium without phytohormones. Seedling fragments fed with MS medium with the addition of auxins and cytokinins formed buds which developed into shoots that rooted in the presence of PBA and IBA. These conditions, however facilitated only the growth of *G. pneumonanthe.* With *G. lutea* the germination of seeds was initiated by the addition of GA_3 (100 mg/l) to an MS medium. In turn, the development of shoots and the induction of roots were only facilitated after the culture was kept for 2 months at 5 °C without light and without phytohormones in the medium.

The micropropagation of *G. lutea* from apical meristems was carried out on an MS medium with BAP (Viola and Franz 1989). Buds appeared when a medium containing kinetin or 2iP was employed. A high rooting index was observed in the presence of IBA or NAA.

The research carried out to date leads to the conclusion that species of *Gentiana* can be propagated via tissue cultures. Regenerated plants synthesize biologically active secoiridoids and hence they can be of pharmaceutical and economic significance. The presence of these compounds also in the above-ground parts of gentians grown in the ground and in in vitro cultures is striking. Suggestion for the utilization of not only the root but also the herb of *G. lutea* as a medicinal material (Nikolaev et al. 1987) seems to be justified.

Acknowledgments. We wish to thank Prof. J. Kozłowski (Institute of Medicinal Plants, Poznań) for the literature data and M. Kawińska (A. Mickiewicz University, Poznań) and J. Budzianowski (K. Marcinkowski Medicinal Academy, Poznań) for checking the language.

References

Bajaj YPS, Furmanowa M, Olszowska O (1988) Biotechnology of the micropropagation of medicinal and aromatic plants. In: Bajaj YPS (ed) Biotechnology in agriculture and forestry, vol 4. Medicinal and aromatic plants I. Springer, Berlin Heidelberg New York, pp 60–103

Bakuridze AD, Dargaeva TD, Patudin AV, Efremow AP, Nikolaeva GG, Brutko LI, Rabinowicz AM (1987a) Metod opredelenija sodierżanija sumy iridoidow u *Gentiana*. Rast Resur 23: 455–458

Bakuridze AD, Dargaeva TD, Nikolaeva GG, Patudin AV, Brutko LI (1987b) Iridoids in plants of the genus *Gentiana* (Gentianaceae). Chim Prirod Soedin 1: 3–11 (in Russian)

Barešová H (1988) *Centaurium erythraea* Rafn: micropropagation and the production of secoiridoid glucosides. In: Bajaj YPS (ed) Biotechnology in agriculture and forestry, vol 4. Medicinal and aromatic plants I. Springer, Berlin Heidelberg New York, pp 350–366

Barralis G, Chadoeuf R, Desmarěst P (1978) New trends of *Gentiana lutea* cultivation. Acta Hortic 73: 303–306

Bellmann G, Jacot-Guillarmod A (1973a) Contribution á la phytochimie du genre *Gentiana*. I. Etude des composes flavoniques et xanthoniques dans les feuilles de *Gentiana lutea* L. Helv Chim Acta 56: 284–294

Bellmann G, Jacot-Guillarmod A (1973b) Contribution á la phytochimie du genre *Gentiana*. II. Etude des cyclitols dans les feuilles de *Gentiana lutea* L. Helv Chim Acta 56: 773–775

Buc TG, Krys ZOP (1975) K biologii prorastanija semjon gorecavki żeltoj. Biul Dl Bòtan Sada AN URSS Nr 96: 52–55; Ref Z-Biol 1975 Nr 11: w 410

Canonica L, Pelizzoni F, Manito P, Jommi G (1961) Structure of gentiopicroside. Tetrahedron 16: 192–200

Chialva F, Frattini C, Martelli A (1986) Öl von *Gentiana lutea*. Z. Lebensm Unters Forsch 182: 212–214

Dragendorff G (1898) Die Heilpflanzen der verschidenen Völker und Zeiten. Enke, Stuttgart, pp 529–530

Dombrowicz E, Świątek L (1987) Kwasy fenolowe w surowcach goryczkowych. Farmacja Pol 43: 639–642

El-Sedawy A, Hattori M, Kobashi K, Namba T (1989) Metabolism of gentiopicroside (gentiopicrin) by human intestinal bacteria. Chem Pharm Bull 37: 2435–2437

Franz C, Fritz D (1975) Anbauversuche mit *Gentiana lutea* und Inhaltsstoffe einiger Ökotypen. Planta Med 28: 289–300

Franz C, Fritz D (1978) Cultivation aspects of *Gentiana lutea* L. Acta Hortic 73: 307–312

Gamborg OL, Miller RA, Ojima K (1968) Nutrient requirements of suspension cultures of soybean root cells. Exp Cell Res 50: 151–158

Garcia J, Chulia AJ (1986) Loganin and new iridoid glucosides in *Gentiana pedicellata*. Planta Med 4: 327–329

Gawłowska J (1986) Pierwsze polowe uprawy goryczki żółtej w Polsce. Wiad Zielar 27(6): 1–3

Ghosal S, Singh AAK, Biswas K (1983) New 6-aryl-2 pyrones from *Gentiana pedicellata*. Planta Med 49: 240–243

Hayashi T, Yamagishi T (1988) Two xanthone glycosides from *Gentiana lutea*. Phytochemistry 27: 3696–3699

Hegi G (1966) Illustrierte Flora von Mitteleuropa, Bd 3. Hanser, München, pp 1979–2047

Hegnauer R (1966) Chemotaxonomie der Pflanzen, Bd 4. Birkhäuser, Basel, pp 176–192

Heymons S, Hölzl J, Weber HC (1986) Va-Mycorrhiza in *Gentiana lutea*, the importance of cultivation and influence on constituents. Planta Med 34th Annu Congr Medicinal Plant Research, Hamburg, p 510

Hostettmann K, Wagner H (1977) Xanthone glycosides. Phytochemistry 16: 821–829

Hostettmann K, Bellmann G, Tabacchi R, Jacot-Guillarmad A (1973) Contribution á la phytochimie du genre *Gentiana*. III Etude des composes flavoniques et xanthoniques dans les feuilles de *Gentiana lutea* L. Helv Chim Acta 56: 3050–3054

Index Kewensis Plantarum Phanerogmarum 1895, Suppl I–XVIII (1906–1987) University Press, Oxford

Inouye H, Nakamura Y (1971) Über die Monoterpenglucoside und verwandte Naturstoffe XIV. Die struktur der beiden stark bitter schmeckenden Glucoside Amarogentin und Amaroswerin aus *Swertia japonica.* Tetrahedron 27: 1951–1966

Inouye H, Yoshida T, Nakamura Y, Tobita S (1968) Die Stereochemie einiger Secoiridoidglucoside und die Revision der Struktur des Gentiopicrosids. Tetrahedron Lett 42: 4429–4432

Inouye H, Ueda S, Nakamura Y (1970) Über die Monoterpenglucoside X. Secoiridoid-glucoside aus *Swertia japonica.* Isolierung von fünf Secoiridoid-glucosiden sowie die Strukturaufklärung des Swerosids, des Swertiamarins und des Gentiopicrosids. Chem Pharm Bull 18: 1856–1865

Inouye H, Ueda S, Nakamura Y, Inoue K, Hayano T, Matsumura H (1974) Über die Monoterpenglucoside und verwandte Naturstoffe XXIV, Trifloroside ein neues Secoiridoidglucosid aus *Gentiana triflora* var. *japonica.* Tetrahedron 30: 571–577

Ishiguro K, Yamaki M, Takagi S (1983) Studies on iridoid-related compounds. III. Gentiopicral, the aglucone of gentiopicroside. Planta Med 49: 208–210

Isiguro K, Yamaki M, Takagi S, Ikeda Y, Kawakami K, Ito K, Nose T (1986) Studies on iridoid-related compounds. IV. Antitumor activity of iridoid aglycones. Chem Pharm Bull 34: 2375–2379

Koshioka M, Takino Y, Hayshi T (1980) Studies on the evaluation of crude drugs. X. II Quantitative determination of bitter compounds in *Swertiae* Herba. Planta Med 38 (Suppl): 64–67

Kozłowski J (1983) Goryczka żółta – wartościowa roślina lecznicza. Wiad Zielar 25(10–11): 16–17

Krupińska A, Skrzypczak L (1989) Oznaczanie amarogentyny metodą densytometryczną. XIV Naukowy Zjazd Polsk Towarzystwa Farm. Wrocław, Streszczenia, p 7/30

Krupińska A, Segiet-Kujawa E, Skrzypczak L, Ellnain-Wojtaszek M (1991) Quantitative determination of amarogentin by TLC densitometry. Sci Pharm 59: 135–138

Lamproye A, Crevecoeur M, Kevers C, Gaspar Th (1987) Multiplication vegetative in vitro de *Gentiana lutea* et de *Gentiana pneumonanthe.* Med Fac Landbuoww Rijkuniv Gent 52: 1255–1257

Luo JP, Lou ZC (1986) TLC-densitometry determination of bitter glycosides in Chinese drug Longdan, Radix Gentianae and its quality evaluation. Yaoxue Xuebao 21: 40–46; I Ph A (1987) 2411788

Maiwald L (1987) Phytotherapie bei Erkrankungen des Magen-Darm-Traktes. Z Phytother 8: 184–185

Menković N, Ralić J, Savin K, Ivanić R (1986) Prilog farmakognozijskom proucavanju nekoliko *Gentiana* – vrsta. Farm Glas 42: 127–135

Meravy L (1987) Phenolic substances in tissue cultures of *Centaurium erythraea.* Biol Plant 29: 81–87

Miura H (1991) *Swertia* spp.: in vitro culture, regeneration, and production of secondary metabolites. In: Bajaj YPS (ed) Biotechnology in agriculture and forestry, vol 15. Medicinal and aromatic plants III. Springer, Berlin Heidelberg New York, pp 451–463

Mpondo EM, Garcia J (1990) Two iridoid glucosides from *Gentiana verna.* Phytochemistry 29: 643–644

Mpondo EM, Garcia J, Chulia AJ (1990a) Secoiridoid glucosides from *Gentiana campestris.* Phytochemistry 29: 1687–1688

Mpondo EM, Garcia J, Cartier G, Pellet G (1990b) 6′-O-β-D-glucosylswertiamarine: a new secoiridoid from *Gentiana alpina.* Planta Med 56: 334

Murashige T, Skoog F (1962) A revised medium for rapid growth and bioassays with tobacco tissue cultures. Physiol Plant 15: 473–497

Nikolaeva GG, Glyzin WI, Mladencewą MS (1983) Xanthones of *Gentiana lutea.* Chim Prirod Soedin 1: 107–108 (in Russian)

Nikolaev SM, Bakuridze AD, Dargaevą TD, Brutko LI, Patudin AV, Schagzhievą GA, Nikolaeva GG (1987) Prospects of the utilisation of the overground part of *Gentiana lutea* L. for medicinal preparations. Izv Sib Otd Akad Nauk URSS Ser Biol Nauk 0(3): 67–70 (in Russian)

Nitsch JP, Nitsch C (1969) Haploid plants from pollen grains. Science 163: 85–87

Paun E, Mihalea A (1977) Probleme actuale si de perspectiva privint plantele medicinale si aromatice. Prod Vegetala Cereale Plante Technice, Bucuresti 29(7): 28–31

Penso G (1983) Index Plantarum Medicinalium Totius Mundi Eorumque Synonymorum. OEMF, Milano, pp 436–437

Pierik RLM (1989) In vitro culture of higher plants. Nijhoff, Dordrecht, pp 305–308

Procyk A (1982) Możliwości uprawy goryczki żółtej. Wiad Zielar 24(1–2): 7–8
Quercia V, Pierini N, Battaglino G (1980) Study of bitter constituents of the gentian root by high pressure liquid chromatography. Acta Hortic 96: 231–244
Rote Liste (1986) Verzeichnis von Fertigarznei-mitteln der Mitglieder des Bundesverbandes der Pharmazeutischen Industrie. Cantor, Aulendorf Württ
Sakamoto I, Morimoto K, Tanaka O, Inouye H (1983) Application of high performance liquid chromatography and field desorption mass spectroscopy to separative analysis of bitter secoiridoid glucosides of Swertiae Herba. Chem Pharm Bull 31: 25–30
Schaufelberger D, Hostettmann K (1985) Analytical and preparative reversed-phase liquid chromatography of secoiridoid glycosides. J Chromatogr 346: 396–400
Schaufelberger D, Domon B, Hostettmann K (1984) Desorption chemical ionization mass spectrometry of bitter glycosides from *Gentiana*. Planta Med 50: 398–403
Schultze J, Franz C (1980) Sugar, bitter substances and essential oils of *Gentiana lutea* L. In dependence of ecotype, plant age and stage of development. Acta Hortic 96: 311–315
Semeniuk P, Griesbach RJ (1987) In vitro propagation of prairie gentian. Plant Cell Tissue Organ Cult 8: 249–253
Skrzypczak L, Wesołowska M (1989) *Blackstonia perfoliata* (L.) Hudson i *Eustoma grandiflorum* Shinn. z kultur in vitro. XIV Naukowy Zjazd Polsk Towarzystwa Farm. Wrocław, Streszczenia, p 7/42; 48 Zjazd Polsk Towarzystwa Botan, Katowice, Streszczenia, p 129
Skrzypczak L, Wesołowska M, Thiem B (1988a) Callus culture and micropropagation of *Eustoma grandiflorum* Shinn. Acta Soc Bot Pol 57: 423–430
Skrzypczak L, Wesołowska M, Thiem B (1988b) Callus culture and micropropagation of *Eustoma grandiflorum* and *Blackstonia perfoliata*. 5th Int Conf Biology of cell cultures and biotechnology, Novosibirsk, Poster session
Sticher O, Meier B (1978) Identifizierung der Bitterstoffe von Radix Gentianae mit Hochleistungs-Flüssigkeitschromatographie (HPLC). Pharm Acta Helv 53: 40–45
Sticher O, Meier B (1980) Quantitative Bestimmung der Bitterstoffe in Wurzeln von *Gentiana lutea* und *Gentiana purpurea* mit HPLC. Planta Med 40: 55–67
Swiątek L, Dombrowicz E, Zadernowski R (1988) Zawartość fenolokwasów w korzeniach goryczki żółtej (*Gentiana lutea* L.). Herba Pol 34: 15–20
Sziszkin BK, Bobrow EG (eds) (1952) Flora URSS, vol 18. AN URSS, Moscow Leningrad, pp 538–620
Takhtajan Al (1980) Outline of the classification of flowering plants (Magnoliophyta). Bot Rev 46: 225–359
Tutin TG, Heywood VH, Burges NA, Moore DA, Valentine DH, Walters SM, Webb DA (1972) Flora Europaea, vol 3. University Press, Cambridge, pp 59–63
Van der Sluis WG, Labadie RP (1981) Fungitoxic activity of the secoiridoid glucoside gentiopicrin (gentiopicroside). Planta Med 42: 139–140
Van der Nat J, van der Sluis WG, Labadie RP (1982) Gentiogenal, a new antimicrobial iridoid derived from gentiopicrin (gentiopicroside). Planta Med 45: 161–162
Van der Sluis WG, Van der Nat JM, Spik AL, Tkeshiro Y, Labadie RP (1983) Gentiogenal, a conversion product of gentiopicrin (gentiopicroside). Planta Med 49: 211–215
Vanhaelen M, Vanhaelen-Fastre R (1983) Quantitative determination of biologically active constituents in medicinal plant crude extracts by thin-layer chromatography densitometry. J Chromatogr 281: 263–271
Verney AM, Debelmas AM (1973) Sur diverses xanthones de *Gentiana lutea* L., *G. purpurea* L., *G. punctata* L., *G. pannonica* Scop. Ann Pharm Fr 31: 415–420
Verotta L (1985) Isolation and HPLC determination of the active principles of *Rosmarinus officinalis* and *Gentiana lutea*. Fitoterapia 56: 25–29
Viola U, Franz C (1989) In vitro propagation of *Gentiana lutea*. 37th Annu Congr on Medicinal plant research, Abstracts of short lectures and poster presentations, Braunschweig, p 115
Wagner H, Vasirian K (1974) Zur Chromatographie der Enzianbitterstoffe. Dtsch Apoth Ztg 114: 1245–1248
Wagner H, Münzing-Vasirian K (1975) Eine chemische Wertbestimmung der Enziandroge. Dtsch Apoth Ztg 115: 1233–1239
Wagner H, Bladt S, Zgainski EM (1984) Plant drug analysis. Springer, Berlin Heidelberg New York, pp 125–136
Weiss RF (1980) Lehrbuch der Phytotherapie. Hippocrates, Stuttgart, pp 54–57

Weiss RF (1988) Phytotherapie bei Magenbeschwerden, Aerztezeitschr Naturheilverf 29: 429–436
Wenzel H (1972) Seltene Pflanzen in der Drogenkammer. Alkohol Ind 85: 24–25
Wesołowska M, Skrzypczak L, Dudzińska R (1985) Rodzaj *Gentiana* L. w kulturze in vitro. Acta Pol Pharm 42: 79–83
Wichtl M (1978) Drogenanalytik und Arzneibuch kritisch betrachtet. Planta Med 34: 113–128
Zenkteler M (ed) (1984) Hodowla komórek i tkanek roślinnych. PWN, Warszawa, p135
Zenkteler M, Zenkteler E (1987) Rozmnażanie *Eustoma grandiflorum* za pomocą kultur in vitro. Ogrodnictwo 24: 19–21
Zimmermann, W, Gaisbauer G, Gaisbauer M (1986) Wirkung von Bitterstoff-Drogen auf das darmassozüerte Immunsystem. Z Phytother 7: 56–64

XIII *Gypsophila paniculata* L. (Baby's Breath): In Vitro Culture and the Production of Gypsogenin Saponins

M. Henry[1]

1 General Account

Gypsophila paniculata saponins have been used for a long time under the generic name of saponin. They have the typical properties of saponins, such as detergent, emulsive, hemolytic, and membrane-toxic substances. Research has been carried out in various biological fields, for example in studies on virus (*Rous sarcoma* virus), on cell membranes (of chicken liver and erythrocyte ghosts) (Dourmashkin et al. 1962), in the preparation of vaccines to enhance the immune response (Freund's adjuvent), or for use as a commercial product, known as Saponin pure white (Merck). This product has also been widely employed as a standard for hemolytic tests in most saponin determinations, and was previously reported to be extracted from roots and rhizomes of *G. paniculata* (Kolodziejski and Stecka 1965). These *G. paniculata* saponins, as well as digitonin, have been studied for their water insoluble complexes with cholesterol (Gögelein and Hüby 1984).

1.1 Distribution and Importance of the Genus *Gypsophila*

The genus *Gypsophila* (from the Greek gypsos: gypsum, calc, and philos: friend) contains 125 species native to the temperate regions of Eurasia (21 species in Europe), living on old walls or on gypseous ground (Fournier 1977). A few of them are encountered in Egypt, Australia, and one species in New Zealand, probably introduced (Mabberley 1989). Among these various species, only a few are of commercial interest, for example, *G. elegans* M. Bieb. (found from Ukraine to Iran) and *G. paniculata* (baby's breath), found from Central Europe to Central Asia are the species most often planted for their medicinal and ornamental interest. The latter species is used for wedding bouquets (Mabberley 1989). *G. patrini* Ser. is used as a copper indicator in Russia, *G. struthium* Loefl. as soap in the Mediterranean Basin and *G. rokejeka* Del. (found in Egypt and the eastern Mediterranean Basin) in halva with sesame seeds and honey (Mabberley 1989). This chapter describes work on *G. paniculata* L.

G. paniculata is well known to contain saponins with various applications of commercial interest. They are used for the composition of photosensitive surfaces

[1] Laboratoire de Botanique, Faculté des Sciences Pharmaceutiques, 31, Allées Jules Guesde, 31000 Toulouse, France

Biotechnology in Agriculture and Forestry, Vol. 21
Medicinal and Aromatic Plants IV (ed. by Y.P.S. Bajaj)

in photography, and of shampoos (Tschesche and Wulff 1973). More recently, these saponins were described as being apparently responsible for spermicidal and antiinflammatory effects (El Bary and Nour 1979; Primorac et al. 1985).

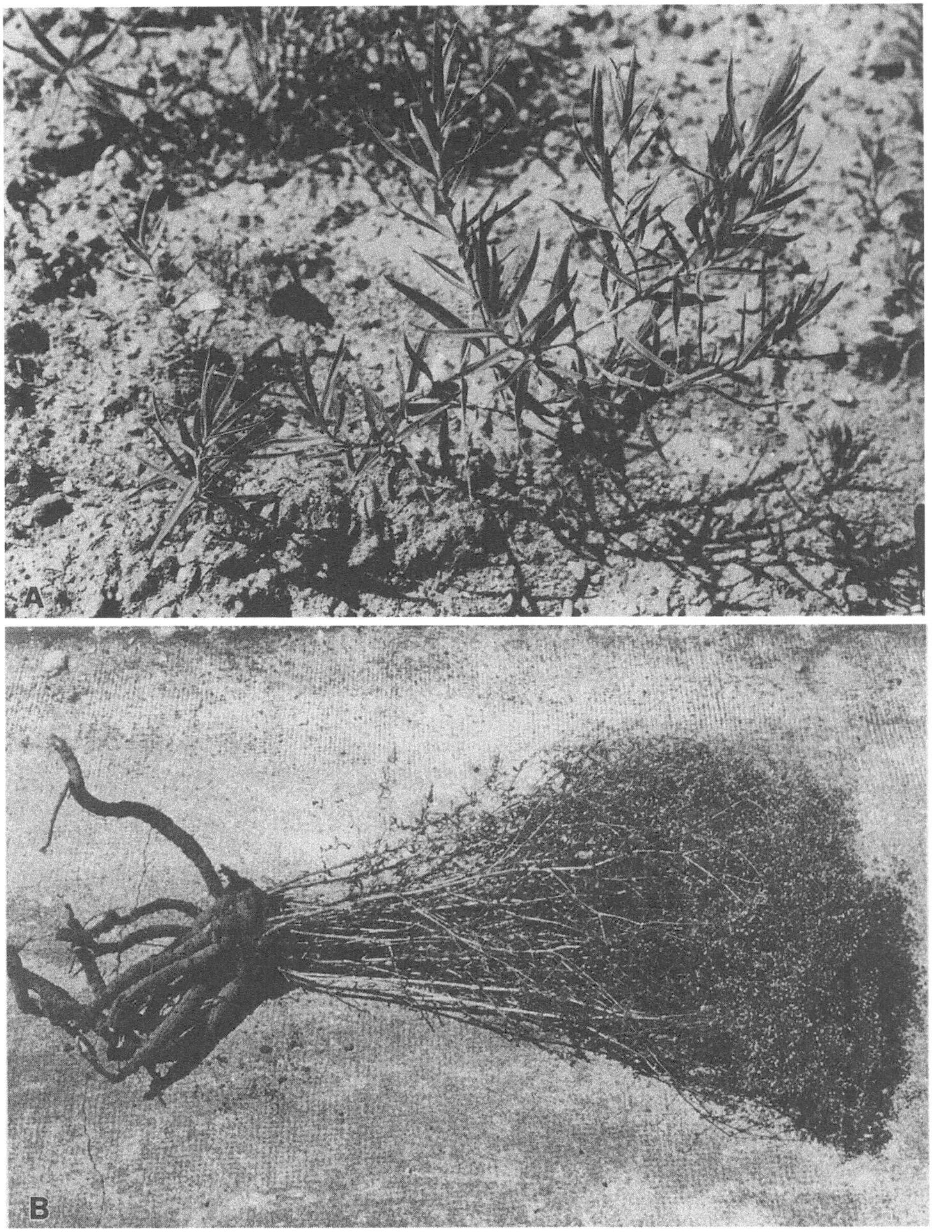

Fig. 1A,B. Plants of *G. paniculata* 6 months (**A**) and 4 years old (**B**) (1.5 m total length) in the Botanical Garden of the Faculty of Sciences of Toulouse

1.2 The Plant: *G. paniculata*

G. paniculata L. (baby's breath) (Fig. 1) is a perennial plant belonging to the family Caryophyllaceae. It is a herb which can reach 1.50 m in height, the most important part being underground. The roots are ramified with long, thick branches. The upper shoot is erect, with leaves reduced to an enlarged and flattened petiole. The elongate bipartite cymes contain many small (2–4 mm), closely gathered, white or pink flowers, which are composed of a short calyx (2 mm) split almost to the bottom. Flowering occurs in July and August (Fournier 1977).

1.3 Saponins of *G. paniculata*

Gypsogenin (Fig. 2) is the main pentacyclic triterpenoidal aglycone of these saponins and was described many years ago (Rûzicka and Giacomello 1936). Substituted on the OH 3 and on the COOH 28 by two chains of different glycosides, these saponins, called bidesmosidic, are among the highest glycosylated. A saponin called gypsoside (Fig. 3) was first extracted from *G. pacifica* (Kochetkov et al. 1963), and found next in *G. paniculata* (Khorlin et al. 1963). This saponin, containing nine glycosides, appears to be the same as the "gypsophila saponin" first described by Van der Haar (1927). However, these highly glycosated saponins break easily during extraction and purification and give shorter saponins. These

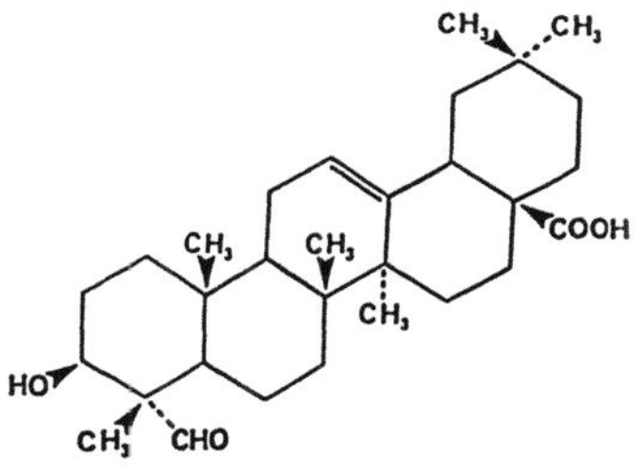

Fig. 2. Structure of gypsogenin

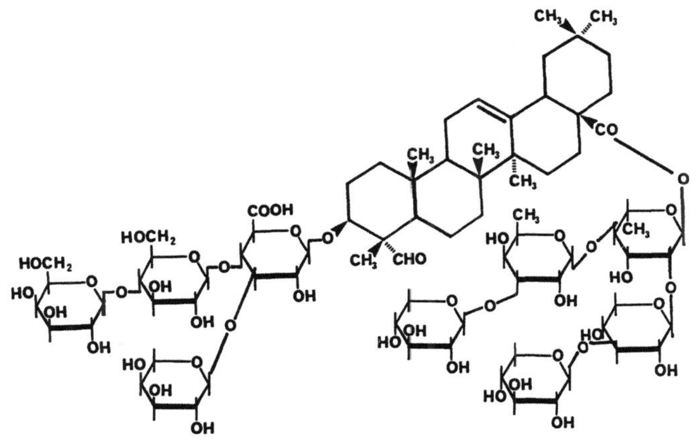

Fig. 3. Structure of gypsoside

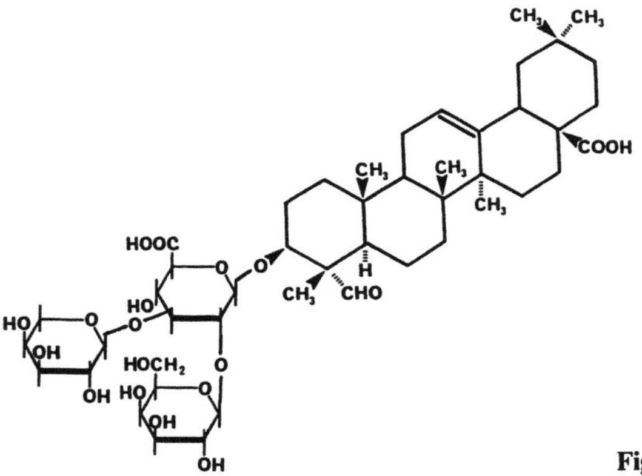

Fig. 4. Structure of gypsogenin 3,O-glycoside

previous structural determinations of *Gypsophila* saponins must be studied carefully, because the difficulties of these first chemical studies do not detract from the great merit of these first saponin chemists.

Recent studies, made with new techniques, have shown the great difficulties involved in the identification of these compounds. For example, Higuchi et al. (1985) discovered the presence of a gypsogenin 3-O-glycoside, [28-hydroxy-23,28-dioxoolean-12-en-3β-yl O-β-D-galactopyranosyl(1 → 2)-α-L-arabinopyranosyl-(1 → 3)-β-D-gluco-pyranosid]luronic acid (Fig. 4) in the commercial product Saponin pure white (Merck). In a systematic approach, they studied the stability of this compound in different chemical conditions of molecule treatment, for example with diazomethane methylation, a very mild method currently employed for saponin structure determination and used to make the molecule apolar (Higuchi et al. 1988). This treatment led to degradation of the sugar-aglycone linkage, proceeding through an epoxide intermediate where the 4α-aldehyde group was shown to be essential. Examined with high-performance liquid chromatography (HPLC) (Mansouri 1989), Saponin pure white (Merck) showed at least nine saponins, none of which is more abundant than the other.

1.4 Quantitative Determination of Saponins in *G. paniculata*

The routine quantitative determination of these saponins is difficult and time-consuming. Tagiev and Ismailov (1986) described a thin layer chromatogrpahic (TLC) method of quantitative determination of a *Gypsophila* saponin in roots of *G. bicolor* after extraction, purification, and measurement of the spot surface, made visible by phosphotungstic acid. Although this method is interesting for a rapid evaluation. it seems to us very imprecise.

The aglycone, gypsogenin, could not be used to quantify the *G. paniculata* saponins because of the weakness of its structure in the acid hydrolysis condition, in spite of HPLC quantitative techniques (see, for example, Lee et al. 1987). Recently, we reported (Henry et al. 1989) an efficient method of determination of *Gypsophila* sp. saponins, by means of their prosapogenin, gypsogenin 3,O-glucuronide [I] (Fig. 5), which is much more stable than gypsogenin. This greater

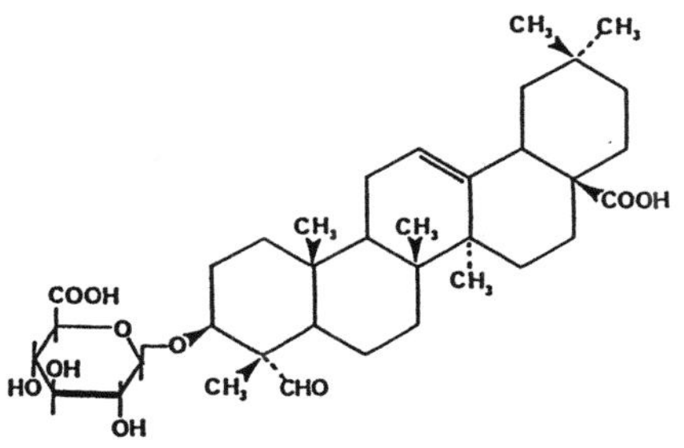

Fig. 5. Structure of gypsogenin 3,O-glucuronide

stability may be explained by a reversible formation of a hemiacetal grouping by the glucuronic acid carboxyl group and the genin 4α-aldehyde group (Kochetkov et al. 1963). [I] is present in most of the saponins isolated and identified from the genus *Gypsophila* (Hiller and Voigt 1977; Adler and Hiller 1985).

In the olean-12-ene group, two 3,O-glucuronides of hederagenin and oleanolic acid obtained from *Lonicera nigra* L. saponins (Domon et al. 1984) eluted with relatively sharp peaks with a classical reversed phase (RP-8) HPLC column. Therefore, the efficiency of the RP-8 column was good enough to perform HPLC separation of these glucuronides under suitable conditions. However, [I] presents quite a different behavior, as was previously observed (Burnouf-Radosevich and Delfel 1986), and with which we were also confronted: [I] under the above classical conditions showed an extensive band broadening with loss of product, which remained attached to the stationary reversed phase column. The only chemical difference between [I] and the two glucuronides of hederagenin and oleanolic acid is the presence of the 4α aldehyde group, which replaces the corresponding CH_2OH or CH_3 groups present in hederagenin and in oleanolic acid, respectively. Thus an attentive study of HPLC efficiency under various conditions of the mobile phase and the stationary reversed phase, combined with a change in the polarity of the molecule of [I], was necessary to carry out a quantitative HPLC. Two methods were used to quantify [I], a reversed phase ion-pair chromatographic method with tetra-n-butylammonium salt (TBA) as counter ion. With a methanol-water solution (75:25) containing 5 mM TBA and adjusted to pH 5.0 with H_3PO_4 0.5 M. [1] (2 mg/ml) was eluted at time appropriated for analysis (13 min with a capacity factor of 5.6) and with a relatively sharp peak and good efficiency (theoretical plate number/meter 22400) with an RP18 end-capped column (5 μm). The HPLC chromatogram of a typical determination is illustrated in Fig. 6.

Generally, the influence of the alkyl chain on retention and efficiency is secondary to the much greater influence of the composition of the mobile phase (Berendesen and De Galan 1980). In this case, however, the influence of the alkyl chain is greater than for other solutes. Under optimal conditions for mobile phase composition, different reversed phases were tested and it was shown that the use of RP-18 increased efficiency by a factor of 7, especially when the remaining silanol groups are well protected by end-capping. But end-capping was not sufficient to give the best efficiency. The state of polarity of the solute is supported by the use of TBA, adjusting the pH with H_3PO_4. With all other conditions identical, the efficiency could be doubled when the pH was adjusted to 5, [I] giving the pairing reagent.

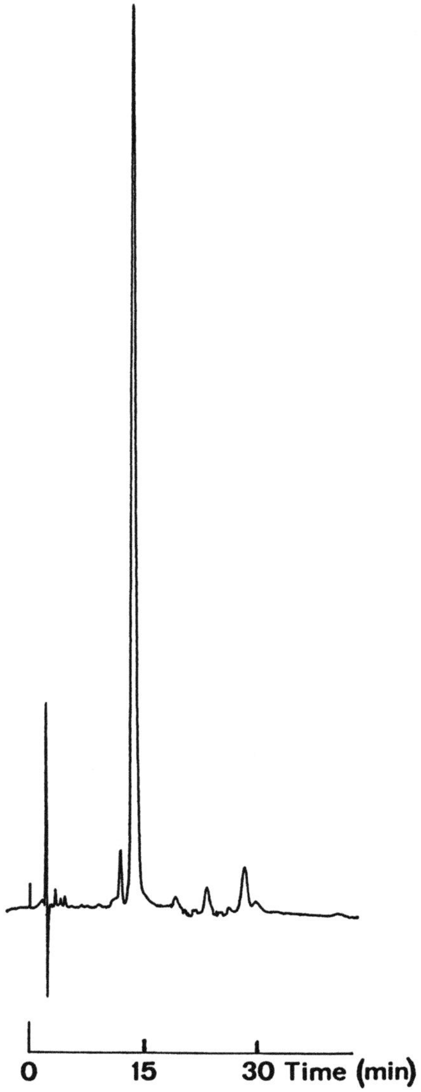

Fig. 6. Reverse phase HPLC chromatogram of a typical determination of [1] (2 mg/ml). UV detection at 206 nm. Sample size: 20 µl. Column: C_{18} (Merck, 5-µm LiChrospher 100 RP-18e, 250 × 4 nm I.D.). Isocratic elution with the following mobile phase: $MeOH/H_2O$ 75/25 with TBA 5 mM, pH adjusted with H_3PO_4 0.5 M. Flow rate: 1 ml/min. (Henry et al. 1989)

An ion-suppression method was also described (Burnouf-Radosevich and Delfel 1986) for analysis of [I], to solve this difficulty, and was apparently efficient, but no values were given of the plate numbers. They used a C-18 column with a 25-min water-acetonitrile gradient in the presence of trifluoroacetic acid (TFA). As TFA has low UV absorbance, detection of [I] could be monitored at 210 nm. This ion-suppression method completes the results presented above with the ion-pair method and also gives a good quantitative efficiency (unpubl.).

In conclusion, contrarily to other glucuronic acid-containing saponins, [I] requires a very apolar support of reversed phase with good end-capping and with apolar protection of the molecule by a pairing reagent or transformation by acidification of the molecule to be eluted under HPLC conditions without band broadening and with a symmetrical peak.

1.5 Determination of [I] in Saponin pure white (Merck)

[I] is a new product that we obtained from Saponin pure white (Merck) by the above method. Recently, Braut-Boucher et al. (1990) showed a cytoprotective effect of [I] towards isolated rat hepatocytes when they were submitted to the toxic effects of galactosamine, CCl_4, or erythromycine. [I] is at present commercialized by Extrasynthese (Lyon, France) as a fine chemical.

The content of [I] in Saponin pure white (Merck) and the reproducibility of the method was measured under the above defined HPLC conditions on increasing concentrations of Saponin pure white (Merck) from 0.5 mg/ml to 30 mg/ml, submitted to subsequent hydrolysis and extraction conditions: boiling with refluxing at 100 °C of a maximum of 20 mg/ml of the Merck saponin mixture in 0.57 M sulphuric acid for at least 1 h. Under these conditions, aglycone (gypsogenin) or byproducts (gypsogenin lactone or degradation products) were not detected. The results showed good correspondence between [I] content versus Saponin pure white (Merck) concentration from 1.5 to 15 mg/ml of Saponin pure white (Merck), with an average of 14.3% of [I] with a coefficient variation of 2.15%. In comparison, roots and rhizomes of *G. paniculata* contain 0.86 to 3.91% of [I] on a dry weight basis (Henry et al. 1990).

2 In Vitro Culture of *G. paniculata*

Examples of in vitro production of triterpenoids, free or bound to sugars, are rare in saponins. The time-consuming and tedious quantitation of these saponins has limited production studies with in vitro cultures. The extraction of *Gypsophila* saponins and, after hydrolysis, their quantitation by HPLC of [I] enabled us to quantify the production of *Gypsophila* saponins with in vitro cultures. Four species of *Gypsophila*: *G. paniculata* L., *G. repens* L., *G. muralis* L., and *G. petraea* (Baumg.) Reichenb., were chosen to compare their ability to produce [I] in vitro cultures.

2.1 Gypsogenin 3,O-Glucuronide Content of Various Types of In Vitro Cultures of Four Species of *Gypsophila*

In June 1986, callus initiation on modified Murashige and Skoog's medium (1962) (called MSA) with four levels of 2,4-D, were performed from the three parts (cotyledon, hypocotyl, and root) of seedlings of four species of *Gypsophila*: *G. paniculata*, *G. repens*, *G. muralis*, and *G. petraea*, germinated under sterile conditions on Knop medium (Pauthe-Dayde 1990; Pauthe-Dayde et al. 1990). Explants of *G. paniculata*, *G. repens*, and *G. petraea* seedlings remained viable on these media except for the cotyledon explant from *G. paniculata*. Root and cotyledon explants

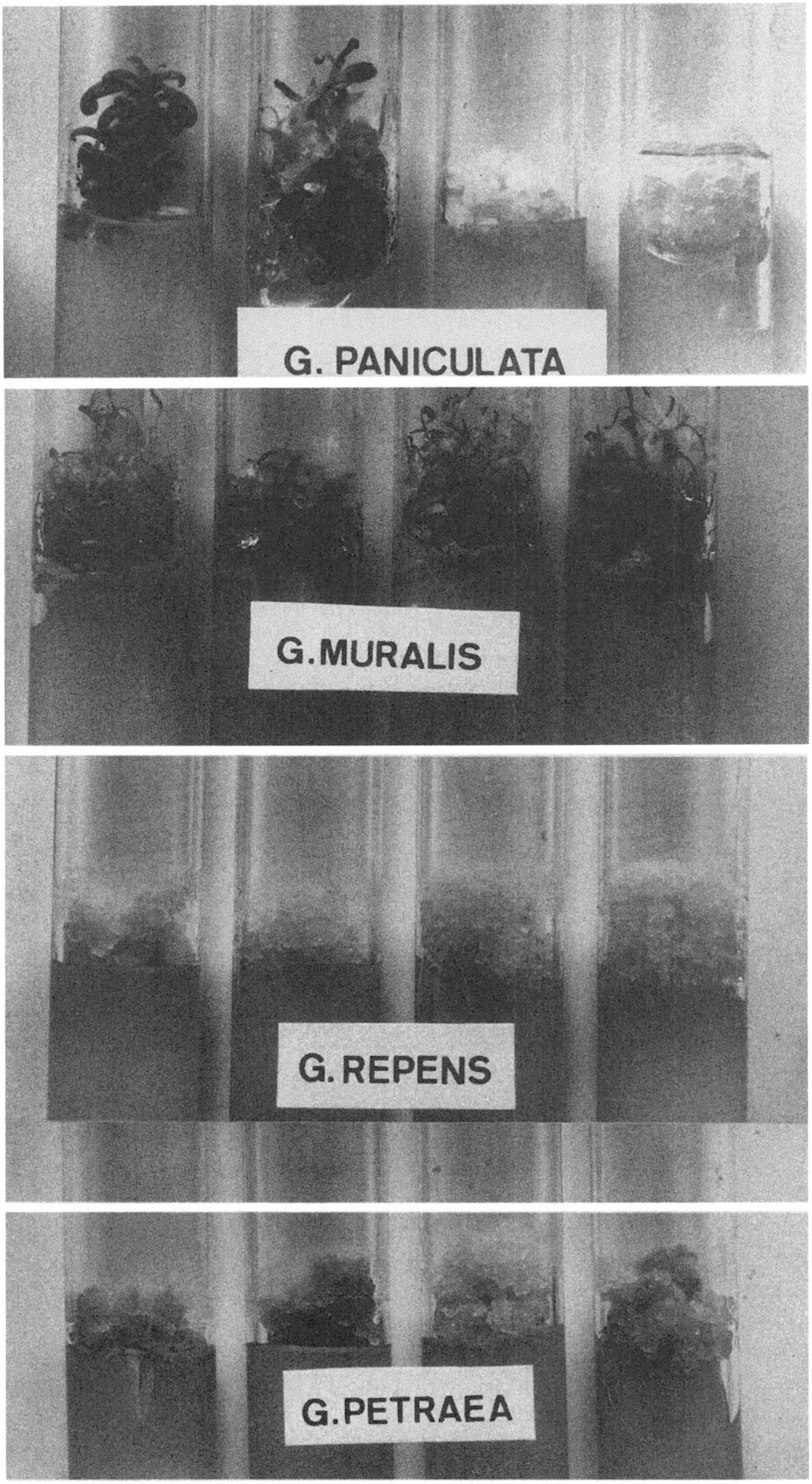

Fig. 7A.

Fig. 7B.

Fig. 7.A. Examples of different callus lines and multiple shoot cultures of the four *Gypsophila* species investigated, photographed at the 15th subculture, 2 weeks after the day of subculture. *G. paniculata*: from *left to right*, two multiple shoot cultures obtained from a hypocotyl explant of respectively plant 5 cultivated on MSB medium and plant 3 on MSA, and two callus cultures obtained from root explants respectively of plant 2 on MSA and plant 1 on B_5. *G. muralis*: from *left to right* four multiple shoot cultures obtained from hypocotyl explants of respectively plant 5 on MSB (*left*) and MSA (*right*) plant 3 on MSB and plant 2 on MSA. *G. repens*: from *left to right* four callus lines obtained from root (*left*) cotyledon (the two in the *center*) and hypocotyl (*right*) of plant 2 on MSA (the two on the *left*) and plant 3 on MSA and plant 1 on B_5 (respectively on the *right*). *G. petraea*: from *left to right* two chlorophyllous callus lines respectively obtained from cotyledon and hypocotyl of plant 1 on MSA medium and two nonchlorophyllous callus lines obtained from cotyledon of plant 3 on MSA and from root explant of plant 2 on MSA respectively. **B.** Multiple shoot culture of *G. paniculata*

of *G. muralis* died, and 40% of the hypocotyl explants gave rise to undifferentiated calli or multiple shoots. Each of these different calli and multiple shoots were transferred onto the same solid medium or onto a further modified Murashige and Skoog's medium (called MSB) or onto B_5 Gamborg et al.'s medium (1968) and then subcultured for 4 years on these media (Fig. 7A). Among the different cultures, those exhibiting fast growth were assayed for [I] content after about five subcultures (3 months after initiation), after 14 subcultures (1 year after initiation), when the cultures seemed to have become stable, and after about 50 subcultures (4 years after initiation). The results are reported in Tables 1 to 4. The behavior of each species cultured under the same conditions was quite different. *G. paniculata* cultures showed the highest [I] contents after 14 subcultures whatever the medium on which they were cultivated. Hypocotyl explants gave rise to multiple shoot cultures with some green calli unlike undifferentiated calli or those obtained with roots, which were yellow-colored. Multiple shoot cultures, being more stable and growing faster than yellow calli, were then chosen for the kinetic studies. The culture obtained with plant no. 5 growing on MSB produced [I] in a good yield at the 14th subculture (1.5 mg/g dry weight), as also the callus culture obtained from plant no. 1 (1.3 mg/g dry weight), and was therefore retained at this date for kinetic

Table 1. Gypsogenin 3,O-glucuronide contents of *G paniculata* in in vitro cultures from different origins (seed, explant) and nature of the subculture medium after 3 months[a] (four to six subcultures), 1 year[a] (14 to 15 subcultures), and 4 years of subculture (about 50 to 70 subcultures)

Plant No.	Explant origin	Subculture		Saponin yield mg/g dry weight
		Medium	No.	
1	Root	B_5	4[b]	1.2
			14[b]	1.3
			49[b]	5.7
2	Root	MSA	5[b]	0.1
			14[b]	0.8
			54[b]	0.1
3	Hypocotyl	MSA	5[c]	0.0
			15[c]	0.6
			55[c]	0.9
3	Hypocotyl	MSB	6[c]	0.3
			14[c]	0.9
			66[c]	1.0
4	Hypocotyl	MSA	5[c]	0.1
			14[c]	0.6
			57[c]	0.6
4	Hypocotyl	MSB	6[c]	0.1
			14[c]	0.6
			67[c]	1.5
5	Hypocotyl	MSB	6[c]	0.3
			14[c]	1.5
5	Hypocotyl	MSB	6[c]	0.2
			14[c]	0.9
			64[c]	0.0

[a] Results previously published in Pauthe-Dayde et al. (1990).
[b] Undifferentiated calli yellow-colored.
[c] Multiple shoot culture with green calli.

experiments. Since then, the callus culture obtained from plant 1 showed much higher [I] production and recently (49th subculture) reached 5.7 mg/g dry weight. The other three species had lower [I] contents, such as *G. muralis* (except for two callus lines which also reached 0.75 and 1.22 mg/g dry weight recently) and *G. petraea*, or they failed to produce [1] after 4 years of subculture (*G. repens*). Only *G. muralis* produced rooted plantlets on MSB (with 0.01 mg/l ANA), and *G. repens* gave rise to unorganized yellow calli whatever the medium. Explants of *G. petraea* gave in vitro cultures only on MSA medium; calli were greenish and covered with small hairy roots during the first subcultures. All these in vitro cultures gave different qualitative HPLC fingerprints of some unidentified compounds in the conditions used to quantify [I] content.

Table 2. Gypsogenin 3,O-glucuronide contents of *G. repens* in in vitro cultures from different origins (seed, explant) and nature of the subculture medium after 3 months[a] (four subcultures), 1 year[a] (14 to 17 subcultures), and 4 years of subculture (about 45 to 60 subcultures).

Plant No.	Explant origin	Subculture		Saponin yield mg/g dry weight
		Medium	No.	
1	Hypocotyl	B_5	15[b]	0.07
2	Root	MSA	4[b]	0.30
			14[b]	0.00
			48[b]	0.17
2	Hypocotyl	MSA	4[b]	0.00
			15[b]	0.00
			58[b]	0.17
2	Cotyledon	MSA	4[b]	0.54
			14[b]	0.00
			54[b]	0.00
3	Cotyledon	MSA	4[b]	0.40
			14[b]	0.12
			45[b]	0.12
3	Cotyledon	B_5	4[b]	0.16
			17[b]	0.03

[a] Results previously published in Pauthe-Dayde et al. (1990).
[b] Undifferentiated calli yellow-colored.

2.2 Growth and [I] Production by *G. paniculata* Multiple Shoot Cultures (Fig. 7B)

Kinetic results on *G. paniculata* cultures on solid medium (Rochd 1990; Pauthe-Dayde et al. 1990) are shown in Fig. 8. A maximal growth rate of 0.0074/h at 25 °C (135 h doubling time) was observed. The stationary phase was reached after 25 days. The dry weight reached 14.3 g/l of MSB solid medium. [I] production increased during the growth period and about 1.5 mg/g of dry material were produced in the stationary phase. As shown in Fig. 8, [I] decreased slightly to reach 1.1 mg/g of dry weight during the exponential phase. Moreover, [I] was not detected in the medium and did not seem to diffuse out of the cells.

When these mutiple shoots were put in the same, but liquid medium and subcultured every 3 weeks, [I] content increased so that after 6 months the first subcultures reached 5.30 $\pm$ 25 mg/g dry weight (determinations carried out at the subculture date during eight following subcultures: 3 months). After the 3 subsequent years, the multiple shoot suspension grew faster. It needed to be subcultured more frequently, every 2 weeks during the last 6 months, and the [I] content decreased to 0.32 mg/g dry weight, all conditions being identical from the beginning of the subcultures. The appearance of the multiple shoot suspension (Fig. 9) did not seem to be changed and, at any moment, normal plantlet regeneration could be obtained by transfer first on the same, but solid medium and, second on the same medium without growth factors.

Table 3. Gypsogenin 3,O-glucuronide contents of *G. muralis* in in vitro cultures from different origins (seed, explant) and nature of the subculture medium after 3 months[a] (two to five subcultures), 1 year[a] (14 to 18 subcultures), and 4 years of subculture (about 40 to 55 subcultures)

Plant No.	Explant origin	Subculture		Saponin yield mg/g dry weight
		Medium	No.	
1	Hypocotyl	MSA	3[b]	0.03
			15[b]	0.33
2	Hypocotyl	MSA	3[b]	0.04
			15[b]	0.72
			42[b]	1.22
3	Hypocotyl	MSB	3[c]	0.00
			14[c]	0.15
			53[c]	0.75
4	Hypocotyl	MSA	5[b]	0.00
			14[b]	0.44
4	Hypocotyl	MSB	5[c]	0.18
			18[c]	0.35
			50[c]	0.02
5	Hypocotyl	MSA	5[b]	0.14
			14[b]	0.61
			43[b]	0.52
5	Hypocotyl	MSB	3[c]	0.06
			15[c]	0.32

[a] Results previously published in Pauthe-Dayde et al. (1990).
[b] Multiple shoot culture with yellow calli.
[c] Rooted plantlets.

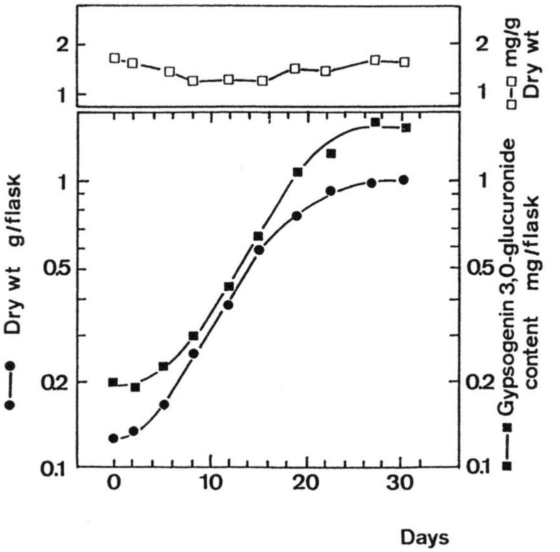

Fig. 8. Culture growth and gypsogenin 3,O-glucuronide production. The contents of four to ten 250-ml (70 ml MSB solid medium) flasks were harvested on the days indicated. The experiment was carried out in duplicate. The dry weight and gypsogenin 3,O-glucuronide content were determined as in *Protocol*. (Pauthe-Dayde et al. 1990)

Table 4. Gypsogenin 3,O-glucuronide contents of *G. petraea* in in vitro cultures from different origins (seed, explant) and nature of the subculture medium after 3 months[a] (four to six subcultures), 1 year[a] (14 to 18 subcultures), and 4 years of subculture (about 40 to 55 subcultures)

Plant No.	Explant origin	Subculture		Saponin yield mg/g dry weight
		Medium	No.	
1	Root	MSA	4[b]	0.07
			14[c]	0.56
			43[c]	0.12
1	Hypocotyl	MSA	4[b]	0.36
			14[c]	0.38
1	Cotyledon	MSA	6[b]	0.00
			15[c]	0.00
2	Root	MSA	4[b]	0.18
			18[c]	0.03
2	Hypocotyl	MSA	4[b]	0.40
			15[c]	0.00
			48[c]	0.14
3	Cotyledon	MSA	4[b]	0.84
			14[c]	0.02
			54[c]	0.15

[a] Results previously published in Pauthe-Dayde et al. (1990).
[b] Hairy-rooted green calli.
[c] Green-yellowish calli.

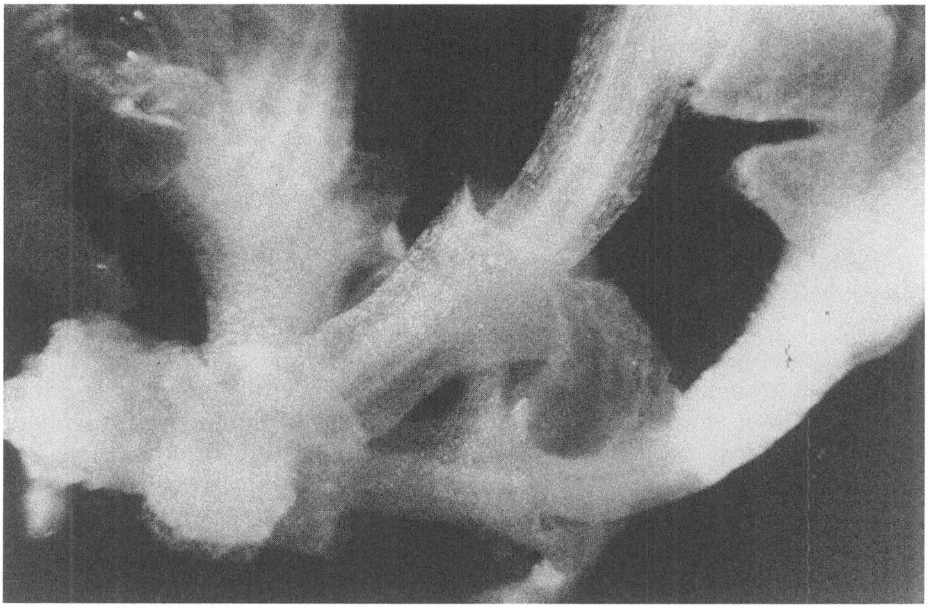

Fig. 9. Multiple shoots of *G. paniculata* obtained in MSB liquid medium (×15)

Table 5. Labeling of gypsogenin 3,O-glucuronide, spinasterol, and spinasteryl glucoside in multiple shoot cultures of *G. paniculata* at different times after feeding with sodium [2-^{14}C]acetate (5.4 μCi, sp. act. 48 mCi/mmol) in each flask of liquid culture (one flask per time interval). (Pauthe-Dayde et al. 1990)

	Gypsogenin 3,O-glucuronide			Spinasterol		Spinasteryl glucoside	
	dpm	%	sp. act. mCi/mmol	dpm	%	dpm	%
Start	2778	0.026	–	485	0.004	381	0.003
30 min	39 880	0.374	0.008	3265	0.027	3835	0.032
1 h	76 195	0.714	0.017	5715	0.047	7587	0.062
2 h	96 335	0.866	0.018	13 038	0.108	13 680	0.114
8 h	126 712	1.140	0.022	31 611	0.260	44 755	0.400
72 h	111 590	1.000	0.023	50 457	0.420	49 855	0.450

2.3 Study on the Turnover of [I] and Comparison with Sterol Biosynthesis

The incorporation of sodium [2-^{14}C]acetate into [1] and into the sterols synthesized by *G. paniculata* was tested (Pauthe-Dayde et al. 1990; Rochd 1990). The main sterols of *G. paniculata* were identified as spinasterol and spinasteryl-glucoside (no other steryl derivatives were detected), as we had previously described in cell suspension cultures of *Saponaria officinalis* (Henry and Chantalat-Dublanche 1985). Time course studies of the incorporation of sodium [2-^{14}C]acetate (Table 5) showed that these compounds were synthesized actively in vitro and that the radioactivity accumulated in the three compounds up to 2 h, and then remained roughly constant until the end of the experiment. The main sterols (spinasterol plus spinasteryl-glucoside) and [1] were labeled to the same extent by 72 h. The distribution of radioactivity between spinasterol and spinasteryl glucoside showed that there was rapid glucosylation of spinasterol. This experiment shows the equal importance of the biosynthetic pathways of these triterpenes in this species, as previously described in *S. officinalis* (Henry and Guignard 1982; Henry and Chantalat-Dublanche 1985).

2.4 Effect of Exogenous Gypsogenin 3,O-Glucuronide on the Biosynthesis of Phytosterols and Endogenous Saponins

Addition of exogenous [I] (16 mg/l) to the in vitro multiple shoot culture (Table 6) at least 8 h before incorporation of [2-^{14}C]acetate, under the same conditions as described above, showed a tenfold inhibition of the saponin biosynthesis without visible decrease of the phytosterol biosynthesis (Rochd 1990). This effect was correlated with a threefold accumulation of squalene. No effect was observed when [I] was added simultaneously with [2-^{14}C]acetate. Eighty percent of [I] penetrated into the cells within 8 h (results obtained with [^{14}C]- labeled [I] and not shown here). Thus, biosynthesis of *G. paniculata* triterpenoidal saponins involves a regulation step located at, or after epoxysqualene β-amyrin cyclase reaction, the first step in the specific triterpenoid saponin branch of their biosynthetic pathway

Table 6. Labeling of gypsogenin 3,O-glucuronide, spinasterol, spinasteryl glucoside, and squalene in multiple shoot cultures of *G. paniculata* at different times after feeding with sodium [2-^{14}C]acetate (5.4 μCi; 48 mCi/mmol) in each flask of liquid culture (one flask per interval). Some of these batch cultures were preincubated with gypsogenin 3,O-glucuronide (16 mg/l) before feeding

Time (h)		Controls	Time of preincubation with gypsogenin 3,O-glucuronide before feeding: 8 h	24 h
Gypsogenin 3,O-glucuronide				
0	dpm	2778	2356	1027
	%	0.026	0.02	0.01
	sp. act. mCi/mmol	–	–	7.15×10^{-3}
0.5	dpm	39880	18047	6043
	%	0.374	0.158	0.06
	sp. act. mCi/mmol	3.34×10^{-3}	–	5.6×10^{-4}
1	dpm	76195	22188	17245
	%	0.714	0.194	0.174
	sp. act. mCi/mmol	5.65×10^{-3}	–	1.17×10^{-3}
2	dpm	139198	43870	14775
	%	1.305	0.385	0.15
	sp. act. mCi/mmol	1.02×10^{-2}	–	1.06×10^{-3}
Spinastéryl-glucoside				
2	dpm	13680	12391	10554
	%	0.114	0.103	0.088

Spinasterol		Controls	0 h	8 h	24 h
2	dpm	10653	9796	6956	8510
	%	0.065	0.060	0.042	0.052
Squalene					
2	dpm	939	1260	1847	2762
	%	5.77×10^{-3}	7.75×10^{-3}	$11.3 \times < 10^{-3}$	17×10^{-3}

(Fig. 10). The independence of the regulation of this pathway and, particularly at its beginning, of this enzyme, was first pointed out by Baisted (1971) in a study on the β-amyrin synthesis in plantlets of pea, where he demonstrated its correlation with plant development.

Recently, a treatment of cell suspension cultures of *Tabernaemontana divaricata* during the growth phase with a *Candida albicans* elicitor preparation both inhibited monoterpene indole alkaloid production within a few hours and caused a rapid accumulation of at least ten pentacyclic triterpenoid phytoalexins (Van der Heijden et al. 1988, 1989). Experiments with cell-free extracts of elicited cells catalyzed the synthesis of squalene and squalene 2,3-oxide and triterpenols (α- and β-amyrin), ursolic and oleanolic acid when they were incubated with either [1-^{14}C]isopentenyl pyrophosphate or [1-$^{3}H_2$] farnesyl pyrophosphate as substrate. Growth inhibition seemed to be due to the inhibition of sterol biosynthesis in order to initiate the biosynthesis of elicited products. Preparations from unelicited cells

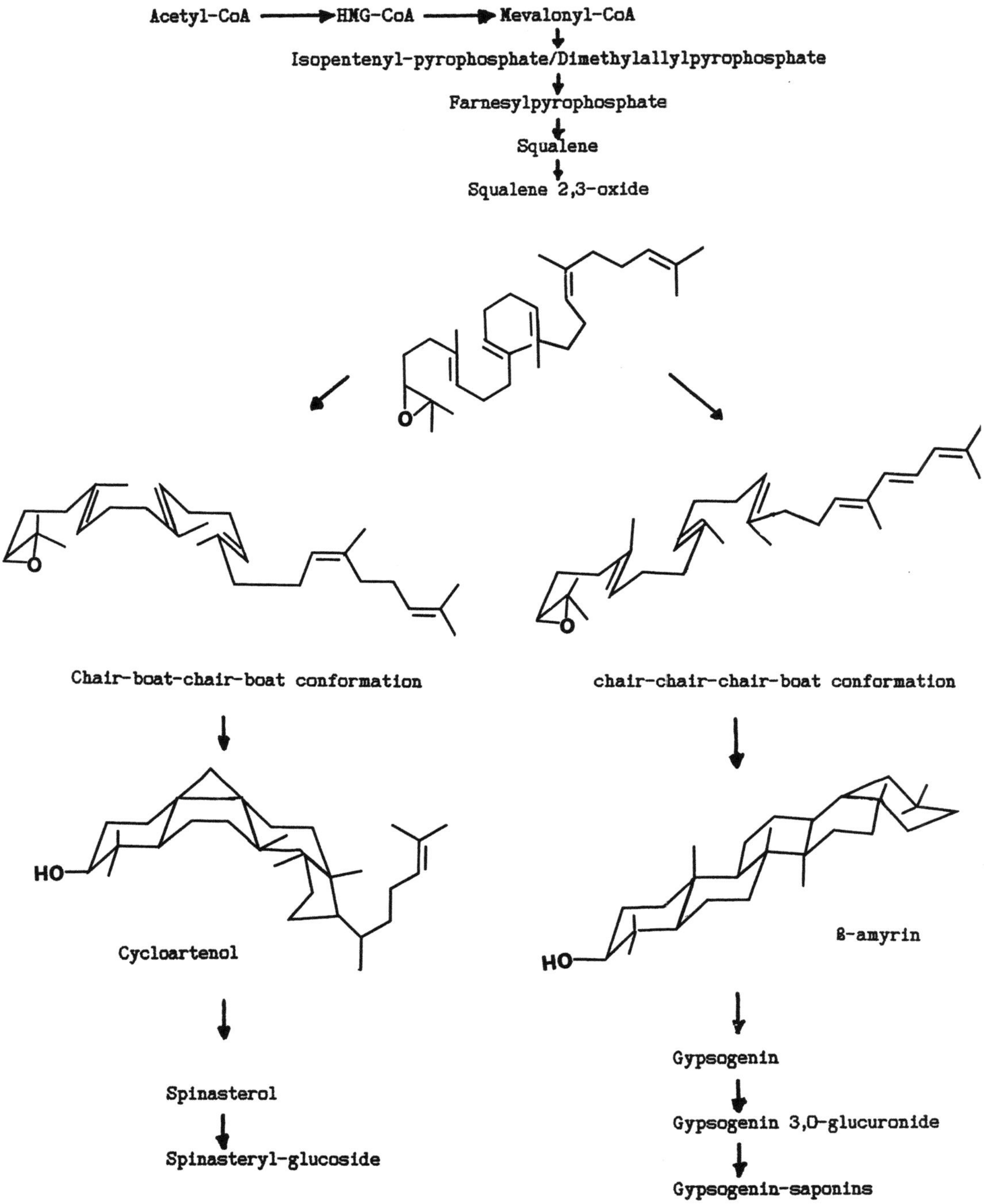

Fig. 10. Summary of the major route leading to the triterpenoid biosynthetic pathway in *G. paniculata*

synthesized squalene and small amounts of cycloartenol only. Feeding experiments indicated that in elicited cultures the synthesis of phytosterols was inhibited at the level of squalene 2,3-oxide:cycloartenol cyclase (Van Der Heijden et al. 1989).

More recently, administration of a specific inhibitor of squalene 2,3-oxide:cycloartenol cyclase caused a marked increase in β-amyrin synthesis in root suspension cultures of *Glycyrrhiza glabra* var. *glandulifera* (Ayabe et al. 1990). In cell suspensions of the same species, surprisingly, this inhibitor did not produce the same effect. An increase in 24-methylenecycloartanol, a compound involved in the later steps of sterol biosynthesis, was noted in unorganized tissues.

These different approaches in the study of the regulation of triterpenoid compounds confirmed the distinct pathways from squalene 2,3-oxide between phytosterols and triterpenoids, and constituted different models on the knowledge of the regulation of this complex pathway. More widely, it can be extended to the interconnections between the terpene pathway, and more narrowly to others such as indole alkaloid compounds via secologanine (a monoterpene).

From this point of view, tissue cultures of *G. paniculata* producing triterpenoidal saponins seem to be a good model to study the biosynthetic behavior of triterpenoidal saponins, to detect the differences in biosynthetic expression between callus or multiple shoot lines in one and the same species *G. paniculata*, or between the different *Gypsophila* species investigated here. For example, the differences during the following subcultures when *G. paniculata* multiple shoot culture is cultivated over a long period and the differences between *G. paniculata* tissues, since roots were described as being the only biosynthetic site in the whole vegetative plant (Henry et al. 1990). This behavior could be compared with other species which are not able to produce genuine triterpenoidal saponins, such as licorice *Glycyrrhiza glabra* or *Centella asiatica* (unpubl.), for example, from which we have established in vitro cultures.

3 Conclusions and Prospects

G. paniculata was shown to synthesize gypsogenin saponins, both in undifferentiated callus and multiple shoot cultures. The other species studied either became inconsistent in producing these saponins (*G. petraea*) or lost the biosynthetic potential after several subcultures (*G. repens*). Such differing biosynthetic behavior may be a model for comparative studies on the regulation of saponin biosynthesis, to enhance saponin production.

From the developmental aspect, the gypsogenin 3,O-glucuronide biosynthesis present in both undifferentiated callus and multiple shoot cultures of *G. paniculata* is all the more surprising as, in the vegetative state of the mature plant, only the roots contain gypsogenin saponins (7.5 mg/g dry weight in 1-year-old plants to 40 mg/g dry weight in 4-year-old plants), and the other parts of the plant are not able to produce any.

4 Protocol

1. Plant Cell Cultures. Seeds of *Gypsophila paniculata*, *G. petraea*, *G. repens*, and *G. muralis* from the Museum d'Histoire Naturelle, Paris, France, were surface sterilized in 10% sodium hypochlorite (10 min twice) and germinated either on Heller's or Knop's medium. Sterile cotyledon, hypocotyl, and root explants were set out on Murashige and Skoog's revised medium (1962) (2 mg/l adenine, 1 mg/l kinetin, 30 g/l glucose, pH 5.7, with different concentrations of 2,4-D: 0.1, 0.25, 0.5, and 1 mg/l).

Primary calli, formed 3 to 6 weeks later, originated in July 1986. These primary cultures were transferred, just after induction, onto different media, Murashige and Skoog's medium (2 mg/l adenine, 1 mg/l kinetin, 30 g/l glucose, pH 5.7) with 0.1 mg/l 2,4-D (MSA) and with 0.01 mg/l NAA (MSB) and Gamborg et al.'s medium B_5 (1968) and subcultured every 2 or 3 weeks on the same medium. Induction and callus cultures were performed at 25 °C under fluorescent light (2000 lx, 12-h photoperiod).

2. Gypsogenin 3,O-Glucuronide [I] Isolation and Assay. The first series of assays was carried out 2 months after initiation and the second 1 year after initiation, when the calli seemed morphologically stable. Tissue fresh weights were immediately determined and saponins were extracted three times, first with pure MeOH and then twice with $MeOH/H_2O$ (20/80; v/v) for 24 h each time at 25 °C. After filtering the methanolic extracts were combined and concentrated under reduced pressure to produce the extract, which was suspended in water and extracted twice with EtOAc. The water solution was submitted to Fractogel TSK HW-4OF chromatography. Elution with a gradient of $MeOH\text{-}H_2O$ containing Na_2CO_3 (0/100 to 50/50, v/v) gave saponins which were revealed by TLC with the Liebermann-Burchard reaction (reddish purple color). The saponins were hydrolyzed with 0.57 M H_2SO_4 for 4 h to liberate [I] in a precipitated acid form. This precipitate was solubilized with EtOAc (4 ml). [I] was purified by precipitation on neutralization with $NaHCO_3$. After filtering, the precipitate was solubilized with MeOH (2 ml) and was ready for HPLC assay.

3. HPLC Procedures. HPLC was performed on a Waters Associates pump model 510 with a stainless steel column (250 × 4 mm I.D.) packed with RP-18e (Merck, 5 µm) for analytical determinations. [I] was quantified at 206 nm with a Waters Associated Lambda-Max Model 481 LC spectrophotometer after elution at room temperature with $MeOH\text{-}H_2O$ (70/30, v/v) containing 5 mM tetra-n-butylammonium hydroxide, pH 5 adjusted with 0.5 M H_3PO_4, at 1.0 ml/min flow rate. The gypsogenin 3,0-glucuronide content in *Gypsophila* sp. was calculated from the peak area by comparison with a known standard.

4. Administration of Radioactive Precursor. Sodium [2-^{14}C]acetate (48 m Ci/mmol) was a product of CEA (Saclay, France). Multiple shoots with green calli of *G. paniculata* were transferred to fresh liquid MSB medium (100 ml) in Erlenmeyer flasks (250 ml each). Plant material in the flasks was mixed on a gyratory shaker at 120 rpm for 15 h. The pH was then adjusted to 4.2–4.3 to enhance acetate incorporation into the cells and 5.4 µCi of [2-^{14}C]NaOAc was added to each flask. At each experimental date, the plant material of one flask was harvested by separating it from the medium through filtration. Separated into two fractions of similar size, the plant material was weighed exactly. The first fraction was lyophilized and used for sterol extraction after addition of unlabeled lyophilized tissue

Table 7. ^{13}C chemical shifts (δ-values) of gypsogenin 3,O-glucuronide in CD_3OD. (Henry et al. 1989)

C_1	39.4	C_{11}	24.3	C_{21}	35.2	C'_1	105.4
C_2	26.0	C_{12}	123.6	C_{22}	33.6	C'_2	76.7
C_3	73.3	C_{13}	145.5	C_{23}	209.3	C'_3	83.9
C_4	56.4	C_{14}	43.2	C_{24}	10.8	C'_4	75.1
C_5	49.7	C_{15}	29.1	C_{25}	16.4	C'_5	77.6
C_6	21.6	C_{16}	24.3	C_{26}	18.0	C'_6	172.7
C_7	34.0	C_{17}	48.6	C_{27}	26.8		
C_8	41.1	C_{18}	43.0	C_{28}	182.0		
C_9	49.1	C_{19}	47.9	C_{29}	33.9		
C_{10}	37.3	C_{20}	31.9	C_{30}	24.8		

(200 mg). Extraction of sterols was carried out three times with cyclohexane. The pooled extracts were concentrated to dryness and 1 ml CH_2Cl_2 was added to solubilize the concentrate. An aliquot of 200 μl was purified with preparative TLC performed on silica gel eluted twice with $CHCl_3$ by comparison with a standard. The band corresponding to free sterols was recovered and assayed for radioactivity. The second fraction was added to fresh unlabeled plant material and was used for [I] and spinasteryl glucoside extraction. Methanolic extraction was performed as described above. Concentrated to dryness, the methanolic extracts were suspended in water acidified with HOAc (pH 2-3). Spinasteryl glucoside was extracted with EtOAc (three times), concentrated to dryness and solubilized in MeOH (2 ml). Purification of the spinasteryl glucoside was performed on 200-μl aliquots by preparative TLC on silica gel eluted with $BuOH/HOAc/H_2O$ (4/1/5) compared to a standard. Spinasteryl glucoside was then assayed for radioactivity. The aqueous phase was treated as described above to extract [I] in MeOH (2 ml), and assayed for radioactivity after HPLC purification.

5. Gypsogenin 3,O-Glucuronide [I]: 28-Hydroxy-23,28-Dioxoolean-12-en-3β-yl O-β-D-Glucopyranosiduronic Acid: M.p. 270–272 °C, $[\alpha]_D = +44$ (c = 10 mg/ml in MeOH), IR (KBr 1%: 3400 cm^{-1} (OH), 1700 cm^{-1} (CO acid), FabMS: with NaI m/z 669 $[M + Na]^+$, 453 $[gypsogenin - H_2O - H^+]^+$, with KI m/z 723 $[M + K]^+$, 453 $[gypsogenin - H_2O - H^+]^+$ $[M = C_{36}H_{53}O_{10}$ (646)], ^{13}C NMR: cf. Table 7 [chemical shifts consistent with those of Higuchi et al. (1985) on gypsogenin methyl ester].

References

Adler C, Hiller K (1985) Bisdesmosidische triterpensaponine. Pharmazie 40: 676–693

Ayabe S, Takano H, Fujita T, Furuya T, Hirota H, Takahashi T (1990) Triterpenoid biosynthesis in tissue cultures of *Glycyrrhiza glabra* var. *glandulifera*. Plant Cell Rep 9: 181–184

Baisted DJ (1971) Sterol and triterpene synthesis in the developing and germinating pea seed. Biochem J 124: 375–383

Berendesen GE, De Galan L (1980) Role of the length of chemically bonded phases and the retention mechanism in reversed-phase liquid chromatography. J Chromatogr 196: 21–37

Braut-Boucher F, Achard-Ellouk S, Pauthe-Dayde D, Henry M, Hoellinger H (1990) Cytoprotective effects of *Gypsophila* saponins towards isolated rat hepatocytes. In: Walker R (ed) Food additive and contaminants, vol 7. Taylor and Francis, London, pp S127–S130

Burnouf-Radosevich M, Delfel NE (1986) High-performance liquid chromatography of triterpene saponins. J Chromatogr 368: 433–438

Domon B, Dorsaz A-C, Hostettmann K (1984) High-performance liquid chromatography of oleanane saponins. J Chromatogr 315: 441–446

Dourmashkin RR, Dougherty RM, Harris RJC (1962) Electron microscopic observations on *Rous sarcoma* virus and cell membranes. Nature 194: 1116–1119

El Bary A, Nour SA (1979) Correlation between the spermicidal activity and haemolytic index of certain saponins. Pharmazie 34: 560–561

Fournier P (1977) Les quatre flores de la France, Corse comprise (Générale, Alpine, Méditerranéenne, Littorale). Lechevalier, Paris, p 1105

Gamborg OL, Miller RA, Ojima K (1968) Nutrient requirements of suspension cultures of soybean root cells. Exp Cell Res 50: 151–158

Gögelein H, Hüby A (1984) Interaction of saponin and digitonin with black lipid membranes and lipid monolayers. Biochim Biophys Acta 773: 32–38

Henry M, Chantalat-Dublanche I (1985) Isolation of spinasterol and its glucoside from cell suspension cultures of *Saponaria officinalis*: ^{13}C-NMR spectra data and batch culture production. Planta Med: 322–325

Henry M, Guignard JL (1982) Obtention d'une suspension cellulaire de *Saponaria officinalis* L. Etude de la croissance et de la production d'acide quillayique. Physiol Veg 20: 201–209

Henry M, Pauthe-Dayde D, Rochd M (1989) Extraction and high-performance liquid chromatographic determination of gypsogenin 3,O-glucuronide. J Chromatogr 77: 413–419

Henry M, Rochd M, Bennini B (1991) Biosynthesis and accumulation of saponins in *Gypsophila paniculata*. Phytochemistry 30: 1819–1821

Higuchi R, Tokimitsu Y, Hamada N, Komori T, Kawasaki T (1985) A new cleavage method for the sugar-aglycone linkage in saponin. Liebigs Ann Chem: 1192–1201

Higuchi R, Tokimitsu Y, Komori T (1988) Study on the mechanism of diazomethane degradation of the sugar-aglycone linkage in gypsogenin 3,O-glycoside. Liebigs Ann Chem: 249–255

Hiller K, Voigt G (1977) Neue Ergebnisse in der Triterpenesaponine. Pharmazie 32: 365–393

Khorlin AYa, Ovodov YuS, Ovodova RG (1963) The identity of gypsoside from *Gypsophila paniculata*. Izv Akad Nauk SSSR Ser Khim 8: 1522–1523 (in Russian)

Kochetkov NK, Khorlin AYa, Ovodov JuS (1963) The structure of Gypsoside-triterpenic saponin from *Gypsophila paniculata* Kom. Tetrahedron Lett 8: 477–482

Kolodziejski J, Stecka L (1965) Occurrence of saponins in various parts of *Gypsophila paniculata* and *Gypsophila altissima*. I. Farm Pol 21: 751–754 (in Polish)

Lee WK, Ryu BG, Kim BK (1987) Determination of gypsogenin by HPLC (high performance liquid chromatography) using 2,4-dinitrophenylhydrazine as a pre-labeling reagent. Yakhak HoeChi 31: 112–115 (in Korean)

Mabberley DJ (1989) The plant book. A portable dictionary of the higher plants. Cambridge Univ Press, Cambridge, p 286

Mansouri L (1989) Extraction et purification des saponines de racines de *Gypsophila paniculata* L. DESU, Univ Toulouse

Murashige T, Skoog F (1962) A revised medium for rapid growth and bioassays with tobacco tissue cultures. Physiol Plant 15: 473–497

Pauthe-Dayde D (1990) Bioproduction et bioconversion de saponines triterpeniques par des tissus de gypsophile et de saponaire cultivés in vitro. Thèse, Université ENSAT, INP Toulouse

Pauthe-Dayde D, Rochd M, Henry M (1990) Triterpenoidal saponin production in callus and multiple shoot cultures of *Gypsophila* sp. Phytochemistry 29: 483–487

Primorac M, Sekulovic D, Antonio S (1985) In vitro determination of the spermicidal activity of plant saponins. Pharmazie 40: 585

Rochd M (1990) Etude de la biosynthèse des saponines et des stérols produits par la plante et par des tissus cultivés in vitro de *Gypsophila paniculata L.* Thèse Université Pharm, UPS Toulouse

Rûzicka L, Giacomello G (1936) Polyterpene und Polyterpenoide CVII. Überführung von Gypsogenin (Albsapogenin) in Oleanolsäure. Helv Chim Acta 19: 1136–1140

Tagiev SA, Ismailov AI (1986) Quantitative determination of gypsoside in roots of *Gypsophila bicolor* (Freyn et Sint) Grossh. Rastit Resur 22: 262–265

Tani T, Kubo M, Katsuki T, Higashino M, Hayashi T, Arichi S (1981) Histochemistry II. Ginsenosides in ginseng (*Panax ginseng* root). J Nat Prod 44: 401–407

Tani T, Katsiki T, Kubo M, Arichi S (1986) Histochemistry IX. Distribution of saikosaponins in *Bupleurum falcatum* root. J Chromatogr 360: 407–416

Tschesche R, Wulff G (1973) Chemie und Biologie der Saponine. Fortschr Chem Org Naturst 3: 461–606

Van Der Haar AW (1927) Saponins and allied compounds. XVII. The gypsophila saponin and the product formed by its hydrolysis, gypsogenin and saccharides. Reg Trav Chim 46: 85–96

Van Der Heijden R, Verheij ER, Schripsema J, Baerheim Svendsen A, Verpoorte R, Harkes PAA (1988) Induction of triterpene biosynthesis by elicitors in suspension cultures of *Tabernaemontana* species. Plant Cell Rep 7: 51–54

Van Der Heijden R, Threlfall DR, Verpoorte R, Whitehead IM (1989) Regulation and enzymology of pentacyclic triterpenoid phytoalexin biosynthesis in cell suspension cultures of *Tabernaemontana divaricata*. Phytochemistry 28: 2981–2988

XIV *Hygrophila erecta*: In Vitro Culture and the Production of Verbascoside

M. Henry[1], J-L. Roussel, and C. Andary[2]

1 Introduction

Verbascoside, or β-(3′,4′-dihydroxy-phenyl)-ethyl-O-α-L-rhamnopyranosyl (1→3)-β-D-(4-O-caffeoyl)-glucopyranoside (Fig. 1), also called "Labiateae tannin", belongs to the phenylpropanoid glycosides. These secondary metabolites, along with other compounds like flavonoids are known to enter as substrates for the enzymes causing browning of damaged plant cells or cells of some species when cultivated in vitro (see, for example, a previous work on rosmarinic acid in this series, De-Eknamul and Ellis 1988).

Verbascoside is the most widespread of the disaccharide caffeoyl esters (Mølgaard and Ravn 1988). It is also known as acteoside (Kitagawa et al. 1984; Kobayashi et al. 1984) and kusaginin (Sakurai and Kato 1983). Moreover, as was shown by Andary and Ibrahim (1986) with labeling experiments of heterosidic caffeoyl esters biosynthesized by plantlets of *Stachys albens* (Lamiaceae), trisaccharide caffeoyl esters are formed from verbascoside, constituting a further or final step in the biosynthetic pathway of the heterosidic caffeoyl esters. These compounds, as well as verbascoside, can be employed as taxonomic markers in Plantaginaceae and Lamiaceae (Andary et al. 1988a, b).

Verbascoside was originally isolated from *Verbascum sinuatum* (Scarpati and Monache 1963), but its chemical structure was exactly defined by Andary et al. in 1982. Actually, verbascoside was said to be present in several families of three orders of the Asteridae (sensu Cronquist 1981): Verbenaceae (Sakurai and Kato 1983; Hänsel and Kallmann 1986; Chanh et al. 1988; Fauvel et al. 1989; Garnier et al. 1989) and Lamiaceae of Lamiales (Andary and Ibrahim 1986; Shimomura et al. 1987; Andary et al. 1988b); Plantaginaceae of Plantaginales (Andary et al. 1988a; Debrauwer et al. 1989); and Oleaceae (Birkofer et al. 1968; Andary et al. 1982; Endo and Hikino 1982; Ellis 1983; Fleuriet et al. 1984; Kitagawa et al. 1984; Andary et al. 1990); Scrophulariaceae (Schilling et al. 1982; Kobayashi et al. 1984; Shoyama et al. 1986), Myoporaceae (Dell et al. 1989), Orobanchaceae (Andary et al. 1982; 1985), Gesneriaceae (Nonaka and Nishioka 1977; Kvist and Pedersen 1986), and Acanthaceae (Andary et al. 1978; Henry et al. 1987), Pedaliaceae (Potterat et al. 1988) of Scrophulariales.

[1]Laboratoire de Botanique, Faculté des Sciences Pharmaceutiques, 31, Allées Jules Guesde, 31000 Toulouse, France
[2]Laboratoire de Botanique, Phytochimie et Mycologie, Faculté de Pharmacie, Avenue Charles Flahault, 34060 Montpellier Cedex, France

Biotechnology in Agriculture and Forestry, Vol. 21
Medicinal and Aromatic Plants IV (ed. by Y.P.S. Bajaj)

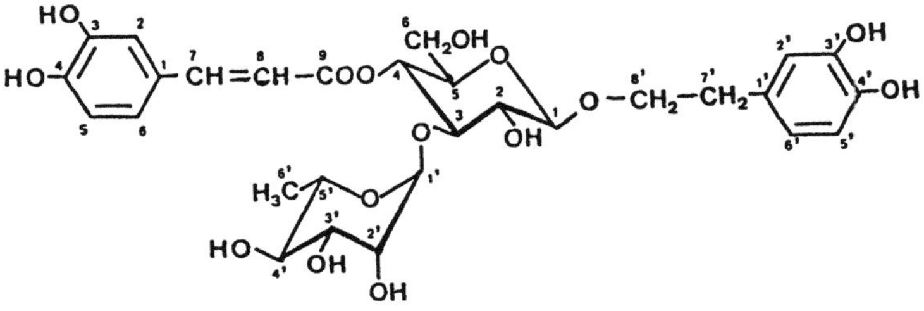

Fig. 1. Structure of verbascoside

Verbascoside and derivatives (forsythiaside, suspensaside, etc.) showed antimicrobial properties (Andary et al. 1981; Kitagawa et al. 1984; Debrauwer et al. 1989) and were described to be in vitro inhibitors of 3′,5′AMP phosphodiesterase and 5-lipoxygenase (Nishibe et al. 1982; Kitagawa et al. 1984; Kimura et al. 1987).

2 Distribution and Importance of the Plant

Hygrophila erecta (Burm. fil.) Hochr. (see Fig. 2) is a widespread plant in Indomalaysia belonging to the family Acanthaceae. The genus *Hygrophila* comprises about 100 tropical species found in wet areas. Some of them are cultivated as an ornamental aquatic plant in aquaria. Others are weeds found in the rice fields from India to Thailand as *H. difformis* (L. f.) Blume (Mabberley 1989).

H. erecta is a perennial herb whose seeds are used by the cosmetic and pharmaceutical industry in central Europe as a skin-regenerating product. As the cells of the central nervous system have the same embryological origin as the skin cells, one of us (M.H.) requested the collaboration of a laboratory specialized in neuronal cultures (Centre de Neurochimie of Strasbourg) to research any effect on neuronal cells. Lipidic extracts showed a striking effect on the morphological differentiation of cerebral neurons of 13-day-old rat embryos in in vitro cultures (Borg et al. 1987). A long-chain *n*-alcohol (with 26 carbon atoms), *n*-hexacosanol was the substance responsible for the growth effect of the glyal and neuronal cells in cultures. It seemed surprising that such a simple long-chain *n*-alcohol might display such activity, but previously another simple, long-chain fatty alcohol containing 30 carbon atoms, the triacontanol, was described to be a growth promotor in the plant kingdom (Ries et al. 1977). We therefore became interested in determining whether *H. erecta* tissue cultures could be induced to produce *n*-hexacosanol. Unfortunately, at present we have not succeeded in producing in vitro *n*-hexacosanol. During the chemical and pharmacological studies on the crude extracts of the plant and the tissue cultures, verbascoside was detected as a major extractable metabolite; we report here its production by callus and cell suspension cultures.

The demand of *H. erecta* in the world market is not possible to estimate because of the confusion in the industrial cosmetic market between the seeds of *H. erecta*

Fig. 2. Flowering inflorescence of *Hygrophila erecta* grown in a greenhouse

and seeds of other species such as *Centella asiatica* employed for the same biological effect.

3 In Vitro Approaches

Verbascoside-producing cell cultures of *Syringa vulgaris* (Ellis 1983), *Rehmania glutinosa* (Shoyama et al. (1986), *Hygrophila erecta* (Henry et al. 1987) and *Eremophila* spp. (Dell et al. 1989) can also be considered as high metabolite-yielding disorganized cell cultures. These studies on the production of verbascoside by plant tissue or suspension cultures had different complementary aims. After the identification of verbascoside as the main constituent of cell suspension cultures of *S. vulgaris*, Ellis (1983) in a first approach examined its biosynthetic pathway by supplying potential radioactive precursors. He showed the importance of tyrosine and tyramine as biosynthetic precursors of the 4-hydroxy- and 3,4-dihydroxyphenylethanol rather than DOPA or dopamine, indicating that 3-hydroxylation is preceded by both decarboxylation and deamination. Shoyama et al. (1986) were more interested in the identification of the caffeoyl glycosides isolated from callus tissue of *Rehmannia glutinosa*. Their chemical study provided substantial information about the physicochemical constants of verbascoside and of three other caffeoyl glycosides, continuing the investigations on the different compounds of this plant to identify the principles responsible for its immunosuppressive property. Studies on *H. erecta* (Henry et al. 1987) and on *Eremophila* spp. were not initially performed to obtain verbascoside, but other compounds: *n*-hexacosanol for *H. erecta* and particular oxygenated diterpenes for *Eremophila* spp. but after establishment of the in vitro cultures, in both cases verbascoside became of great interest with the particularly high yield of the cultures.

3.1 Establishment of Cultures of *H. erecta*

3.1.1 Material and Methods

Callus was initiated from stem explants of adult plants and from hypocotyl of seedlings germinated under sterile conditions. All plant material had the same geographical origin: Burma.

Induction of disorganized tissues, callus, and suspension cultures was performed at 25 °C and maintained in the dark. The sterilization of the seeds was carried out by rinsing with 70% alcohol and then by soaking twice in a 10% solution of sodium hypochlorite for 15 min, followed by several washings with sterile water. Sterilization of the stems was done the same way, however only for 5 min in the 10% solution of sodium hypochlorite.

3.1.2 Media Composition and Culturing

Sterile hypocotyl explants were obtained from surface-sterilized seeds germinated on Heller's medium. Stems segments of adult plants were cut from 1 year-old plants grown in a greenhouse. Hypocotyl explants and surface-sterilized stems segments were set out on Nitsch and Nitsch's medium (1956) containing growth hormones as follows: 0.1 mg/l 2,4-dichlorophenoxyacetic acid; 2 mg/l adenine; the pH was adjusted at 5.5. This medium permitted the formation of primary calli 3 weeks later. Those primary cultures were transferred just after induction on Gamborg et al. medium 1B5 (1968) and were subcultured every 3 weeks on the same medium. Stabilization of the cell lines was different between the two origins of plant material. If stem initiations gave rise rapidly to undifferentiated and stable light-gray cultures after 6 months of subculture, either on solid (Fig. 3) or on liquid medium, hypocotyl initiations needed 2 years of subculture mostly on liquid medium to establish a continuous line before any study could take place.

Cell suspension cultures were obtained from the agar cultures by transferring callus to a liquid medium of the same composition, but without agar. This transfer did not present any difficulties in the dissociation of the calli. Cells were maintained in suspension and subcultured in 250-ml flasks containing 80 ml fresh medium and mixed on a gyratory shaker at 120 rpm.

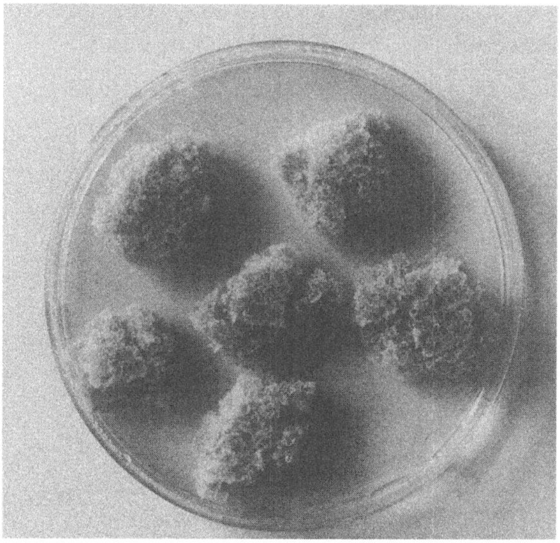

Fig. 3. Callus culture of *Hygrophila erecta* obtained on a solidified Nitsch and Nitsch's medium (1956) containing 0.1 mg/l 2,4-dichlorophenoxyacetic acid; 2 mg/l adenine. pH adjusted to 5.5; subcultured every 3 weeks on medium 1B5

Suspension cultures were vacuum filtered (200–300 Hg) on a glass filter, and fresh weights were immediately determined. Cell numbers were counted on an aliquot of fresh culture after dissociation by boiling with a mixture of 10% chromic anhydride in 2 N H_2SO_4 for 10 min. Cells were lyophilized to determine the dry weights.

3.1.3 Verbascoside Isolation and Assay

The technique, described previously (Henry et al. 1987), consisted of three extractions of the lyophilized cells or solubilization of the suspension media concentrated to dryness with hot 70% MeOH. Methanolic extracts or solutions were analyzed by densitometry of fluorescence on cellulose TLC after elution and spraying Neu's reagent (2-aminoethyl diphenyl borinate in 1% MeOH) to determine the verbascoside content.

3.2 Production of Verbascoside by *H. erecta*

In 1984, 25 callus initiations were performed from stem explants of eight different plants of *H. erecta* (Henry et al. 1987). These initiations gave rise rapidly to undifferentiated and stable light-gray calli. Thin-layer chromatography screening on these different calli showed the production of a main component, verbascoside. In plants, leaves are generally the richest in verbascoside content, stems being the poorest (Andary et al. 1978). In *H. erecta*, the dried leaves showed 2.9% dry matter of verbascoside, the dried stems contained 0.85% dry matter. The verbascoside content of the different callus lines measured on the calli after 3 weeks of culture (Table 1) displayed only one high producing cell line with 1% dry matter, three cell

Table 1. Verbascoside contents of 25 callus cell lines of *H. erecta* compared to the 8 different plants from which they were initiated

Plants	Verbascoside content in the leaves (g% dry wt.)	Cell lines	Verbascoside content in the cells (g% dry wt.)
1	2.3	1a	0.10
		1b	0.10
		1c	0.01
		1d	0.04
2	5.2	2a	0.01
		2b	0.01
		2c	0.02
		2d	0.07
		2e	0.01
3	3.3	3a	0.07
		3b	0.17
4	2.0	4a	1.00
5	5.2	5a	0.02
		5b	0.02
6	4.4	6a	0.19
		6b	0.10
7	2.4	7a	0.09
		7b	0.04
		7c	0.20
8	2.9	8a	0.01
		8b	0.01
		8c	0.02
		8d	0.02
		8e	0.04
		8f	0.02

lines containing 0.1–0.2% dry matter, six containing 0.05–1% dry matter and 15 of them producing less than 0.05% dry matter of verbascoside. There was apparently no correlation between the verbascoside contents of the different callus lines and those of the leaves of the plants from which they were initiated. Among these callus lines, three of them produced other unidentified caffeoyl glycosides.

In 1981, we obtained the first cell line from hypocotyls of seedlings germinated under sterile conditions. This primary culture was also obtained and subcultured on solid 1B5 medium and then transferred to the corresponding liquid medium to obtain a cell suspension. The suspension was subcultured every 3 weeks for 2 years to establish a continuous line before any further study. Cell growth and verbascoside content measurements were then performed. The kinetic results are presented in Fig. 4. Cell growth showed a maximal rate of 0.012/h at 25 °C, i.e., a doubling time of 82 h. Stationary phase was reached in 15 days. The cell number rose from 7×10^3 cells/ml to 1.2×10^6 cells/ml and the dry matter reached 10 g/l of cell suspension. Verbascoside production increased during the growth period and about 2–2.6 g per 100 g of dry matter was produced in the stationary phase. As

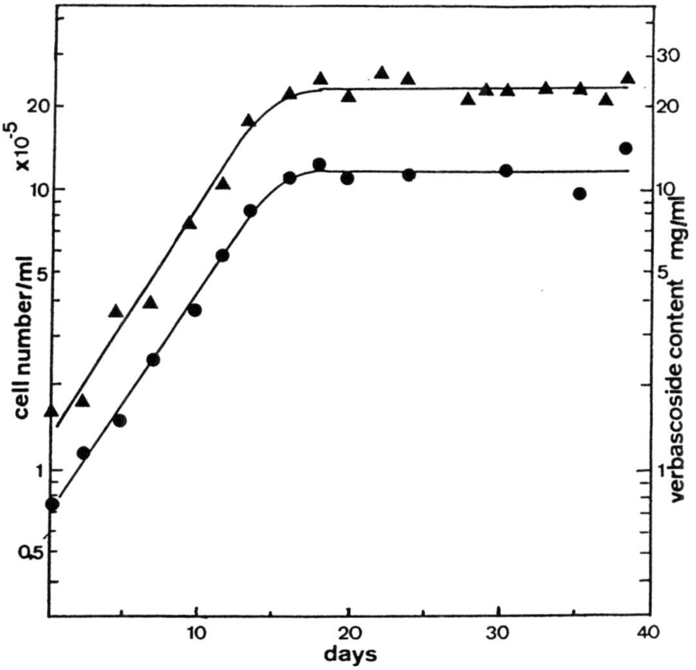

Fig. 4. Culture growth and verbascoside production. The contents of triplicate 250-ml (62-ml suspension) flasks were harvested by filtration on the days indicated. The cell number (*circles*) and verbascoside content (*triangles*) were determined as described in the text

shown in Fig. 4, the verbascoside content stayed remarkably constant in the cells during the growth phase. Moreover, verbascoside was not detected in the medium and it did not seem to diffuse out of the cells.

H. erecta cell suspension could not grow in the light. In a few hours, its creamy grey color changed to a darkish color and the cell growth dropped drastically. In contrast, in the dark the cells could be maintained in the stationary phase (Fig. 4) for more than 20 days without damage. One explanation could be the effect of ultraviolet rays in the natural light on the induction of phenoloxidases. Verbascoside was then oxidized giving different darkish products that damaged the cells. This phenomenon is now well known (Hahlbrock and Grisebach 1975; De-Eknamul and Ellis 1988) and also usually observed with the whole plant of *H. erecta* when it is poorly dried and preserved.

The different callus lines obtained from explant of adult plant had a lower verbascoside content than the cell suspension culture obtained from hypocotyl explants. It is known that seedlings in the hypocotyl state are richer in this caffeoyl compound than the more adult plants (Andary and Ibrahim 1986). To test the stability of verbascoside, production, three assays were conduced on the same cell suspension at 6 month intervals, but no change was observed in the verbascoside contents of the cells of the batch culture.

4 Conclusion

Botanical confusion in the identification of the seeds of different species used by the cosmetic industry promoted research on *Hygrophila erecta*. Pharmacological studies on lipid extracts in the plant showed a growth factor of neuronal cells in cultures, *n*-hexacosanol. Up to the present, in vitro cell cultures of *Hygrophila erecta* have failed to produce *n*-hexacosanol. Despite this failure, verbascoside, a disaccharide caffeoyl ester, was detected in great amounts in the biomass of the cell suspension. Such results gave rise to the increased interest in the in vitro cultures of this plant which can be considered a substantial source of verbascoside, as well as in other species of the three orders of Asteridae where this compound is mainly encountered.

5 Protocol

5.1 Establishment of Tissue Cultures

To obtain cell lines able to produce great amounts of verbascoside, use plantlets grown aseptically as explant source. Allow seeds to germinate on classical medium, for example, on Knop's or Heller's medium at 25 °C, after surface sterilization in 10% NaOCl for 15 min, removed by correct rinsing with sterile distilled water. Hypocotyls are the best explants for obtaining callus easily. Maintain germination in complete darkness for a few weeks using hypocotyl fragments of sufficient length.

Set out sterile hypocotyl explants on Nitsch's revised solid medium (Nitsch and Nitsch 1956) (0.1 mg/l 2,4-dichlorophenoxyacetic acid; 2 mg/l adenine; pH 5.5; 7 g/l agar). After ca. 3 weeks, primary calli will appear. Transfer them just after induction on Gamborg's medium 1B5 and subculture these calli every 3 weeks on the same medium in a 25 °C chamber exclusively in the dark to avoid any initiation of darkish products; otherwise, cell growth will drop drastically and the callus culture will be lost in a few weeks. Most explants will proliferate and produce soft, creamy gray callus.

5.2 Cell Suspension Cultures

Transfer fast-growing callus into a small volume of the same 1B5 medium without agar and put on a shaker under the same environmental conditions (dark exclusively). Subculture the suspension every week by adding 20–50 ml log-phase cells to 330 ml fresh medium in a 1-l Erlenmeyer flask. Stress such as light, stopping shaker agitation for a few hours, abnormal changes in temperature, etc., which will induce damage by producing toxic products, should be avoided.

References

Andary C, Ibrahim RK (1986) Biosynthetic capacity of *Stachys* seedlings for verbascoside and related caffeoyl derivatives. Z Naturforsch 41c: 87–90

Andary C, Pellecuer J, Soediro I, Privat G (1978) Identification et dosage du verbascoside, ester hétérosidique de l'acide caféique, chez *Acanthus mollis* (Acanthacées). Pharm Mediterr Tolosae 13: 237–249

Andary C, Roussel JL, Motte ME, Rascol JP, Privat G (1981) Activité antifongique comparée de divers esters de l'acide dihydroxy-3,4-cinnamique. Crytogam Mycol 2: 119–130

Andary C, Wylde R, Laffite C, Privat G, Winternitz F (1982) Structure of verbascoside and orobanchoside caffeic acid sugar esters from *Orobanche rapum-genistae*. Phytochemistry 21: 1123–1127

Andary C, Privat G, Wylde R, Heitz A (1985) Pheliposide et arenarioside, deux nouveaux esters hétérosidiques de l'acide caféique isolés de *Orobanche arenaria*. J Nat Prod 48: 778–783

Andary C, Motte-Florac ME, Gargadennec A, Wylde R, Heitz A (1988a) The caffeic esters of the genus *Plantago*, identification and chemotaxonomic value. Plant Med Phytother 22: 17–22

Andary C, Rascol JP, Puech S, Roussel JP, Privat G (1988b) Les esters de l'acide cafeique dans la chimiotaxinomie des *Teucrium* de la section *polium* (Lamiaceae). Can J Bot 66: 1007–1012

Andary C, Tahronch S, Marion C, Wylde R, Heitz A (1992) Caffeic glycoside esters from *Jasminum nudiflorum* and some related species. Phytochemistry 31: 885–886.

Birkofer L, Kaiser C, Thomas U (1968) Acteosid und Neoacteosid; Zückerester aus *Syringa vulgaris* (L.). Z Naturforsch 23b: 1051–1058

Borg J, Toazara J, Hietter H, Henry M, Schmitt G, Luu B (1987) Neurotrophic effect of naturally occurring long-chain fatty alcohols on cultured CNS neurons. Febs Lett 213: 406–410

Chanh PH, Koffi Y, Chanh APH (1988) Comparative hypotensive effects of compounds extracted from *Lippia multiflora* leaves. Planta Med 54: 294–296

Cronquist A (1981) An integrated system of classification of flowering plants. Columbia Univ Press, New York

Debrauwer L, Maillard C, Babadjamian A, Vidal-Ollivier E, Laget M, Salmona G, Afzal-Raffi Z (1989) Study in the chemical constituents of *Plantago cynops* L. and antibacterial evaluation of verbascoside. Pharm Acta Helv 64: 183–187

De-Eknamul W, Ellis BE (1988) Rosmarinic acid: production in plant cell cultures. In: Bajaj YPS (ed) Biotechnology in agriculture and forestry, vol 4. Medicinal and aromatic plants I. Springer, Berlin Heidelberg New York, pp 310–329

Dell B, Elsegood CL, Ghisalberti EL (1989) Production of verbascoside of *Eremophila* spp. Phytochemistry 28: 1871–1872

Ellis BE (1983) Production of hydroxyphenylethanol glycosides in suspension cultures of *Syringa vulgaris*. Phytochemistry 22: 1941–1943

Endo K, Hikino H (1982) Structures of forsythoside C and D, antibacterial principles of *Forsythia suspensa* fruits. Heterocycles 19: 2033–2036

Fauvel MT, Gleye J, Andary C (1989) Verbascoside: a constituent of *Clerodendrum inerme*. Planta Med 55: 493–582

Fleuriet A, Macheix J-J, Andary C, Villemur P (1984) Mise en évidence et dosage par chromatographie liquide à haute performance du verbascoside dans le fruit de six cultivars d'*Olea europaea*. L. C R Acad Sci Paris 299(III): 253–256

Gamborg OL, Miller RA, Ojima K (1968) Nutrient requirements of suspension cultures of soybean root cells. Exp Cell Res 50: 151–158

Garnier J, Mahuteau J, Plat M, Sastre C (1989) Esters caféiques de deux verbénacées des Antilles. Plant Med Phytother 22: 17–22

Hahlbrock K, Grisebach H (1975) Biosynthesis of flavonoids. Flavonoids 2: 866–915

Hänsel R, Kallmann S (1986) Verbascoside, a main constituent of *Verbena officinalis*. Arch Pharm (Weinheim) 319: 227–230

Henry M, Roussel J-L, Andary C (1987) Verbascoside production in callus and suspension cultures of *Hygrophila erecta*. Phytochemistry 26: 1961–1963

Kimura Y, Okuda H, Nishibe S, Arichi S (1987) Effects of caffeoylglycosides on arachidonate metabolism in leukocytes. Planta Med 53: 148–153

Kitagawa S, Tsukamoto H, Hisada S, Nishibe S (1984) Studies on the chinese crude drug Forsythia fructus. VII. A new caffeoyl glycoside from *Forsythia viridissima*. Chem Pharm Bull 32: 1209–1213

Kobayashi H, Karasawa H, Miyase T, Fukushima S (1984) Studies on the constituents of *Cistanche herba*. III. Isolation and structure of new phenylpropanoid glycosides, cistanosides A and B. Chem Pharm Bull 32: 3009–3014

Kvist LP, Pedersen JA (1986) Distribution and taxonomic implications of some phenolics in the family Gesneriaceae determined by EPR spectroscopy. Biochem Syst Ecol 14: 385–405

Mabberley DJ (1989) The plant book. A portable dictionary of the higher plants. Cambridge Univ Press, Cambridge, p 286

Mølgaard P, Ravn H (1988) Evolutionary aspects of caffeoyl ester distribution in Dicotyledons. Phytochemistry 27: 2411–2421

Nishibe S, Okabe K, Tsukamoto H, Sakushima A, Hisada S (1982) The structure of Forsythiaside isolated from *Forsythia suspensa*. Chem Pharm Bull 30: 1048–1050

Nitsch JP, Nitsch C (1956) Auxin-dependent growth of excised *Helianthus tuberosus* tissues. Am J Bot 43: 839–851

Nonaka G, Nishioka I (1977) Bitter phenylpropanoid glycosides from *Conandron ramoidioides*. Phytochemistry 16: 1265–1267

Potterat O, Msonthi Y, Hostettman K (1988) Four iridoid glucosides and a phenylpropanoid glycoside from *Sesamum angolense*. Phytochemistry 27: 2677–2679

Ries SK, Wert V, Sweeley CC, Leavitt RA (1977) Triacontanol: a new naturally occurring plant growth regulator. Science 195: 1339–1341

Sakurai A, Kato T (1983) A new glycoside, kusaginin isolated from *Clerodendron trichotomum*. Bull Chem Soc Jpn 56: 1573–1574

Scarpati ML, Monache D (1963) Isolation from *Verbascum sinuatum* of two new glucosides, verbascoside and isoverbascoside. Anal Chem 53: 356–367

Schilling G, Hügel M, Mayer W (1982) Verbascoside and isoverbascoside from *Paulownia tomentosa* Steud. Z Naturforsch 37B: 1633–1635

Shimomura H, Sashida Y, Ogawa K (1987) Iridoid glucosides and phenylpropanoid glycosides in *Ajuga* species of Japan. Phytochemistry 26: 1981–1983

Shoyama Y, Matsumoto M, Nishioka I (1986) Four caffeoyl glycosides from callus tissue of *Rehmania glutinosa*. Phytochemistry 25: 1633–1636

XV *Leontopodium alpinum* Cass. (Edelweiss): In Vitro Culture, Micropropagation, and the Production of Secondary Metabolites

I.L.I. Hook[1]

1 Introduction

1.1 Distribution of the plant

The genus *Leontopodium* (Compositae; Inulae; Gnaphaliinae *sensu amplo*) consists of between 30 and 40 species (Handel-Mazzetti 1927) found growing in mountainous areas of Japan (Nishikawa 1985), Asia (Hartmann 1972; Voroshilov 1979; Shmetova and Shcherbova 1982; Khokhryakov 1983; Lee et al. 1985; Park et al. 1987; Khatoon and Ali 1988) and Europe (Sugar 1971; Bicchi et al. 1975). A single species, *L. alpinum* Cassini, is considered to represent the genus in Europe (Wagenitz 1979) and the once separate *L. nivale* (Ten) Huet ex Hand.-Mazz. is now regarded as a subspecies, i.e. *L. alpinum* subsp. *nivale* (Ten) Tutin, stat. nov. (Heywood 1973). The plant is protected by national legislation in Austria, Germany, Italy and Liechtenstein. *L. alpinum*, commonly known as edelweiss, is a perennial plant with a branching rootstock and fibrous roots (Fig. 1a). Aerial structures exhibit wide morphological diversity (Siljak et al. 1974). Foliage leaves are linear to lanceolate in shape, 3–8 mm wide and pubescent. Inflorescence stalks develop from June to August and grow 2–45 cm high. The characteristically star-shaped "flower" varies in diameter from 1.5–10.5 cm and consists of an inflorescence made up of up to 12 densely aggregated capitula, which are subtended by an involucre of hairly leaves (Hess et al. 1972; Tutin et al. 1976; Wagenitz 1979; Fig. 1b). *L. alpinum* is traditionally found growing on limestone formations at altitudes up to 3140 m but can be easily cultivated on moist, humus-rich soil (Wagenitz 1979).

1.2 Conventional Propagation Methods and Commercial Value

Edelweiss plants are grown without difficulty from seed obtained from either wild or cultivated sources. Larger plants are more readily obtained from subdivisions of the senescing root stock made during winter and planting these in a humus-rich soil. When placed in a sunny location, flowering will occur from June of the same year onwards. This vegetative propagation ensures uniformity of the plants grown (Fig. 1c). Development of plants from seeds risks the production of hybrids, ready

[1] Department of Pharmacognosy, School of Pharmacy, Trinity College, Dublin 4, Ireland

Biotechnology in Agriculture and Forestry, Vol. 21
Medicinal and Aromatic Plants IV (ed. by Y.P.S. Bajaj)

Fig. 1. **a** Example of cultivated *L. alpinum* plant; **b** inflorescence of cultivated *L. alpinum* showing a variable number of densely aggregated capitula; **c** *L. alpinum* plants growing in peat-based compost

hybridization being a problem considered to be associated with the genus *Leontopodium*. Although classified as a medicinal plant (Chiej 1984), Edelweiss is rarely used as such. The major commercial uses are currently purely decorative, it being offered for sale as a fresh alpine-garden plant or for dried flower arrangements (Borntrager and Schlemmer 1988). Trading is, however, severely restricted in Europe due to the protected status of the plant.

1.3 Secondary Metabolites in Species of *Leontopodium* and Their Medicinal Value

L. alpinum is listed as a poisonous plant (Gessner and Orzechowski 1974) and several reports refer to it as a herbal folk remedy, for example, in the Ukranian-Carpathian mountains (Pizov and Bartkov 1973) and in Poland, where it has been used in the treatment of cancer (Hartwell 1968). More commonly, the aerial parts of Edelweiss plants have been referred to as being useful in the treatment of intestinal disorders such as diarrhea, dysentery and colic (Hoppe 1958; Chiej 1984). The active constituents have not been identified; in general, relatively little research has been carried out into the secondary metabolites of *Leontopodium* species. However, flavonoid glycosides such as cosmosiin and polustrin have been isolated from *L. ochroleucum* Beauv. (Dashbalyn and Glyzin 1978) and luteolin and its glucosides from *L. alpinum* Cass. (Tira et al. 1970) and *L. ochroleucum* (Dashbalyn and Glyzin 1978). Various hydroxyacinnamic acid esters have also been found to occur, for example, 3,4-dihydroxycinnamic and vanillic acids in *L. leontopodioides* (W) Beauv. (Zhao et al. 1984) and 3-O-caffeoylquinic (chlorogenic acid) and 3,4-di-O-caffeoylquinic acids in *L. alpinum* (Hennessy et al. 1989). Organic extracts of aerial structures of *L. ochroleucum* Beauv. var. *campestre* (Ledb.) Grub contained various terpenoids (Huneck and Knapp 1986), while *L. alpinum* and *L. espeletia* have been found to produce alkanes, predominantly C^{31} (Bicchi et al. 1975; Paternostro et al. 1973) and the ubiquitous sterols campesterol, stigmasterol and sitosterol (Paternostro et al. 1973; Hennessy et al. 1989).

2 In Vitro Approaches

Only two references, other than that to be discussed below, concerning the in vitro culture of *Leontopodium* species have been reported in the literature. Russian workers (Lapchik et al. 1984) developed optimal tissue culture and micropropagation techniques for various endangered plants, including *L. alpinum*, but details of these are unavailable in English. *L. coreanum* was successfully micropropagated by Korean workers (Park et al. 1987) using MS salt medium (Murashige and Skoog 1962) containing BAP (0.2 mg/l) and NAA (0.01 mg/l) for shoot multiplication and half-strength GD medium (Gresshoff and Doy 1974) with BAP (0.3 mg/l) and NAA (0.01 mg/l) for root development.

2.1 Micropropagation of *L. alpinum*

Apical buds from mature, senescing plants or seeds were used as source material. These were surface sterilized by treatment with a 10% aq. solution of commercial sodium hypochlorite concentrate for 20 min, rinsing with sterile distilled water. Shoots developed and were placed for rooting on an agar-solidified (7 g/l) MS salt medium containing NAA (0.1 mg/l), kinetin (0.05 mg/l), thiamine HCl (0.4 mg/l), mesoinositol (80 mg/l), casein hydrolysate (100 mg/l) and sucrose (30 g/l) (Fig. 2). Rooted plantlets were subdivided and subcultured at 4–6 week intervals into 20 ml fresh medium in culture tubes and grown in continuous bright fluorescent light at 25 °C. Each parent plant could be divided at subculture into between 3 and 14 plantlets. Rooted plantlets were transplanted into a peat based compost in propagating trays, maintained under humid conditions for 2–3 weeks and then grown to maturity. One plant type (Type 1), originating from a single seed, has been continuously micropropagated using the conditions outlined above, since Dec. 1984. These plants when mature are indistinguishable from the original parent plant from which the seed material was derived (Fig. 1b).

Plants of Type 2 originated in Dec. 1985 from callus developed from a micropropagated Type-1 plant. Callus maintained on the same culture medium began to show cytodifferentiation and spontaneously developed shoots (Fig. 3A). These subsequently flowered under the culture room conditions outlined and have

Fig. 2. Micropropagated plant (Type 1) in growth culture tube. Four weeks growth on modified MS salt medium as specified in Sect. 2.1

Fig. 3. A Spontaneous development of shoots on callus culture of *L. alpinum*. Four weeks growth on modified MS salt medium as specified in Sect. 2.1; **B** micropropagated plant (Type 2) flowering under culture room conditions. Four weeks growth on modified MS salt medium as specified in Sect. 2.1; **C** mature micropropagated plant (Type 2) after 4-months growth under garden conditions (transplanted from growth culture tubes into peat compost May 1986)

continued to do so to the present day (Fig. 3B). Some rooted plants were transplanted into potting compost and grown to maturity under garden conditions. These plants are identical in appearance to the parent plant, although much smaller (Fig. 3C) but differ by producing flowers between August and October instead of June and August.

2.2 Establishment of Tissue Cultures

Callus cultures were developed from leaves of sterile, micropropagated plants (Type 1) placed on an agar-solidified (9 g/l) MS salt medium containing 2,4-D (0.22 mg/l), NAA (0.18 mg/l), glycine (2 mg/l), nicotinic acid (0.5 mg/l), pyridoxine HCl (0.5 mg/l), mesoinositol (200 mg/l), thiamine HCl (0.5 mg/l) and sucrose (30 g/l). Cultures grew in continuous dim fluorescent light at 25 °C. Suspension cultures were initiated by transferring callus portions into the liquid modified MS medium noted above. Flasks were shaken on an orbital shaker (100 rpm, 1 cm throw). Suspensions were grown as batch cultures in 250-ml flasks and subcultured at monthly intervals by transferring 10 ml of culture into 90 ml fresh medium or 4 g fresh weight of separated cells into 100 ml medium. Cultures yielded approx 14 g dry cell weight per litre of medium per growth cycle. The suspension cultures were relatively homogeneous with respect to cellular differentiation, tending to darken in colour from yellowish-grey to dark brown towards the end of the monthly growth cycle (Fig. 4).

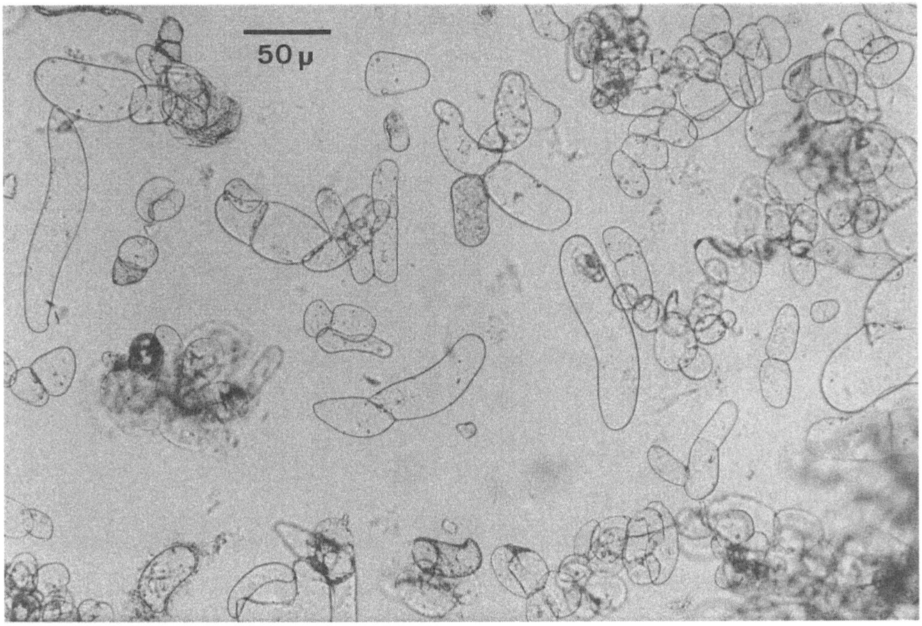

Fig. 4. Suspension culture of *L. alpinum*

2.3 Establishment of Transformed ("Hairy") Roots

A sterile micropropagated plant (Type 1) was inoculated with a 36-h activated kanamycin-resistant strain of *Agrobacterium rhizogenes*. After 7 weeks, hairy roots showing negative geotropism had been established and were picked off (Fig. 5a). After sterilization with ampicillin (500 mg/l) these were subcultured at 4–6-week intervals into an agar-solidified medium containing MS salts, thiamine HCl (0.4 mg/l), mesoinositol (100 mg/l) and sucrose (30 g/l). After 4 months rapidly growing roots were transferred to the liquid agar-free medium noted above. Culture conditions were as those for suspension cultures (see Sect. 2.2). Five hairy root lines were established and are now maintained in liquid batch culture (2–4 g inoculum per 300-ml medium per 500-ml flask) for a 6-week growth period (Fig. 5b). Growth can be measured as fresh weight increase and expressed as the factor by which inoculum weight has increased at harvest. From Table 1 it is apparent that the root lines exhibit variability in growth, with HR line 1 consistently showing highest productivity. TLC evaluation of methanol extracts of these transformed roots showed them to produce a spectrum of secondary metabolites identical to the roots of the parent plant. Quantitative metabolite analyses and confirmation of transformation have yet to be carried out.

Fig. 5. a Transformed ("hairy") roots of *L. alpinum* showing negative geotropism, after 7-weeks growth in a modified MS medium as specified in Sect. 2.3; **b** hairy root line (HRI) growing as batch culture in liquid modified MS medium as specified in Sect. 2.3 after 6 weeks growth

Table 1. Biomass production by hairy root lines of *L. alpinum*

Hairy root lines	Growth factor		
	(Total fresh wt. ÷ inoculum fresh wt.)		
	Harvest (i)	Harvest (ii)	Harvest (iii)
HR1	10	18	23.3
HR2	8.3	9.9	7
HR3	3.8	4.4	3.4
HR4	6.9	6.4	5.7
HR5	–	1.9	3.3

2.4 Extraction, Quantitation, and Production of Secondary Metabolites by *L. alpinum* Plants and Cultures

2.4.1 Extraction and Quantitation

Plant or cell culture material was dried in a fan oven (<40 °C). Powdered material was refluxed twice with methanol for 30 min; filtered and combined filtrates evaporated to dryness. The residue was resuspended in methanol and applied to a column of Sephadex LH 20. This was eluted with methanol, fractions were collected, monitored by TLC and like fractions combined. The major compounds were further purified by PLC (System I) and identified by ^{1}H-NMR and ^{13}C-NMR spectroscopy and by comparison with published data (Hennessy et al. 1989). TLC was performed on precoated silica gel plates (60 F_{254}) with System I: *n*–BuOH-HOAc–H_2O (4:1:5 upper phase); System III: EtOAc–HCOOH–HOAc–H_2O (100:11:11:27) and System II: Et_2O–EtOAc (3:2) used with $KHSO_4$-silica gel (1:14). Quantitation was by HPLC carried out on a Spherisorb 10 ODS, 250 × 4.6-mm column, a solvent system of AcCN-HOAc-H_2O (18:1:82), at a flow rate of 2 ml/min, with detection by UV at 340 nm. Two major compounds were separated and identified as hydroxycinnamic acid esters (Fig. 6): compound *1* as 5-O-caffeoyl-quinic acid (Syn. chlorogenic acid, 3-O-caffeoylquinic acid (IUPAC 1976) and compound *2* as 4,5-di-O-caffeoylquinic acid (Syn. 3,4-dicaffeoylquinic acid) (IUPAC 1976).

2.4.2 Isolation from Cultivated Plants

The two major hydroxycinnamic acid esters, compounds *1* and *2* were found to occur in all edelweiss plants, especially the aerial plants. Dried leaves from micro-propagated plants (Type 1), matured in peat compost under garden conditions and harvested at the preflowering stage were found to contain 0.45% of *1* and 0.55% of *2*. Roots from the same plants yielded only 0.05% of *1* and 0.07% of *2* (Hennessy et al. 1989). *L. alpinum* types cultivated under identical conditions and collected at the flowering stage showed higher levels of chlorogenic acid in leaves, with dicaffeoyl-quinic acid occurring mainly in the flower heads (Fig. 7).

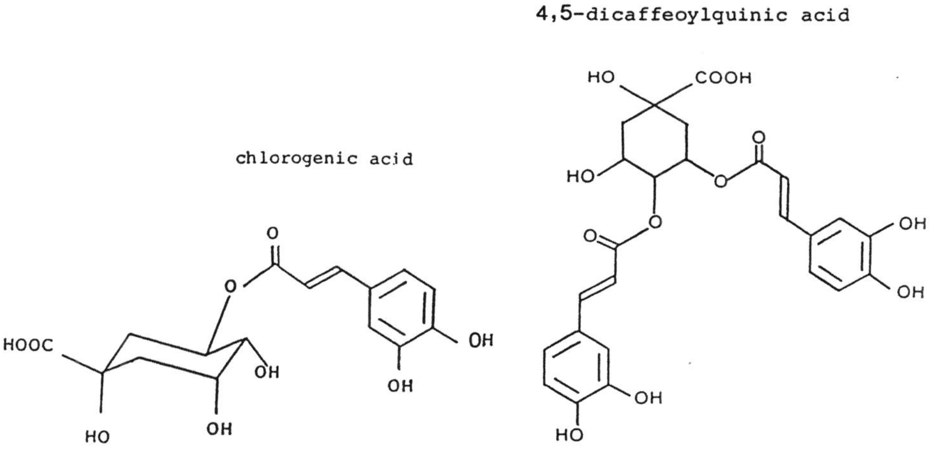

Fig. 6 Major metabolites isolated from *L. alpinum*. (Hennessy et al. 1989)

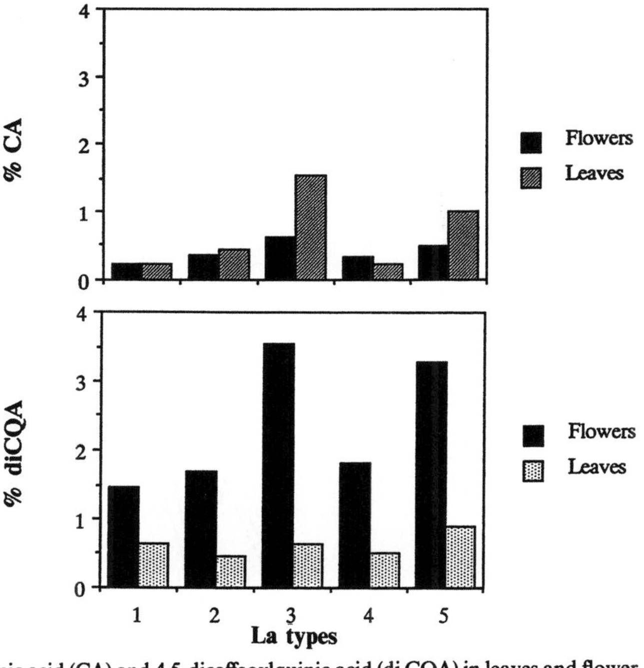

Fig. 7. Distribution of chlorogenic acid (CA) and 4,5-dicaffeoylquinic acid (di CQA) in leaves and flower heads of cultivated *L. alpinum* (La) types

2.4.3 Isolation from Suspension Cultures and Enhancement of Metabolite Yields

Dried suspension cultures produced approx. 2% of both chlorogenic and 4,5-dicaffeoylquinic acids under the culture conditions specified in Sect. 2.2. A time-course experiment (Fig. 8) showed the production of both hydroxycinnamic acid

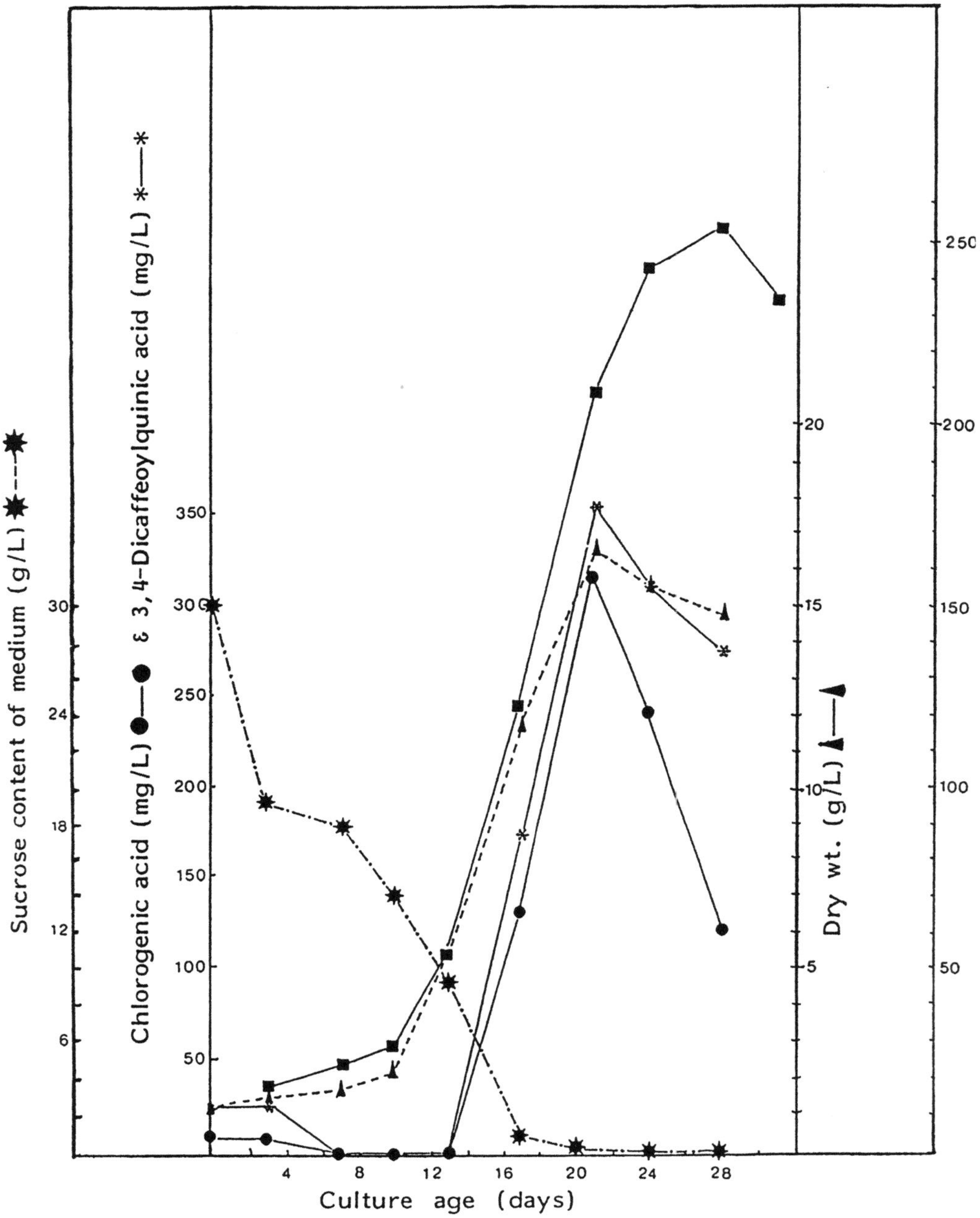

Fig. 8. Time-course experiment of *L. alpinum* cells grown as batch suspension cultures (50-ml medium in 100-ml flasks). Determinations were carried out in triplicate, mean results are shown

esters to be growth associated, their highest concentrations being reached at the time of maximum dry cell-weight production, which in turn was coincident with the depletion of sucrose from the medium. Growth and secondary metabolite formation in cell suspension cultures are known to be affected by manipulation of the concentration of media components, especially sucrose and phytohormones (Zenk et al. 1975; De-Eknamkul and Ellis 1988). Figure 9 shows that by increasing the initial sucrose concentration from 3 to 6%, the biomass yield was more than doubled, while the percentage content of hydroxycinnamic esters also increased. High sucrose levels are known to enhance production of phenolic metabolites, e.g. rosmarinic acid, in suspension cultures (De-Eknamkul and Ellis 1988). Figures 10 and 11 show the influence of phytohormones on growth and metabolite yields in suspension cultures of *L. alpinum*. The optimal growth range was found to be very narrow for both 2,4-D and NAA. Of the concentrations examined, a concentration of 0.2 mg/1 proved best for both hormones, with concentrations above this being strongly inhibitory. Similar results have been reported in the literature for biomass and phenolic metabolite production (Zenk et al. 1975; De-Eknamkul and Ellis 1988). Immobilization and entrapment of plant cells have been shown to be promising and potentially valuable techniques for the production and enhancement of plant secondary metabolites. There are several entrapment techniques available, e.g. alginate beads, polyurethane foam, etc. (Brodelius 1985). For *L. alpinum* cultures, chemical wood pulp and polyurethane foam were evaluated as immobilizing agents. These were washed, dried, divided into small pieces and added to the medium prior to sterilization (0.5 g/50 ml). Culture and growth conditions were as previously described. The suspension cells invaded and were retained within the pores of the foam pieces (Fig. 12). Results of the immobilization experiment shown in Fig. 13 indicate a significant increase in dry cell weight and the

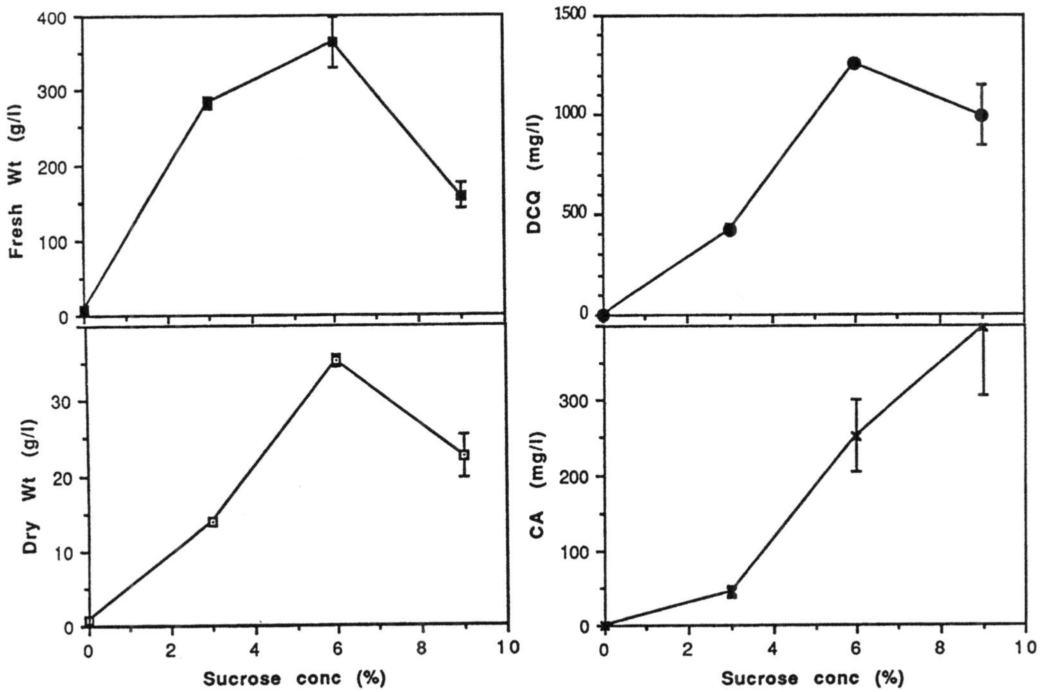

Fig. 9. Effect of initial sucrose concentration in medium on biomass weight and metabolite production. Experiments were carried out using 50 ml of medium in 100-ml flasks and performed in triplicate. Mean results are shown

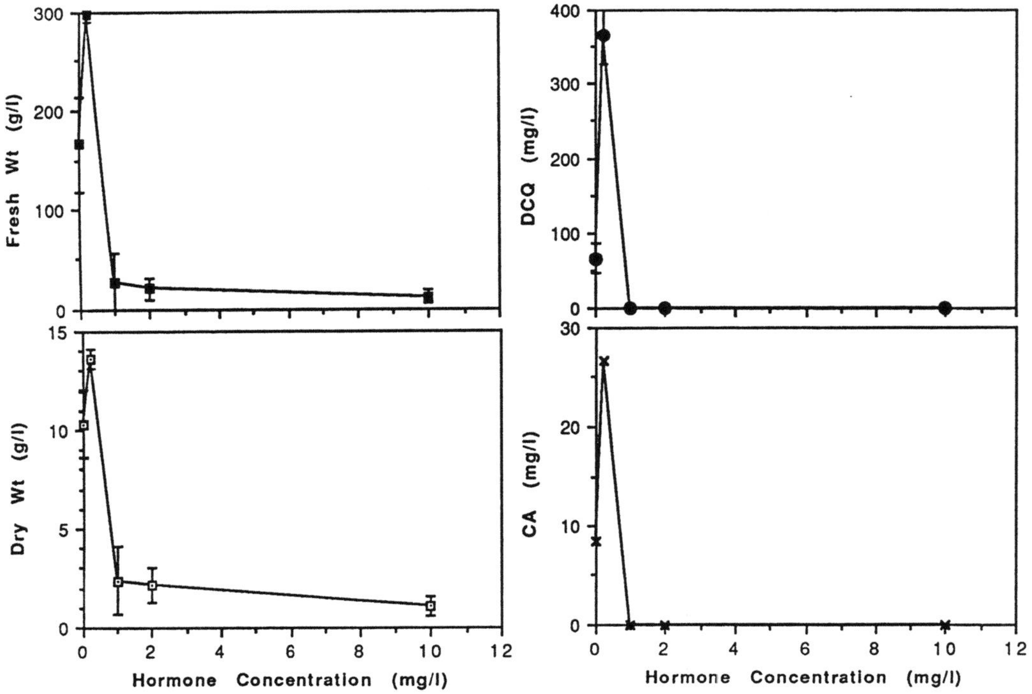

Fig. 10. Effect of initial 2,4-D concentration in medium on biomass weight and metabolite production. Experiments were carried out using 50 ml of medium in 100-ml flasks and performed in triplicate

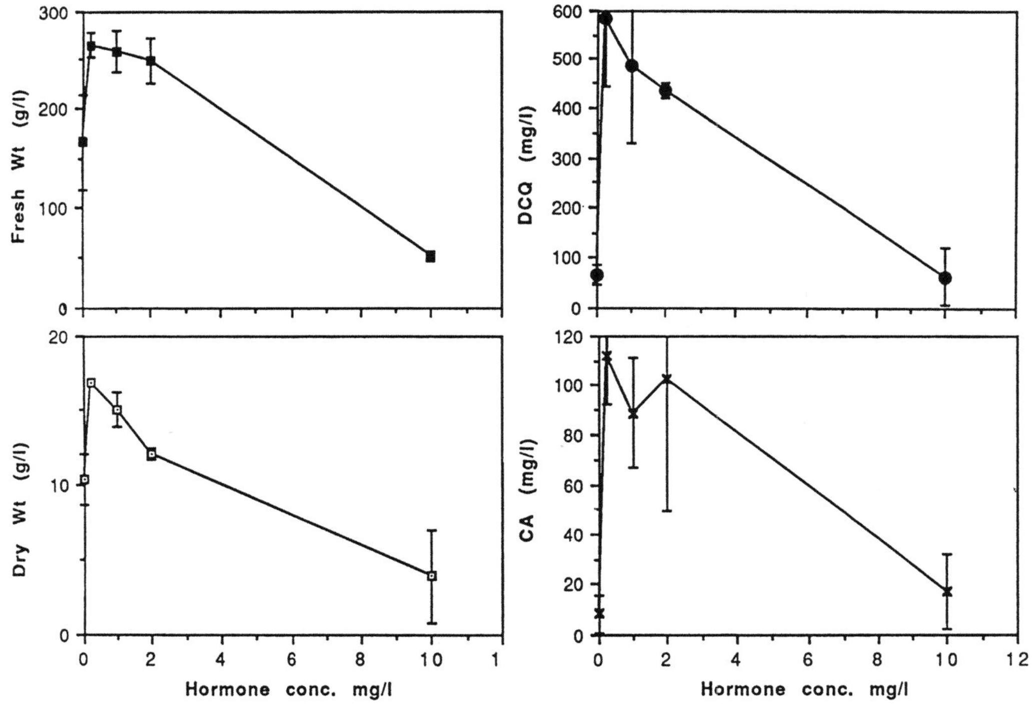

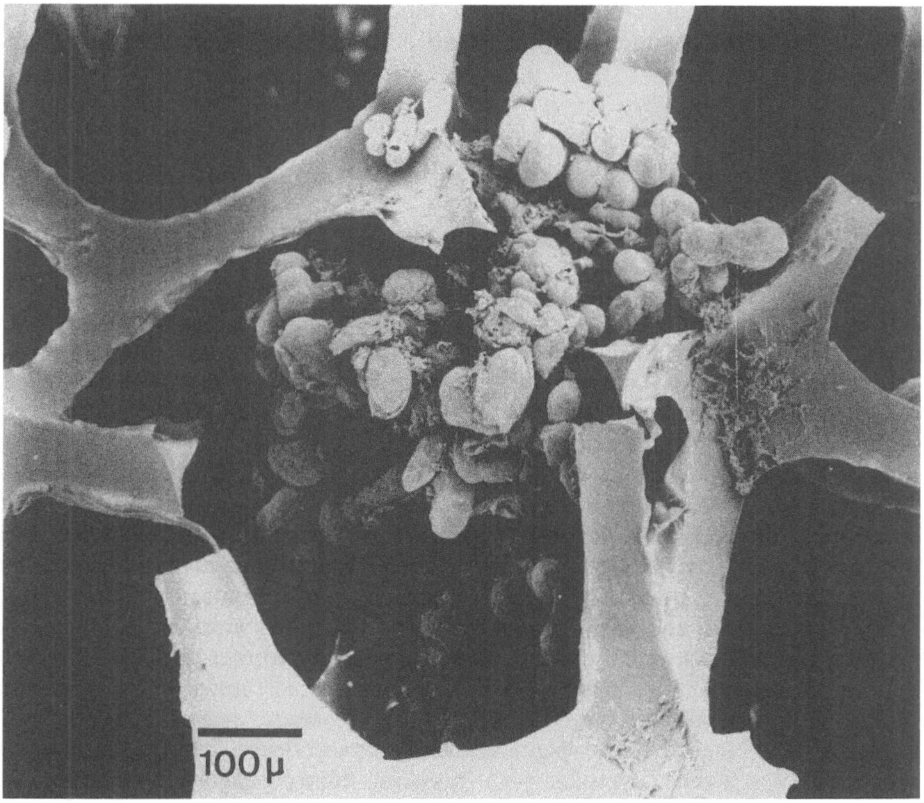

Fig. 12. Scanning-electron photomicrograph of section through polyurethane foam showing immobilized cells of *L. alpinum*. Courtesy of C. Reid, the Electron Microscope Unit, Trinity College Dublin

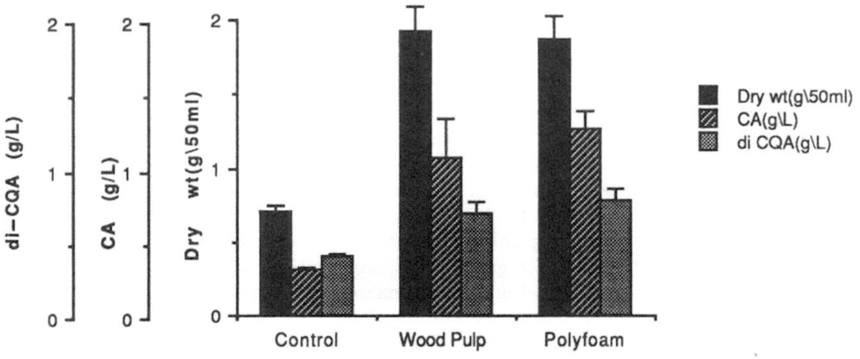

Fig. 13. Effect of immobilization with wood pulp and polyurethane foam on biomass weight and metabolite production. Experiments were carried out using 50 ml of medium in 100-ml flasks and performed in triplicate. Mean results are shown

Fig. 11. Effect of initial NAA concentration in medium on biomass weight and metabolite production. Experiments were carried out using 50 ml of medium in 100-ml flasks and performed in triplicate. Mean results are shown yield of both major metabolites. This product enhancement was the result of a greater biomass production rather than a stimulation in the percentage of metabolites formed.

3 Conclusions and Prospects

Although *L. alpinum* cannot be considered as a traditional herbal remedy of major importance, it can now be thought of as a medicinal plant with potential biological activities. The hydroxycinnamic acid esters, i.e. chlorogenic acid and 4,5-dicaffeoylquinic acid, isolated from its aerial structures have also been found to occur in the more important medicinal plants *Echinacea* species (Cheminat et al. 1988), *Pluchea* species (Martino et al. 1979; Scholz et al. 1990) and *Artemisia* species (Kimura et al. 1985a, b). These plants have been traditionally used for the treatment of a variety of conditions, including inflammation, stomach pains, diarrhea, and liver disorders. The caffeoylquinic acids isolated from them have been found to be biologically active, possessing antimicrobial, immunostimulatory, anti-inflammatory, antioxidant and antiviral activities. Phytochemical work on *L. alpinum* is ongoing and will examine the intact plant, suspension cultures and "hairy" roots for the presence of other potentially active constituents known to occur in the Inulae, namely, sesquiterpenes (Bokadia et al. 1986), sesquiterpene lactones (Bohlmann et al. 1978), flavonoid pigments and polyacetylenes. Due to the limited supply of commercially available plant material, micropropagation and other in vitro techniques will enable sufficient material to be produced for research purposes and possible commercial exploitation.

4 Protocol

For general micropropagation purposes, edelweiss seeds proved to be the best source material, as they were easiest to surface sterilize. The modified MS salt medium (specified in Sect. 2) was satisfactory under the culture conditions used for the establishment of roots. These were well developed after 4 weeks growth, allowing subculture or transplantation of the rooted plantlets. Plants of Type 1 proved to be hardiest and most amenable to garden transplantation.

Evaluation of cell suspension cultures for optimized secondary metabolite (hydroxycinnamic acid esters) production showed highest concentrations were obtained when the modified MS medium (specified in Sect. 2.2) contained sucrose at a concentration of 60/l instead of 30 g/l and cells were cultivated for 21 days instead of 4 weeks. Immobilization techniques could be used to stimulate the metabolite production, but the conditions used would need further refinement.

References

Bicchi C, Nano GM, Tira S (1975) *n*-Paraffin components of some Gnaphalieae. Planta Med 28(4): 389–391

Bohlmann F, Mahanta PK, Jakupovik J (1978) New sesquiterpene lactones from *Inula* species. Phytochemistry 17(7): 1165–1172

Bokadia MM, Macleod AJ, Mehta BH (1986) The essential oil of *Inula racemosa*. Phytochemistry 25(12): 2887–2888

Bornträger + Schlemmer (1988) Commercial catalogue. B + S Offstein, Germany

Brodelius P (1985) Immobilised plant cells – preparation and biosynthetic capacity. In: Woodward J (ed) Immobilised cells and enzymes, a practical approach. IRL Press, Oxford Washington DC pp 127–145

Cheminat A, Zawatzky R, Becker H, Brouilliard R (1988) Caffeoyl conjugates from *Echinea* species: structures and biological activity. Phytochemistry 27(9): 2787–2794

Chiej R (1984) The MacDonald encyclopeidia of medicinal plants. MacDonald, London

Dashbalyn TS, Glyzin VI (1978) Flavonoid glycosides of the edelweiss (*Leontopodium ochroleucum*). Khim Prir Soedin (Tashk) No 6: 807

De-Eknamkul W, Ellis BE (1988) Rosmarinic acid: production in plant cell cultures. In: Bajaj YPS (ed) Biotechnology in agriculture and forestry, vol 4, Medicinal and aromatic plants I. Springer, Berlin Heidelberg New York, pp 310–329

Gessner O, Orzechowski G (1974) Gift – und Arzneipflänzen von Mitteleuropa. Carl Winter, Heidelberg

Gresshoff PM, Doy CH (1974) Derivation of a haploid cell line from *Vitis vinifera* and the importance of the stage of meiotic development of anthers for haploid culture of this and other genera. Z Pflanzenphysiol 73: 132–141

Handel-Mazzetti H (1927) Systematische Monographie der Gattung *Leontopodium*. Beih bot Centralbl 44(II): 1–178

Hartmann H (1972) On the vegetation of Karakorum, Part 2. Vegetatio 24(1–3): 91–157

Hartwell JL (1968) Plants used against cancer. Lloydia 31(2): 71–170

Hennessy D, Hook I, Sheridan H, McGee A (1989) Hydroxycinnamic acid esters from cell suspension cultures and plants of *Leontopodium alpinum*. Phytochemistry 28(2): 489–490

Hess HE, Landolt E, Hizel R (1972) Flora der Schweiz, Band 3. Birkhauser Basel Hamburg

Heywood VH (1973) Flora europaea notulae systematicae and floram europaeam spectantes No 14. Bot J Linn Soc 67: 275–283

Hoppe HA (1958) Drogenkunde 7th edn Cram de Gruyter, Hamburg

Huneck S, Knapp HD (1986) Components of further Compositae from Mongolia. Pharmazie 41(9): 673

IUPAC (1976) Nomenclature of cyclitols. Biochem J 153: 23–31

Khatoon S, Ali SI (1988) Chromosome numbers in Compositae from Pakistan. Candollea 43(2): 455–466

Khokhryakov AP (1983) New plant species from the northeast of the Asian continent. Byull Mosk O VA Ispyt Prir Otd Biol 88(5): 95–97

Kimura Y, Okuda H, Okuda T (1985a) Studies on the activities of the tannins and related compounds from medicinal plants and drugs VI: Inhibitory effects of caffeoyl quinic acids on histamine release from rat peritoneal mast cells. Chem Pharm Bull 33(2): 690–696

Kimura U, Okuda H, Okuda T (1985b) Studies on the activities of tannins and related compounds from medicinal plants and drugs VII: Effects of extracts of leaves of *Artemisia* species and caffeic acid and chlorogenic acid on lipid metabolic injury in rats fed peroxidised oil. Chem Pharm Bull 33(5): 2028–2034

Lapchik VF, Gleba DM, Stupnitskii VA, Tsa VM (1984) Tissue culture for the rapid and effective propagation of rare and endangered medicinal and ornamental plant species. Okhrana, Izuchenie i Obogashchenie Rastitel nogo Mira (No 11): 97–100

Lee KE, Lee WT, Kim JH (1985) Studies on growing conditions of plants of the genus *Leontopodium* distributed in Korea. J Korean Soc Hortic Sci 26(1): 59–65

Martino VS, Debenedetti SL, Coussio J (1979) Caffeoylquinic acids from *Pterocaulon virgatum* and *Pluchea sagittalis*. Phytochemistry 18: 2052

Murashige T, Skoog F (1962) A revised medium for rapid growth and bioassays with tobacco tissue culture. Physiol Plant 15: 473–497

Nishikawa T (1985) Chromosome counts of flowering plants of Hokkaido Japan – 9. J Hokkaido Univ Educ Sect IIB 36(1): 25–40

Park JI, Shim SY, Kim JH, Lee BC, Moon HK (1987) In-vitro rapid propagation of edelweiss *Leontopodium coreanum*. Res Rep Inst Genet 0(22): 128–131

Paternostro MP, Passannanti S, Venturella P, Bellino A (1973) Alkanes and sterols from some *Sideritis*, *Stachys*, *Leontopodium espeleta* and *Cistus* species. Atti Accad Sci Lett Arti Palermo Parte 1 vol Date 1971–1973, 32: 39–43

Pizov V YU, Bartkov YA F (1973) Rare medicinal plants of the subalpine and alpine zones of the Ukranian-Carpathian mountains of USSR and their protection. Farm Zh 28(2): 92–94

Scholz E, Heinrich M, Hunkler D (1990) New caffeic acid esters from *Pluchea symphytifolia* and their antimicrobial activity. Int Symp, Biology and chemistry of active natural substances, Bonn, July 17–22, Planta Med 56(6): 648

Shmetova NS, Shcherbova MA (1982) Unique community with *Leontopodium palibinianum*. SFSR USSR. zv Sib Otd Akad Nauk SSSR Ser Biol Nauk 0(2): 55–57. (in Russian)

Siljak S, Cartier D, Gorenflot R (1974) Introduction to the study of *Leontopodium alpinum* Cass: variability in morphology and chromosome number in the natural population. Hebd Seances Acad Sci Ser D Sci Nat 278(23): 2905–2908

Sugar I (1971) Mount Ucka – a new locality for *Leontopodium alpinum* var *krasense* D in Croatia. Acta Bot Croat 30: 153–156

Tira S, Galeffi C, Di Modica G (1970) Flavonoids of Gnaphalieae: *Leontopodium alpinum*. Experientia 26(11): 1192

Tutin TG, Heywood VH, Burges NA (eds) (1976) Flora europaea, vol 4. Cambridge Univ Press, Cambridge

Voroshilov VN (1979) *Leontopodium blagoveshczenskyi* – new species of edelweiss. Byull Mosk O VA Ispyt Prir Otd Biol 84(4): 102–104

Wagenitz G (1979) *Leontopodium alpinum*. In: Hegi G (ed) Flora Mitteleuropa, Band IV, Teil 3. Paul Perey, Berlin, pp 131–136

Zenk MH, El-Shagi H, Schulte U (1975) Anthraquinone production by cell suspension cultures of *Morinda citrifolia*. Planta Med 28 (suppl): 79–101

Zhao Q, Lu J, Xu D (1984) Chemical constituents of Lao Tou Cao (*Leontopodium leontopodiodes* Beauv). Zhongcaoyao 15(3): 103–104

XVI *Nerium* spp.: In Vitro Culture and the Production of Secondary Metabolites

D.H. Paper[1, 2] and G. Franz[1]

1 Introduction

The genus *Nerium* consists of two species *Nerium oleander* L. (Fig. 1) and *Nerium odorum* Sol. (= *Nerium odorum* Lam. = *Nerium indicum* Mill.). They are distributed in the Mediterranean region, Arabia, the Near East and East India. In many other areas they are cultivated as ornamental plants (Hegi 1927). Various parts of the plants are reputed to be therapeutic agents in the treatment of swellings, leprosy, eye and skin diseases (Dymock et al. 1891; Chopra et al. 1956). The leaves mainly possess cardiotonic and antibacterial properties and are utilized as counter-poisons against snake venom (Dymock et al. 1891; Chopra et al. 1956). Because of these applications and their pharmacological effects, many researchers have investigated the various compounds of *Nerium* spp. The substances which have been found in the leaves, root bark and seeds belong to the flavonoids, saponins, cardenolides, pregnanes and hexadecanoic acid derivatives. The structures of these components will be described in this chapter.

2 Main Components of *Nerium* spp.

2.1 Flavonoids

The leaves of *Nerium oleander* L. contain rutin, kaempferol-3-rhamnosidoglucoside (Fig. 2) (Hörhammer et al. 1956, Görlich 1961).

2.2 Cardenolides

Cardenolides have been found in the leaves, root bark, stem bark and the seeds of *Nerium oleander* L. and *Nerium odorum* Sol. (Tables 1–3). The compounds consist of an aglycone which is glycosidically linked to one or more sugars (Fig. 3).

[1] Institute of Pharmaceutical Biology, University of Regensburg, Universitätsstr. 31, W-8400 Regensburg, FRG
[2] Address for correspondence: St. Georgs-Platz 8, W-8451 Rieden, FRG
Abbreviations: Ac = Acetate, Dgn = D-Diginose, Dgt = D-Digitalose, GB = Gentiobiose, Glc = D-Glucose, Oln = L-Oleandrose, Rhm = L-Rhamnose, Srm = D-Sarmentose

Biotechnology in Agriculture and Forestry, Vol. 21
Medicinal and Aromatic Plants IV (ed. by Y.P.S. Bajaj)

Fig. 1. A plant of *Nerium oleander* in a flowering state

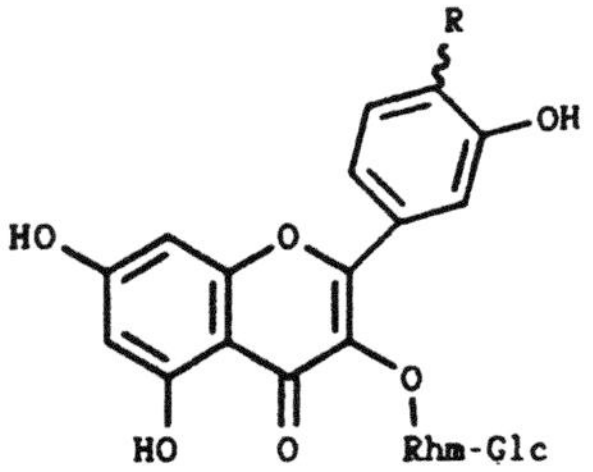

R	flavonoid
H	kaempferol-3-rhamnosidoglucoside
OH	rutin

Fig. 2. Flavonoids in *Nerium oleander* L.

Table 1. Monoglycosides

Name	Aglycone	Sugar	Plant species	Plant organ	Reference
Digitoxigenin-glucoside	Digitoxigenin	Glc	*Nerium oleander*	Leaf	Paper (1989) Paper and Franz (1989)
Odoroside H	Digitoxigenin	Dgt	*Nerium oleander*	Seed	Jäger et al. (1959)
			Nerium odorum	Bark	Yamauchi et al. (1976b)
Digitoxigenin--α-L-oleandroside	Digitoxigenin	Oln	*Nerium odorum*	Leaf	Abe and Yamauchi (1978)
Odoroside A	Digitoxigenin	Dgn	*Nerium oleander*	Seed	Jäger et al. (1959)
			Nerium odorum	Bark	Yamauchi et al. (1976b)
Digistroside	Digitoxigenin	Srm	*Nerium oleander*	Seed	Jäger et al. (1959)
Uzarigenin-digitaloside	Uzarigenin	Dgt	*Nerium odorum*	Root	Yamauchi et al. (1976b)
Odoroside B	Uzarigenin	Dgn	*Nerium odorum*	Bark	Yamauchi et al. (1976b)
Strospeside	Gitoxigenin	Dgt	*Nerium oleander*	Seed	Jäger et al. (1959)
Desacetyl-nerigoside	Gitoxigenin	Dgn	*Nerium oleander*	Seed	Jäger et al. (1959)
Desacetyl-oleandrin	Gitoxigenin	Oln	*Nerium oleander*	Seed Leaf	Jäger et al. (1959) Görlich (1961)
Desacetyl-cryptogran-doside A	Gitoxigenin	Srm	*Nerium oleander*	Seed	Jäger et al. (1959)
Oleandrigenin-glucoside	Oleandrigenin	Glc	*Nerium oleander*	Leaf	Paper (1989) Paper and Franz (1989)
			Nerium odorum	Leaf	Yamauchi et al. (1975)
Neritaloside	Oleandrigenin	Dgt	*Nerium oleander*	Seed	Jäger et al. (1959)
Oleandrin	Oleandrigenin	Oln	*Nerium oleander*	Leaf Seed	Görlich (1961) Jäger et al. (1959)
Nerigoside	Oleandrigenin	Dgn	*Nerium odorum*	Leaf	Abe and Yamauchi (1979)
			Nerium oleander	Seed	Jäger et al. (1959)
Cryptogran-doside A	Oleandrigenin	Srm	*Nerium oleander*	Seed	Jäger et al. (1959)
Oleaside B	Oleagenin	Dgt	*Nerium oleander*	Leaf	Abe and Yamauchi (1979)
Oleaside A	Oleagenin	Dgn	*Nerium oleander*	Leaf	Abe and Yamauchi (1979)

Table 1. Continued

Name	Aglycone	Sugar	Plant species	Plant organ	Reference
Adynerin	Adynerigenin	Dgn	*Nerium odorum*	Leaf	Yamauchi et al. (1973)
			Nerium oleander	Leaf	Görlich (1961)
5α-Adynerin	5α-Adynerigenin	Dgn	*Nerium odorum*	Leaf	Abe and Yamauchi (1978)
Δ^{16}-Dehydro-adynerigenin-digitaloside	Δ^{16}-Dehydro-adynerigenin	Dgt	*Nerium odorum*	Leaf	Yamauchi et al. (1973)
Δ^{16}-Dehydro-adynerin	Δ^{16}-Dehydro-adynerigenin	Dgn	*Nerium odorum*	Leaf	Abe and Yamauchi (1979) Yamauchi et al. (1973)
Δ^{16}-Anhydro-strospeside	Δ^{16}-Anhydro-gitoxigenin	Dgt	*Nerium oleander*	Seed	Jäger et al. (1959)
Δ^{16}-Anhydro-desacetylneri-goside	Δ^{16}-Anhydro-gitoxigenin	Dgn	*Nerium oleander*	Seed	Jäger et al. (1959)
Δ^{16}-Anhydro-desacetylcrypto-grandoside A	Δ^{16}-Anhydro-gitoxigenin	Srm	*Nerium oleander*	Seed	Jäger et al. (1959)
Kaneroside	2α-Hydroxy-8,14β-epoxy-5β-card 16:17, 20:22-dienolide	Dgn	*Nerium oleander*	Leaf	Siddiqui et al. (1987b)
			Nerium odorum	Leaf	Siddiqui et al. (1987b)
Neriumoside	2α,14β-Dihydroxy-5β-carda-16:17 20:22-dienolide	Dgn	*Nerium oleander*	Leaf	Siddiqui et al. (1987b)
			Nerium odorum	Leaf	Siddiqui et al. (1987b)
Neriaside	Neriagenin	Dgn	*Nerium odorum*	Leaf	Yamauchi and Abe (1978)

Glc (1——>6) Glc (1——>4) hexose (1——>3) aglycone

hexose = Glc, Dgt, Dgn, Oln, Srm

Fig. 3. Basic structure of the cardenolides of *Nerium* spp.

A great variety of aglycones have been discovered in *Nerium* (Fig. 5). Distinct from most of the cardenolides derived from *Digitalis* spp., the sugar moiety contains at most only one deoxysugar which is connected directly to the aglycone. These sugars are in some cases glucose, but in the main cardenolides of *Nerium* this sugar is either a 6-deoxysugar or a 2,6-dideoxysugar which is methylated at C′-3 (Fig. 4). Either a glucose, in the case of the diglycosides, or the disaccharide gentiobiose, in the case of the triglycosides, can be attached to position C′-4 of the deoxysugar. These glucoses could rapidly be cleaved by endogenous plant

Table 2. Diglycosides

Name	Aglycone	Sugar	Plant species	Plant organ	Reference
Odoroside D	Digitoxigenin	Dgn-Glc	*Nerium odorum*	Bark	Yamauchi et al. (1976b)
Uzarigenin-digitaloside-glucoside	Uzarigenin	Dgt-Glc	*Nerium odorum*	Root	Yamauchi et al. (1976b)
Bioside K	Uzarigenin	Dgn-Glc	*Nerium odorum*	Bark	Yamauchi et al. (1976b)
Glucosyl-nerigoside	Oleandrigenin	Dgn-Glc	*Nerium odorum*	Leaf	Yamauchi et al. (1975)
Glucosyl-neritaloside	Oleandrigenin	Dgt-Glc	*Nerium odorum*	Leaf	Yamauchi et al. (1975)
Glucosyl-oleandrin	Oleandrigenin	Oln-Glc	*Nerium odorum*	Leaf	Yamauchi et al. (1975)
Oleaside D	Oleagenin	Dgt-Glc	*Nerium oleander*	Leaf	Abe and Yamauchi (1979)
Oleaside C	Oleagenin	Dgn-Glc	*Nerium oleander*	Leaf	Abe and Yamauchi (1976)
Neriumoside A-2	3β,21-Dihydroxy-5β-carda-8,14,16,20(22)-tetraenolide	Dgt-Glc	*Nerium odorum*	Root	Yamauchi et al. (1976b)
Neriumoside B-2	3β-Hydroxy-5β-carda-8,14,16,20(22)-tetraenolide	Dgt-Glc	*Nerium odorum*	Root	Yamauchi et al. (1976b)
Δ^{16}-Dehydro-adynerigenin digitaloside-glucoside	Δ^{16}-Dehydro-adynerigenin	Dgt-Glc	*Nerium odorum*	Leaf	Yamauchi et al. (1975)

[a]Glucosyl-neritaloside = 16-O-acetyl-digitalinum verum.

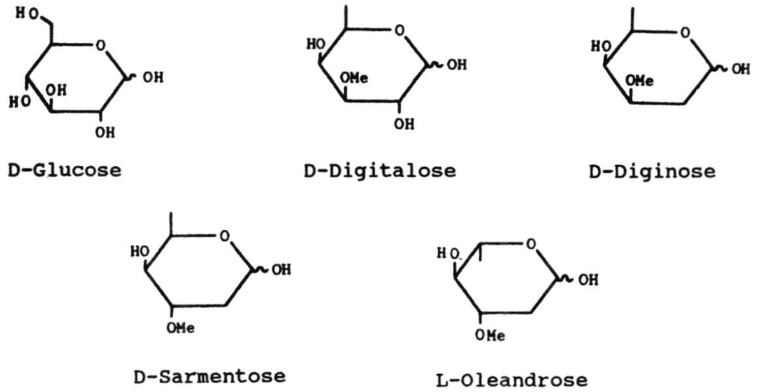

Fig. 4. Sugars occurring in the cardenolides of *Nerium* spp.

Table 3. Triglycosides

Name	Aglycone	Sugar	Plant species	Plant organ	Reference
Odoroside G	Digitoxigenin	Dgt-GB	*Nerium odorum*	Leaf	Abe and Yamauchi (1979) Yamauchi et al. (1975)
GB-odoroside A[a]	Digitoxigenin	Dgn-GB	*Nerium odorum*	Leaf	Abe and Yamauchi (1979) Yamauchi et al. (1975)
Odoroside K	Uzarigenin	Dgn-GB	*Nerium odorum*	Bark	Yamauchi et al. (1976b)
16-O-acetyl-neogitostin	Oleandrigenin	Dgt-GB	*Nerium odorum*	Root Leaf	Yamauchi et al. (1976b) Abe and Yamauchi (1979)
GB-nerigoside[b]	Oleandrigenin	Dgn-GB	*Nerium odorum*	Leaf	Abe and Yamauchi (1979)
GB-oleandrin	Oleandrigenin	Oln-GB	*Nerium odorum*	Leaf	Yamauchi et al. (1975)
Oleaside F	Oleagenin	Dgt-GB	*Nerium oleander*	Leaf	Abe and Yamauchi (1979)
Oleaside E	Oleagenin	Dgn-GB	*Nerium oleander*	Leaf	Abe and Yamauchi (1979)
GB-adynerin	Adynerigenin	Dgn-GB	*Nerium odorum*	Leaf	Yamauchi et al. (1975)
GB-Δ^{16}-adynerin	Δ^{16}-Adynerigenin	Dgn-GB	*Nerium odorum*	Leaf	Yamauchi et al. (1975)
Neriumoside A-1	3β-21-Dihydroxy-5β-carda-8,14,20,(22)-tetraenolide	Dgt-GB	*Nerium odorum*	Root	Yamauchi et al. (1976a)
Neriumoside B-1	3β-Hydroxy-5β-carda-8,14,16,20(22)-tetraenolide	Dgt-GB	*Nerum odorum*	Root	Yamauchi et al. (1976a)
Neriumoside C-1	–„–	Dgn-GB	*Nerium odorum*	Leaf	Yamauchi et al. (1976a)

[a] GB-odoroside A = digitoxigeninneritrioside.
[b] GB-nerigoside = oleandrigeninneritrioside.

enzymes. Therefore, cardenolides in the freshly harvested plant are mostly triglycosides. These are decomposed as a result of the drying procedure. Drying at 80 °C resulted in the hydrolysis of the di- and triglycosides to monoglycosides. The genuine triglycosides are preserved if the plant material is dried at room temperature (Yamauchi and Ehara 1972).

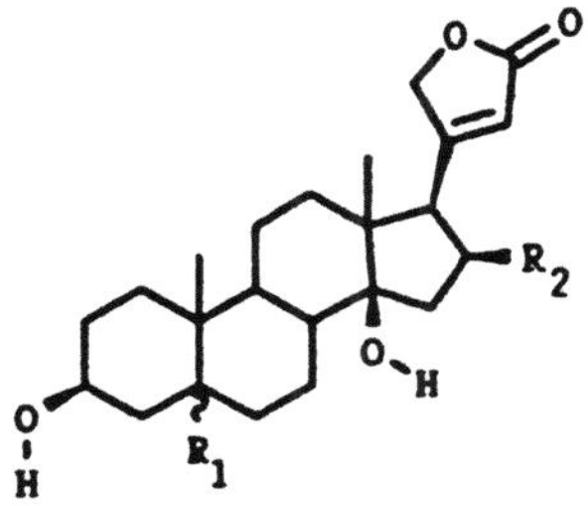

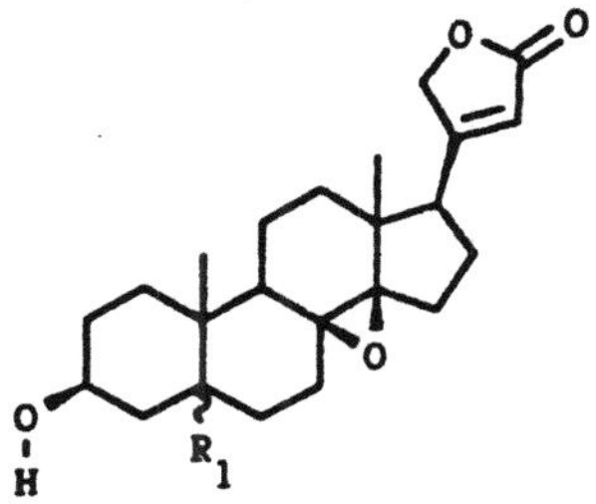

R_1	R_2	aglycone
αH	H	Uzarigenin
ßH	H	Digitoxigenin
ßH	OH	Gitoxigenin
ßH	OAc	Oleandrigenin

R_1	aglycone
αH	5α-Adynerigenin
ßH	Adynerigenin

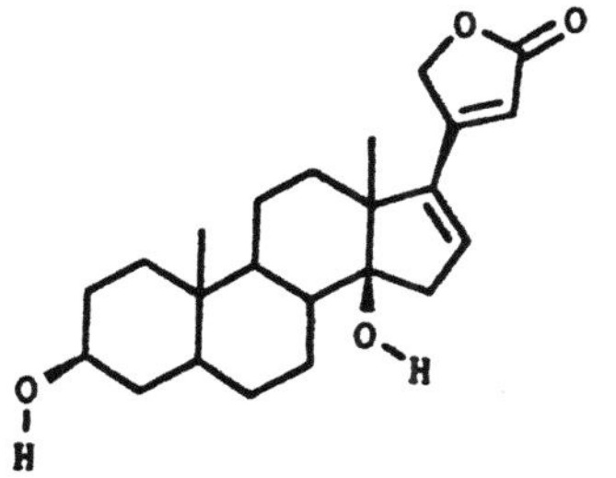

Δ16-Dehydrodigitoxigenin

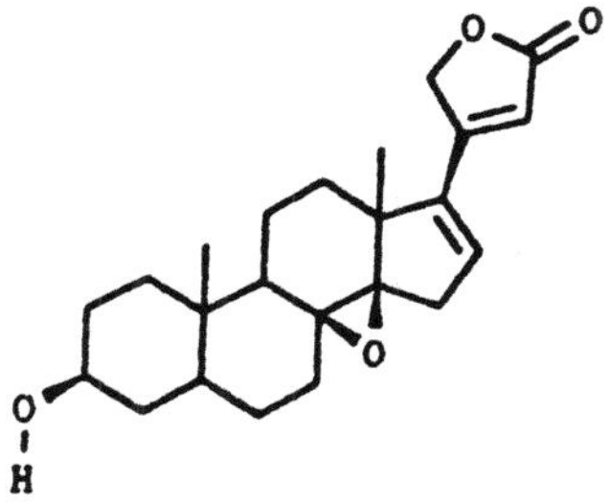

Δ16-Dehydroadynerigenin

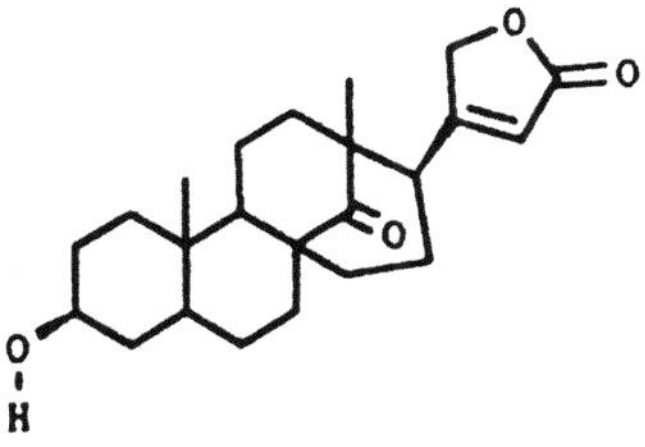

Oleagenin

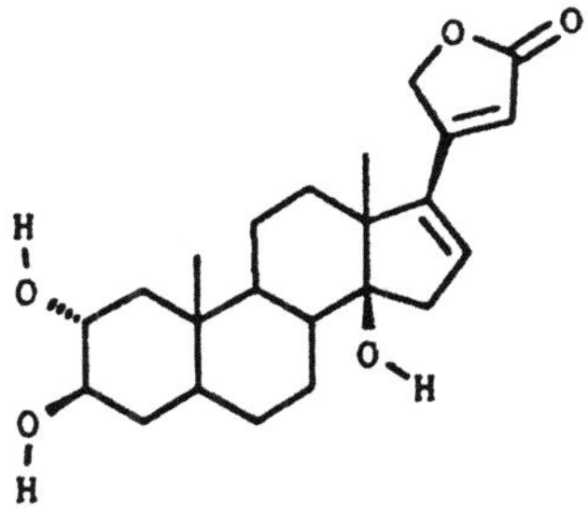

2α,14ß-hydroxy-5ß-carda-16:17,20:22-dienolide

Fig. 5. Aglycones of the cardenolides in *Nerium* spp.

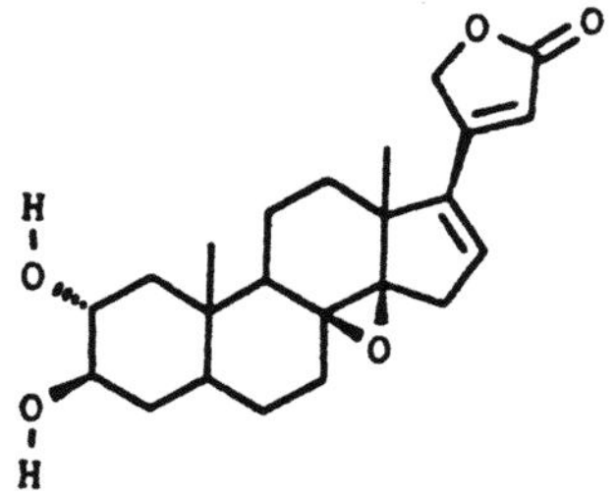

2α-hydroxy-8,14ß-epoxy-5ß-carda-16:17,20:22-dienolide

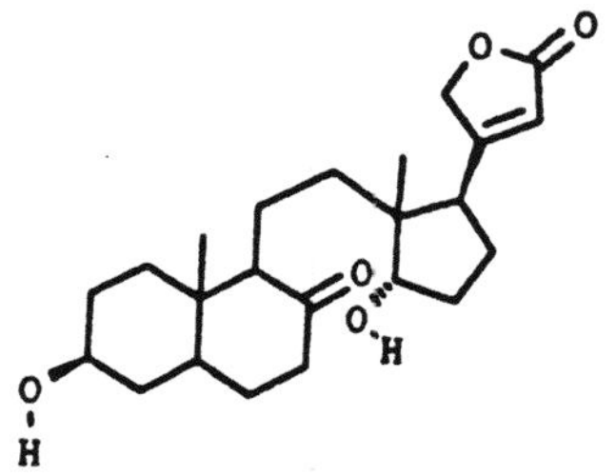

Neriagenin

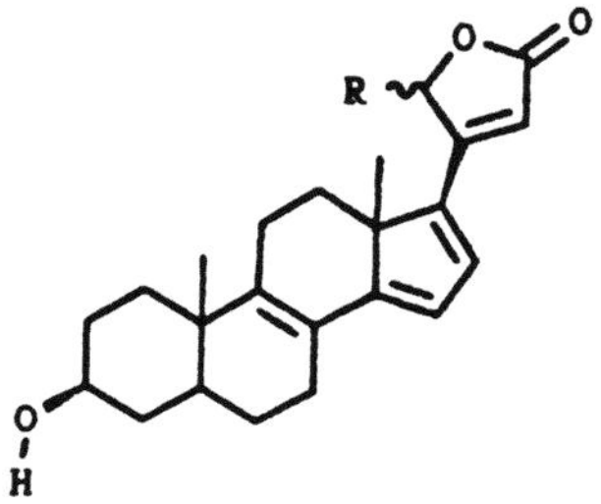

R	aglycone
H	3ß-hydroxy-5ß-carda-8,14,16,20(22)-tetraenolide
OH	3ß,21,dihydroxy-5ß-carda-8,14,16,20(22)-tetraenolide

Fig. 5. Continued

2.3 Pregnanes

2.3.1 Pregnanaglycones

Neridienone A, neridienone B, 12β-hydroxy-pregna-4,6-diene-3,20-dione, 12β-hydroxy-pregn-4-en-3,20-dione, 12β-hydroxy-16α-methoxy-pregna-4,6-diene-3,20-dione (Fig. 6) were found in the root bark of *Nerium odorum* Sol. (Abe and Yamauchi 1976).

2.3.2 Pregnenolone Glycosides

Also, four pregnenolone glycosides have been identified in the root bark of *Nerium odorum* Sol. (Yamauchi et al. 1972; Fig. 7)

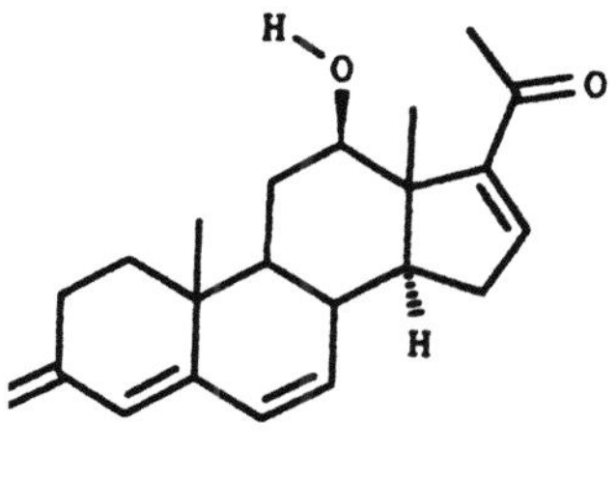

Neridienone A

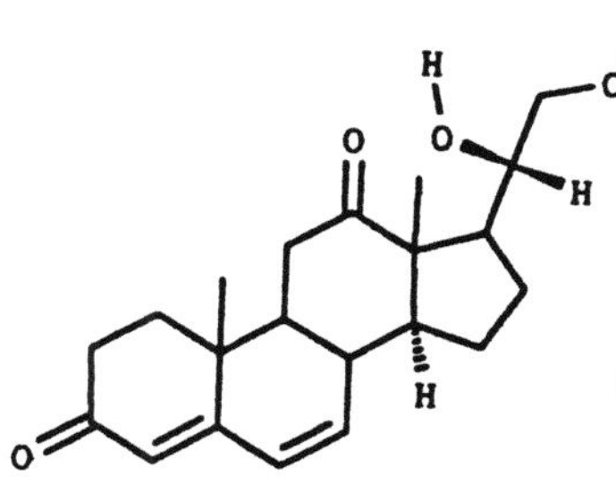

Neridienone B

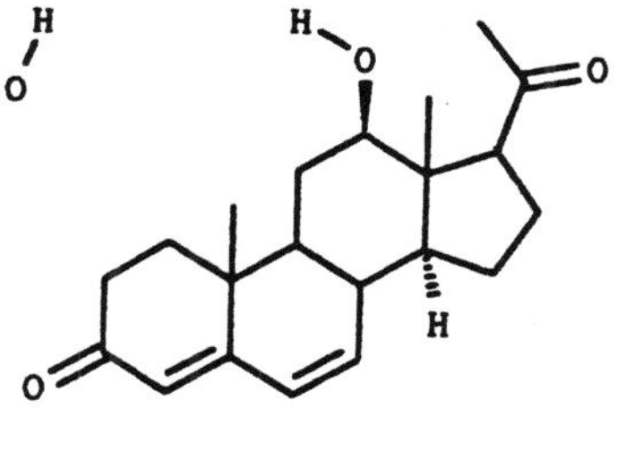

12ß-hydroxy-pregna-
-4,6-diene-3,20-dione

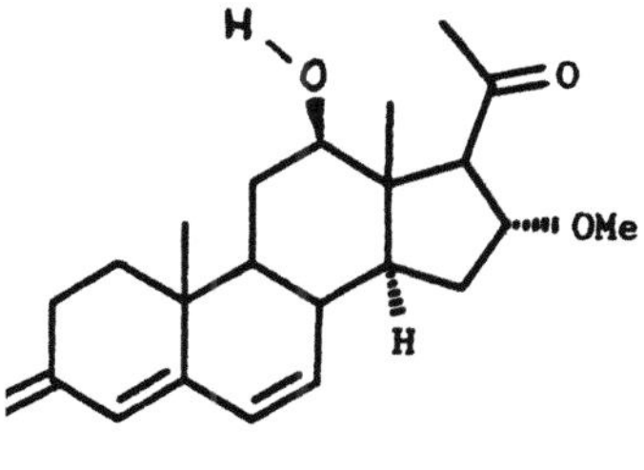

2ß-hydroxy-16α-methoxy-
pregna-4,6-diene-
3,20-dione

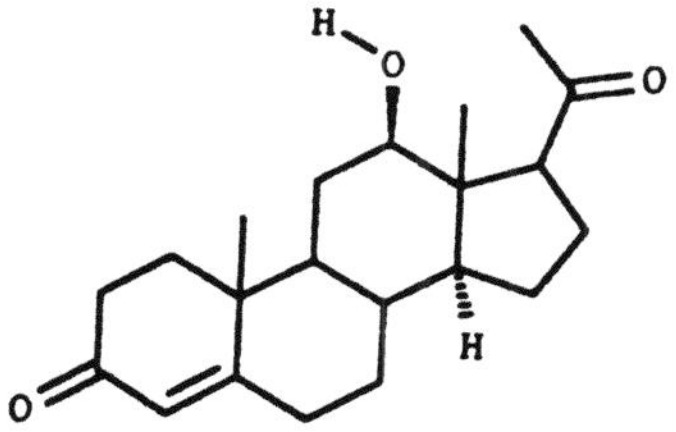

12ß-hydroxy-pregn-4-en-3,20-dione

Fig. 6. Pregnanaglycones in *Nerium odorum* Sol.

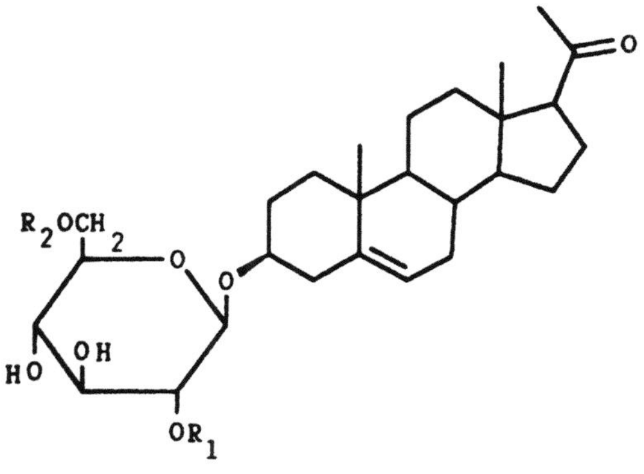

R_1	R_2	Name
H	H	Pregnenolon-ß-D-glucopyranoside
H	Glc	Pregnenolon-ß-D-glucopyranosyl-(1→6)- -D-glucopyranoside
Glc	H	Pregnenolone-ß-D-glucopyranosyl-(1→2)- -D-glucopyranoside
Glc	Glc	Pregnenolone-bis-ß-D-glucopyranosyl- (1→2,1→6)-ß-D-glucopyranoside

Fig. 7. Pregnenolone glycosides of *Nerium odorum* Sol.

2.4 Triterpenes

From the leaves of *Nerium oleander* L. and *Nerium odorum* Sol., betulin, betulinic acid, kanerin, kanerocin, oleanderol, oleanolic acid and ursolic acid have been isolated and identified (Fig. 8; Siddiqui et al. 1988, 1989a, b).

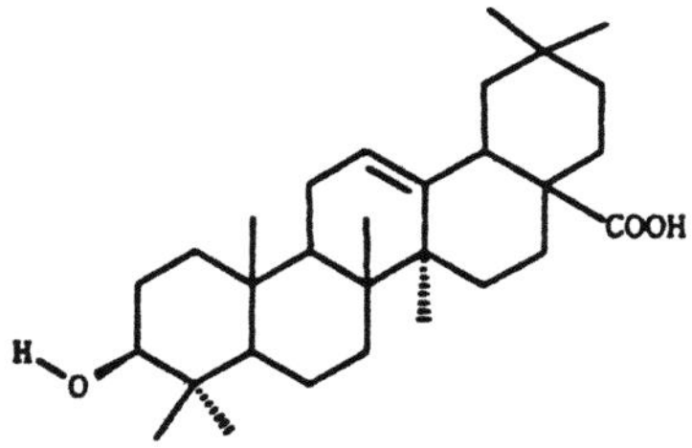

Oleanolic acid

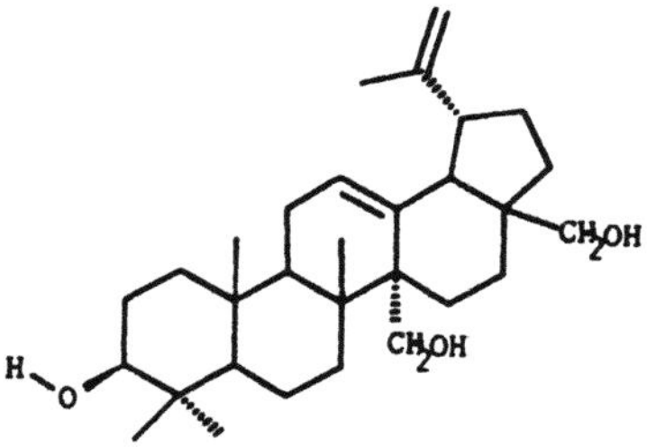

Oleanderol

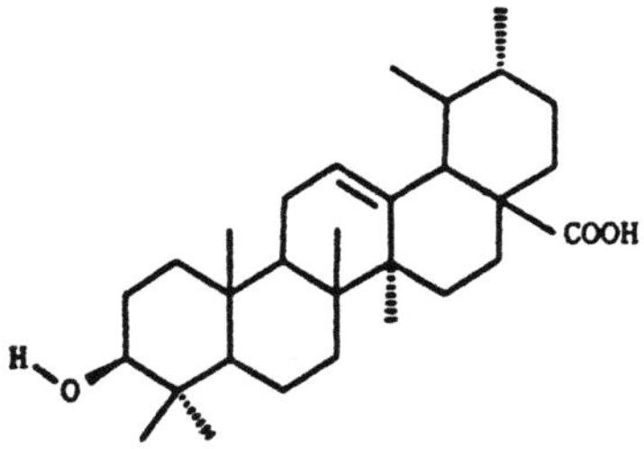

Ursolic acid

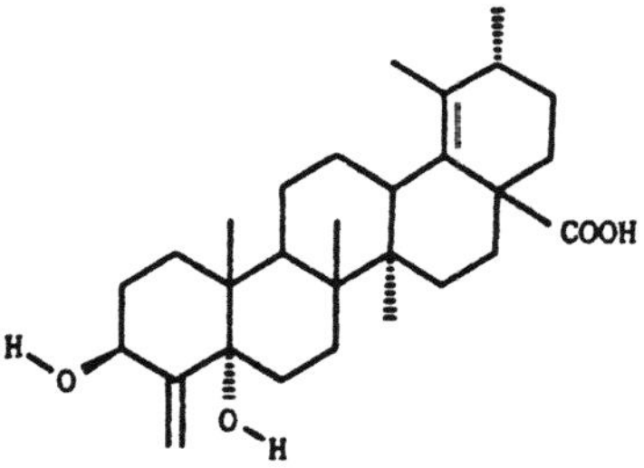

Kanerin

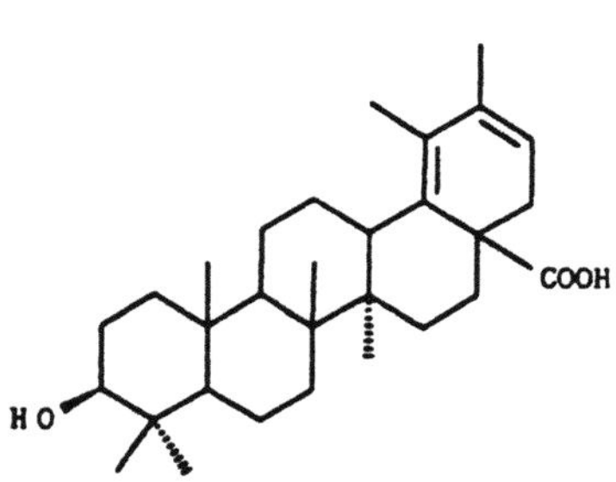

Kanerocin

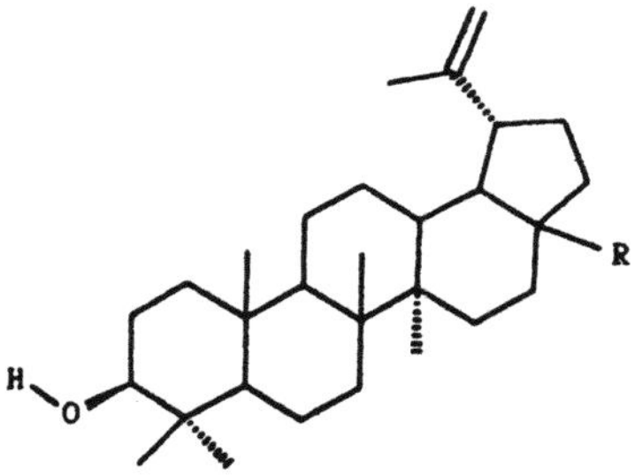

R	Aglycone
CH_2OH	Betulin
COOH	Betulinic acid

Fig. 8. Triterpene aglycones in *Nerium* spp.

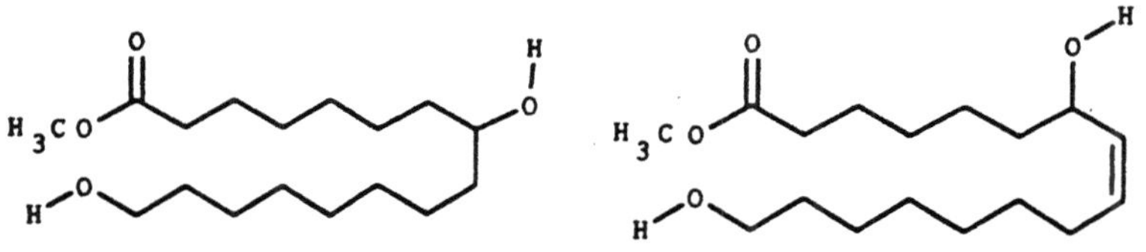

Nerifol Neriumol

Fig. 9. Hexadecanoic acid derivatives in *Nerium odorum* Sol.

2.5 Hexadecanoic Acid Derivatives

Neriumol and nerifol have been found in the leaves of *Nerium odorum* Sol. Some of these derivatives (Fig. 9) are known for their antibiotic effects (Siddiqui et al. 1987a).

3 Cell Cultures of *Nerium* spp.

3.1 Growth Conditions and Formation of Secondary Compounds

First attempts to establish cell cultures of *Nerium oleander* L. were not very successful, due to the formation of polyphenols which resulted in a negative influence upon the callus growth rate. In a modified Murashige and Skoog medium, however, containing a high concentration of 2,4-D and PVP cell cultures derived from leaves and stem of *Nerium oleander* L., resulted in good growth (Paper 1989). In these cell cultures no cardenolides or metabolites containing 2,6-dideoxysugars were detectable (Paper 1989). Only the flavonoids rutin and kaempferol-3-rhamnosidoglucoside, which are present in the genuine plant material, could be identified.

3.2 Biotransformation Experiments with Suspension Cultures of *Nerium oleander* L.

Starting from callus cultures, suspension cultures were established. These are a helpful tool for biotransformation experiments in order to study the biosynthesis of cardenolides and 2,6-dideoxysugars. Further, the cell cultures should be a potential enzyme source to modify administered substrates, especially cardenolides and steroids.

These experiments demonstrated that cell cultures of *Nerium oleander* L. are able to glycosylate, oxidize and isomerize steroids and cardenolide aglycones added to the culture medium (Paper and Franz 1989).

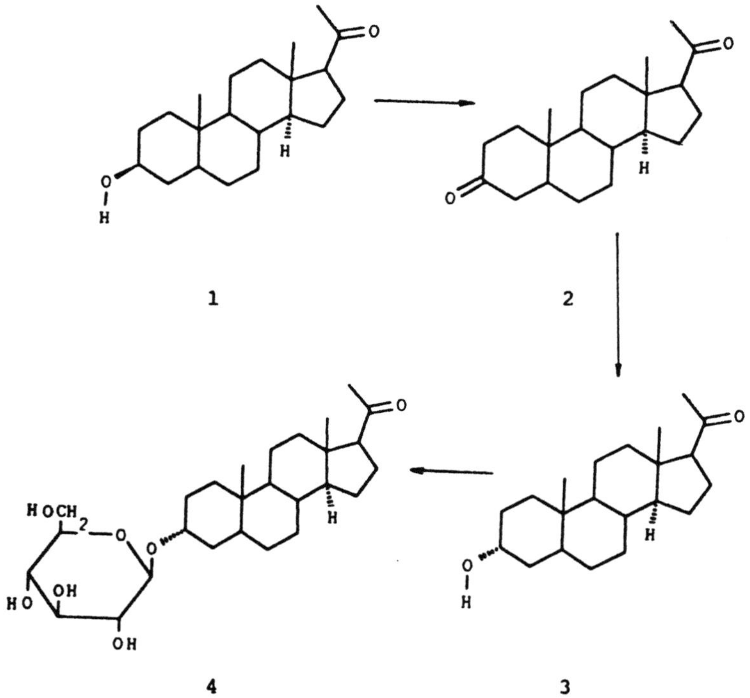

Fig. 10. Biotransformation by cell suspension cultures of *Nerium oleander* L.: 5β-pregnan-3βol-20-one (*1*) to 5β-pregnan-3,20-dione (*2*), 5β-pregnan-3αol-20-one (*3*), 5β-pregnan-3αol-20-one glucoside (*4*)

5β-Pregnan-3βol-20one was transformed to 5β-pregnan-3,20-dione, 5β-pregnan-3αol-20-one and 5β-pregnan-3αol-20-one glucoside (Fig. 10). Digitoxigenin was converted into digitoxigenone, epi-digitoxigenin, digitoxigenin glucoside and epi-digitoxigenin glucoside (Fig. 11). Oleandrigenin was transformed into oleandrigenin glucoside, gitoxigenin glucoside and Δ^{16}-dehydro-digitoxigenin glucoside (Fig. 12). Neither 2,6-dideoxysugar-transferase activity nor cleavage of 2,6-dideoxysugars or 6-deoxysugars could be detected. However, it is noteworthy that a large amount of the cardenolide aglycones were converted into their respective glucosides. Therefore, it should be possible to utilize cell cultures of *Nerium oleander* L. as a source for the isolation of a 5β-steroid-glucosyltransferase (Paper 1989; Paper and Franz 1989).

4 Summary

As shown above, the genus *Nerium* contains a large variety of different chemical compounds and has been utilized for different applications. Cell cultures of *Nerium oleander* L. derived from stem and leaves produce flavonoids, but cardenolides and

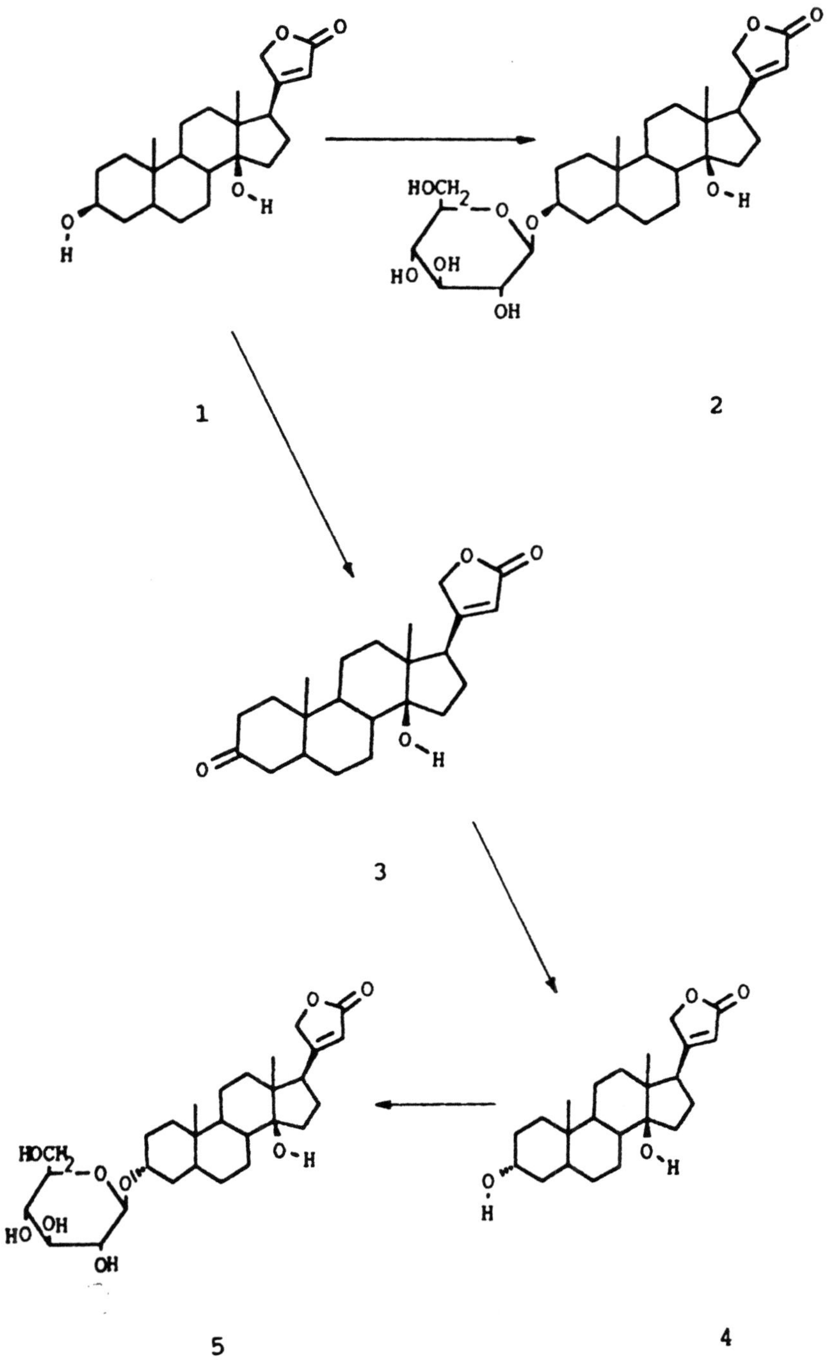

Fig. 11. Biotransformation of digitoxigenin (*1*) to digitoxigenin glucoside (*2*) or digitoxigenone (*3*), epi-digitoxigenin (*4*) and epi-digitoxigeninglucoside (*5*)

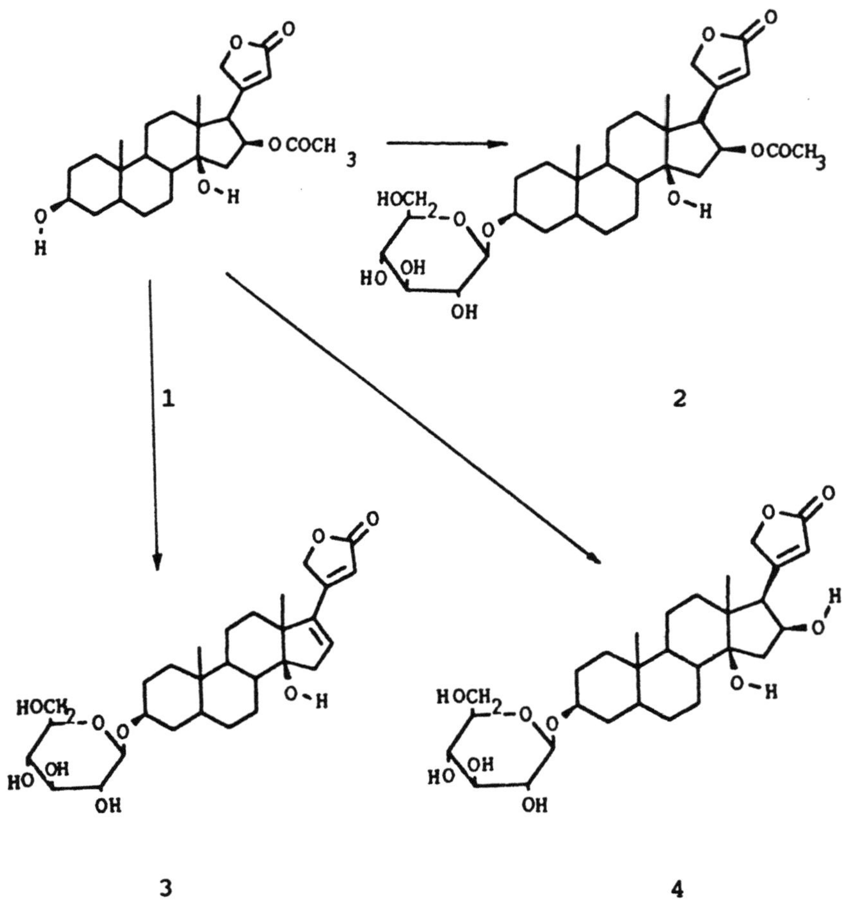

Fig. 12. Biotransformation of oleandrigenin (*1*) to oleandrigenin glucoside (*2*) or digitoxigenin glucoside (*3*) or Δ^{16}-dehydrodigitoxigenin glucoside (*4*)

compounds containing 2,6-dideoxysugars could not be detected. Biotransformation experiments with suspension cultures revealed that they are able to glucosylate 5β-steroids with a 3β-hydroxy group in fair amounts.

It is obvious that many data are documented about *Nerium* secondary metabolites. Since there have been reports about *Nerium oleander* L. extracts against cancer (Özel 1988), a group of *Nerium* compounds, which up to now was of minor interest, was brought into focus, i.e., the water-soluble polysaccharides (Carbik et al. 1990; Müller et al. 1991). Some polysaccharides of varying structure from other biological sources are known to have an effect on the immune system. Therefore, the search for active compounds in *Nerium* will still be interesting and exciting in the future.

5 Protocol

Callus cultures of *Nerium oleander* L. prepared from the stem and leaves were grown on agar plates containing a modified Murashige and Skoog medium with the following composition:

Mineral salts	mg/l	Organic components	mg/l
NH_4NO_3	1650	Glycine	2
KNO_3	1900	Nicotinic acid	0.5
$CaCl_2 \cdot 2H_2O$	440	Myo-inositol	100
$MgSO_4 \cdot 7H_2O$	370	Pyridoxine-HCl	0.5
KH_2PO_4	170	Thiamine-HCl	0.1
NaFeEDTA	36.7	Folic acid	0.5
H_3BO_4	6.2	Biotin	0.5
$MnSO_4 \cdot 4H_2O$	22.3	Sucrose	20000
$ZnSO_4 \cdot 4H_2O$	8.6	Coconut water	100 ml/l
KJ	0.83	PVP	1250
$Na_2Mo_4 \cdot 5H_2O$	0.25	2,4-D	6
$CuSO_4 \cdot 5H_2O$	0.025	Agar	10000
$CoCl_2 \cdot 6H_2O$	0.025		

adjusted to pH 5.8

The medium for the suspension cultures contained the same components except agar and coconut water. These cultures were grown in 250-ml Erlenmeyer flasks containing 50 ml medium. They were shaken at 100 rpm at 28 °C. The callus cultures of *Nerium oleander* L. produce the flavonoids rutin and kaempferol-3-rhamnosidoglucoside.

References

Abe F, Yamauchi T (1976) Pregnanes in the root bark of *Nerium odorum*. Phytochemistry 15: 1745–1748

Abe F, Yamauchi T (1978) Digitoxigenin oleandroside and 5α-adynerin in the leaves of *Nerium odorum* (*Nerium* 9). Chem Pharm Bull 26(10): 3023–3027

Abe F, Yamauchi T (1979) Oleasides; novel cardenolides with an unusual framework in *Nerium* (*Nerium* 10). Chem Pharm Bull 27(7): 1604–1610

Carbik I, Baser HKC, Wagner H (1990) Immunologically active polysaccharides from the aqueous extract of *Nerium oleander*. Poster 3/47, Int Joint Symp Ges Arzneipflanzenforsch, Am Soc Pharmacogn, Assoc Francaise Enseignement Recherche Pharmacogn and Phytochem Soc Eur in Bonn, 17–22.7.90

Chopra RN, Nayer SL, Chopra IC (1956) Glossary of Indian medicinal plants. CSIR, New Delhi, p 175

Dymock W, Warden CJH, Hooper D (1891) Pharmacographia indica, vol 2. Inst of Health and Tibbi Research, Republ under the auspices of Hamdard National Foundation of Pakistan, p 398

Görlich B (1961) Chemische Wertbestimmung von Oleanderextrakt. Planta Med 9: 442–455

Hegi G (1927) Illustrierte Flora von Mitteleuropa, Bd V/3, pp 2056–2058

Hörhammer L, Wagner H, Luck R (1956) Zur Kenntnis der Flavonglycoside von *Nerium oleander* L. Arch Pharm 289/61: 613

Jäger H, Schindler O, Reichstein T (1959) 105. Die Glycoside der Samen von *Nerium oleander* L. Helv Chim Acta 17: 977–1013

Müller B, Roβkopf F, Paper DH, Kraus J, Franz G (1991) Polysaccharide aus *Nerium oleander*: Struktur und biologische Aktivität. Pharmazie 46/9: 657–663

Özel HZ (1988) *Nerium oleander* ekstreleri NOI ve NOO nun habis tümörlelrinde etkisine ait klinik neticeler. February, TDA-Yayini, Maltepe, Ankara, Turkey, p 66

Paper DH (1989) Untersuchungen zur Biosynthese der 2,6-Didesoxyzucker an Ganzpflanzen und Zellkulturen von *Nerium oleander* L. PhD Thesis, University of Regensburg

Paper D, Franz G (1989) Glycosylation of cardenolide aglycones in the leaves of *Nerium oleander*. Planta Med 55: 30–34

Siddiqui S, Begum S, Hafeez F, Siddiqui BS (1987a) Isolation and structure of neriumol and nerifol from the leaves of *Nerium odorum*. Planta Med 53: 47–49

Siddiqui S, Hafeez F, Begum S, Siddiqui BS (1987b) Isolation and structure of two cardiac glycosides from the leaves of *Nerium oleander*. Phytochemistry 26(1): 237–241

Siddiqui S, Hafeez F, Begum S, Siddiqui BS (1988) Oleanderol, a new pentacyclic triterpene from the leaves of *Nerium oleander*. J Nat Prod 51(2): 229–233

Siddiqui S, Begum S, Siddiqui BS, Hafeez F (1989a) Kanerin and 12,13-dihydroursolic acid, two new pentacyclic triterpenes from the leaves of *Nerium oleander*. J Nat Prod 52(1): 57–62

Siddiqui S, Siddiqui BS, Begum S, Hafeez F (1989b) Kanerocin: a new triterpene from the leaves of *Nerium oleander*. Planta Med 55: 292–293

Yamauchi T, Ehara Y (1972) *Nerium* l. Drying conditions of leaves of *Nerium odorum*. Yakugaku Zaschi 92: 155

Yamauchi T, Hara M, Mithashi K (1972) Pregnenolone glucosides of *Nerium odorum*. Phytochemistry 11: 3345–3347

Yamauchi T, Mori Y, Ogata Y (1973) Δ^{16}-Dehydroadynerigenin glycosides of *Nerium odorum*. Phytochemistry 12: 2737–2739

Yamauchi T, Takata N, Mimura T (1975) Cardiac glycosides of the leaves of *Nerium odorum*. Phytochemistry 14: 1379–1382

Yamauchi T, Abe F, Takahashi M (1976a) Neriumosides, cardenolide pigments in the root bark of *Nerium odorum*. Tetrahedron Lett 14: 1115–1116

Yamauchi T, Takahashi M, Abe F (1976b) Cardiac glycosides of the root bark of *Nerium odorum*. Phytochemistry 15: 1275–1278

Yamauchi T, Abe F (1978) Neriaside, a 8,14-seco-cardenolide in *Nerium odorum*. Tetrahedron Lett 21: 1825–1828

XVII *Picrasma quassioides* Bennet (Japanese Quassia Tree): In Vitro Culture and Production of Quassin

A.H. Scragg[1] and E.J. Allan[2]

1 Introduction

The genus *Picrasma* contains about six species and belongs to the family *Simaroubaceae*. A wide variety of biologically active compounds have been isolated from members of this family, the majority of which are quassinoids. The wood from these trees has been used as a traditional medicine and lately some quassins have been reported to have important antileukemic properties (Cordell 1978).

Picrasma quassioides (D. Don) Bennett, (*Simaba quassioides* D. Don; *P. ailanthoides* Planch.) is a lofty, deciduous tree whose distribution is reported throughout the tropics, with most commercial supplies being obtained from the West Indies. It does, however, also grow in more temperate regions (Clarke 1976). The slender tree grows to 20–40 ft. and is characterized by its young bark, which is reddish-brown and marked with yellow spots. The glossy green leaves are pinnate, bearing several long, ovate leaflets, which turn to a dramatic scarlet color in the autumn (Fig. 1).

P. quassioides, like some other simaroubaceous species, produce bitter drugs which have been used extensively in folk medicine. Quassia wood was originally derived from the bushy tree *Quassia amara*, but is now also harvested from *P. quassioides*.

After felling, the trunk and main branches are logged, the bark removed and the wood chipped. These wood chips are then dried in kilns before being transported for chemical extraction (Ayensu 1980). The bitter principles of these trees form a family of compounds known as the quassinoids or simaroubolides (Polonsky 1973a, b).

Quassia wood is still sometimes used as an enema in the treatment of threadworm and as a dusting powder to destroy lice and other parasites (Ayensu 1980). As an organic insecticide, it is effective against sawfly caterpillars and aphids (Hills 1982). Quassia wood is also commercially extracted to produce the bitter agent, quassin. Commercial supplies of quassin obtained in the United Kingdom have been shown to consist of the chemicals quassin, neoquassin and 18-hydroxyquassin (Fig. 2). Quassin is used as a food flavouring agent, being 1670 times more bitter than caffeine and is used as a bittering agent in the soft drinks industry (pers. comm).

[1] Department of Applied Science, The University of the West of England, Cold Harbour Lane, Frenchay, Bristol BS16 1QY, United Kingdom
[2] Department of Agriculture, University of Aberdeen, 581 King Street, Aberdeen AB9 1UD, Scotland

Biotechnology in Agriculture and Forestry, Vol. 21
Medicinal and Aromatic Plants IV (ed. by Y.P.S. Bajaj)

Fig. 1. *Picrasma quassioides* plant

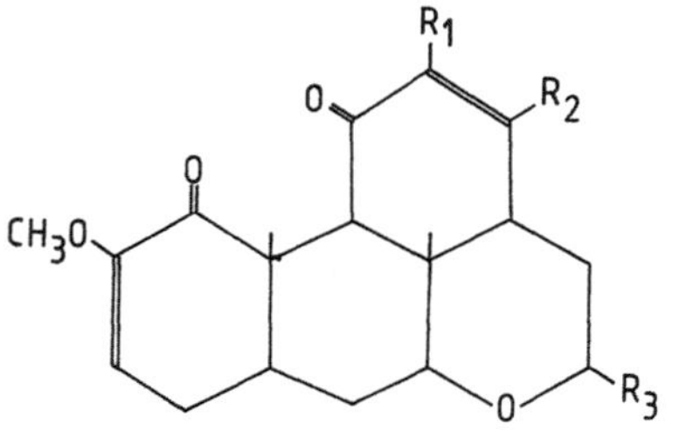

	R_1	R_2	R_3
quassin	OCH_3	CH_3	O
neoquassin	OCH_3	CH_3	HOH
18-hydroxyquassin	OCH_3	CH_2OH	O

Fig. 2. The structure of quassin, neoquassin and 18-hydroxyquassin

2 In Vitro Approaches

Reports of the development of callus and suspension cultures are limited for the species *Picrasma*, although other simaroubaceous plants such as *Ailanthus altissima* (Anderson et al. 1983) have been cultured. Callus culture has been developed from stem cuttings of *Picrasma quassioides* (Scragg and Allan 1986) using B5 medium (Gamborg et al. 1968) supplemented with coconut milk (CM). Suspension cultures were developed, without much difficulty, from the callus cultures by omitting the CM (Scragg and Allan 1986; Allan et al. 1988). The suspension culture used glucose as a carbon source, but a second line using sucrose was also used. At the start of the suspension culture, the cells formed a large number of aggregates and had a slow growth rate, but with time the aggregate size reduced and the growth rate improved (Allan et al. 1988).

2.1 Establishment of Cultures

2.1.1 Callus Culture

It proved difficult to obtain either seeds or leaves of *P. quassioides* and no source was found for the related species *P. excelsa*. However, dormant stem cuttings of *P. quassioides* (D. Don) Benn. were obtained from the Royal Botanic Gardens, Kew. The stem explants were kept in water (2 weeks) until buds formed and these were used to initiate callus. The buds were surface sterilized by washing in ethanol, followed by shaking in 10% (v/v) bleach (Domestos, Lever Ltd., UK) for 15 min.

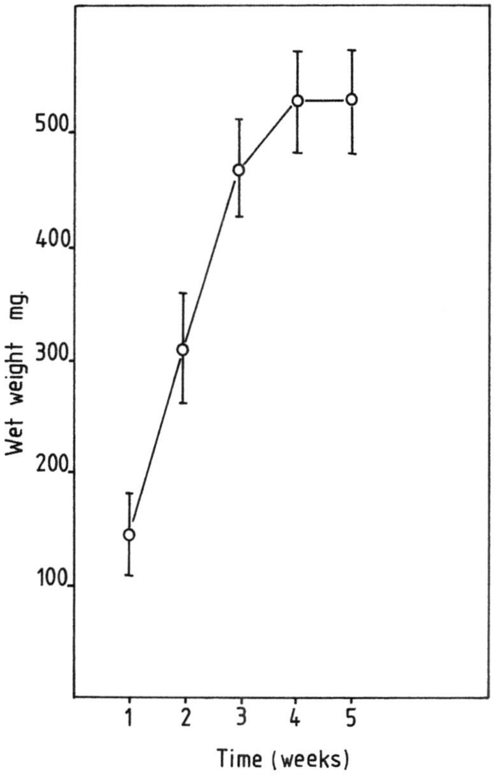

Fig. 3. The growth of *Picrasma quassioides* callus. The wet weight values shown are the mean of five determinations

After several washings in sterile distilled water, the buds were transferred to solid B5 medium, 40 ml in a 60-ml Sterlin pot. The B5 medium was supplemented with 10% (v/v) deproteinized CM, 2% (w/v) glucose, 1 mg/l 2,4-D, 0.1 mg/l kinetin and 0.8% agar. This medium was used for all callus experiments. The explants were incubated at 25 °C in subdued light. The *P. quassioides* explants developed a dark, hard, slow-growing callus but after 3 months of culture, a pale yellow-green callus formed on the dark callus. This callus was excised and subcultured every 3–4 weeks.

The growth of *P. quassioides* callus is shown in Fig. 3, which gave a doubling time, td. (wet wt.) of 8.4 days and growth rate of 0.08/day.

Carbon Source. The callus was developed using glucose as the carbon source, but this was replaced by four other carbon sources and the results are shown in Table 1. The callus grew well in all five carbon sources, with the best values obtained with sucrose and maltose.

Table 1. The growth of *Picrasma quassioides* callus on various carbon sources[a]

Carbon source	Wet weight (mg)	Dry weight (mg)
Glucose	380 ± 12	13
Sucrose	596 ± 83	14.8
Galactose	362 ± 72	13.3
Maltose	520 ± 51	14.7
Fructose	544 ± 54	13.3

[a] The weights are the mean of five estimates after 4 weeks growth.

Table 2. The effect of varying the auxins and cytokinins on the growth of *Picrasma quassioides* callus[a]

Cytokinin	Auxin[b]			
	2,4-D	IAA	NAA	IBA
Kinetin	+++	−	+++	−
Kinetin riboside	+++	−	−	−
Zeatin	++++	+	+++	+++
Zeatin riboside	++++	−	−	+++
IpA	++++	−	−	++
IpAR	+++	−	−	−
6 Benzyldenine	+++	−	++	−

[a] The table is a composite of two experiments where the auxin or cytokinin was varied while the other remained fixed over 5 different concentrations. The figures yielded are the mean of 5 determinations and the full data are given in Scragg and Allan (1986).

[b] ++++ = Very good growth; +++ = good growth, + = poor growth; − = no growth.

Growth Regulators. 2,4-dichlorophenoxyacetic acid (2,4-D), naphthalene acetic acid (NAA), kinetin and zeatin can have a profound effect on growth and secondary product accumulation. The effect of various concentrations (0, 0.5, 1, 1.5, 2 mg/l) of four auxins at a fixed concentration (0.5 mg/l) of cytokinins on both wet and dry weights of callus after 4 weeks growth has been determined. The reverse experiment was also carried out where the concentration of cytokinins was varied (0, 0.25, 0.5, 0.75, and 1 mg/l) and the auxin fixed at 1 mg/l. A combination of the results of these two series are shown in Table 2. It was clear that 2,4-D supported growth with all cytokinins used, whereas IAA, NAA, and IBA only supported growth with a limited number of cytokinins.

2.1.2 Suspension Culture

A suspension culture was initially developed from the callus using the basal medium minus coconut milk and agar. A 4-week old callus (~500 mg) was added to 20 ml of medium which was shaken at 25 and 150 rpm on an orbital shaker (New Brunswick, G-10) in the light. After 1 week incubation, a further 50 ml of medium was added, and after a further week the suspension was subcultured using a 1:5 ratio (20 ml added to 100 ml fresh medium). The suspension developed readily and the growth pattern is shown in Fig. 4, where it had a doubling time of 5.4 days and growth rate of 0.13/day. In view of the results with the growth regulators, a second

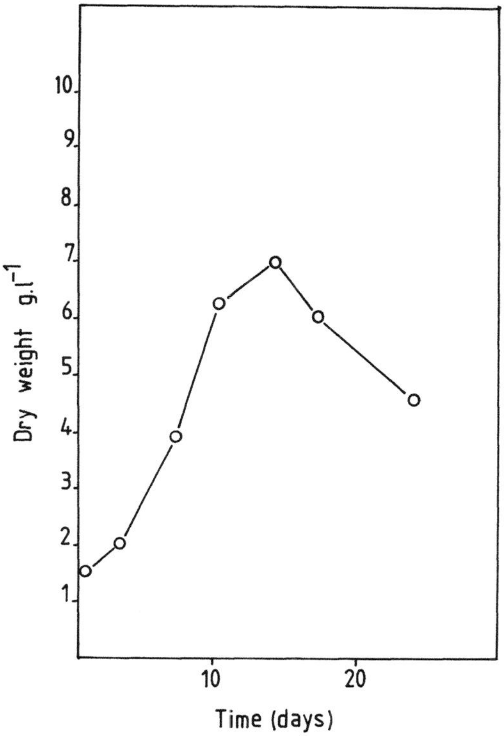

Fig. 4. The growth of *Picrasma quassioides* suspension culture in B5 medium containing 1 mg/l 2,4-D and 0.5 mg/l kinetin

suspension culture was initiated on B5 medium containing 2% glucose, 1 mg/l indolebutyric acid (IBA) and 0.5 mg/l N^6 (v^2 isopentenyl) adenine riboside (IPAR). This medium was used for all subsequent growth experiments. 1–2 g wet weight of callus was transferred to 50 ml of medium and a further 50 ml added after 14 days. As before, the cultures were incubated at 25 °C, in subdued light, at 150 rpm on an orbital shaker. The suspensions were subcultured at 14-day intervals using an inoculation ratio of 3.10 (v/v) on some 1.5 g wet cells. In contrast to the first suspension, growth of the cultures was initially very poor and highly aggregated (Fig. 5A). The aggregated nature of the culture gave high errors when estimating growth, but transfer of the culture to fluted flasks for 17 subcultures reduced aggregation and improved growth (Fig. 5B). The full cultural history and variation in biomass yields and growth rates is given in Fig. 6 and Table 3.

Inoculation Density. The effect of inoculation density on the growth of *P. quassioides* suspension cultures has been investigated at the 35th subculture and at the 53–60th subculture. Both results are shown in Table 4 and the results of the 53–60th subculture in Fig. 7. Inoculation levels of 0.5–1 g wet wt. per 100 ml

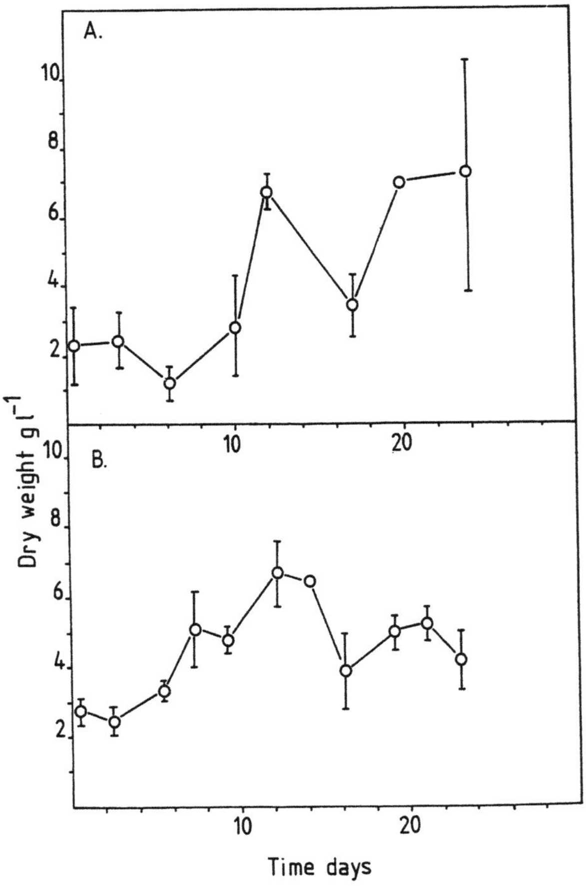

Fig. 5A, B. The growth of *Picrasma quassioides* suspension culture in B5 medium containing 1 mg/l IBA and 0.5 mg/l IPAR. Both figures show growth of the same cell line **A** before and **B** after subculture in fluted flasks to reduce cell aggregation

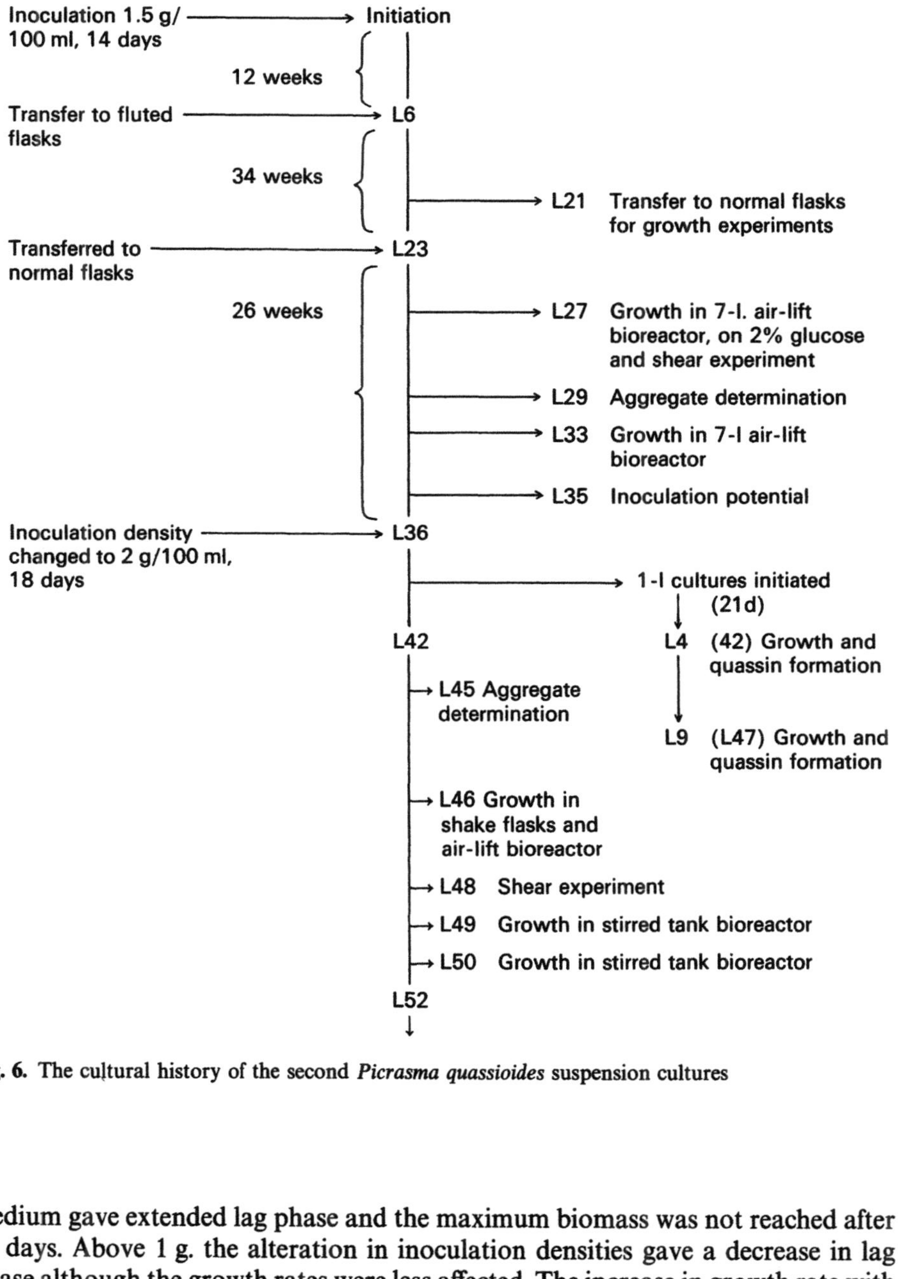

Fig. 6. The cultural history of the second *Picrasma quassioides* suspension cultures

medium gave extended lag phase and the maximum biomass was not reached after 34 days. Above 1 g. the alteration in inoculation densities gave a decrease in lag phase although the growth rates were less affected. The increase in growth rate with the length of time in culture can be seen with the increase in growth rate between the two determinations.

Age of Inoculum. Often the age of the inoculated cells can affect the growth of plant cell cultures. Figure 8 shows the effect of inoculation age on the growth of *P. quassioides*. Inoculation of cells of 7, 11 and 14 days had very similar growth

Table 3. Growth rate, doubling time and quassin formation by *Picrasma quassioides* suspension cultures

Inoculum age time (days)	Growth rate µ/days	Doubling time (days)	Maximum biomass g/l;	time (days)	Quassin µg/g;	
100-ml Cultures						
L6	0.058	12	7.4	24	0	0
L21	0.102	6.8	6.8	12	0	0
L27	0.142	4.89	9.1	21	0	0
L35	0.138	5.03	8.1	21	0	0
L43	0.143	4.85	10	18	165	10
L48	0.239	2.9	11.1	14	30	31
1000-ml Cultures						
L4 (L42)	0.152	4.56	13.62	15	0	0
L9 (L47)	0.123	5.63	13.57	19	145	23

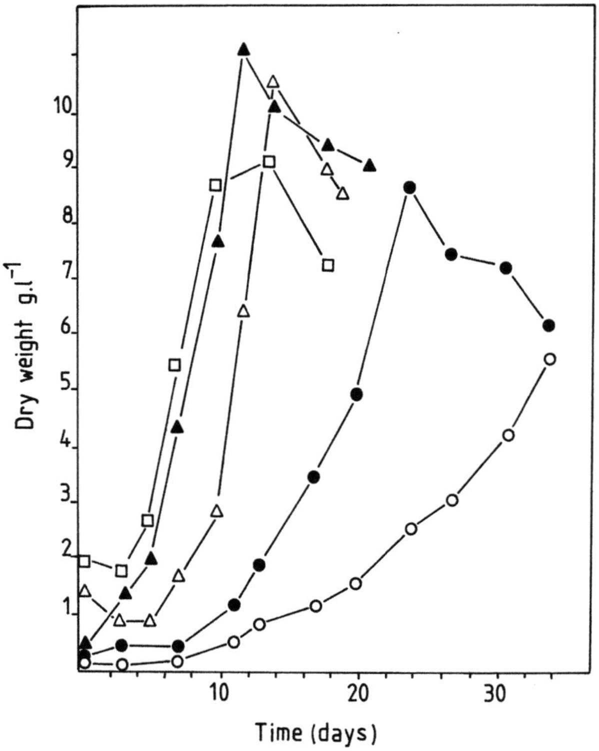

Fig. 7. The growth of *Picrasma quassioides* suspension cultures using various inoculation densities. Grams wet weight per 100 ml medium; ○, 0.5 g; ●, 1 g; △, 1.5 g ▲, 3 g; □, 4 g

rates (Table 5) of 0.25–0.28/day (doubling times 2.48–2.81 days), whereas at 18 days the growth rate was somewhat reduced (0.21/day; td = 3.27). The maximum biomass levels varied, but no correlation was seen with the age of the inoculum.

Carbon Sources. Suspension cultures were grown on a range of carbon sources and the results are shown in Fig. 9. As expected, best growth was obtained with glucose

Table 4. The effect of inoculation density on the growth of *Picrasma quassioides*

Inoculation level (g/wet wt. per 100 ml)	Growth rate (μ/day)	Doubling time (td, days)	Maximum biomass (g/l)
Subculture 35			
1	0.07	9.47	8.6
1.5	0.16	4.36	8.3
2	0.14	5.03	8.1
3	0.14	5.03	10.5
4	0.12	5.61	9.8
Subculture 53–60			
0.5	0.08	9.11	5.5[a]
1	0.14	5.11	8.6
1.5	0.24	2.96	10.5
3	0.26	2.67	11.1
4	0.22	3.12	9.1

[a] Culture did not reach a maximum.

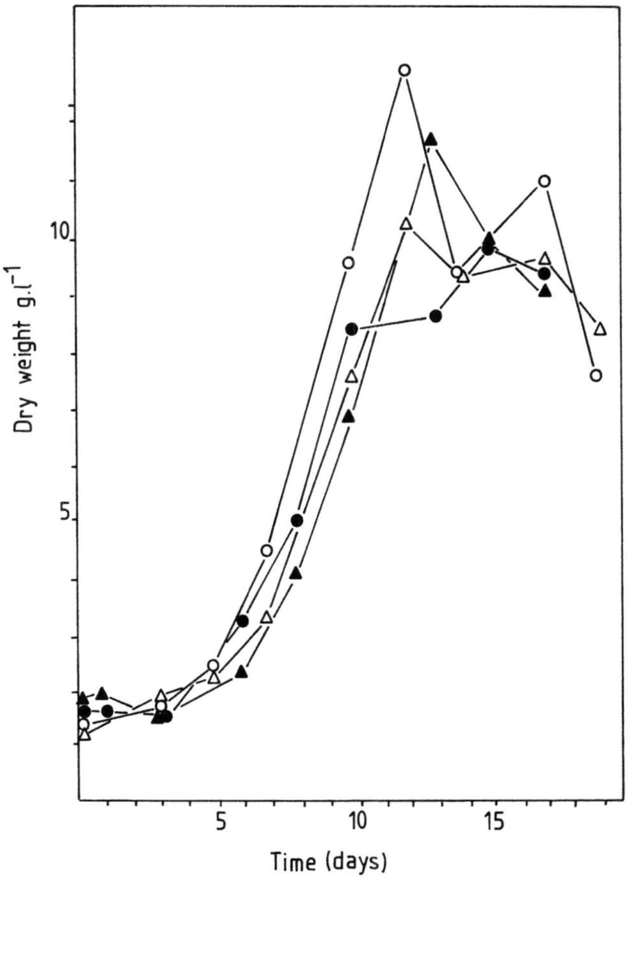

Fig. 8. The effect of inoculation age on the growth of *Picrasma quassioides* suspension culture. Inoculation age; ○, 7 days; ●, 11 days; △, 14 days; ▲, 18 days

Table 5. The effect of inoculum age on the growth of *Picrasma quassioides* suspension cultures

Age at inoculation (days)	Growth rate (μ/days)	Doubling time (td, days)	Maximum biomass (g/l)
7	0.28	2.48	12.5
11	0.25	2.77	9.8
14	0.25	2.81	10.2
18	0.21	3.27	11.5

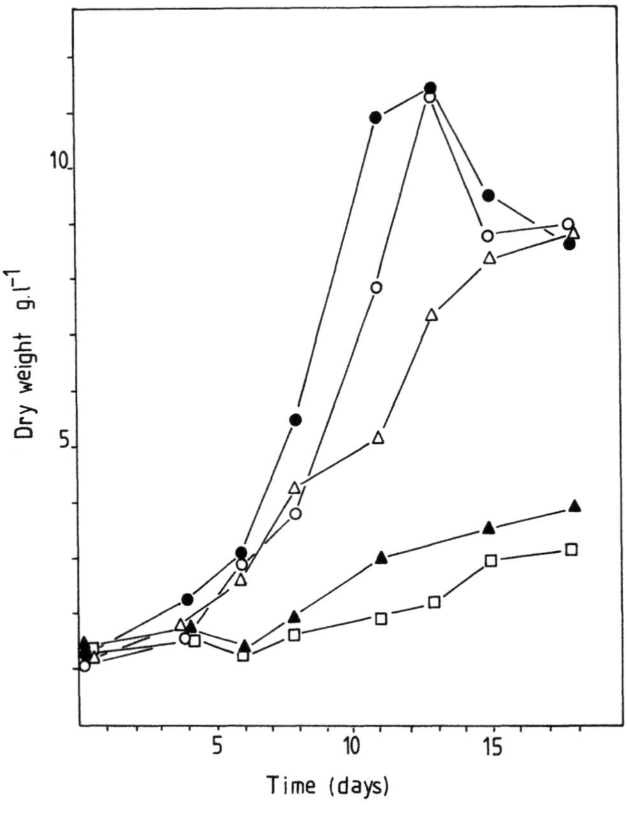

Fig. 9. The effect of various carbon sources on the growth of *Picrasma quassioides* suspension cultures. ○, Sucrose; ●, glucose; △, galactose; ▲, maltose; □, lactose; all at 2% (w/v)

and sucrose. Galactose supported a lower growth rate and biomass and maltose and lactose gave very little growth.

Variation in Carbon Source (Fig. 10). With 2% glucose, a growth rate of 0.14 days^{-1} (td = 4.89 days) was obtained with a maximum biomass of 9.1 g/l (Table 6). The biomass yields were much larger when the glucose was increased to 5, 10 and 15%. The biomass increased to 25 g/1 at day 28 with 10% glucose which was reduced to 23 g/1 with 15% glucose. The maximum growth rate was found with

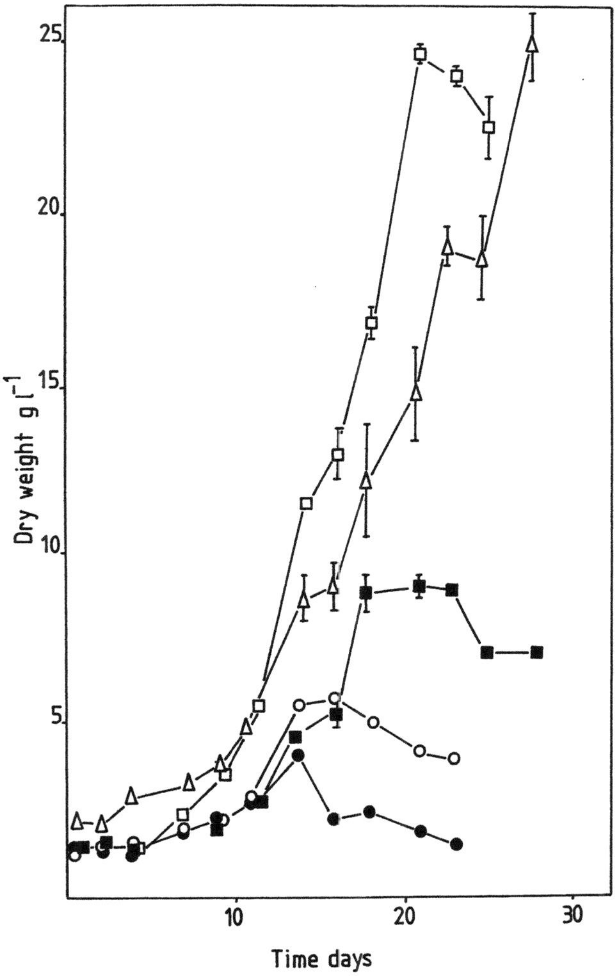

Fig. 10. The growth of *Picrasma quassioides* suspension culture with various concentrations of glucose (w/v). ●, 0.5%; ○, 1%; ■, 2%; □, 5%; △, 10%

Table 6. The effect of glucose levels on the growth of *Picrasma quassioides* suspension cultures

Glucose levels (%)	Growth rate (μ/day)	Doubling time (td, days)	Maximum biomass (g/l)
0.5	0.11	6.3	4.2
1	0.13	5.3	5
2	0.14	4.95	9
5	0.17	4.1	25
10	0.086	8.1	25[a]

[a] Maximum not reached.

5% glucose ($\mu = 0.17$; td = 4.1 days) which was reduced at higher concentrations of glucose.

Effect of Temperature. The *P. quassioides* suspension cultures were normally maintained at 25 °C and the effect of changing the incubation temperature is shown in Fig. 11. Growth was poor at 20 °C, but there was only a small reduction in growth rate at 30 °C.

Large-Scale Culture. The growth of *P. quassioides* suspension cultures in volumes of above 100-ml or in bioreactors was difficult initially, with the cells losing viability rapidly. By using a higher inoculation density (50 g wet weight cells into 1 liter medium) growth was achieved; the results are shown in Table 3. When the cultivation of *P. quassioides* was attempted in airlift bioreactors, the cells failed to grow, although viability was maintained for some time. However, by using an inoculum grown in 1-liter volumes, growth was obtained in a 7-l airlift bioreactor (Fig. 12). Long lag phases were found and the growth rate was lower (0.12/day;

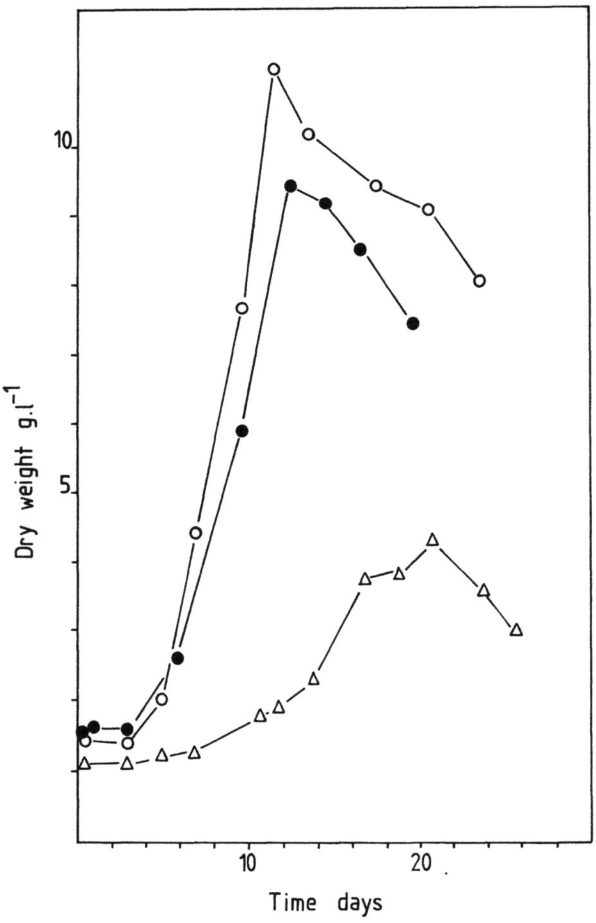

Fig. 11. The effect of temperature on the growth of *Picrasma quassioides* suspension cultures. ○, 25 °C; ●, 30 °C; △, 20 °C

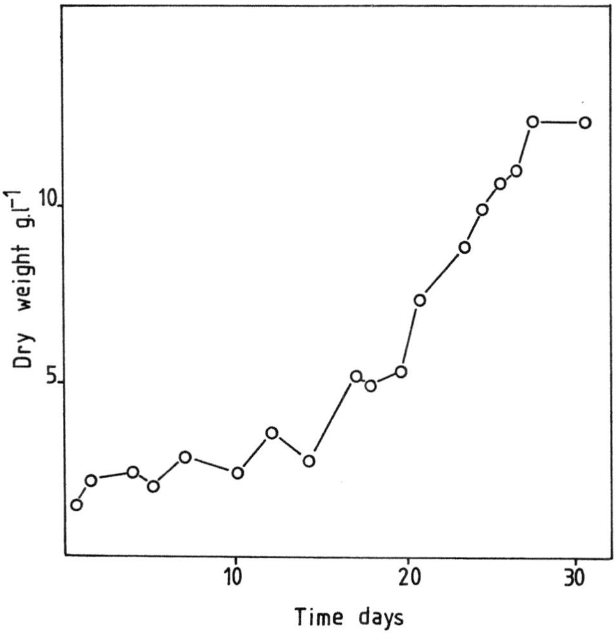

Fig. 12. The growth of *Picrasma quassioides* suspension cultures in a 7-l airlift bioreactor (Morris et al. 1985) using medium containing 5% (w/v) glucose. Temperature was 15 °C and the aeration was a 1 l/min (0.14 vvm, i.e. volume of gas/volume of bioreactor/min)

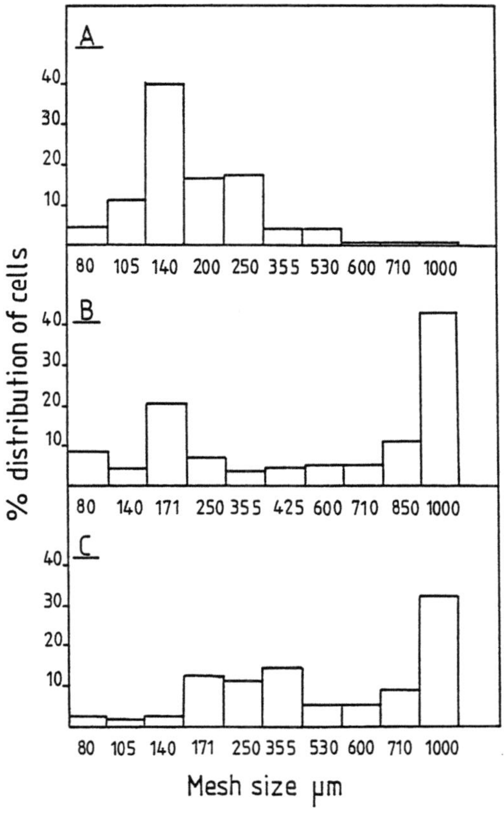

Fig. 13A–C. Cell aggregate distributions for cultures of *Picrasma quassioides* grown in various vessels. **A** 14-day-old culture from a 7.1 airlift bioreactor (L48); **B** 17-day-old culture from a 1-l shake-flask culture (L35); **C** 14-day-old culture from a 100-ml shake-flask culture. The size of aggregates was determined by washing the cultures through a range of meshes of known pore size td = 5.85 day) than that found in shake flasks. The size distribution of the suspension grown in the airlift bioreactor has been compared with aggregates found in shake flasks (Fig. 13). At 14 days in the bioreactor, the culture had aggregates between 105 and 200 μm in the main (40%), whereas in 2-l and 100-ml shake flasks, aggregates of above 1000 μm were in evidence.

Shear Sensitivity. Plant cells have been regarded as sensitive to shear, and this has been one of the reasons that airlift bioreactors have been adopted for the growth of plant cells. However, little experimental evidence was available on the shear sensitivity of *P. quassioides* culture. Suspension cells were subjected to 1000 rpm and 440 rpm (167/s or 73/s shear rate) in a 3-l stirred tank bioreactor for up to 5 h. The cells were removed at intervals and tested for their ability to grow in shake flasks. A suspension culture taken at subculture 27 was exposed to the lower impeller speed (440 rpm) and as can be seen in Fig. 14, lost viability after 2.5 h of exposure. When the culture was tested some 1 year later (subculture 48) at 1000 rpm, the culture proved to be shear stress tolerant. The shear stress tolerance appears to be related to the increase in growth rate of the culture and possibly a reduction in aggregate sizes. The increased tolerance to shear encouraged the use

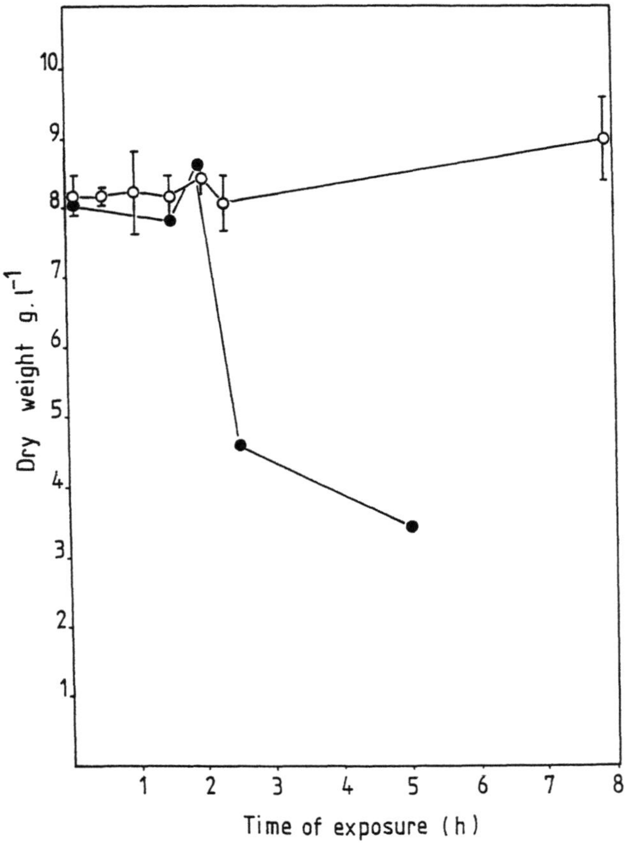

Fig. 14. The loss of viability of *Picrasma quassioides* cultures when exposed to shear stress in a 3-l stirred-tank bioreactor. Viability is expressed by the dry weight achieved after 14 days growth in shake flasks. ●, Culture treated at 440 rpm [shear rate (av) 73/s] at subculture 27; ○, culture treated at 1000 rpm (average shear rate 167/s) at subculture 48

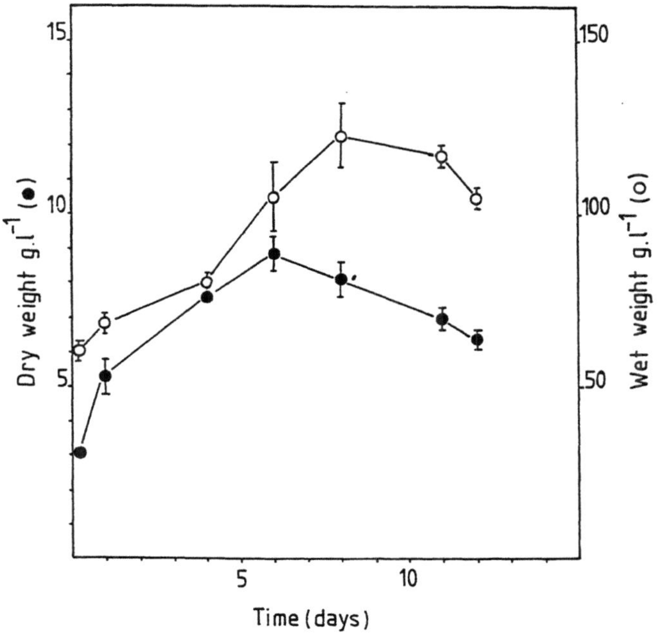

Fig. 15. The growth of *Picrasma quassioides* in a 3-l stirred tank bioreactor. One liter of 14-day-old cells was inoculated into 2 l of fresh medium. The incubation was at 25 °C, the aeration rate 100 ml/min and agitation was at 150 rpm. ○, Wet wt. g/l; ●, dry wt. g/l

of a stirred tank for the growth of *P. quassioides* suspension cultures. The growth of the culture in a 3-l stirred-tank bioreactor is shown in Fig. 15, which had a growth rate of 0.2/day (td = 3.5 day) with a final biomass yield of 8.8 g/l after 6 days.

2.2 Extraction and Accumulation of Quassin

2.2.1 Extraction

Picrasma is a member of the Simaroubaceae, which are the source of quassinoids, of which quassin is of commercial value. Commercial quassin is a mixture of quassin, its hemiketal neoquassin and 18-hydroxyquassin (Fig. 2). Initially, quassin was extracted from freeze-dried callus or suspension cells by soxhlet extraction with methanol. The methanol extract was analyzed by HPLC using a reverse phase μBondapak C18 (Waters Associates) column with a solvent system of methanol, water and n-heptane sulphonate (nhs) at 45:55:5 and a gradient of 45–65% methanol.

Detection of peaks was by UV absorbance at 254 and 280 nm. This method of quassin investigation was used with a TLC method where Kieselgel GF plates (Merck Ltd.) are used in a solvent of 3% methanol in chloroform. Quassin was detected by quenching at 254 nm and turns brown/black on heating at 60 °C for 10 mins after spraying with 0.5% cerric sulphate in 2M sulphuric acid. The presence of

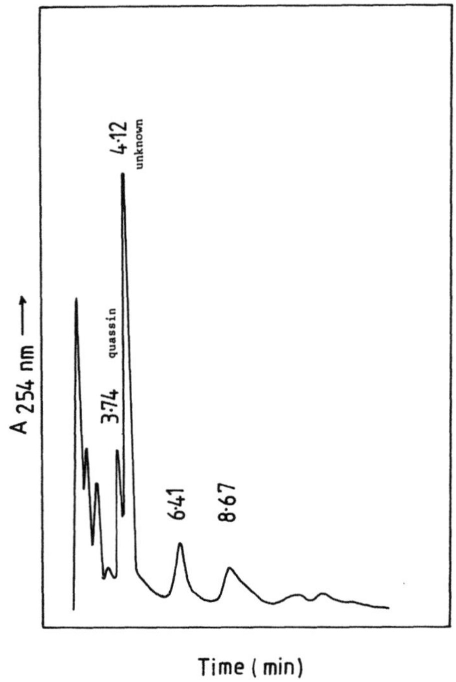

Fig. 16. A typical HPLC analysis pattern of an extract from a *Picrasma quassioides* suspension culture. Separation was by normal phase HPLC

a peak close to the position expected for quassin (Fig. 16) prompted the use of another method of HPLC analysis. A normal phase system was used with a cartridge (8 mm × 10 cm) packed with 5 µm silica with a solvent system of 2% methanol in dichloromethane. Detection was absorbance at 250 nm. Purified quassin, neoquassin, was confirmed by mass spectroscopy in extracts (Fig. 17). From feeding experiments it was determined that simple methanol extraction was insufficient to extract all quassin and this was replaced by dichloromethane containing 10% (v/v) methanol.

Changes in extraction and HPLC allowed the isolation and identification of the large fluorescent compound extracted with quassia. By comparison with authentic samples, that was identified as the alkaloid 1-methoxycanthin-6-one (Fig. 18).

2.2.2 Quassin in Cultures

The callus cultures grown under various concentrations of growth regulators (Table 2) were analyzed for the presence of quassin and the results are shown in Table 7. Quassin was detected mainly in calli grown in the presence of 2,4-D, but the highest values were found in the presence of IBA and zeatin riboside. With the suspension cultures developed in the presence of IBA and IPAR, no quassin was detected in the early culture, but by subculture 48, growth rate had increased and low levels of quassin were detected (Fig. 19). Figure 20 shows the formation of both quassin and 1-methoxycanthin-6-one during the growth of *P. quassioides* suspension cultures.

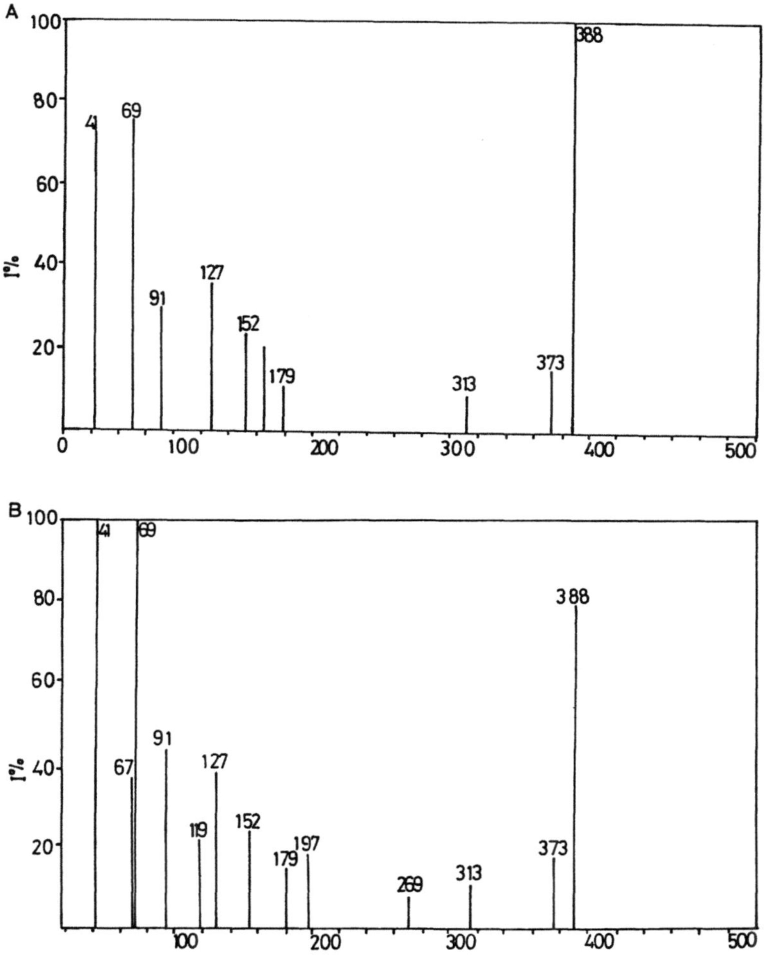

Fig. 17A, B. Mass spectra of **A** purified quassin and **B** pooled extracts of *Picrasma quassioides* cell suspension purified by HPLC

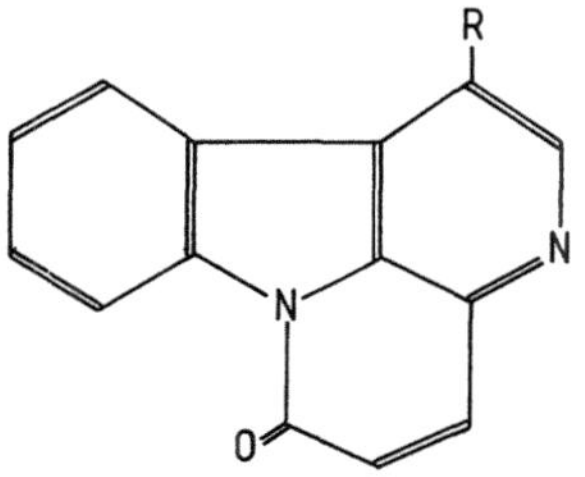

	R
canthin-6-one	H
1-hydroxycanthin-6-one	OH
1-methoxycanthin-6-one	OCH_3

Fig. 18. Structure of canthin, 1-methoxycanthin-6-one and 1-hydroxycanthin-6-one

Table 7. Quassin accumulation in *Picrasma quassioides* callus grown on B5 medium supplemented with various auxins and cytokinins[a]

Cytokinins	Auxin[b]			
	2,4-D	IAA	NAA	IBA
Kinetin	+	−	−	−
Kinetin riboside	+	−	−	−
Zeatin	++	−	−	++
Zeatin riboside	−	−	−	+++
IpA	++	−	−	++
IpAR	++	−	−	−
6 BA	−	−	+	−

[a] Values of quassin expressed as μg/g dry wt.
[b] − = No quassin; + = Trace–10; ++ = 10–100; +++ = 100–200.

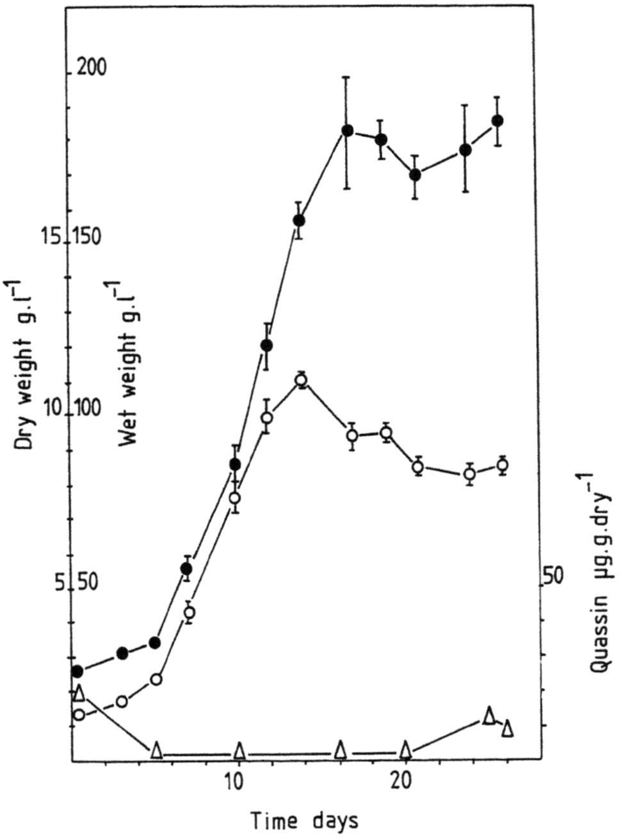

Fig. 19. Growth and quassin accumulation by a suspension culture of *Picrasma quassioides* at subculture 48. ○, Wet wt. g/l; ●, dry g/l; △, quassin μg/g dry wt

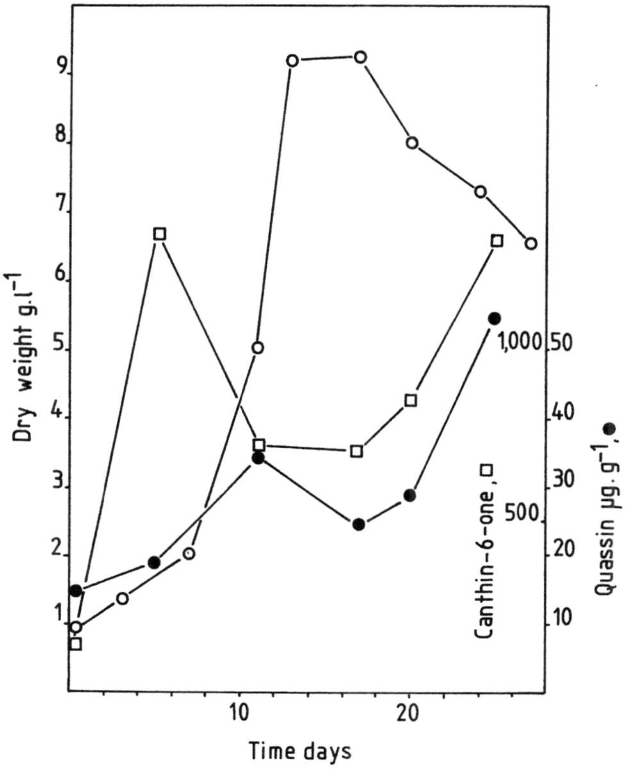

Fig. 20. The growth, canthin-6-one and quassin accumulation in a suspension culture of *Picrasma quassioides*. ○, dry wt. g/l; ●, quassin µg/g dry wt.; □, canthin-6-one µg/g dry wt.

Although *Picrasma* plants have been shown to produce a wide range of interesting compounds, quassin is the only commercial product to date. At present, although *P. quassioides* cultures can be grown in bioreactors, the yield of quassin is too low to make the process economical.

3 Summary and Conclusions

Callus and suspension cultures have been successfully initiated and maintained from *P. quassioides* (D. Don) Bennett. Callus was formed from germinating buds cultivated on Gamborgs B5 medium supplemented with 1 mg/l 2,4-D, 0.1 mg/l kinetin and 10% (v/v) coconut milk. Callus was in fact slow to develop, but after a few weeks in culture a soft friable callus was obtained. Despite this, suspension cultures proved to be difficult to maintain, as the cells were very sensitive to shear and tended to form large aggregates. These problems were alleviated by growing the cells in fluted flasks, which gradually reduced aggregate size, while shear sensitivity decreased with culture age. A large variety of growth conditions have been studied and a fast-growing, shear-resistant cell line has been obtained.

It is difficult to select the best form of explant, as only stem cuttings were available; seeds would perhaps be the best choice. The medium of choice for callus was B5 containing 10% coconut milk, 2% glucose, 1 mg/l 2,4-D, and 0.1 mg/l kinetin; and for the suspension cultures, B5 containing 2% glucose, 1 mg/l IBA and 0.5 mg/l IPAR. The conditions for growth of both callus and suspension were 25 °C, in subdued light, with shaking at 150 rpm for the suspension cultures. Quassin was detected in suspensions cultured in the above medium but only after sometime in culture, therefore it is difficult to specify a medium for quassin accumulation.

In addition to studies on growth and maintainance of the *P. quassioides* cell line, investigations have involved methods for the extraction and detection of the sesquiterpenoid quassin from these tissue cultures. These methods are described and the cultures examined for quassin accumulation under the wide range of environmental conditions reported. Quassin, albeit at low concentrations, has been detected in suspension cultures growing in B5 medium containing 2% glucose, 1 mg/l indolebutyric acid and 0.5 mg/l N^6 (Δ^2 isopentyl) adenine riboside. This result offers the possibility for future selection of higher yielding cell lines which, combined with the fact that these cells can be easily grown in 3-l continuously stirred tank bioreactor, promises some potential for the production of quassin from tissue culture systems.

References

Allan EJ, Scragg AH, Pugh K (1988) Cell suspension culture of *Picrasma quassioides*: the development of a rapidly growing, shear resistant cell line capable of quassin formation. J Plant Physiol 132: 176–183

Anderson LA, Harris A, Phillipson JD (1983) Production of cytotoxic canthin-6-one alkaloids by *Ailanthus altissima* plant cell cultures. J Nat Prod 46: 374–378

Ayensu ES (ed) (1980) Jungles. Jonathon Cape, London

Clarke DL (ed) (1976) Trees ,and shrubs hardy in the British Isles, 8th edn, revised vol III. Bean & Murray, London, 199 pp

Cordell GA (1978) Anticancer agents from plants. In: Reinhold L, Harborne JB, Swain T (eds) Progress in biochemistry, vol 5. Pergamon Press, Oxford New York, pp 273–316

Gamborg OL, Miller RA, Ojima K (1968) Nutrient requirements of suspension cultures of soybean root cells. Exp Cell Res 50: 151–258

Hills LD (ed) (1982) Fruit, pest and disease control the organic way. Henry Doubleday Res Assoc, Essex, England

Morris P, Scragg AH, Smart NJ, Stafford A (1985) Secondary product formation by cell suspension cultures. In: Dixon RA (ed) Plant cell culture. A practical approach. IRL Press, Oxford, pp 127–167

Polonsky J (1973a) Chemistry and biogenesis of the quassinoids (simaroubolides). Recent Adv Phytochem 6: 31–64

Polonsky J (1973b) Quassinoid bitter principles. Fortschr Chem Org Naturst 30: 101–150

Scragg AH, Allan EJ (1986) Production of the triterpenoid quassin in callus and cell suspension cultures of *Picrasma quassioides*. Bennett. Plant Cell Rep 5: 356–359

XVIII *Polygonum hydropiper* L. (Water Pepper): In Vitro Culture and the Production of the Aphid-Antifeedant Polygodial

D.V. Banthorpe[1]

1 General Account

1.1 Distribution and Importance

Polygonum hydropiper L. (family Polygonoceae) is a member of a genus of some 175 species. It is a semi-erect (25–75 cm) annual herb with a branched stem and lance-shaped leaves, carrying its greenish-pink flowers in slender racemes (Fig. 1). The species is widespread in most parts of Europe, temperate Asia, and North America, and it also occurs at scattered sites in North Africa. Across its main range it is abundant in the verges of ponds and ditches and on waterlogged grasslands and water meadows. *P. hydropiper* is not grown commercially but has found an exceptionally impressive range of uses in folk medicine and also as a culinary herb, and this has led to the adoption of a rich variety of apt local names, e.g. fireweed, arsemart and smartweed are examples of some 20 English regional names in addition to the accepted vernacular name of "water pepper". The flower heads have little odour but all the aerial parts have a bitter acrid taste and contain vesicant compounds that blister the skin upon repeated handling (Clapham et al. 1952).

Medicinal use of *P. hydropiper* goes back to Dioscorides (ca. 60 A.D.) and tinctures of foliage are used as diuretics, diaphoretics, and to arrest gynecological bleeding (Wisniewski and Szlaski 1970). Water effusions are astringent and anti-inflammatory and have been claimed efficacious for colds, coughs, dysentery, gout, sores and ulcers, toothache, dropsy, menstrual pains, and hemorrhoids. Foliage can be applied to arrest bleeding or be secreted in beds to kill fleas, cf. German synonyms *Flohkraut* and *Flohpfeffer* (Grieve 1931; Grigson 1975; Launert 1981). Linneaus recorded that the herb yielded a yellow dye with an alum mordant. The peppery taste of the foliage has led to widespread culinary use in the Far East, in particular as a garnish for fish dishes.

These varied biological effects, if not anecdotal, must be a consequence of the secondary metabolism of the herb. The essential oil from foliage (ca. 0.43% dry wt.) contains monoterpenoids with a variety of skeletons, e.g. 1,4-cineole, car-3-ene, α and β-pinenes, fenchone, borneol, camphor, linaloyl acetate etc., but is otherwise unexceptional (Yankov and Damyanova 1969). More noteworthy is the sesquiterpenoid component of the flower heads. The main compound (ca. 0.05% wet wt.) is

[1] Chemistry Department, University College London, 20 Gordon St., London WC1H-0AJ, United Kingdom

Biotechnology in Agriculture and Forestry, Vol. 21
Medicinal and Aromatic Plants IV (ed. by Y.P.S. Bajaj)

Fig. 1. *Polygonum hydropiper* L. (Courtesy of Dr. G.S. Morris, University College London)

(−)-polygodial (Fig. 2, *1*) accompanied by its hydroxy-derivative (−)-warburganal (Fig. 2, *2*) in small amounts. The former (≡9β-drim-7-en-11, 12-dial) has also been isolated from liverworts and named tadeonal (Barnes and Loder 1962; Fukuyama et al. 1982). Both polygodial and warburganal are responsible for the hot taste of the plant (Kubo and Ganjian 1981; Askawa et al. 1988). Polygodial has been assigned widespread physiological activities. It is a fungicide, and as it acts by damaging cell membranes, it also facilitates the entry of antibiotics and thus is synergistic for them (Kubo and Taniguchi 1988; Taniguchi et al. 1988). It also acts as a mutagen (Sterner et al. 1987); an allergen (Stampf et al. 1982); a plant growth inhibitor, preventing germination of rice at 100 ppm (Asakawa and Takemoto 1979); is moluscidal, piscicidal and phytotoxic (Asakawa et al. 1988); and is a viral-genome inactivator (Tokuda and Matsumoto 1989). Best studied is its antifeedant activity towards aphids (Dawson et al. 1986) and armyworms (Kubo et al. 1976). Thus, leaves of *Zea mays* dipped into 0.1 ppm solutions of the dialdehyde are uneaten and the insects perish of starvation. Warburganal is also antifungal (Taniguchi et al. 1984) and is an antifeedant for armyworms (Kubo et al. 1976) and aphids (Asakawa et al. 1988). Other drimane derivatives isolated from foliage

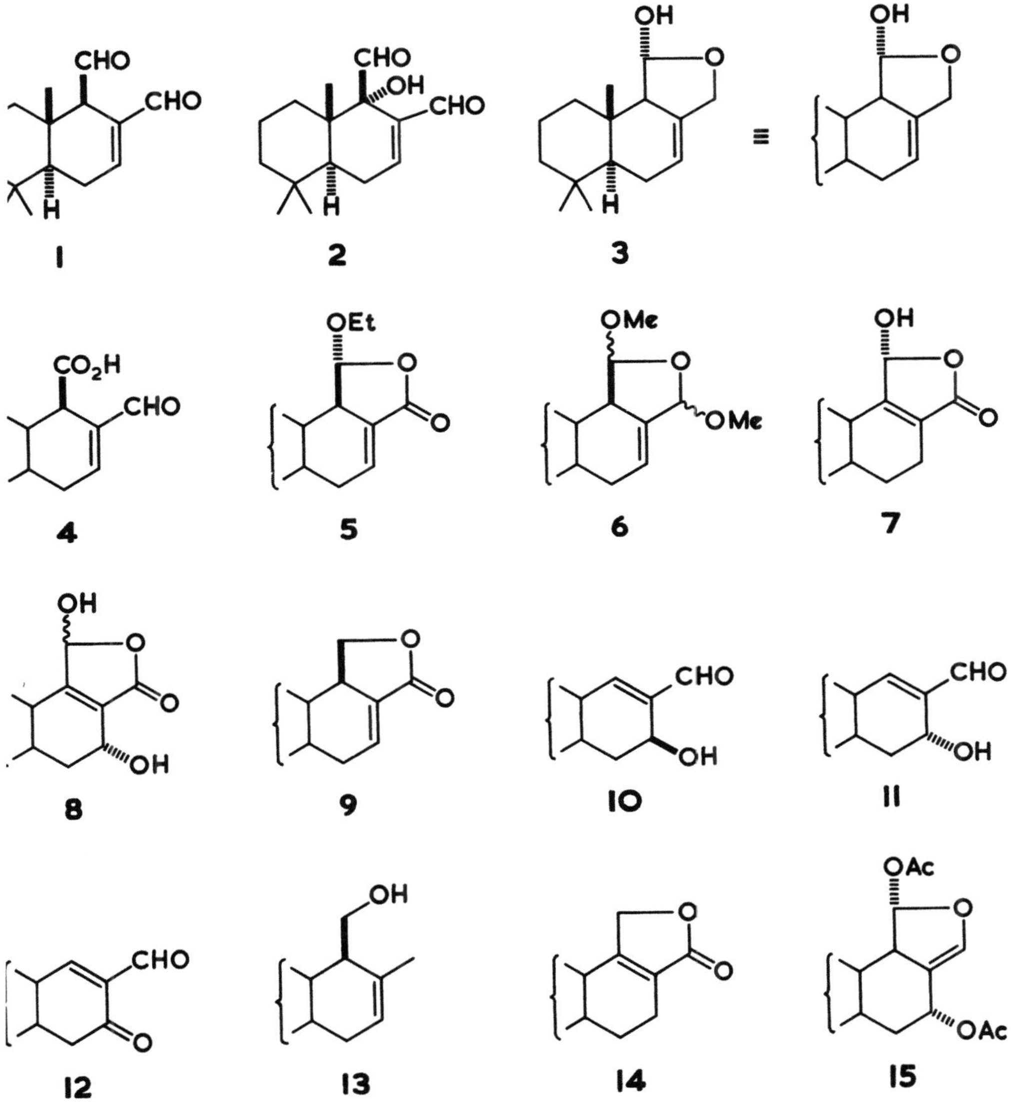

Fig. 2. Drimane-type sesquiterpenoids isolated from *P. hydropiper*. *1* (−)-Polygodial; *2* (−) warburganal; *3* isopolygodial; *4* polygonic acid; *5* 11-ethoxycinnanolide; *6* polygodial acetal; *7* valdiviolide; *8* fueqin; *9* cinnanolide; *10* isopolygonal; *11* polygonal; *12* polygonone; *13* drimenol; *14* concertifolin; *15* oleopupuane

(usually in small amounts) are (see Fig. 2) isopolygodial (*3*); polygonic acid (*4*: which shows weak anti-coagulant activity); 11-ethoxycinnanolide (*5*); polygodial acetal (*6*); valdiviolide (*7*); fuegin (*8*); cinnanolide (a fungicide); and drimenol (*13*); in addition to the nor-sesquiterpenoids (C_{14}-compounds) isopolygonal (*10*); polygonal (*11*); and polygonone (*12*) (F. Fukuyama et al. 1982; Y. Fukuyama et al. 1985). Seeds contain polygodial, together with several of the above and concertifolin

(*14*, a plant growth regulator; Asakawa and Takemoto 1979), and up to 4% linolenic acid (Ahmed et al. 1987).

Foliage also contains anthocyanidins and anthocyanins (Yoshitama et al. 1987), flavonoids including guercetin and kaempferol, and glycosides such as isorhamnetin and rhamnazin (Quadrat-i-Khuda et al. 1965; Fukiyama et al. 1983; Yusif and Blinova 1984; Vysochina et al. 1987); caffeic, gallic, sinapic and many other aromatic acids (Swiatek and Dombrowicz 1987); and other aromatics (Yankov and Damyanova 1977); β-sitosterol and a variety of high molecular-mass linear hydrocarbons and waxes (Yankov and Damyanova 1970); and tannins (Shalanova et al. 1983). Of especial interest are isocoumarins that possess anti-inflamatory activity (Furata et al. 1986) and an unidentified fraction of extract from foliage that is a potent anti-fertility agent for rats (Garg and Mathur 1972).

1.2 Extraction and Purification of Components

The biologically active sesquiterpenoids may be obtained from foliage by ether extraction followed by liquid chromatography on SiGel and application of standard HPLC techniques (Fukuyama et al. 1985). A specific method for polygodial is to extract the dried leaves with MeOH–H_2O (70:30 v/v), addition of NaCl, recovery with Et_2O and alumina chromatography to give the product: oil, [α]-131 (EtOH), λ_{295}(ε49000: EtOH) (Barnes and Loder 1962). Polygodial may also be separated by GC as its adduct with pyrrole (Brooks and Watson 1985). Isopolygodial (*3*) is believed to be an artifact of the isolation procedure (Cortes and Oyarzun 1981). Polygodial occurs in certain nudibranches as a protected form, oleopupuane (*15*), which is decomposed during isolation (cf. San Feliciano et al. 1985; Cimino et al. 1988): it is not clear to what extent this situation occurs in higher plants. A neat separation method for flavonol aglycones involves chromatography on polyamide with elution by aqueous ethanol (Yusif and Blinova 1984).

1.3 Commercial Applications

Polygodial would seem ripe for exploitation in order to protect agricultural and horticultural crops, not only in view of its antifeedant properties which could prevent aphid infestation, but the consequent prevention of the distribution of plant viruses by the insects. The known different responses of aphidian and lepidopteran larvae to the compound would also auger well. However, despite extensive studies on the dialdehyde and analogues (cf. Dawson et al. 1986; Asakawa et al. 1988) few concrete control proposals have been reported (Res. Corporation Patent 1981). The compound is volatile and unstable and its small but significant phytotoxicity and fish toxicity may well restrict its use in pest control. It seems likely that future control systems may rely on several compounds, in low concentrations, acting synergistically; polygodial may well become a component of such an integrated and environmentally acceptable approach (Gibson et al. 1982; Dawson et al. 1986). Pharmaceuticals containing warburganal for use in combating fungal infections have been patented (Matsumoto and Kageyama 1988). Use of

Polygonum hydropiper as a cocrop to provide protection for aphid-susceptible crops such as beans and potatoes has not been reported: possibly the damp conditions suitable for successful cultivation of the former cannot easily be achieved in cultivated environments.

2 Biotechnological Approaches

2.1 Prospects

Polygodial is the subject of an interesting speculation concerning gene transfer (cf. Pickett 1985). The compound is believed to be formed in only a few steps from 2E,6E-farnesyl pyrophosphate (*16*; FPP; Fig. 3) *via* drimenol (*13*); and as *16* is ubiquitous in higher plants, the possibility exists of isolation of the gene responsible for the single step *16* → *13* and its transference to a commercially important crop plant. The oxidative steps that convert *13* into polygodial also involve genes that could be isolated and transferred, but the hope would be that these steps are mediated by nonspecific enzymes – perhaps the components of a metabolic grid – that would occur in the host species. Such genetic manipulation is the more attractive as, unlike the situation for, e.g., monoterpenoids and alkaloids, specific secretory and storage structures for polygodial are not present in *P. hydropiper* and the antifeedant is stored in vacuoles throughout the aerial parts. Consequently, structural genes for the formation of trichomes would not have to be cointroduced.

The most convenient source of biomass for studies of gene transfer is tissue culture: both for the isolation of the gene and for its reintroduction into the host. Consequently we have recently studied both callus and suspension cultures of *P. hydropiper* (Banthorpe et al. 1989).

2.2 Establishment of Cell Cultures

The only reported study on tissue culture of *P. hydropiper* yielded poorly characterized calli that were claimed to produce various amino acids (Minoda and Kodama

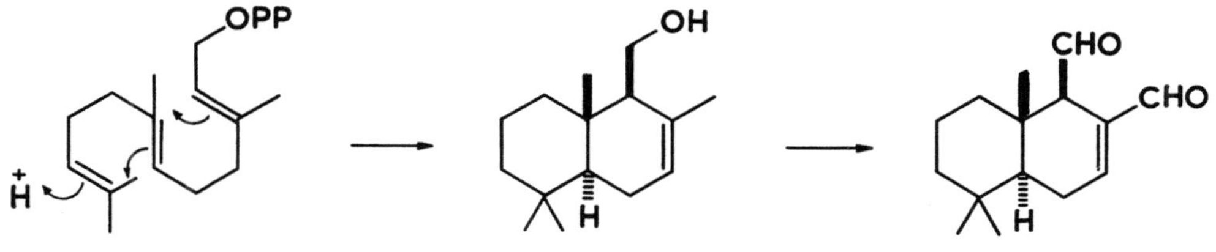

16 **13** **1**

(OPP = pyrophosphate group)

Fig. 3. Biogenesis of (−)-polygodial (*1*) and drimenol (*13*) from 2E,6E-farnesyl pyrophosphate (*16*)

1977). We found that the use of the medium developed by these workers or most other conventional media, including even those supplemented with 2,4-D, led to very slow callus initiation followed by very restricted growth and the cultures usually did not survive the first subculture. However, use of Murashige and Skoogs' medium supplemented with BAP (0.5 mg/l), NAA (2 mg/l) and gibberellin A_3 (GA_3; 10 mg/l) caused ready initiation and admirable subsequent growth from explants of stem, foliage, or hypocotyl of seedling. Use of GA_3 was essential and was noteworthy in that this additive has rarely been used in tissue culture studies. At 25 °C under constant illumination (ca. 600 lx; λ max 580 nm) usually 70–90% of the explants had formed adequate callus within ca. 2–4 weeks, and passage on the same medium was made thereafter ca. every 14–28 days, depending on the source of the explants, for some 6–12 subcultures before production of sesquiterpenoids stabilized.

The callus lines established appeared morphologically and biochemically stable for up to 2 years (when experiments ceased). Nearly always the tissue was white and friable and consisted largely of parenchyma-like cells with some tracheid formation, but a few lines developed into hard greenish tissue that was inferior in its biosynthetic activity for production of sesquiterpenoids. Coarse cell suspensions were established from the above lines in media minus the agar support and maintained in similar conditions on a gyrorotary shaker (120 rpm). Now subculture was carried out at 7-day intervals for 3 months before assay. These suspensions at subculture contained ca. 10^6 cells/cm^3, mainly in large aggregates of ca. 2 mm diameter. Attempts to produce fine cell suspensions led to biosynthetically inert material. The culture lines were unstable with respect to metabolite production during the early subcultures: this may be because sublines of different types of cells, with differing growth rates, have to undergo a settling-down period to establish a more or less equilibrium population which then shows stable metabolism. This was the reason for the protracted periods of culture before assays were systematically carried out.

2.3 Products Produced by Culture Lines

Callus was extracted after 12 passages and then after subsequent subcultures, and the products of interest were characterized by comparison with authentic standards using TLC (three systems), GC and GC-MS. The crude extracts predominantly comprised waxes and compounds of high molecular mass that were not identified, but they also contained the phytosterols, stigmasterol and sitosterol, as well as polygodial, drimenol and 2E,6E-farnesol. The time profiles of accumulation of the drimanes in a typical callus line are presented in Fig. 4. The maximum yield occurred at ca. day 8 of the culture cycle and the profile was similar when lines with longer (e.g. 28-day culture periods) were assayed. At day 8 of the cycle, the drimanes comprised ca. 20% of the oil extracted from the callus together with about 1% farnesol.

An assay across the culture period of numerous calli of differing provenance, and of cell suspensions showed similar peaks for production of polygodial and its parents, although for suspensions the maximum yield was now at ca. day 4. Several reasons may be proposed to account for these fluctuations in products. One factor

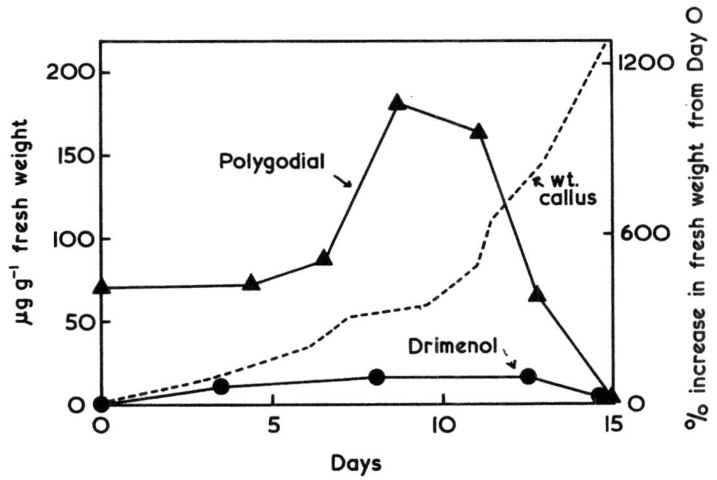

Fig. 4. Yields of polygodial and drimenol at different times in the culture cycle of callus from *P. hydropiper* (unpublished data obtained at UCL by J.T. Brown and G.S. Morris). Similar profiles are presented by Banthorpe et al. (1989)

may be the occurrence of extensive necrosis in the cell cultures which never entered a true stationary phase. Another may be the instability of products which may mask the true rates of synthesis (polygodial is known from controls to be fairly unstable at room temperature). Another may reflect the possibility that in the absence of adequate storage structures within the cultured cells, the product may be metabolized in vivo or excreted into the medium and there "biotransformed". It is important to draw a distinction between the capacity for synthesis and the capacity to accumulate a metabolite: the latter depends on the balance between synthetic and degradative processes. This distinction is rarely made in studies on cell cultures. Whatever the reason for the accumulation-time profiles that we repeatedly observed, the results stress the need to monitor production of secondary metabolites in general across the whole of the cell cycle. This is rarely done.

Polygodial mainly (96–99% of total in plant ca. 10% wt./fresh wt.) occurs in the flower-heads of the parent plant: the balance is in the leaves and then only can be detected during the flowering period. These observations differ from the implications concerning location in the flowers in previous papers. At optimum, cultures accummulated levels of polygodial at ca. 0.25 and 29%, respectively, of those in the flower head and foliage. Callus did not secrete detectable amounts of secondary metabolites into the supporting medium, but 20–50% of the assayed product was secreted from the cell suspensions. These levels appear low but it must be remembered that the synthetic ability of unorganized culture material is being compared with that of the specialized synthetic and storage organs of the field-grown plant.

Monoterpenoids and pigments (flavonoids, anthocyanidins etc.) were not detectably produced by our cultures. This is understandable as there is now a considerable body of evidence that sesquiterpenoids are synthesized and stored at different sites (in intact plants) from these other metabolites.

2.4 Biotransformations by Cell Suspension Cultures

The cell suspensions were capable of metabolizing exogeneously supplied drimenol and polygodial (ca. 5 mg per 500 cm^3 culture per 24-h period). A variety of products were formed including *2*, *3*, and *4* (Banthorpe and Branch, unpublished).

2.5 Studies Directed Towards Gene Transfer

Preparation of cell-free extracts revealed that the biosynthetic activity for polygodial formation from FPP (*16*) resided almost entirely within the flower heads of the field-grown plant. It has been anticipated that tissue cultures would provide a more convenient source of biomass for studies on enzyme purification, ultimately directed at gene transfer; and this was indeed the case. Extracts from both callus and cell suspensions made at the period of the culture cycle that yielded main accumulations of polygodial and drimenol were some 70% efficient in synthesis of polygodial compared with extracts prepared from the flower heads and assayed under identical conditions and contained 45–75% of the protein content of the latter (at max. ca. 0.15 mg/ml). Thus, tissue cultures could certainly be used for enzyme purifications.

Enzyme extracts from some 5 kg callus was processed in three steps to yield ca. 45 μg of drimenol synthetase, the enzyme that catalyzes the step *16* → 13, Fig. 3. The final product represented a 2850-fold purification and yielded enzyme with activity 9.2 μkat/kg. The final product was shown by electrophoretic screening to contain ca. 10% synthetase. The key step which allowed such a rapid, efficient purification was the use in the final stage of Phosphocellulose-Cellex which introduced a purification factor of ca. 400-fold. These studies are continuing.

3 Conclusions

Clearly, tissue culture studies are unlikely to be competitive for the commercial production of polygodial and its relatives. In addition to *P. hydropiper*, this compound occurs in several other natural sources that could be exploited (Loder 1962; Anderson et al. 1977; McCallion et al. 1982; Calle et al. 1985; Cimino et al. 1985; Kioy et al. 1989); and several total and partial chemical syntheses are available (Mori and Watanabe 1986; Razmilic et al. 1987; Nakaro and Martin 1989). However, gene transfer whereby commercially important plants are endowed with the ability to synthesize and secrete the anti-feedant compound seems an encouraging approach and the first steps have been laid in the above described establishment of suitable biomass and the partial purification of the appropriate synthetase.

A great deal of work is also necessary to understand the sources of the other biological activities of extracts of *P. hydropiper*. An important line of work could follow the claimed use of polygodial as a virus genome inactivator, as this could lead to the compounds, use as a viricide and perhaps as an antitumour compound.

4 Protocol

The following optimized protocol resulted in callus lines that gave high accumulations of (−)-polygodial reported in Sect. 2.3.

1 Explants (ca. 1 × 0.5 × 0.5 cm) of stem or foliage of mature plants of *P. hydropiper* were used. Material (ca. 4 × 0.1 cm) of the hypocotyl of seeds freshly germinated under sterile condition was also suitable.

2. After surface sterilization by conventional methods the explant was cultured on Murashige and Skoogs' standard medium supplemented with BAP (0.5 mg/l), NAA (2 mg/l) sucrose, (30 g/l) and (crucially) gibberellin A_3 (GA_3; 10 mg/l) and was made up with agar (0.9% w/v) and taken initially to pH 5.5. After maintenance at 25 °C under constant illumination (ca. 600 lx; Philips "warm white" tubes, max 580 nm), most explants formed sufficient callus for subculture after 2 to 4 weeks.

3. Serial passages were then carried out on the same medium and under the same conditions every 14 days.

4. Assays (GC-MS) on material extracted from the early subcultured material indicated erratic accumulations of polygodial; but after ca. 12 passages, the level stabilized to the values indicated in Sect. 2.3. Such cell lines were then stable for at least 2 years, when work ceased.

5. In addition to the requirement for gibberellin to be present in the medium to allow initiation and maintenance of the cultures, two highly significant observations were made. Firstly, the variation of accumulation of the sesquiterpenoid across the culture cycle (Fig. 4); and secondly, the period of instability in such accumulation (see above). Such effects have rarely been evaluated in previous studies on secondary metabolism in plant cell cultures but, if general, could have a crucial importance in the selection, cloning, and harvesting of high-producing cell lines.

Acknowledgement. This work was funded by Applied Technologies (Cambridge) Ltd., Cambridge, UK, and was carried out by Drs. Jacqueline T. Brown and G.S. Morris.

References

Ahmad S, Ansari MH, Ahmad M, Osman SH (1987) Studies on minor seed oils. Fett Wiss Technol 89: 154–156; Chem Abst 107: 9253

Anderson NH, Bissonette P, Liu CB, Shunk B, Ohta Y, Tseng RL, Moore A, Honeck S (1977) Sesquiterpenoids of nine European liverworts. Phytochemistry 16: 1731–1751

Asakawa Y, Takemoto T (1979) New nor-sesquiterpenoid aldehyde and sesquiterpenoid hemiacetals from seed of *Polygonum hydropiper*. Experientia 35: 1420–1421

Asakawa Y, Pickett JA, Ley S (1988) Activity of drimane-antifeedants and related compounds against aphids and comparative biological effects and chemical reactivity of (+) & (−)-polygodial. J Chem Ecol 14: 1844–1855

Banthorpe DV, Brooks CJW, Brown JT, Lappin GJ, Morris GS (1989) Synthesis and accumulation of polygodial by tissue cultures of *Polygonum hydropiper*. Phytochemistry 28: 1631–1633

Barnes CS, Loder JW (1962) Structure of polygodial; a new sesquiterpene dialdehyde from *Polygonum hydropiper*. Aust J Chem 15: 322–327

Brooks CJW, Watson DG (1985) Analytical separation of enantiomeric polygodials by gas chromatography of pyrrole derivatives. J Chromatogr 347: 455–457

Calle J, Guerrero MT, Rivera A (1985) Isolation of polygodial from bark of *Drymis granadensis*. Rev Colomb Quim 14: 27–30; Chem Abst 107: 214871

Cimino G, De Rosa S. Destefano S, Mamore R, Sodano G (1985) Structure, biosynthetic origin and defence properties of terpenes from dorid nudibrach *Dendrodonis grandiflora*. Tetrahedron 41: 1093–1100

Cimino G, Sodano G, Spinella A (1988) Occurrence of olepupuane in two mediterranean nudibranches. A protected form of polygodial. J Nat Prod 51: 1010–1011

Clapham AR, Tutin T, Warburg EF (1952) Flora of the British Isles. Cambridge Univ Press, Cambridge, pp 697–700

Cortes MM, Oyarzun LM (1981) Tadeonal and isotadeonal from *Drimys winteri*. Fitoterapia 52: 33–35

Dawson GW, Griffths DC, Smart LE, Hassanali A, Pickett JA, Plumb RT, Pye BJ, Woodcock CM (1986) Antifeedants – a new concept for control of barley yellow dwarf virus in winter cereals. Proc Brit Crop Protect Conf; Pestic Div 1001–1008; Chem Abst 106: 171111

Fukuyama F, Sato T, Asakawa Y, Takemoto T (1982) A potent cytotoxic warburganal and related drimane-type sesquiterpenoids from *Polygonum hydropiper*. Phytochemistry 21: 2895–2898

Fukuyama F, Sato T, Miura I, Asakawa Y, Takemoto T (1983) Hydropiperoside – a novel coumaryl glycoside from roots of *Polygonum hydropiper*. Phytochemistry 22: 549–552

Fukuyama Y, Sato T, Miura I, Asakawa Y (1985) Drimane-type sesquiterpenoid and non-sesquiterpenoids from *Polygonum hyropiper*. Phytochemistry 24: 1521–1524

Furata T, Fukuyama Y, Asakawa Y (1986) Polygonolide – an isocourmarin from *Polygonum hydropiper* possessing anti-inflammatory activity. Phytochemistry 25: 517–520

Garg SK, Mathur VS (1972) Effect of chromatographic fractions of *Polygonum hydropiper* root on fertility in female albino rats. J Reprod Fertil 29: 421–423

Gibson RW, Rice AD, Pickett JA, Smith MC, Sawicki RM (1982) Effect of repellant dodecanoic acid and polygodial on the acquisition of ion, semi and prevalent plant virus by the aphid *Myzus persicae* Ann Appl Biol 180: 55–59

Grieve M (1931) A modern herbal. Jonathan Cape, London, pp 743–746

Grigson G (1975) Englishmans flora. Paladin, St Albans, pp 460–462

Kioy D, Gray AT, Waterman PG (1989) Further drimane sesquiterpenoids from bark of *Camella winterana*. J Nat Prod 52: 174–177

Kubo I, Ganjian I (1981) Insect-antifeedant terpenes that are hot-tasting to humans. Experientia 37: 1063–1064

Kubo I, Taniguchi M (1988) Polygodial – an antifungal potentiator. J Nat Prod 51: 22–29

Kubo I, Lee YW, Pettei M, Pilkiewicz F, Nakanishi K (1976) Potent army worm antifeedants from the E. African *Warburgia* plant. JCS Chem Commun: 1013–1014

Launert E (1981) Edible and medicinal plants of Britain and N Europe. Hamlyn, New York, pp 114–115

Loder J (1962) Occurrence of the sesquiterpenoids polygodial and guaiol in leaves of *Drimys lanceolata* Aust J Chem 15: 389–390

Matsumoto A, Kageyama S (1988) Trichophyton-inhibiting pharmaceuticals containing waburganal; Jpn Pat 63: 156718; Chem Abst 109: 176376

McCallion RF, Cole ALJ, Walker JR, Blunt JW, Munro HMG (1982) Polygodial – an anti-candida agent from *Pseudowintera colorata* from New Zealand. Planta Med 44: 134–138

Minoda T, Kodama T (1977) Amino acids from plant calluses. Jpn Pat 85: 85200; Chem Abstr 88: 205557

Mori K, Watanabe H (1986) Synthesis of both enantiomers of polygodial – an unsaturated antifeedant sesquiterpenoid. Tetrahedron 42: 273–281

Nakaro T, Martin A (1989) A very short, highly efficient synthesis of natural (−)-warburganal. JCS Res Synop: 52–53

Pickett JA (1985) Production of behaviour-controlling chemicals by crop plants. Philos Trans R Soc Lond 310B: 235–239

Quadrat-i-Khuda M, Khalique A, Khuda HAM (1965) Examination of *Polygonum hydropiper*, constituents of the plant. Sci Res 2: 135–142

Razmilic I, Lopez J, Sierra J (1987) Alternative partial synthesis of (−)-polygodial. JCS Chem Commun; 95–103

Res Corporation (1981) Antifeedants: compositions containing antifeedants and the use thereof. Brit Pat 1590174; Chem Abstr 95: 127448

San Feliciano A, Barrero AF, Miguel del Corral JM, Gordaliza M, Medarde M (1985) Dialdehyde formation during isolation of extracts from *Linaria* species. Tetrahedron 41: 671–685

Shalanova GG, Yakoulev Al, Churilov Gl (1983) Chemical composition of above-ground part of three species of *Polygnum*. Rastit Resur 19: 526–529

Stampf JF, Benezra C, Askakawa Y (1982) Experimentally-induced allergic contact dermatitis to a natural sesquiterpenoid dialdehyde, polygodial. Arch Dermatol 274: 277–278

Sterner O, Carter RE, Nilson LM (1987) Mutagenic activity of unsaturated dialdehydes. Mutat Res 188: 169–174

Swiatek L, Dombrowicz E (1987) Phenolic acids in medicinal plant drugs from *Polygonum* species. Farm Pol 43: 420–423; Chem Abst 108: 156523

Taniguchi M, Adachi T, Oi S, Kimura A, Katsumura S, Isoe S, Kubo I (1984) Structure-activity relationship of the warburgia sesquiterpenoid dialdehyde. Agric Biol Chem 48: 73–78

Taniguchi M, Yano Y, Tada E, Ikenishi K, Oi S, Horoguchi H, Hashimoto K, Kubo I (1988) Mode of acton of polygodial – an antifungal sesquiterpene dialdehyde. Agric Biol Chem 52: 1409–1414

Tokuda H, Matsumoto T (1989) Warburganal as a virus-genome inactivator. Jpn Pat 89: 29307; Chem Abst 111: 167386

Vysochina Gl, Kulpina TG, Berezovskaya TP (1987) Flavonoid composition of some species of *Polygonium* in the Siberian flora. Rastit Resur 23: 229–234

Wisniewski W, Szlaski J (1970) Stability of anti-hemorrahagic herbal extracts. Farm Pol 26: 239–242; Chem Abst 75: 40336

Yankov LK, Damyanova L (1969) Components of *Polygonum hydropiper* I. Composition of the essential oil. Pharmazie 24: 696–699

Yankov LK, Damyanova LD (1970) Components of *Polygonum hyropiper* II, Hydrocarbons, waxes and higher aliphatic alcohols. Pharmazie 25: 199–201

Yankov L, Damyanova L (1977) Study of aromatic hydrocarbon composition of *Polygonum hydropiper*. Dokl Bolg Akad Nauk 30: 1713–1716; Chem Abstr 88: 186132

Yoshitama K, Nishino H, Ozawa H, Satatani M, Okabe Y, Ishikura N (1987) Distribution pattern of anthocyanidins and anthocyanins in polygonaceous plants. Bot Mag 100: 143–149

Yusif KS, Blinova KF (1984) Flavonol aglycones of *Polygonum hydropiper*. Khim Prir Soedin: 658–659

XIX *Ptelea trifoliata* (Quinine Tree, Hop Tree): In Vitro Culture and the Production of Alkaloids and Medicinal Compounds

G. PETIT-PALY, M. MONTAGU, J. TRÉMOUILLAUX-GUILLER, J.C. CHÉNIEUX, and M. RIDEAU[1]

1 Introduction

Ptelea trifoliata L. (Rutaceae) is a bush of North American origin that has been cultivated in Europe since the eighteenth century. Pharmacological properties (particularly bacteriocidal and cytotoxic activities) are due to the presence of coumarins and quinoline alkaloids.

1.1 Botany and Distribution

Ptelea trifoliata's common names include: quinine tree, potato chip tree, and hop tree (the latter being the most widely used today); in Spanish, *Cola de Zorillo*; in French *Ptelea à 3 feuilles, trèfle de Virginie, Orme de Samarie* – this last name was first used in France around 1800 and is still widely used (Petit-Paly 1988).

Ptelea trifoliata L., described by Linnaeus in 1753, is extremely variable in its morphology and chemical composition. This explains the description of numerous varieties which have often been raised to the rank of species. The most recent revision of the genus *Ptelea* is by Bailey (1962) who recognizes only three species: *P. trifoliata* L., *P. crenulata* Greene, and *P. aptera* Parry, although he subdivides *P. trifoliata* into five subspecies and ten varieties.

The *Ptelea* species are deciduous bushes, 3–4 m tall, with trifoliate aromatic leaves (Fig. 1). A large number of detailed descriptions exist (Baillon 1872; Gray 1888; Sargent 1892; Bailey 1962). There have been many studies of the floral morphology, most recently that of Ambrose et al. (1985) which clearly confirms the dioecious nature of this species, with only 2% of otherwise male plants producing a few hermaphrodite flowers and fruits. These fruits are two-celled samaras (Gray 1849; Ulrich 1942). The seeds germinate weakly; this is certainly linked to the presence of inhibitors in this species (Mac Leod and Murphy 1977).

The present habitat of *Ptelea trifoliata* is central and southern parts of North America, where it is often localized, and hence designated as "rare" or "endangered" in Ontario and in four states of America (Ambrose et al. 1985). *P. trifoliata* seems to have originated in southern Mexico where certain populations have persisted in the earliest colonized areas, while others have migrated toward the

[1] Plant Biotechnology Laboratory, Faculty of Pharmacy, University of Tours, 2 bis Bd. Tonnellé, 37042 Tours Cedex, France

Biotechnology in Agriculture and Forestry, Vol. 21
Medicinal and Aromatic Plants IV (ed. by Y.P.S. Bajaj)

Fig. 1A–D. *Ptelea trifoliata* L. (Rutaceae). **A** 30-year-old tree in the Tours Botanical Gardens (France); **B** full-blossom flowers; **C** two-celled samaras; **D** view of a plantlet obtained after germination in the light

north in the soils made available after glacial retreat (Ambrose et al. 1985). Geological and climatic fluctuations since the tertiary era account for the great number of varieties and forms, more or less linked by hybridization. In a chemotaxonomic study on the genera *Ptelea*, Bailey et al. (1971a) used the coumarin and alkaloid patterns of 150 specimens as a means of determining the relationship between different subspecies and varieties. We recently analyzed the dihydrofuro[2,3-b]quinolinium alkaloid patterns of trees belonging to natural populations in Ontario. They were sometimes quite different: studying these patterns alongside morphology would possibly yield a better characterization of these populations (Petit-Paly et al. 1989b).

1.2 Medicinal Value

Pharmaceutical uses of *Ptelea* have been known through the reports of an English army doctor, Schoepf, who noted that the Canadian populations used the leaves as a antihelminthic and for healing of wounds (Schoepf 1787).

The most frequently reported medicinal properties concern the tonic, stomachic, and aperitive effects of the root bark, which is more active in this respect than flowers, fruit, or leaves (Mulvey and Zalewski 1969). The antipyretic properties of root bark, used in Louisiana before the introduction of quinine, account for the common name of "quinine tree". Spencer et al. (1947) found that this root bark also showed weak antimalarial properties, but trials using material from our collection – trees growing in the Botanical Garden in Tours – showed no antimalarial activity although antipyretic activity was observed (unpublished).

The antimicrobial and tuberculostatic properties reported by King and Lloyd (1886) have been confirmed by more recent studies: various alkaloids and coumarins with bacteriocidal properties have been isolated, in particular quaternary alkaloids such as pteleatinium (Mitscher et al. 1971, 1975a, b) and ptelefolonium (Rideau et al. 1979).

Other physiological properties of coumarins and alkaloids are listed in Table 1. The plant possesses toxicity proportional to the alkaloid content of its organs, stem bark being the most toxic and seeds the least (Minker et al. 1976). Also, dermatitis has been reported after contact with the leaves and this reaction is almost certainly linked to the presence of coumarins and some alkaloids (Table 1).

Ptelea trifoliata is currently used in treatment of gastric disorders due to its spasmolytic activity: it is included, for this reason, in some French homoeopathic preparations (Vannier and Poirier 1976). It was listed in the American Pharmacopoeia from 1878 to 1941. It remains a folk remedy in many Eastern European countries and in India.

1.3 Conventional Practices for the Preparation. Production of Medicinal Compounds

Acclimatized in Europe very early, *P. trifoliata* is now cultivated throughout the world. Cultivation is possible under all conditions of soil and exposure, and sowing is carried out directly after harvest, the dormant seeds needing 2–4 months at 1 °C (Schroeder 1937). Other cultivation methods used include layering and grafting under frames on young seedlings.

The Hungarian team of Reisch, Szendrei and Novák, known for their work on *Ruta graveolens* (see Petit-Paly et al. 1989a, for a review), began the analysis of *P. trifoliata* in 1969, resulting in the identification of a large number of alkaloids and coumarins responsible for the majority of the biological activities. Parallel research on antimicrobial substances by Mitscher et al. (1971, 1975a, b) and that on plant growth inhibitors by Garestier and Rideau (1972) have resulted in the isolation of many alkaloids, particularly quaternary ones. Some of them (ptelefolonium and hydroxyluninium) have a prominent cytotoxic effect both on KB tumorous cells and Wi 38 human cells (André et al. 1981). More recently, our group isolated two new dihydrofuro[2,3-b]quinolinium alkaloids: ptelecultinium

Table 1. Physiological activities of some coumarins and alkaloids[a]

Compounds	Activities	References
Coumarins	Growth inhibitor on roots	Andreae (1952)
Furocoumarins		
Bergapten	Spasmolytic	Minker et al. (1979)
	Photosensitizer	Musajo and Rodighiero (1971), Nikolin (1971)
	Growth inhibitor	Sinha-Roy and Chakraborty (1976)
	Molluscicid	Soine (1964)
	Fish toxic	Torres (1950)
Isopimpinellin	Growth inhibitor	Sinha-Roy and Chakraborty (1976)
	Molluscicid	Soine (1964)
	Fish toxic	Torres (1950)
Imperatorin	Photosensitizer	Kumar et al. (1976), Soine (1964)
(+) Imperatorin oxide	Photosensitizer	Kumar et al. (1976)
Furoquinoline alkaloids, 2-quinolone, 4-quinolone		
Dictamnine	Photosensitizer	Towers et al. (1981), Mizuta and Kanamori (1985)
	Uterus contraction	Gastaldo (1973)
	Vasoconstriction	Kovalenko (1946)
	Heart contraction	Kovalenko (1946)
Kokusaginine	Ganglioplegic	Kovalenko (1946)
Skimmianine	Hypothermic	Polievtsev et al. (1972)
	(+ iron bactericid)	Mitscher et al. (1975b)
Ptelefoline	Hypotensive	Minker et al. (1979)
	Thermoregulation inhibitor	Minker et al. (1979)
Ptelefolone	Thermoregulation inhibitor	Minker et al. (1979)
Dihydrofuro [2,3-*b*] *quinolinium alkaloids*		
Ptelefolonium	Spasmolytic	Minker et al. (1976)
	Ganglioplegic	Szendrei et al. (1973)
	Growth inhibitor	Chénieux et al. (1976)
	Cytotoxic	André et al. (1981)
	Bactericid	Rideau et al. (1979)
Isoptelefolonium	Growth inhibitor	Chénieux et al. (1976)
Hydroxyluninium	Growth inhibitor	Chénieux et al. (1976)
	Slightly cytotoxic	André et al. (1981)
Pteleatinium	Bactericid	Mitscher et al. (1971, 1975a)

[a] Most of the structures of coumarins and quinoline alkaloids also present in *R. graveolens* were reported in Petit-Paly et al. (1989a), those of quaternary alkaloids are reported in Table 3b.

and ptelefolidonium (Petit-Paly et al. 1987, 1989b; Petit-Paly 1988). Various flavonoids (Novák et al. 1970) and essential oils (Earle et al. 1960) were also obtained.

The organs or tissue cultures that produce these metabolites are summarized in Tables 2 and 3a–d. The chemical variability encountered in parallel with the morphological variations most likely explains the differences observed in pharmacological activity between extracts prepared from different sources of *P. trifoliata*. In fact, two of the alkaloids reported to be responsible for the major physiological properties of the extracts, ptelefolonium and kokusaginine, are completely absent in some samples.

Table 2. Some coumarins isolated from *P. trifoliata*[a]

	Roots	Stems	Leaves	Flowers	Fruits
Coumarins					
Auraptene	6	4		5	5
Scopoletin		4			
Scopoletin glucoside		4			
Furocoumarins					
Bergapten		3	3		
Isopimpinellin		3	3		
Phellopterin		3	3		
Imperatorin		6			
Imperatorin oxide		3, 8	3		8
(−) Byakangelicin		3	3		
Dihydrofurocoumarins					
(+) Marmesin	6	3	3, 7		
(−) Marmesinin	6	1	2, 7		

[a] Structures were reported in Petit-Paly et al. (1989a). 1, Bailey and Bailey (1965); 2, Bailey et al. (1971a); 3, Dreyer (1969); 4, Körösi et al. (1974a); 5, Novák et al. (1970); 6, Reisch et al. (1975a); 7, Szendrei et al. (1973); 8, Weinstein et al. (1972).

Table 3b. Dihydrofuro [2, 3b] quinolinium alkaloids[a]

	Roots	Stems	Leaves	Flowers	Fruits
1-*Furoquinoline alkaloids*					
Dictamnine	6				
Pteleine	4				
Kokusaginine	1, 2, 4, 5, 6, 9		4, 5, 6		4
Skimmianine	1, 2, 4, 5, 6, 9, 10	7	4, 5, 6	12	4
7-isopentenyloxy-γ fagarine					3
2-*Pyranoquinoline alkaloids*					
N-methylisoflindersine	11				
Neohydroxylunine		7	8		
Pteleflorine					12

[a] Some structures of these compounds were reported in Petit-Paly et al. (1989a). 1, Bailey and Bailey (1965); 2, Bailey et al. (1971a); 3, Dreyer (1969); 4, Frolova et al. (1964); 5, Kowalska (1963); 6, Kowalska and Borkowski (1966); 7, Mitscher et al. (1975a); 8, Mitscher et al. (1975b); 9, Mooken et al. (1969); 10, Mulvey and Zalewski (1969); 11, Reisch et al. (1975a); 12, Reisch et al. (1975b).

Table 3b. Dihydrofuro [2, 3b] quinolinium alkaloids[a]

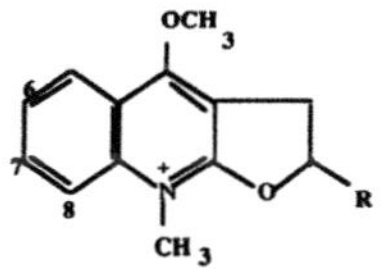

	6	7	8	R	Roots	Stems	Leaves	Flowers	Tissue culture
Pteleatinium			OH			——4——			
Ptelefolonium	OCH_3		OCH_3			—1, 2, 3, 5, 8, 9, 11—			10
Isoptelefolonium		OCH_3	OCH_3			——9, 11——			
Ptelecultinium			OCH_3		11	in seedlings only 12			10
Luninium		$O–CH_2–O$						7	
Hydroxyluninium		$O–CH_2–O$				—5,6,9,11—			
Ptelefolidonium		$O–CH_2–O$				——11——			

[a] 1, Garestier and Rideau, (1972); 2, Körösi et al. (1974b); 3, Körösi et al. (1976); 4, Mitscher et al. (1971); 5, Mitscher et al. (1975a); 6, Mitscher et al. (1975b); 7, Novák et al. (1970); 8, Reisch et al. (1973); 9, Rideau et al. (1979); 10, Petit-Paly et al. (1987); 11, Petit-Paly et al. (1989b); 12, Petit-Paly et al. (1990).

Table 3c. 4-Quinolone

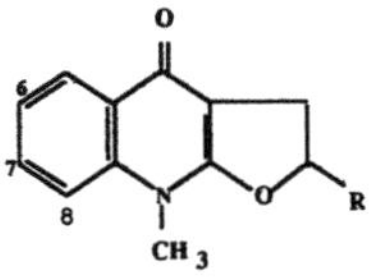

	6	7	8	R	Roots	Stems	Leaves	Flowers
(+) Balfourodin			OCH_3					8
Ptelefolon	OCH_3		OCH_3		1,2		6	4,6
Lunin		O–CH_2–O						5
(+) Hydroxylunin		O–CH_2–O					8	3,4,5
Ptelefolidon		O–CH_2–O						7

[a] 1, Bailey et al. (1971a); 2, Bailey et al. (1971b); 3, Körösi et al. (1976); 4, Novák et al. (1970); 5, Reisch et al. (1969); 6, Reisch et al. (1970); 7, Reisch et al. (1975b); 8, Szendrei et al. (1973).

Biogenetically, furoquinolines are derived from anthranilic acid. Key intermediates are 2,4-dihydroxyquinoline and 3-dimethylallyl-4 methoxy-2 quinolone. Recently, it was found that an S-adenosyl L-methionine-anthranilic acid N-methyl transferase and an N-methyl anthranilic acid "activating" enzyme are involved in the first steps of quinoline alkaloid biosynthesis in Rutaceae (Baumert et al. 1990). Only one study of the dimerization of the pyranoquinoline, N-methylflindersine, has been conducted on *P. trifoliata* (Grundon and Rutherford 1985). It may be of interest to note that coumarins and quinolines both derived biogenetically from anthranilic acid.

2 In Vitro Culture Approaches

The tissue culture of *Ptelea trifoliata* has not, to our knowledge, been previously reported. Our interest in these cultures derives from the fact that we found dihydrofuroquinolinium alkaloids presenting cytotoxic activities in this species (Rideau et al. 1979). Moreover, we wanted to compare the intraspecific variability within the natural populations of trees with the somaclonal variations currently observed in plant tissue cultures. In a first set of experiments, a number of cell lines were prepared by culturing stem explants from different trees (Fig. 2). In a second set of experiments, protoplasts prepared from one of these cell lines were cultured in order to obtain a population of protoclones (i.e. protoplast-derived cell lines; Fig. 3).

Table 3d. 2-Quinolone[a]

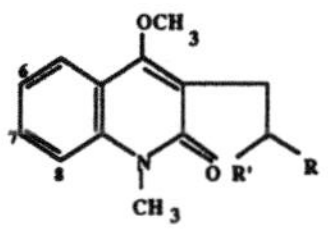

	6	7	8	R	R′	Roots	Stems	Leaves	Flowers	Fruits
Ptelefolinol	OCH_3		OCH_3		OH		2	2		
Ptelefolin	OCH_3		OCH_3		OH		2	2	4,5	
Ptelefolin methyl ether	OCH_3		OCH_3		OCH_3			7	7,9	
4,6,8 Trimethoxy 3′3′ dimethyl allyl 2-N methylquinolone	OCH_3		OCH_3			8				
Isoptelefolin		OCH_3	OCH_3		OH					6
Lunidin		$O–CH_2–O$			OH			5	9	
Lunidonin		$O–CH_2–O$			=O			3	3	
Hydroxylunidin		$O–CH_2–O$			OH		1,2	1,2	4,5	
Hydroxylunidonin		$O–CH_2–O$			=O			3		9
Pteleprenin		$O–CH_2–O$					1	1		
Pteleolin		$O–CH_2–O$		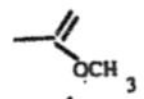			7		9	
Ptelefolidin		$O–CH_2–O$			OH		—1,2,6—			
Ptelefolidin methyl ether		$O–CH_2–O$			OCH_3		—1,3,6—			
6-Hydroxy methoxy lunidin	OCH_3	$O–CH_2–O$			OH		—2,3—			
6-Methoxylunidonin	OCH_3	$O–CH_2–O$			=O		—3—			
Ptelecortin		$O–CH_2–O$				8	1,7			
Ptelefructin		$O–CH_2–O$	OCH_3		OH					4,6

[a] 1, Körösi et al. (1974a); 2, Körösi et al. (1976); 3, Mitscher et al. (1975a); 4, Novák et al. (1970); 5, Reisch et al. (1969); 6, Reisch et al. (1970); 7, Reisch et al. (1972); 8, Reisch et al. (1975a); 9, Reisch et al. (1975b).

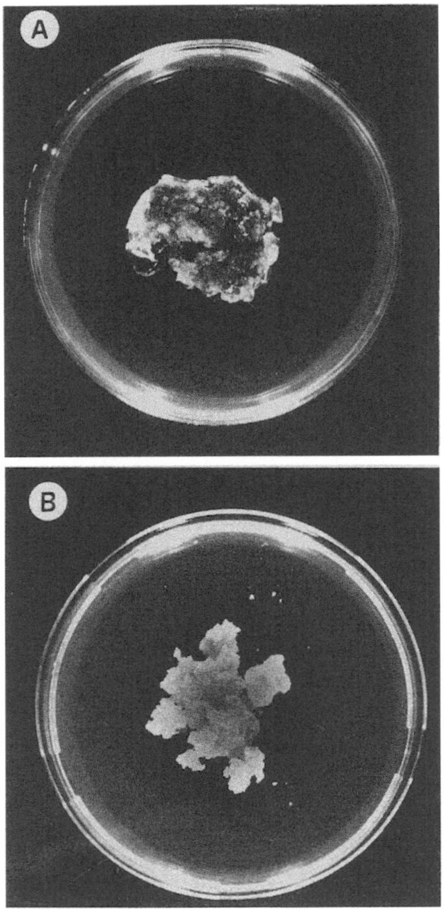

Fig. 2A, B. Stem explant-derived cell lines of *P. trifoliata*. **A** dense chlorophyllous line maintained on MT, NAA 5.3 μM, K 0.47 μM without coconut milk; **B** pale yellow friable line maintained on MT, 2,4-D 4.5 μM, K 0.47 μM

2.1 Stem Explant-Derived Cell Lines

Young stems of *Ptelea trifoliata* were collected in May 1982 from 30-year-old trees grown in the Botanical Garden in Tours. They were treated with 1.75% sodium hypochlorite (prepared from commercial bleach) for 5 min, followed by 70% ethanol for 1 min and 3.5% sodium hypochlorite for 5 min. Surface-sterilized stems were washed three times with sterile distilled water, then cut into segments (5 mm length).

A number of culture media containing various auxins and cytokinins were used (Petit-Paly 1988). Using the Murashige and Tucker (MT) agar medium (1969) supplemented with 0.47 μM kinetin (K), either 4.5 μM 2,4-dichlorophenoxyacetic acid (2,4-D) or 5.3 μM α- naphthalene acetic acid (NAA) and 5% coconut milk, well-growing primary callus cultures were obtained. They were subcultured on the same medium supplemented with coconut milk. This substance was progressively suppressed during the following subcultures. The media were autoclaved at 0.5 bar for 20 min. The cultures were kept under continuous white light at 24 ± 1 °C.

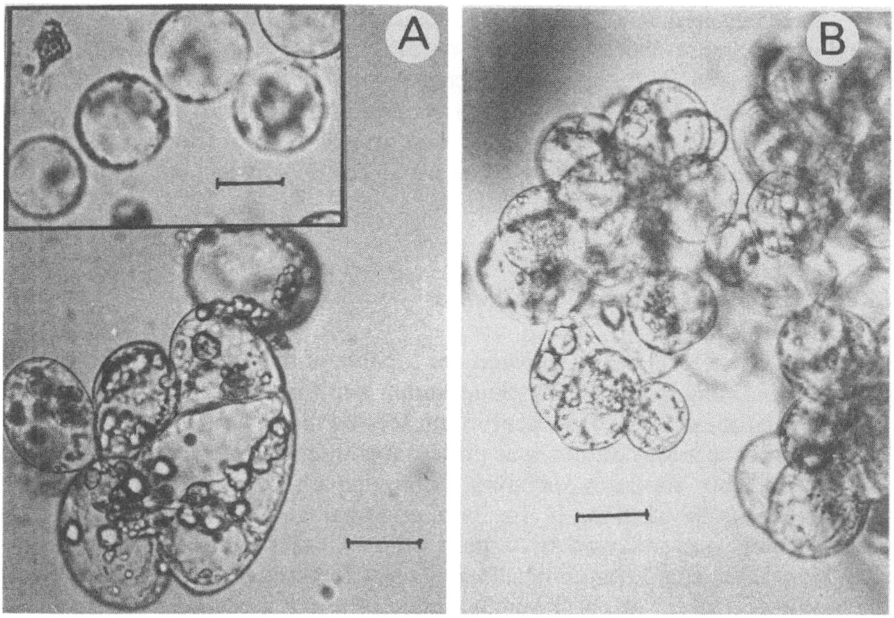

Fig. 3A, B. Protoplast-derived cell lines. **A** first cell division after 7 days of culture. *Insert* freshly isolated protoplasts. *Bar* = 50 μm. **B** 2-month-old microcolonies. *Bar* 50 μm

2.2 Protoplast-Derived Cell Lines

Protoplasts were isolated from 2-week-old callus cultures. Callus (0.5 g fresh weight) was harvested and incubated in 5 ml of an enzyme solution consisting of 1% (w/v) cellulase Onozuka R.10, 0.5% (w/v) macerozyme R.10 (Yakult Biochemicals Co. Ltd., Mishinomiya, Japan) in a 0.034 M $CaCl_2$, $2H_2O$ solution containing 0.5 M mannitol. The pH was adjusted to 5.5. Incubation was carried out at 27 °C in the dark. After 16 h, the digested mixture was filtered through a nylon sieve (100 μm pore size). Gentle centrifugation (100x g, 3 min) produced a pellet enclosing the protoplasts together with some cellular debris; this pellet was taken up in a 20% sucrose solution allowing intact protoplasts to float. The protoplast band was removed and pelleted twice in $CaCl_2$, $2H_2O$ (0.2 M) then in mannitol (0.5 M). The protoplasts were then resuspended in the following culture medium; Murashige and Tucker medium supplemented with mannitol (0.35 M); glucose (0.25 M); NAA (5.37 μM); 2,4-D (4.5 μM); K (0.47 μM).

Protoplasts were cultivated in Parafilm sealed Corning dishes (3.5 cm diameter), with 0.5 ml culture medium. Protoplast density was adjusted to 10^4–10^5/ml. Cultures were incubated at 27 °C in the dark for 1 week, then transferred to continuous white light. Protoplasts divided within 8 days, giving microcolonies which progressively turned brown and necrosed after 2 months. However, we were able to obtain protoplast-derived cell lines of *P. trifoliata* by using the following nurse culture method. The nurse cells – 1-month-old *Catharanthus roseus* protoplast-derived microcolonies (10^5/ml) obtained from a tumorous line – were cultured in a Millicell CM chamber (Millipore). This one was introduced in a Corning dish containing 1-month-old *P. trifoliata* protoplast cultures. Both *C. roseus* and *P. trifoliata* microcolonies were cultured in the medium mentioned above. After 3–4 months in coculture the *Ptelea trifoliata* protoclones which had reached 0.5–1 mm diameter were transferred onto solid medium. Thus, a large population of protoclones was obtained, 60 of which were submitted to alkaloid analysis.

2.3 Extraction and Structure of Alkaloids

The tissue cultures of *Ptelea trifoliata* are capable of biosynthesizing furoquinoline alkaloids and dihydrofuro[2,3-b]quinolinium and furoquinoline alkaloids.

A new alkaloid, named ptelecultinium (Petit-Paly et al. 1987), isolated for the first time from in vitro cultures, was present together with ptelefolonium (Table 3b) and several other alkaloids, including skimmianine (Table 3a) (unpublished).

The change in content of the two principal alkaloids, ptelecultinium and ptelefolonium, was analyzed by a spectrofluorodensitometric method (Montagu et al. 1986) after separation of the alkaloids by overpressure layer chromatography (Pothier et al. 1990). Table 4 shows that ptelecultinium was always absent in stems, whereas it was always present in the derived tissue cultures. Its concentration reached 20 mg/g dry weight in the highest-ptelecultinium accumulating protoclone. We also found that dense chlorophyllous cell lines showed higher alkaloid content than the friable and weakly pigmented ones (Fig. 2B; results not shown).

Table 4. Ptelecultinium (Pc) and ptelefolonium (Pf) content (mg/g dry wt.) in plant material

	Ptelecultinium					Ptelefolonium			
	n[a]	x	SD[b]	Pc_{min}	Pc_{max}[c]	x	SD[b]	Pf_{min}	Pf_{max}[c]
Mature trees									
Stems	61		Nondetectable			2.3	1.2	0	6.7
Leaves	61		Nondetectable			3.2	2.1	0	7.2
Flowers	26		Nondetectable			1.7	1.4	0.25	3.8
Callus cultures									
– From stem explants (8 years old)	32	0.4	0.3	0.06	1.8	0.1	0.1	0	0.7
– From protoplasts (6 months old)	60	3.7	5.6	0	22	1.8	3.1	0	15.1

[a] Total number of samples.
[b] $x \pm SD$, Mean value ± standard deviation.
[c] Pc_{min}, Pf_{min}: minimum alkaloid yield.
Pc_{max}, Pf_{max}: maximum alkaloid yield.

3 Conclusions and Prospects

3.1 In Vitro Culture of *Ptelea* Is Possible But Difficult

Most of the stem explants developed primary callus. This required the addition of coconut milk to the culture media. However, it was not possible to obtain cell lines from all the trees. Certain trees are better suited to in vitro culture than others. Thus, the importance of the genotype of origin for in vitro is shown for *P. trifoliata*. This observation should be considered in conjunction with the difficulty in obtaining cell lines from *P. trifoliata* protoplasts.

3.2 Dihydro [2,3-b] Quinolinium Is Accumulated in *P. trifoliata* Tissue Cultures

The different cell lines and protoclones showed important genotypic variations in growth, pigmentation, friability as well as in alkaloid content. A new alkaloid, ptelecultinium, was present in all cell cultures, irrespective of the tree of origin: although absent in stems, it accumulated rapidly in the primary explants, while the other quaternary alkaloids which were present in the explant progressively decreased. It appears, therefore, that alkaloid metabolism can change very rapidly once the explant is no longer under the regulatory control of the whole plant. Ptelecultinium was mostly dominant in the in vitro cultures often associated with ptelefolonium. This alkaloid pattern is similar to that of roots and cotyledons where ptelecultinium has been isolated (Petit-Paly et al. 1990). This correlation in chemical composition between in vitro cultures and roots has been observed in two members of Rutaceae, *Choisya ternata* and *Ruta graveolens* (unpublished) and also by Wink (1989).

Apparently, the in vitro *P. trifoliata* cultures are unable to reach the level of chemical differentiation of the cells of stems or flowers, and therefore offer a choice of material for the analysis of the early stages of metabolite biogenesis.

3.3 Prospects

Future developments are possible in several directions:

1. Further knowledge of the quinoline alkaloid biogenesis pathway;
2. Optimization of the production of quinoline alkaloids by the use of elicitors;
3. Search of new dihydrofuroquinoline alkaloids by the use of somatic fusions between protoplasts obtained from different genotypes;
4. Further knowledge of pharmacological properties of the isolated dihydrofuroquinolinium alkaloids.

4 Protocol

Extraction of Dihydrofuroquinolinium Alkaloids from Tissue Cultures of *Ptelea trifoliata*

Three-week-old callus culture was harvested and freeze-dried (570 g fresh weight; 50 g dry wt). The callus powder was extracted (2x) with 0.5 l methanol, for 48 h at 4 °C. The organic layers were pooled and concentrated under vacuum. The residue was dissolved in 50 ml distilled water and extracted with 50 ml diethyl ether (3x). The aqueous phase was concentrated under vacuum and fractionated on a silica gel column (3.2 × 40 cm, 50 g silica gel 60, Merck 7754) with $CHCl_3$/MeOH (75:25). Fractions of 10 ml were collected (120 fractions all together), and fractions 60–70, 71–82, and 83–120 were evaporated to dryness. The residues were subjected to TLC in the solvent system ethyl acetate/formic acid/distilled water (10:1:1).

Fractions 60–70: A blue fluorescent compound (254 nm) showed an Rf value of 0.47 which was identical to that of ptelefolonium and a positive reaction with Dragendorff's reagent. It was further purified by preparative TLC (silica gel G60, Merck 7731; solvent system as above), eluted from silica gel with methanol, precipitated in diethyl ether with perchloric acid. This substance (22 mg) was identified as ptelefolonium perchlorate.

Fractions 71–82 contained a blue-purple fluorescent compound under 254 nm showing a Dragendorff positive reaction (Rf = 0.41) and a Folin reactive compound (Rf = 0.38). Both compounds were separated on a Sephadex LH20 column (2.7 × 50 cm; 40 g) with an aqueous HCl solution (10^{-3} M) as eluent: 65 fractions of 10 ml were collected. Fractions 55–60 which contained the alkaloid were combined and freeze-dried. Ptelecultinium chloride (16 mg) was obtained (Petit-Paly et al. 1987).

References

Ambrose JD, Kevan PG, Gadawski RM (1985) Hop tree (*Ptelea trifoliata*) in Canada: population and reproductive biology of a rare species. Can J Bot 63: 1928–1935

André S, Lasselain MJ, Pareyre C, Deysson G, Chénieux JC, Rideau M (1981) Etude de quelques propriétés biologiques d'alcaloïdes dihydrofuroquinoléiques isolés de Rutacées. Ann Pharm Fr 39: 255–266

Andreae WA (1952) Effect of scopoletin on indoleacetic acid metabolism. Nature 170: 83–84

Bailey VL (1962) Revision of the genus *Ptelea* (Rutaceae). Brittonia 14: 1–45

Bailey HE, Bailey VL (1965) Phylogenetic studies on the genus *Ptelea* (Rutaceae). Chromatography as an aid to clarification of taxonomic relationships between varieties of *Ptelea trifoliata* subsp. *angustifolia* and subsp. *pallida*. Lloydia 28: 27–43

Bailey HE, Mooken JD, Bailey VL (1971a) Studies on the geographical distribution of some alkaloids and coumarins in leaf extracts of *Ptelea* species. Lloydia 34: 377–382

Bailey HE, Mooken JD, Bailey VL (1971b) The use of TLC profiles in studies on *Ptelea* (Rutaceae). Herba Hung 10: 49–54

Baillon H (1872) Histoire des plantes, vol 4. Hachette, Paris, pp 373–511

Baumert A, Crèche J, Rideau M, Chénieux JC, Gröger D (1990) Anthranilic acid-specific enzymes in alkaloid-producing tissue cultures of *Choisya ternata* and *Ruta graveolens*. Plant Physiol Biochem 28: 587–592

Chénieux JC, Rideau M, Verchère C, Viel C (1976) Propriétés inhibitrices de la croissance cellulaire de quelques alcaloïdes furoquinoléiques isolés de Rutacées. CR Acad Sci D 283: 101–104

Crèche J, Guiller J, Andreu F, Gras M, Chénieux JC, Rideau M (1987) Variability in a population of *Choisya ternata* strains issued from a single individual. Phytochemistry 26: 1947–1953

Dreyer DL (1969) Coumarins and alkaloids of the genus *Ptelea*. Phytochemistry 8: 1013–1020

Earle FR, Glass CA, Glenda C, Geisinger-Wolff IA (1960) Search for new industrial oils. J. Am Oil Chem Soc 37: 440–447

Engler A (1931) Rutaceae. In: Engler A, Prantl K (eds) Die natürlichen Pflanzenfamilien, Bd 19a Wilhelm Englemann, Leipzig, pp 187–359

Frolova VI, Kuzovkov AD, Kibal'chich PN (1964) Alkaloids of *Ptelea trifoliata*. Structure of pteleine. Zh Obshch Khim 34: 3499–3505 (in Russian)

Garestier R, Rideau M (1972) Isolement d'un alcaloïde inhibiteur de la croissance chez *Ptelea trifoliata* L. (Rutacées). CR Acad Sci 274: 2651–2654

Gastaldo (1973) Compendium della flora officinale italiana. Fitoterapia 46: 61–79

Gray A (1849) The genera of the plants of the United States, vol 2. Putnam GP, New York, pp 149–150

Gray A (1888) Contributions to American botany Proc Am Acad Arts Sci 23: 223–227

Greene EL (1905) Some *Ptelea* segregates. Torreia 6: 99–100

Greene EL (1906) The genus *Ptelea* in the western and south-western United States and Mexico. Contrib US Natl Herb 10: 49–78

Grundon MF, Rutherford MJ (1985) Dimerization of N-methylflindersine. J Chem Soc Perkin Trans 1: 197–198

King J, Lloyd J (1886) The American dispensatory. Wilstach Baldwin, Cincinnati, pp 679–681, 1111–1112

Körösi J, Papp E, Szendrei K, Novák I, Reisch J, Minker E (1974a) Kinolon-Alkaloidok és kumarinakizolálása a *Ptelea trifoliata* L. Bol Herba Hung 13: 23–32

Körösi J, Szendrei K, Novák I, Reisch J, Minker E (1974b) A *Ptelea trifoliata* kvaterner alkaloidjai. I. A Fókomponensek izolálása. Herba Hung 13: 17–22

Körösi J, Szendrei K, Novák I, Reisch J, Minker E (1976) A *Ptelea trifoliata* Kvaterner alkaloidjai. II. A Komponensek vizsgálata indirekt modszevrel. Herba Hung 15: 9–17

Kovalenko VN (1946) Pharmacochemical and pharmacological properties of alkaloid dictamnine from *Dictamnus albus turkestaniens*. Farmatsiya 9: 20 (in CA 1947 41: 6989b)

Kowalska M (1963) Alkaloidy w Liściach i Korzeniach *Ptelea trifoliata* L. Acta Pol Pharm 20: 458

Kowalska M, Borkowski B (1966) Alkaloidy w Lisciach i Korzeniach *Ptelea trifoliata* L. (Rutaceae). Acta Pol Pharm 23: 295–304

Kumar R, Banerjee SK, Handa KL (1976) Coumarins of *Heracleum canescens* and *Heracleum pinnatum*. Sources for dermal photosensitizing agents. Planta Med 30: 291–293

Mc Leod KW, Murphy PG (1977) Germination ecology of *Ptelea trifoliata*. Am Midl Nat 97: 363–372

Linnéron C (1737) *Hortus cliffortianus* (Amsterdam) *Ptelea* 36

Linnéron C (1753) *Species Plantarum* T.I.P. 118 Holmiac Imp. Laurentii SAL VII

Minker E, Koltai M, Blazsò G, Szendrei K, Novák I, Körösi J, Reisch J (1976) Contribution to the pharmacology of the water soluble constituents from *Ptelea trifoliata* L. Herba Hung 15: 93–101

Minker E, Bartha Cs, Koltai M, Rózsa Zs, Szendrei K, Reisch J (1979) Pharmacological study of the antispasmodic principles isolated from *Ruta graveolens*. Planta Med 36: 255–256

Mitscher LA, Bathala MS, Beal JL (1971) Antibiotics from higher plants: pteleatinium chloride, a new quaternary quinoline alkaloid from *Ptelea trifoliata* with antitubercular and antiyeast activity. J Chem Soc Chem Commun: 1040

Mitscher LA, Bathala MS, Clark GW, Beal JL (1975a) Antimicrobial agents from higher plants. The quaternary alkaloids of *Ptelea trifoliata*. Lloydia 38: 109–116

Mitscher LA, Bathala MS, Clark GW, Beal JL (1975b) Antimicrobial agents from higher plants. The antimicrobially inactive components of *Ptelea trifoliata*. Lloydia 38: 117–124

Mizuta M, Kanamori H (1985) Mutagenic activities of dictamnine and γ fagarine from *Dictamni* radicis cortex (Rutaceae). Mutat Res 144: 221–225

Montagu M, Levillain P, Chénieux JC, Rideau M (1986) Dosage d'alcaloides dihydrofuro [2,3-b] quinoléinium dans des tissus in vitro par chromatographie sur couche mince de gel de silice et fluorodensitométrie. J Chromatogr 351: 144–149

Mooken JD, Bailey VL, Bailey HE (1969) Occurrence of furoquinoline alkaloids in *Ptelea trifoliata*. Lloydia 32: 526–527

Mulvey RK, Zalewski CJ (1969) Kokusaginine: plant sources and its isolation from *Ptelea trifoliata* root bark. Econ Bot 23: 75–81

Murashige T, Tucker DPH (1969) Growth factor requirements of *Citrus* tissue cultures. In: Proc 1st Int *Citrus* Symp, Riverside, California, vol. 3, pp 1155–1161

Musajo L, Rodighiero G (1971) Photosensitizing properties of furocoumarins. Herba Hung 10: 79–94

Nikolin K (1971) Klinische Untersuchungen mit einem aus *Ruta graveolens* hergestellten, Bergapten- und Xanthotoxin-enthaltenden Präparat bei an Vitilago-leidenden Kranken. Herba Hung 10: 141

Novák I, Szendrei K, Pàpay V, Minker E, Koltai M (1970) Hazai Rutaceae-Fajok vizsgálata *Ptelea trifoliata* L. Herba Hung 9: 23–31

Petit-Paly G (1988) Les alcaloïdes quaternaires dihydrofuroquinoléiques à activité cytotoxique de *Ptelea trifoliata* L. (Rutacée). Comparaison de la variabilité intraspécifique (in vivo) et de la variabilité somaclonale (in vitro). Mise en évidence de molécules nouvelles Thèse Doctorat d'Etat ès Sci Pharm, Tours, 189 pp

Petit-Paly G, Montagu M, Viel C, Rideau M, Chénieux JC (1987) Dihydrofuro [2,3-b] quinolinium alkaloids in cultured cells of *Ptelea trifoliata*. Isolation of a new alkaloid Ptelecultinium. Plant Cell Rep 6: 309–312

Petit-Paly G, Ramawat KG, Chénieux JC, Rideau M (1989a) *Ruta graveolens*: in vitro production of alkaloids and medicinal compounds. In: Bajaj YPS (ed) Biotechnology in agriculture and forestry, vol. 7. Medicinal and aromatic plants II. Springer, Berlin Heidelberg New York, pp 488–505

Petit-Paly G, Montagu M, Mérienne C, Viel C, Rideau M, Chénieux JC (1989b) New alkaloids isolated from *Ptelea trifoliata* L. Planta Med 55: 209–210

Petit-Paly G, Montagu M, Viel C, Chēnieux JC, Rideau M (1990) Occurrence of ptelecultinium in *Ptelea trifoliata* L. (Rutaceae). Pharmazie 45: 698

Polievtsev NP, Evdokinova N, Sultanov MB (1972) Effect of *Haplophyllum* alkaloids on animal body temperatures. Farmakol Alkaloidov Ikh Proizvod 41–47 (Ca 1974, 80: 103867 u)

Pothier J, Petit-Paly G, Montagu M, Galand N, Chénieux JC, Rideau M, Viel C (1990) Separation of quinolizidine and dihydrofuro [2,3-b] quinolinium alkaloid by over pressured layer chromatography. J Planar Chromatogr 3: 356–358

Reisch J, Szendrei K, Novák I, Minker E, Pàpay V (1969) Inhaltsstoffe der Blüten von *Ptelea trifoliata*: Arctigenin-Methyläther, (+)-Hydroxilunin und Ptelefolin. Tetrahedron Lett 43: 3803–3806

Reisch J, Szendrei K, Pàpay V, Minker E, Novák I (1970a) Über die Strukturen weiterer Chinolon-2 Alkaloide aus *Ptelea trifoliata*. Tetrahedron Lett 22: 1945–1948

Reisch J, Szendrei K, Pàpay V, Novák I, Minker E (1970b) Ptelefolin, ein chinolon-4 Alkaloid aus *Ptelea trifoliata*. Tetrahedron Lett 38: 3365–3368

Reisch J, Szendrei K, Novák I, Minker E, Körösi J, Csedo K (1972) Drei neue Chinolon-2 Alkaloide aus *Ptelea trifoliata*. Tetrahedron Lett 5: 449–452

Reisch J, Mirhom YM, Körösi J, Szendrei K, Novák I (1973) O-Methyl ptelefolonium als *Ptelea* "Leital alkaloid". Phytochemistry 12: 2552–2553

Reisch J, Körösi J, Szendrei K, Novák I, Minker E (1975a) Cumarine und Chinolin-Alkaloide aus der Wurzelrinde von *Ptelea trifoliata*. Phytochemistry 14: 1678–1679

Reisch J, Körosi J, Szendrei K, Novák I, Minker E (1975b) 7,8-Methylendioxy-chinolin-Derivate als neue Alkaloide der Blüten von *Ptelea trifoliata*. Phytochemistry 14: 2722–2723

Rideau M, Verchère C, Hibon P, Chénieux JC, Maupas P, Viel C (1979) Alcaloïdes dihydrofuro-quinoléiques de quelques Rutaceae. Isolement, structure, propriétés biologiques. Phytochemistry 18: 155–169

Sargent GHS (1892) The silva of North America, vol. I. Riverside Press, Cambridge, pp 75–77

Schoepf D (1787) Materia medica americana. Potissimum regni vegetabilis. Erlangae Sumtibus *10 IAC* Palmii. In: Bull Lloyd Libr Bot Pharm Mat Med (1903)3: 15

Schroeder EM (1937) Germination of fruits of *Ptelea* species. Contrib Boyce Thompson Inst 8: 355–359

Sinha-Roy SP, Chakraborty DP (1976) Psoralen, a powerful germination inhibitor. Phytochemistry 15: 2005–2006

Soine TO (1964) Naturally occurring coumarins and related physiological activities. J Pharm Sci 53: 231

Spencer CF, Koniuszy FR, Rogers EF, Shavel J, Easton NR, Kaczka EA, Kuehlf A, Philips RF, Walti A, Folkers K (1947) Survey of plants for antimalarial activity. Lloydia 10: 145–174

Standley PC (1923) Trees and shrubs of Mexico. Contr US Natl Herb 23: 517–848

Szendrei K, Novák I, Petz M, Reisch J, Bailey HE, Bailey VL (1973) Coumarins and alkaloids from *Ptelea trifoliata* ssp. *pallida* var *confinis*. Lloydia 36: 333–337

Torres JC (1950) Estudo farmacognostico de la hoja de Ruda (*Ruta graveolens*). Farmacognosia 10: 275–361

Towers GHN, Graham EA, Spenser ID, Abramowski Z (1981) Phototoxic furanoquinolines of the Rutaceae. Planta Med 41: 136–142

Trémouillaux-Guiller J, Andreu F, Crèche J, Chénieux JC, Rideau M (1987) Variability in tissue culture of *Choisya ternata*. Alkaloid accumulation in protoclones and aggregate clones obtained from established strains. Plant Cell Rep 6: 375–378

Trémouillaux-Guiller J, Kodja H, Andreu F, Créche J, Rideau M, Chénieux JC (1988) Variability in tissue cultures of *Choisya ternata*. Comparing alkaloid production in cell lines obtained by various strategies. Plant Cell Rep 7: 456–458

Ulrich R (1942) Observations sur la croissance de quelques fruits. Rev Sci 80: 24–30

Vannier L, Poirier J (1976) Précis de Matière médicale homéopathique. Doin, Paris 369 pp

Waterman PG (1975) Alkaloids of the Rutaceae: their distribution and systematic significance. Biochem Syst Ecol 3: 149–180

Waterman PG (1986) A phytochemist in the African rain forest. Phytochemistry 25: 3–17

Weinstein B, Craig AR, Fuller LW, Kang JB, Mc Breen SA (1972) Furanocoumarins from *Poncirus trifoliata* and *Ptelea baldwinii*. Phytochemistry 11: 1530

Wilson P (1911) Notes on Rutaceae. Some characters in *Ptelea* and *Taravalia*. Bull Torrey Bot Club 38: 296–297

Wink M (1989) Genes of secondary metabolism: differential expression in plants and in vitro cultures and functional expression in genetically transformed microorganisms. In: Kurz WGW (ed) Primary and secondary metabolism of plant cell cultures. Springer, Berlin Heidelberg New York, pp 239–251

XX *Rheum* Species (Rhubarb): In Vitro Culture and the Production of Sennosides

S. SHIBATA[1]

1 Introduction

Rhubarb, the rhizome and root of *Rheum* spp. (Polygonaceae), has been used since ancient times as an important drug in the East and West. It was described in Dioscorides' *De Materia Medica* as Ra (ρα), designating its native place, the Volga (Ra River) basin. It was said to be effective for disorders of stomach and intestine, as well as for pains in spleen, liver, kidney, abdomen, bladder, and chest (Madaus 1976). In traditional Chinese medicine, rhubarb (*Da-Huang* in Chinese) has been used as a major component of some prescriptions for the treatment of blood stasis, in which it produces mildly purgative, antiinflammatory, and sedative effects. In western countries, rhubarb is mostly employed as a purgative drug in folk medicine.

Several *Rheum* species are recognized as being the original rhubarb plants, from which *Rheum palmatum* L., *Rh. tanguticum* Maxim., and *Rh. officinale* Bail. are recommended for medicinal use. All these species with palmately or elliptically lobed large leaves are native in cool, high-altitude districts in Tibet, Chianhai, Yunnan, and Si-Chuang Provinces, China. A Korean species, *Rh. coreanum* Nakai, possesses similarly shaped leaves to those of Chinese origin. The rhizome and root of *Rheum* sp. growing in lower lands, *Rh. undulatum* L., *Rh. rhaponticum* L., and *Rh. emodi* Wall., which have round-edged leaves, are inferior for medicinal use, while the stem of *Rh. rhaponticum* is edible (Xiao et al. 1959).

Rheum plants of superior quality are not native to Japan. The author introduced the seeds of these species from abroad in 1949 to cultivate at an experimental farm attached to the University of Tokyo and Shinshu University in the Nobeyama Highland (1300 m a.s.l.) in Nagano Pref., Japan. The cultivation of *Rh. palmatum*, *Rh. tanguticum*, *Rh. officinale*, and *Rh. coreanum* was successful in adapting them to the climate and soil conditions there (Shibata and Kobayashi 1983).

A hybrid of *Rh. palmatum* and *Rh. coreanum* was produced from the strains grown on the Nobeyama farm by the research laboratory of Takeda Chemical Industry Co., Ltd. to cultivate on a large scale at Ikeda-machi, Hokkaido, the northern most island of Japan, where the temperature in mid-summer is cool enough to suit the practical cultivation of the plant (Matsuoka and Hatta (1970).

[1]Laboratory of Natural Medicinal Materials, c/o Minophagen Co, 3rd Tomizawa Bldg, 4th floor, Yotsuya 3-2-7, Shinjuku-ku, Tokyo 160, Japan

Biotechnology in Agriculture and Forestry, Vol. 21
Medicinal and Aromatic Plants IV (ed. by Y.P.S. Bajaj)

Fig. 1. Shinshu-daiwo (*Rheum palmatum* × *Rh. coreanum*). (Courtesy of Dr. T. Iwasa, Kyoto Experimental Farm, Takeda Chemical Industry Co., Ltd.)

1.1 Purgative Principles of Rhubarb

Rhubarb is commonly used as a purgative, and the anthraquinones and their glycosides, which are usually contained in various species of rhubarb, have been considered as the active principles. Sennosides which were considered by Stoll et al. (1950) as the purgative principles of *Senna* leaves were also isolated from rhubarb by Miyamoto et al. (1967) and Oshio et al. (1974, 1978). Sennosides A–F are bianthrone glucosides consisting of 2 mol of rheinanthrone or rhein- and aloe-emodin-anthrone. Recently, rheinosides A–D, which are monomeric rheinanthrone glucosyl derivatives, have also been isolated as the purgative principles (Yamagishi et al. 1987).

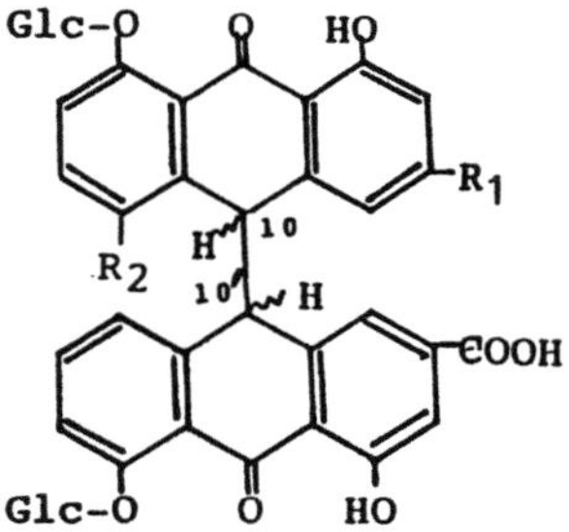

	R_1	R_2	10–10′
Sennoside A	COOH	H	threo
Sennoside B	COOH	H	erythro
Sennoside C	CH_2OH	H	threo
Sennoside D	CH_2OH	H	erythro
Sennoside E	COOH	OC–COOH	threo
Sennoside F	COOH	OC–COOH	erythro

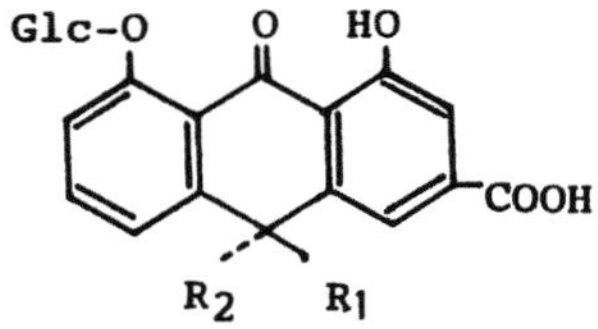

	R_1	R_2
Rheinoside A	Glc	OH
Rheinoside B	OH	Glc
Rheinoside C	Glc	H
Rheinoside D	H	Glc

1.2 Other Constituents of Rhubarb

Several other types of compounds have been isolated from several species of rhubarb. Until recently rhaponticin-type stilbenes had been believed to be the constituents of inferior types of rhubarb, but now they have been found more or less in every type of rhubarb, and perhaps will prove to be the treatment for stasis of blood in Chinese medicine (Nonaka et al. 1977).

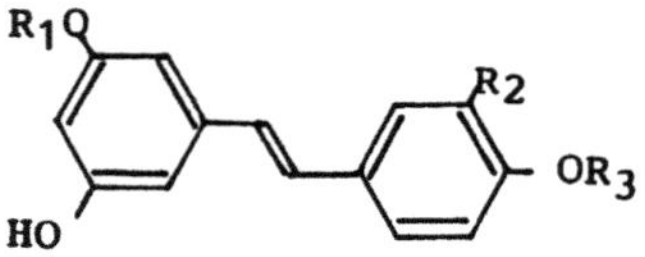

	R_1	R_2	R_3
Rhaponticin	Glc	OH	Me
Desoxyrhaponticin	Glc	H	Me
Piceid	Glc	H	H

The naphthalene derivatives, torachrysone 8-O-glucoside and 8-O(6′-oxalyl) glucoside, and 6-hydroxymusizin 8-O-glucoside were found in several types of rhubarb. Torachrysone was first isolated from *Cassia tora* seeds (Leguminosae). Eight kinds of chromone derivatives have been isolated from Mati-dahuang (*Rh. officinale* ?), but a very minute amount in Ya-huang (another type of *Rh. officinale*).

Phenylbutanone derivatives have recently been found in rhubarb. Lindleyin, which was first isolated from *Aconium lindleyi* (Crassulaceae) growing in the Canary Islands, has been isolated from rhubarb with a maximum yield of 3%. It shows fairly strong antiinflammatory activity like aspirin and phenylbutazone (Nonaka et al. 1981).

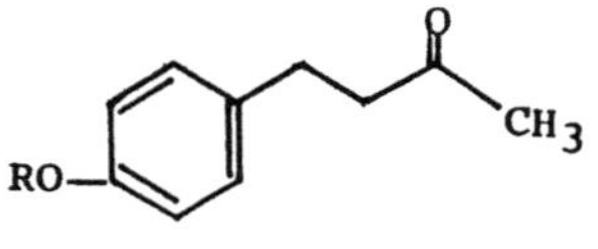

Lindleyin
R = Galloyl——^{2}Glc –
Isolindleyin
R = Galloyl——^{6}Glc –

Rhubarb contains a large amount of nonhydrolyzable catechin-tannins as well as hydrolyzable tannins. The structures of these tannins have been described by Nonaka and Nishioka (1983) and Nonaka et al. (1983), and highly polymerized rhatannin I and II were characterized as suppressing the increase on blood urea nitrogen (BUN), which appears in chronic renal disorders. Rhatannins are polymerized galloylepicatechins as formulated below:

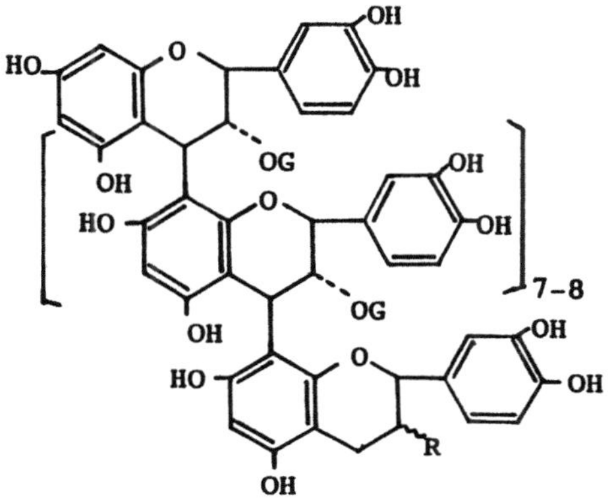

Rhatannin I
R = – OG
Rhatannin II
R = OH
G = galloyl

1.3 Mechanism of Purgative Action of Rhubarb

The multiple functions of rhubarb have thus been explained by the presence of various types of compounds, but the purgative effect is still recognized as the major function of rhubarb of higher quality. The oral administration of sennosides A and C is effective in producing a purgative action, whereas the intravenous injection of sennosides is ineffective. Therefore, it is obvious that sennosides have no direct purgative action, but are made effective by microbial conversion into rheinanthrone in the large intestine (Sasaki et al. 1979). However, the oral administration of rheinanthrone is not effective due to its instability in the digestive organ.

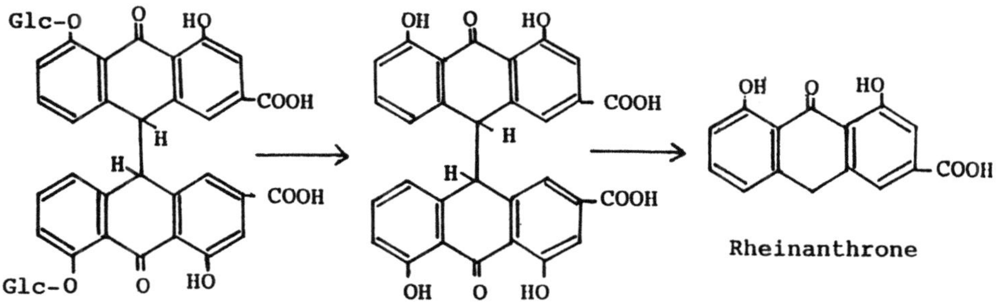

Sennoside A Sennidin A Rheinanthrone

Pretreatment with chloramphenicol decreased the purgative effect of sennosides A and C by the suppression of intestinal microbe activity. The purgative action of rheinanthrone is related to the promotion of biosynthesis and release of prostaglandin E_2, because the inhibitors of prostaglandin biosynthesis, indomethacin,

SC19220, and PPP (polyphloretin phosphate), suppress the purgative activity of rheinanthrone (Benbler and Juan 1979; Benblar and Kallar 1985). The anthraquinones coexisting in rhubarb which have antimicrobial action may control the activity of intestinal microbes, resulting in a mild purgative effect.

2 The Quantitative Analysis of Sennosides in Rhubarb by High Performance Liquid Chromatography (HPLC)

A colorimetric analysis of the binding anthrone or thin layer chromatography has been used to determine the amount of sennosides in rhubarb, but the results are not satisfactory (Stahl 1970; Brendl and Schneider 1974). Nishizawa et al. (1984) reported the total HPLC analysis of water-soluble and ethyl acetate-soluble fractions of rhubarb extracts.

For the purpose of the microquantitative determination of sennosides, the author and his coworkers employed HPLC, using an amino-binding silica-gel column which effectively absorbs the principles having carboxylic functional groups (Ohshima and Takahashi 1983). As the commercially available NH_2-binding silica gel absorbed sennosides too firmly, a column of $N(CH_3)_2$-binding silica gel (Senshu Pak, SN-342N) was employed using tetrahydrofuran (THF)/water/acetic acid (8 : 2 : 1) as the mobile phase to efficiently separate sennosides A, B, C, and E. (Fig. 2)

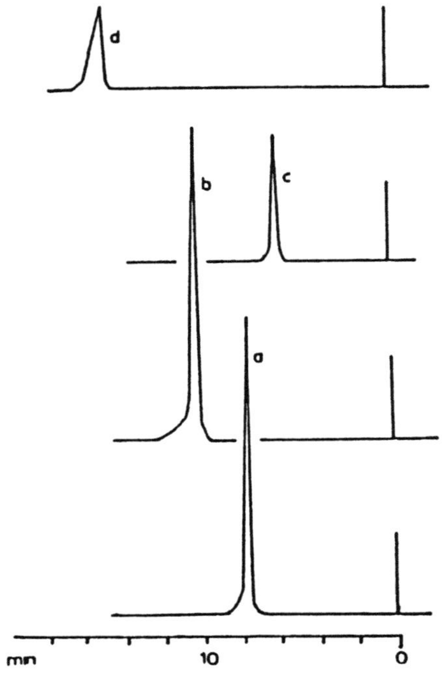

Fig. 2. HPLC of standard samples of *a* sennoside A; *b* sennoside B; *c* sennoside C; *d* sennoside E. Column: Senshu-Pak SN-342N; mobile phase, THF/water/acetic acid (8:2:1); flow rate, 3.5 ml/min; detection, UV 280 nm. (Ohshima and Takahashi 1983)

Table 1. The contents of sennoside A and B in rhubarb

Rhubarb species	Sennoside A (%)	Sennoside B (%)
Rheum palmatum L.	1.30	0.58
Rh. tanguticum Maxim	0.94	0.43
Rh. coreanum Nakai	1.50	0.81
Rh. coreanum × *Rh. palmatum* (Shinshu-Daiwo)	1.30	0.51
Kinmon I	0.16	0.05
Kinmon II	1.43	0.37
Kinmon III	0.52	0.11
Ya-huang (Gawo)	0.97	0.42
Mati-dahuang	0.70	0.17
Wa-daiwo (*Rh. undulatum*)	—	—

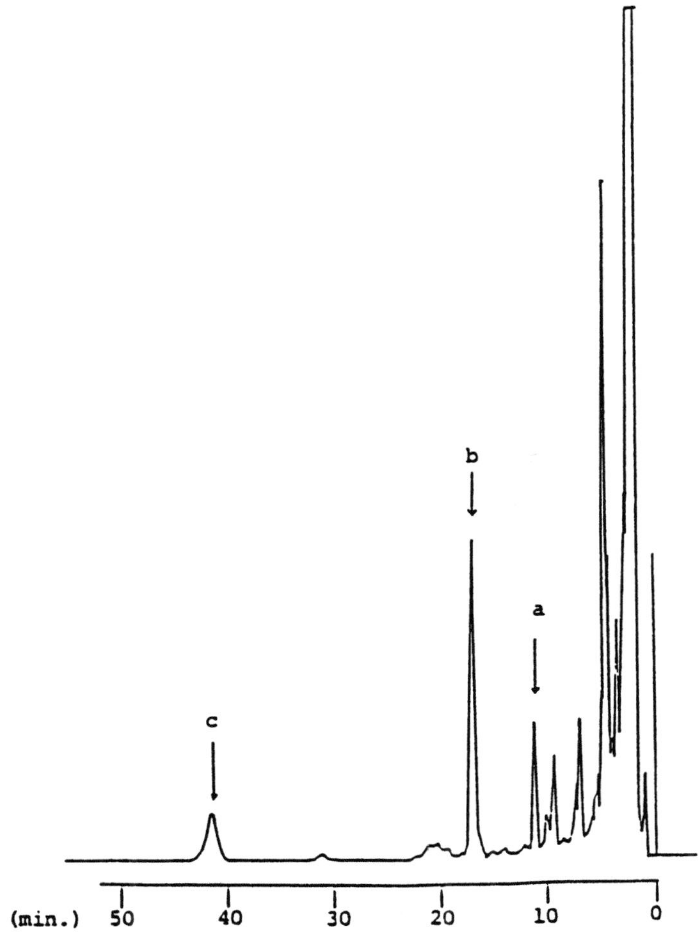

Fig. 3. HPLC elution profile of methanolic extracts of rhubarb. Column, Senshu-Pak SN 112N (4.6 × 150 mm i.d.); mobile phase, THF/H_2O/AcOH (160:40:7); flow rate, 1 ml/min; detection, UV 360 nm. *a* Sennoside C; *b* sennoside A; *c* sennoside B

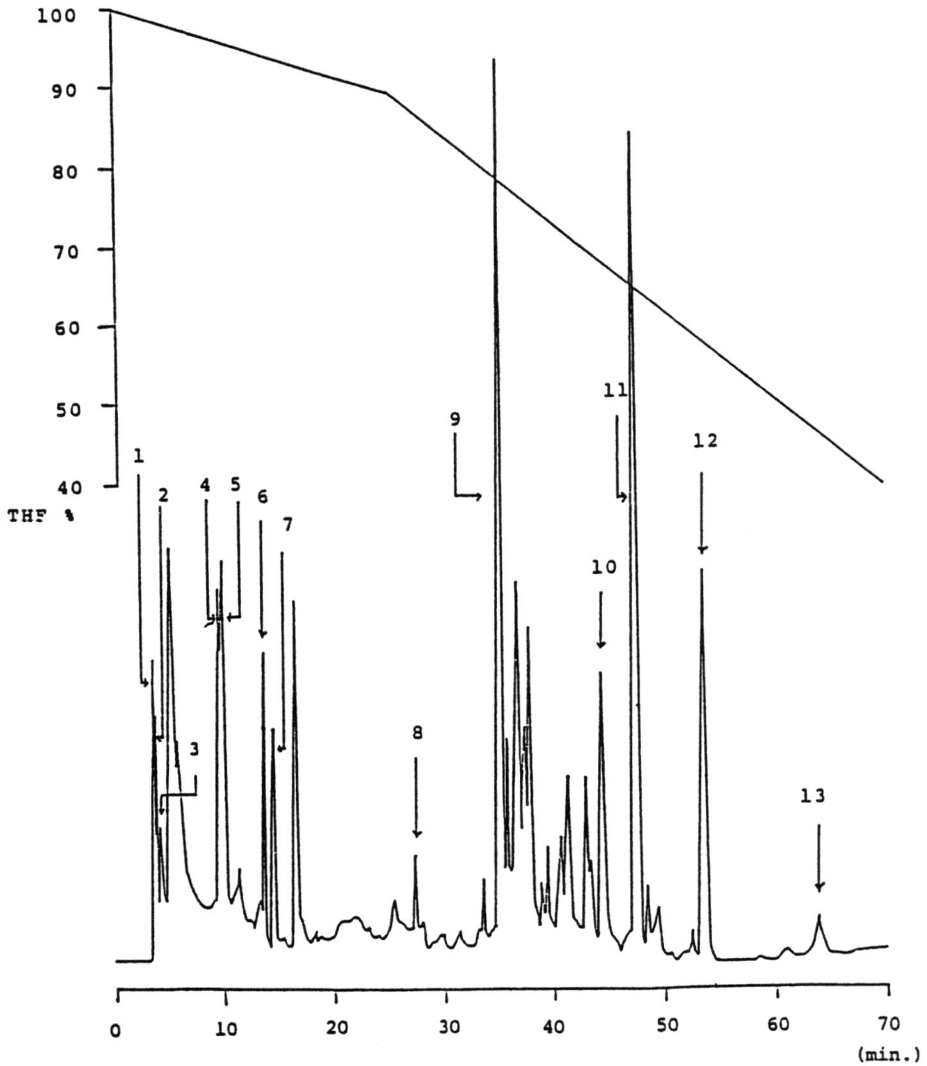

Fig. 4. HPLC elution profile of methanolic extracts of rhubarb. Column, Senshu-Pak SN 112N (4.6 × 150 mm i.d.); mobile phase, THF-15% acetic acid gradient; flow rate, 1 ml/min; detection, UV 361 nm. *Peak 1* Chrysophanol, physcion mixture; *2* emodin; *3* aloe-emodin; *4* chrysophanol-1-glc, -8-glc mixture; *5* aloe-emodin-8-glc; *6* citreorosein; *7* emodin-1-glc; *8* rhein; *9* rhein-8-glc; *10* sennoside C; *11* sennoside A; *12* sennoside B; *13* sennoside E

Experimental. SP 8700 (Spectra physics Co.); NS (Nihon-seimitsu Co.); Detector: UVILOG 5-111 (Oyo-bunko Co.); MPCD-350 (Union-giken Co.); recorder: SS-205F (Seconik); column: Senshu Pak Aquasil SN-322N (8 mm × 150 mm i.d.); solvents: analytical grade (Kanto Kagaku Co.) and distilled water.

The recovery percentage of sennoside A and B from the column was 95.8 and 94.5, respectively.

For the HPLC detection of sennosides in extracts of rhubarb, other, coexisting constituents should be removed by pretreatment using Sep-Pak C_{18} in which tannins are retained. After the anthraquinone derivatives are eluted out with chloroform/methanol (8:1), a sennoside mixture is obtained by elution with methanol/water/acetic acid (27:73:1). The mixture is loaded on a column of Senshu Pak SN-342N (25 cm × 80 mm i.d.) and eluted with THF/water/acetic acid (8:2:1), detecting with UV 280 nm.

In simplifying the above procedure, a practical method has been developed using a UV wavelength of 360 nm for the detection of sennosides unaffected by the coexisting anthraquinones and tannins, showing UV absorption in the range 340–360 nm. Pretreatment is excluded in this new method to shorten the time of analysis (Fig. 3).

Using the improved HPLC method, the rhubarbs of different botanical species cultivated in the experimental fields and the marketing varieties were analyzed to determine the content of sennoside A and B. The cultivated species showed the contents of sennoside A to be 0.9–1.5% and sennoside B to be 0.4–0.8% dry matter, which correspond to those of the commercial materials of high quality.

The author and his coworkers have developed a total analysis of rhubarb constituents by HPLC on a dimethyl amino column using a gradient solvent system consisting of THF-15% acetic acid. The total analysis of methanolic extracts of rhubarb on this HPLC system showed a well-defined separation of the constituents from chrysophanol to sennoside E (Fig. 4).

Applying a multiwavelength photomeric detector, the total HPLC analysis of rhubarb extracts has been demonstrated by three-dimensional and contour map displays, in which sennosides A, B, C, and E are shown by well-defined peaks with characteristic UV absorptions at 330–370 nm (Takahashi et al. 1983).

3 Tissue Culture of Rhubarb for the Production of Sennosides

The tissue culture of rhubarb was first reported by Furuya et al. (1975) to reveal the formation of chrysophanol and emodin. Thereafter, several papers were published on the tissue culture of *Rheum rhaponticum* L. (see Roggemans and Boxus 1988) and *Rheum palmatum*, showing the formation of anthraquinone glycosides but not sennosides (Rai 1978; Dusek et al. 1986, 1988, 1989; Lal 1989).

3.1 Induction of Callus

The author and his coworkers demonstrated the formation of sennosides along with anthraquinone derivatives by tissue culture induced from the seedling of a stable hybrid of *Rheum palmatum* and *Rh. coreanum* (Shinshu-Daiwo) (Ohshima et al. 1988). The selection of callus cells and the control of cultivation conditions are strictly required to achieve sennoside formation.

The callus was induced from the seeds of Shinshu-Daiwo. The seeds, supplied by the Kyoto Experimental Farm, Takeda Chemical Industry Co., Ltd., were

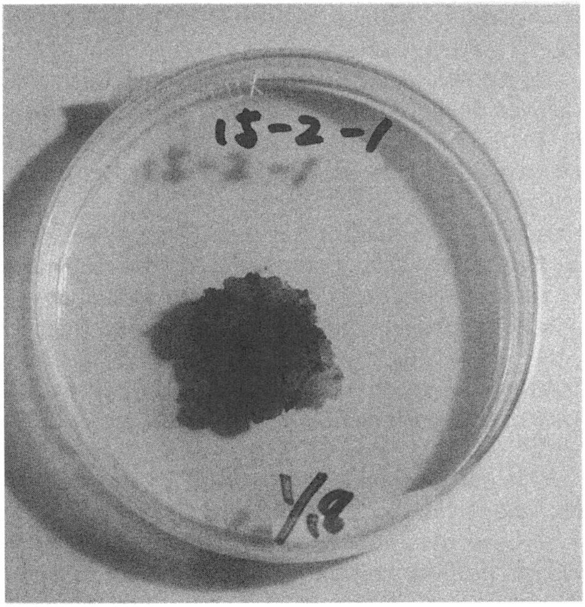

Fig. 5. Callus induced from Shinshu-daiwo

treated at 4 °C for 3 weeks and placed on absorbent cotton soaked with water for 24 h. For the germination, a sucrose (20 g/l)-agar (9 g/l) mixture in water medium, divided into 20 ml each in the 100-ml conical flasks, was employed. Pretreatment of the seeds at a low temperature (4 °C) was essential for germination. The seeds, sterilized with sodium hypochlorate, were sown on the above medium and incubated at 21 °C, resulting in 65% germination. The seedlings grew to 2–3 cm length, and were cut into small pieces and put on the Murashige-Skoog (MS) medium containing sucrose (20 g/l), 2,4-dichlorophenoxyacetic acid (2,4-D; 1 ppm), kinetin (K; 0.1 ppm), and agar (9 g/l). After 2–3 weeks, callus was formed in 100% of the cases after incubation in the dark at 21 °C. The plant hormone combination, indole acetic acid (IAA; 1 ppm) and K (0.1 ppm), or α-naphthalene acetic acid (NAA; 1 ppm) and K (0.1 ppm), were not effective in inducing callus.

3.2 Proliferation of Callus

In higher plants, the secondary metabolism is generally active in the differentiated cells, whereas in the dedifferentiated callus cells, the primary metabolism is activated to achieve proliferation, and conversely the secondary metabolite production is decreased. Therefore, for the production of plant products by callus cultivation, it is necessary to find an optimal condition promoting both primary and secondary metabolisms.

By the comparative studies on the effects of plant hormones, 2,4-D was found to be more effective than IAA in the proliferation of callus. In the case of IAA, the root differentiation from the callus was observed at 7–8 weeks after transplantation (Fig. 6).

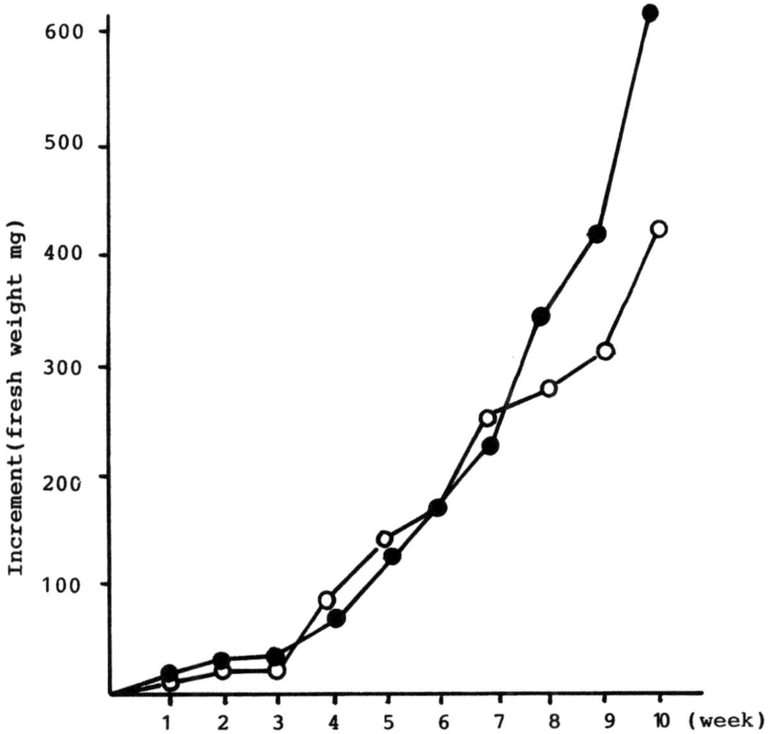

Fig. 6. Effects of plant hormones on the proliferation of rhubarb callus. -●- 2,4-D (1 ppm) and K (0.1 ppm); -○- IAA (1 ppm) and K (0.1 ppm) in MS medium containing 2% sucrose. (Ohshima et al. 1988)

The effects of sugars on the proliferation of callus were studied, as illustrated in Fig. 7. Maltose gave the best result in the proliferation of callus among the other carbon sources so far tested such as glucose and sucrose.

For the large-scale cultivation of rhubarb callus, suspension culture has been attempted in order to find an optimum condition. The best condition in proliferation of the callus was given by the suspension culture in the medium containing IAA (1 ppm) and N-(2-chloro-4-pyridyl)-N′-phenylurea (4-PU-30) (1 ppm) at week 4, which showed 5.5-fold greater increase in weight than the original weight of the callus, and the rate of increase was about twofold greater than the best rate of growth obtained by the stationary culture of the callus on the medium containing 2,4-D (1 ppm) and K (0.1 ppm) (Fig. 8).

3.3 Secondary Metabolites Produced in Callus

The callus of rhubarb was cultivated on the MS medium containing 2,4-D (1 ppm) and K (0.1 ppm) in the dark at 21 °C for 4 weeks. The callus harvested was lyophilized, and then successively extracted with chloroform, NaCl-THF, and

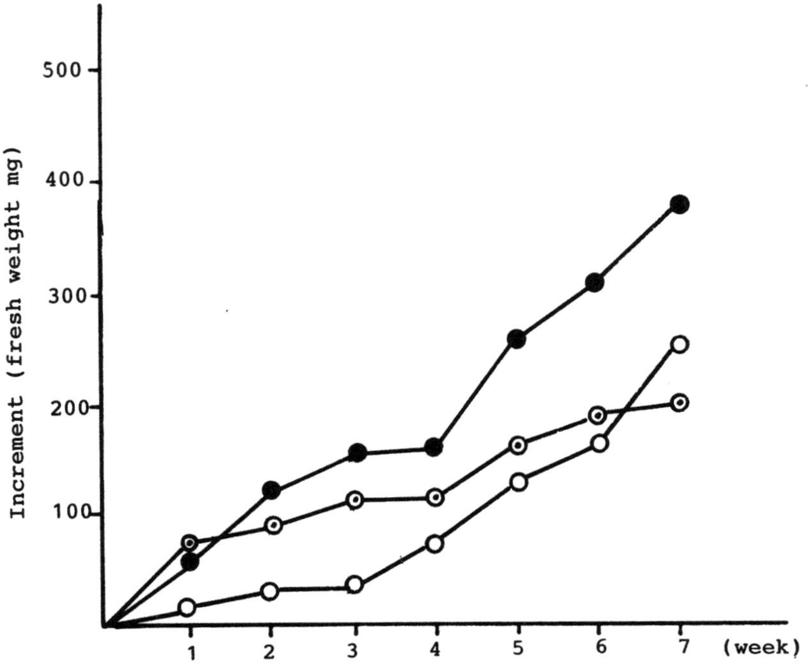

Fig. 7. Effects of sugar components on the proliferation of *Rheum* callus (-⊙- 2% Glucose; -●- 2% maltose; -○- 2% sucrose in MS medium containing IAA (1 ppm) and K (0.1 ppm). (Ohshima et al. 1988)

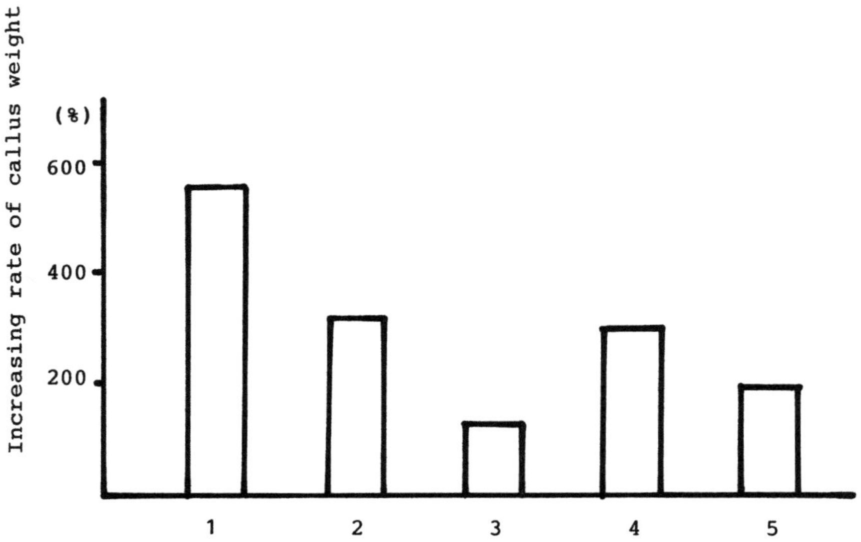

Fig. 8. Effects of plant hormones and the mode of cultivation on the growth of rhubarb callus. *1* IAA (1 ppm), 4-PU-30 (1 ppm), and maltose in MS medium on suspension culture. *2* NAA (1 ppm), 4-PU-30 (0.1 ppm) and maltose in MS medium on suspension culture. *3* IAA (1 ppm), 4-PU-30 (0.1 ppm) and maltose in MS medium on suspension culture. *4* 2,4-D (1 ppm), K (0.1 ppm) and maltose in MS agar medium on stationary culture. *5* IAA (1 ppm), K (0.1 ppm) and maltose in MS agar medium on stationary culture. *1–5* Incubated in the dark at 21 °C for 4 weeks. (Ohshima et al. 1988)

oxalic acid-THF. The chloroform extract of callus gave a similar HPLC profile with that given by the rhubarb extract (Fig. 9), showing the presence of free anthraquinones, chrysophanol (0.47%), physcion (0.4%), emodin (0.7%), aloe-emodin, and rhein. The yields were determined using a reference calibration curve.

The HPLC profiles of THF extracts of the callus and rhubarb showed the presence of rhein-8-glucoside (0.09%) (Fig. 10).

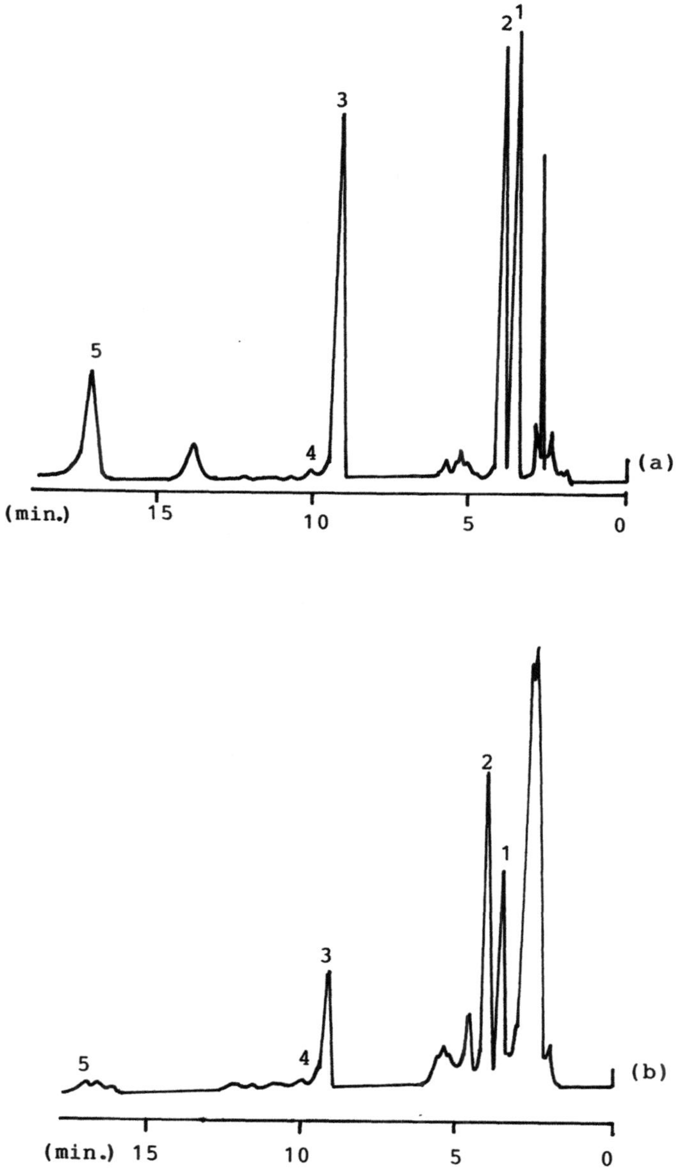

Fig. 9a, b. HPLC profile of chloroform extracts of **a** rhubarb and **b** callus induced from **a**. Column, Senshu-Pak silica gel (8 × 250 mm i.d.); mobile phase, n-hexane/acetic acid (4:1); flow rate, 2.5 ml/min; detection UV 280 nm. *Peak 1* chrysophanol; *2* physcion; *3* emodin; *4* aloe-emodin; *5* rhein. (Ohshima, et al. 1988)

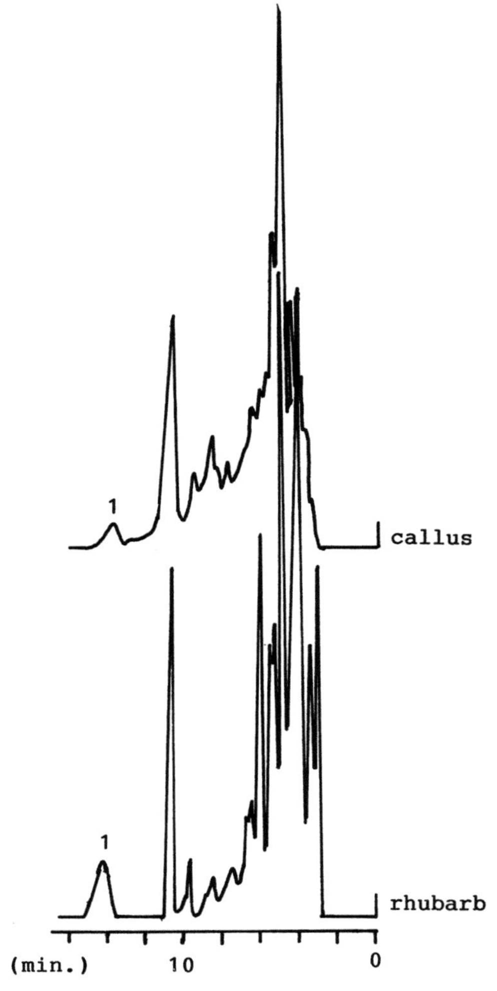

Fig. 10. HPLC profile of THF extracts of callus induced from callus of rhubarb and from rhubarb. *1* Rhein-8-glucoside. (Ohshima, et al. 1988)

The presence of sennosides in the oxalic acid-THF extract of rhubarb callus was preliminarily proved by thin layer chromatography (TLC) using n-PrOH/ethyl acetate/H_2O (4:4:3) as the developing solvent. The spot of sennoside A on the TLC was cut off and eluted to measure optical rotatory dispersion (ORD), which indicated a cotton effect of *threo*-type sennoside A (Fig. 11).

The presence of sennoside A in the elute was also proved by means of paper electrophoresis using a pyridine-acetic acid buffer solution at pH 6.5, under 2500 V, 2.5 mA/cm^2 for 70 min to give a flow rate of 9.5 cm, identical to that given by the reference sample of sennoside A.

In 70% methanolic extract of the rhubarb callus, the presence of sennoside B was also proved, along with sennoside A with HPLC on a Senshu-Pak SN 117N column using THF/H_2O/acetic acid (8:2:1) as the mobile phase (Fig. 12).

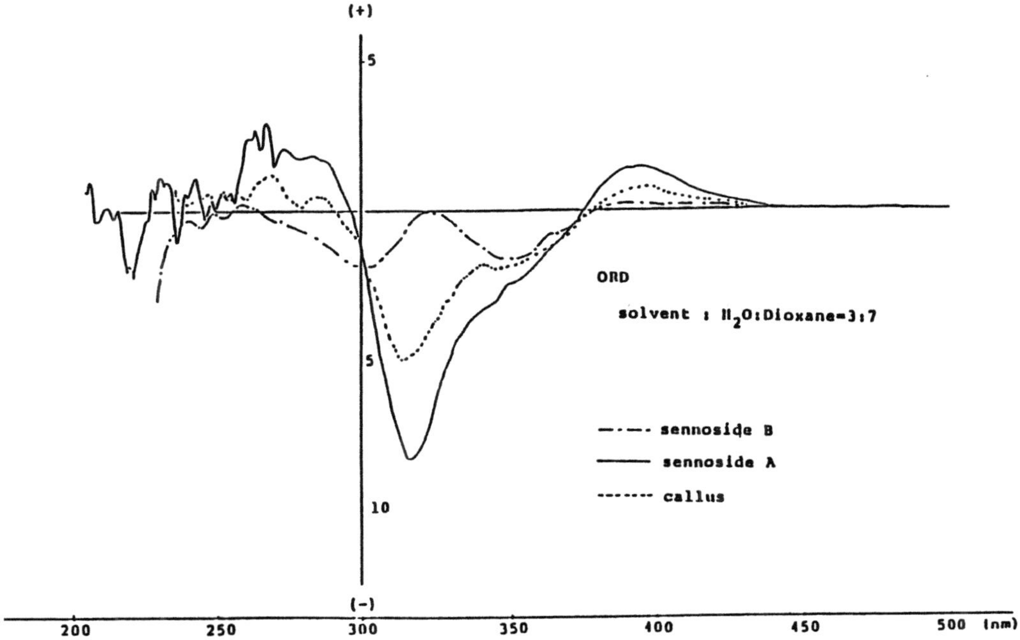

Fig. 11. ORD curves of sennosides produced by *Rheum* callus and the explant

3.4 Formation of Sennosides by the Callus Cultures

Sennoside formation in the callus of rhubarb has been studied to find the optimum conditions for growth by using several combinations of plant hormones and carbohydrates as the carbon source. Sennoside B was found in the dark-colored callus and the rootlets differentiated from the callus, which sennoside A was formed in all the callus cultures induced from the seedlings of Shinshu-daiwo. Sennoside A was formed with a comparatively higher yield by the cultivation on the MS medium containing IAA (1 ppm), K (0.1 ppm), and maltose. Shaking the culture of the callus gave better results than stationary culture in the formation of sennosides, with a two- to ten fold greater yield (Table 2).

For the promotion of sennoside production in the rhubarb callus, elite cells were selected from the callus. The callus was smashed into small pieces with a glass rod in sterilized distilled water in a test tube; these pieces were then inoculated on MS agar medium with plant hormones added and incubated for 4 weeks in the dark at 21 °C. The colonies growing on the agar plates were separated to inoculate them again on a fresh MS medium containing plant hormones, and transferred to the fresh media every month to attain a callus size of 2–3 cm in diameter. The harvested callus was lyophilized and extracted to determine the contents of sennosides by the HPLC-procedure, as mentioned previously.

The selected callus strains induced from rhubarb seedlings (Shinshu-daiwo) are tabulated to show their characters in the production of sennosides A and B.

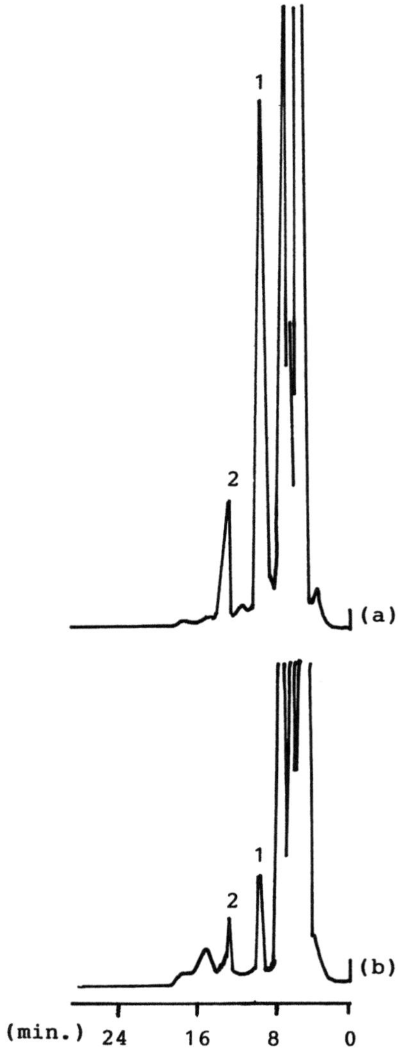

Fig. 12a, b. HPLC elution profile of methanolic extracts of **a** rhubarb and **b** callus induced from **a**. Column, Senshu-Pak SN 112N (4.6 × 150 mm i.d.); mobile phase, THF/H_2O/AcOH (8:2:1); flow rate, 1 ml/min; detection, UV 360 nm *Peak 1* Sennoside A; *2* Sennoside B. (Ohshima, et al. 1988)

Table 2. Cultivation conditions of rhubarb callus and the formation of sennosides

Medium (ppm)	Mode of culture[a]	Sennoside (%) A	Sennoside (%) B
MS IAA (1), K (0.1) sucrose, agar (white callus)	st	0.001	—
MS IAA (1), K (0.1) maltose	sh	0.005	0.005
MS IAA (1), K (0.1) sucrose, agar (root)	st	0.005	0.0003
MS 6-Benzyladenine(BA) (5) K (0.1), maltose	sh	0.002	0.007
MS 4PU-Cl (1), sucrose, agar (white callus)	st	0.0002	—
MS 4PU-Cl (1), sucrose, agar (brown callus)	st	0.001	0.001
MS 4PU-Cl (1), K (0.1) maltose	sh	0.003	0.008

[a] st, Stationary; sh, shaking.

The strains 1091 and 141224, producing neither sennoside A nor sennoside B, have been morphologically transformed into white stem like tissues and later differentiated into stem and leaves which produced chlorophyll upon cultivation under light (4000 lx). Strain 13112, producing sennosides A and B in higher yields, consists of typical dedifferentiated callus cells of dark color.

The selected callus cells of rhubarb were subjected to shaking culture in MS medium, with some plant hormone combinations added.

The experiment showed that the selection of elite callus cells induced from rhubarb seedlings is an important factor controlling the potency of sennosides biosynthesis. The formation of sennosides, the purgative principles of rhubarb, by callus cultivation has been established, but the yield of active principles is still not enough. Therefore, further investigation along this line is required to further practical application.

Table 3. The yields of sennosides A and B by the selected callus induced from rhubarb seedlings (Shinshu-daiwo). (Ohshima et al. 1988)[a]

Callus strain (No.)	Sennoside A	Sennoside B
1001	0.0010[b]	0.0006[b]
1044	0.0006	–
1041	0.0002	–
1035	0.0002	–
1032	0.0014	0.0006
13112	0.0083	0.0040
141223	0.0050	0.0071
1091	–	–
10321	0.0050	0.0017
141224	–	–

[a] Culture medium: MS maltose-agar medium containing IAA (1 ppm) and 4PU-30 (1 ppm).
[b] Content (%) in the dried callus.

Table 4. Sennosides formation by the shaking culture of selected callus cells of rhubarb. (Ohshima et al. 1988)

Callus strain (No.)	Medium[a]	Sennoside A (%)	Sennoside B (%)
2001	a	0.0030	0.0029
2002	a	0.0017	0.0020
3003	b	0.0023	0.0041
4000	c	0.0017	0.0017
4001	c	0.0048	0.0041
4002	c	0.0023	0.0017

[a] Medium a: MS, IAA (1 ppm), 4PU-30 (1 ppm), maltose;
b: MS, IAA (1 ppm), 4PU-30 (0.1 ppm), maltose;
c: MS, NAA (1 ppm), 4PU-30 (1 ppm), maltose.

4 Summary and Conclusion

Callus cultivation of rhubarb has been investigated to find the optimal conditions for the production of purgative principles, i.e., sennosides.

A stable hybrid of *Rheum palmatum* and *Rheum coreanum* was employed as the explant which induced callus of the best quality. The callus formed in 100% of the cut seedlings on the Murashige-Skoog (MS) medium containing sucrose, 2,4-D, and kinetin in the dark at 21 °C.

The cultivation of the selected callus in the MS medium containing maltose (20 g/l), IAA (1 ppm), and 4PU-30 (1 ppm) in the dark resulted in the best production of sennosides A and B.

Acknowledgments. The author wishes to thank Dr. Y. Ohshima, who devoted himself to the tissue culture experiments, and Dr. K. Takahashi, who pursued HPLC analyses, for their cooperation at Meiji College of Pharmacy, Tokyo. Thanks are also due to Dr. M. Goto, who was at that time at the Kyoto Experimental Farm, Takeda Chemical Industry Co., Ltd., and Mr. Y. Kobayashi, Experimental Station for Medicinal Plant Studies attached to the Faculty of Pharmaceutical Sciences, University of Tokyo, for their cooperation in the cultivation of rhubarb plants and in supplying plant materials for the present study.

References

Benbler E, Juan H (1979) Effect of ricinoleic acid and other laxatives on net water flux and prostaglandin E release by the rat colon. J Pharm Pharmacol 31: 681–685

Benbler E, Kollar G (1985) Stimulation of PGE_2 synthesis and water and electrolyte secretion by senna anthraquinones inhibited by indomethacin. J Pharm Pharmacol 37: 248–251

Brendel WD, Schneider D (1974) Wertbestimmung von Folia und Fructus Sennae sowie deren Zubereitungen. Planta Med 25: 63–67

Dusek J, Zahradnicek M, Hubik J (1986) Effect of the conditions of cultivation on the growth of the tissue culture of the rhubarb root and the production of secondary metabolites. Cesk Farm 35: 313–317

Dusek J, Zahradnicek M, Duskova J (1988) Effect of metathion E50 on *Rheum palmatum* in field and tissue cultivations. Pharmazie 43: 786–787

Dusek J, Sicha J, Duskova J (1989) Influence on the production of anthracene derivatives in tissue culture of *Rheum palmatum* by a modification of the cultivating medium or potential precursors. Cesk Farm 38: 210–213

Furuya T, Ayabe S, Noda K (1975) Chrysophanol and emodin from callus tissue of rhubarb (*Rheum palmatum*). Phytochemistry 14: 1457

Lal N, Ahuja PS (1989) Propagation of Indian rhubarb (*Rheum emodi* Wall.) using shoot-tip and leaf explant culture. Plant Cell Rep 8: 493–496

Madaus G (1976) *Rheum*. In: Madaus G (ed) Lehrbuch der biologischen Heilmittel, Vol 3. Georg Olms, Hildesheim New York, pp 2300–2306

Matsuoka T, Hatta R (1970) Studies on cultivation of Shinshu-Daio. J Takeda Res Lab 29: 776–787

Miyamoto M, Imai S, Shinohara M, Fujioka S, Goto M, Matsuoka T, Fujimura H (1967) Studies on purgative substances I. Isolation of sennoside A, one of the most active principles from rhubarb. Yakugaku Zasshi (J Pharm Soc Jpn) 87: 1040–1043 (in Japanese)

Nishizawa S, Yamagishi T, Kashiwada Y, Nonaka G, Nishioka I (1984) Total analysis of rhubarb principles with HPLC. Comparative studies on rhubarbs on market. Abstr 104th Annu Meet Pharm Soc Jpn (Sendai), Tokyo p 193

Nonaka G, Nishioka I (1983) Tannins and related compounds X. Rhubarb (2). Isolation and structures of a glycerol gallate, gallic acid glucoside gallates, galloylglucoses and isolindleyin. Chem Pharm Bull 31: 1652–1658

Nonaka G, Minami M, Nishioka I (1977) Studies on rhubarb (Rhei Rhizoma) IV. Naphthalene glycosides. Chem. Pharm Bull 25: 2708–2712

Nonaka G, Nishika I, Nagasawa T, Oura H (1981) Tannins and related compounds I. Rhubarb (1). Chem Pharm Bull 29: 2862–2870

Nonaka G, Ezaki E, Hayashi K, Nishioka I (1983) Flavonol glucosides from rhubarb and *Rhaphiolepis umbellata*. Phytochemistry 22: 1659–1661

Ohshima Y, Takahashi K (1983) Separation of sennosides. J Chromatogr 258: 292–296

Ohshima Y, Takahashi K, Shibata S (1988) Tissue culture of rhubarb and isolation of sennosides from the callus. Planta Med: 20–24

Oshio H, Imai S, Fujioka S, Sugawara T, Miyamoto M, Tsukui M (1974) Investigation of rhubarbs III. New purgative constituents, Sennosides E and F. Chem Pharm Bull 22: 823–831

Oshio H, Naruse Y, Tsukui M (1978) Quantitative analysis of the purgative components of rhubarb and senna. Chem Pharm Bull 26: 2458–2464

Rai PP (1978) The production of anthraquinones in callus cultures of *Rheum palmatum*. Lloydia 41: 114–116

Roggemans J, Boxus P (1988) Rhubarb (*Rheum rhaponticum* L.). In: Bajaj YPS (ed) Biotechnology in agriculture and forestry, vol 6. Crops II. Springer Berlin Heidelberg New York, pp 339–350

Sasaki K, Yamauchi K, Kuwano S (1979) Metabolic activation of sennoside A in mice. Planta Med 37: 370–378

Shibata S, Kobayashi S (1983) Trial cultivation of rhubarb in Japan. Gendai Toyoigaku (Mod Orient Med) 4: 68–71 (in Japanese)

Stahl E (1970) Chromatographische und mikroskopische Analyse von Drogen. Gustav Fischer Stuttgart, p 68

Stoll A, Becker B, Helfenstein A (1950) Die Konstitution der Sennoside. Helv Chim Acta 33: 313–336

Takahashi K, Kaizuka H, Ohshima Y (1983) Application of three-dimensional UV-absorbance-HPLC patterns for the analysis of plant extracts. J Chromatogr 268: 522–527

Xiao BG Chem B-Z, Li J-D (1959) Dahuang. In Lau Z-C (ed): Zung Yao Zhi, vol 1. Renmin Weisheng Publ, Beijing, pp 24–35

Yamagishi T, Nishizawa S, Nonaka G, Nishioka I (1987) New laxative constituents of rhubarb isolation and characterization of rheinosides A, B, C and D. Chem Pharm Bull 35: 3132–3138

XXI *Scopolia* spp.: In Vitro Culture and the Production of Scopolamine and Hyoscyamine

H.J. Scholten[1,2], S. Batterman[1], and J.F. Visser[1]

1 Introduction

Scopolia comprises a number of species which are a rich source of tropane alkaloids. In the literature the following species and synonyms have been mentioned:

S. acutangula Wu et Chen	
S. atropoides	
S. carniolica Jacq.	Asia, Eastern Europe
S. hladnikiana	
S. japonica Maxim	Japan
S. lurida	
S. parviflora Nakai	Korea
S. physaloides Dun	
S. sinensis	China
S. stramonifolia	Central Himalaya
S. tangutica	West China

Synonyms are:

S. atropoides = *S. hladnikiana* = *S. carniolica*
S. stramonifolia = *S. lurida* = *Anisodus luridus* = *Anisodus stramonifolius*
S. physaloides = *Physalis virginiana*
S. acutangula = *Anisodus acutangulus*
S. tangutica = *Anisodus tanguticus*
S. parviflora = *S. japonica* var. *parviflora*

Zheng (1989) gives a fourth species of *Anisodus*: *A. mairei*. It is not known whether this species has been described as a *Scopolia* species.

S. carniolica from West Asia was naturalized in Europe. As far as is known, in the last century *S. carniolica* was cultivated to some extent in the Netherlands and in Lithuania. There was little demand for cultivated plants, however, since plant

[1] University of Groningen, University Center for Pharmacy, Department of Pharmacognosy, A. Deusinglaan 2, 9713 AW Groningen, The Netherlands.
[2] Agricultural University, Department of Horticulture, Haagsteeg 3, 6708 PM Wageningen, The Netherlands

Biotechnology in Agriculture and Forestry, Vol. 21
Medicinal and Aromatic Plants IV (ed. by Y.P.S. Bajaj)

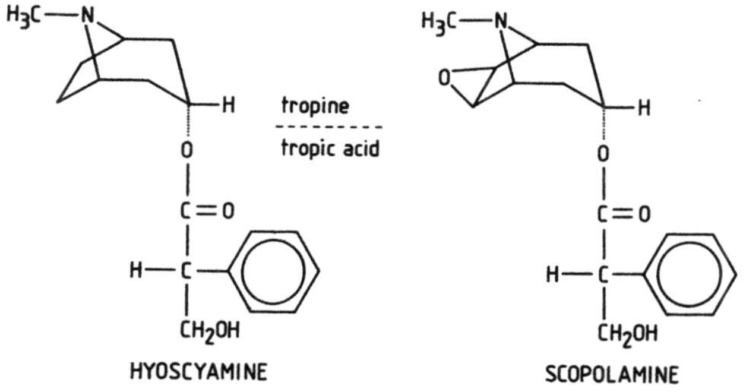

Fig. 1. Hyoscyamine and scopolamine. Tropine and tropic acid indicate the position of the precursors in the molecule

material could easily be supplied by collecting plants in nature. Both the leaves (Herba Scopoliae) and the roots (Radix Scopoliae) were used.

S. parviflora (*S. japonica* var. *parviflora*) is very similar to *S. japonica*, not only morphologically but also with respect to alkaloid composition and pharmacological activity. From both plants, rhizome extracts are used as antispasmodics. The rhizome of *S. parviflora* contains 0.34% dry wt. total alkaloid (Tabata et al. 1972).

The family Solanaceae consists of about 68 genera. From 21 of these, over 30 different tropane alkaloids have been described. In *Scopolia*, depending on the organ examined, hyoscyamine or scopolamine can be found (Fig. 1). Roots of intact plants contain considerable amounts of cuscohygrine, pseudotropine, and tigloidine. Other nontropane alkaloid components reported are pyridin alkaloids, rutin, kaempferol-3-rhamnoglucosid, astragalin, chlorogenic acid, and aesculetin (Hegnauer 1973; Frohne and Jensen 1979; Evans 1989). The coumarin scopoletin and its monoglucoside, scopoline, are characteristic for Solanaceae. In the leaves of adult plants of *S. carniolica* we measured 0.19% dry wt. hyoscyamine and 0.13% scopolamine.

The tropane alkaloids scopolamine and hyoscyamine are used in medicine for their parasympatholytic and anticholinergic activity. The chemical synthesis of tropane alkaloids has been described (Iida et al. 1985), but for economic reasons these alkaloids are still extracted from plant material. The solanaceous plants containing tropane alkaloids, especially the plants belonging to the genera *Atropa*, *Datura*, *Duboisia*, *Hyoscyamus*, and *Scopolia*, are known for their relative high scopolamine and hyoscyamine content (0.04–0.6% dry wt., on a dry weight basis; Evans 1989).

2 In Vitro Culture Studies

2.1 Anther culture

Wernicke and Kohlenbach (1975) were able to induce embryogenesis in three species. In 0.3% of *S. carniolica* anthers embryogenesis was induced. This percentage could be increased to 3% by applying liquid medium; 3% of the anthers of

S. lurida were also active. *S. physaloides* showed the best response, with 17% of the anthers yielding plantlets. In these plantlets the ploidy level was established to be haploid. Unfortunately, no information was given on the alkaloid content and the growth characteristics of the adult plants. It would be interesting to determine whether doubled haploids of *Scopolia* might be a useful tool in selection procedures for high alkaloid-producing plants.

2.2 Cell Culture

Cell cultures of *S. parviflora* have been initiated from stem and rhizome explants on an LS basal medium with 10^{-5} M IAA or 10^{-6} M 2,4-D (Tabata et al. 1972). IAA proved to be less effective in initiating callus proliferation than 2,4-D unless supplemented with kinetin. The cultures were grown in the dark. *S. parviflora* plants were very similar to plants of *S. japonica.* However, callus cultures of *S. parviflora* clearly differ in appearance. These cultures exhibited strong morphological activity, especially root formation. Romeike (1975) used MS medium with 9×10^{-6} M 2,4-D to establish callus cultures from stem explants of *S. lurida.* Cheng and Cheng (1978) and Zheng (1989) used LS medium for *S. acutangula.* Light appeared to have a positive effect on growth, as did NAA and higher concentrations of 2,4-D. They found kinetin to be slightly negative. Cheng et al. (1987, 1989) also used LS medium with 2.3×10^{-6} M 2,4-D to start callus cultures from seedlings of *A. tanguticus.* The cells could be subcultured in a liquid LS medium with 4.5×10^{-6} M 2,4-D, 10.7×10^{-6} M NAA, and 3% sucrose. Their cultures were grown in the dark. In our laboratory at the University of Groningen we have initiated callus cultures from leaf explants of *S. carniolica* on a B5 medium with 4.5×10^{-6} M BAP, 9×10^{-6} M 2,4-D and 4% sucrose. For maintainance, the cultures were grown with 16 h light (1500 lx). Like all other cultures of *Scopolia*, our cultures were grown at 25 °C.

2.3 Root Culture

A general feature of tropane alkaloid-producing plants is that the alkaloid content of undifferentiated tissue is only a small proportion of that measured in intact plants. When cell cultures are regenerated to plants, the pattern of secondary products can differ from that of the original plants, but the concentrations of hyoscyamine and scopolamine are in the same range as those of the intact plant (Kitamura et al. 1985). It is not surprising that part of the research efforts have been focused on the biotechnological production of biomass with differentiated tissues, namely roots, since roots are also the site of hyoscyamine synthesis. In general, roots grow very slowly and the yield of biomass is too low to be of commercial interest, at least with the alkaloid concentrations normally found in them. By inoculation with *Agrobacterium rhizogenes*, a root-inducing (Ri) plasmid can be integrated in the plant genome which causes adventitious, hairy roots arising from the inoculum site. These hairy roots grow vigorously on a hormone-free basal medium and can be subcultured in a liquid medium, even in large-scale fermentors. Mano et al. (1986) reported the transformation of *S. japonica* root segments with

A. rhizogenes and the production of alkaloids in the hairy roots. The growth rate of the roots was best on hormone-free Heller medium supplemented with FeNaEDTA and 3% sucrose. The roots were grown at 25 °C. With the basal medium of Schenk-Hildebrandt similar results were obtained. Several root clones were selected and screened for growth characteristics and alkaloid content. The pattern of alkaloid content varied between clones and differed from the original roots. A high scopolamine-producing clone S1 and a high hyoscyamine-producing clone S22 were isolated. The medium composition greatly influenced the alkaloid content of the roots. Also the medium supplement, casamino acids, could increase the scopolamine content of S1. Under optimum medium conditions, S1 contained 0.5% dry wt. scopolamine, which is three times the content of the untransformed rhizome. Clone S22 contained as much as 1.3% dry wt. hyoscyamine, which is eight times the rhizome content. These results suggest that it is worthwhile to screen for high-producing clones and to select the optimum medium composition. The results presented by Zheng (1989) with *Anisodus acutangulus* show that such variant cell lines can be maintained by cryopreservation. It is clear that for commercial exploitation of the root cultures, the concentrations have to be raised to considerably higher levels, in any case higher than the 4% dry wt. measured in plants of *H. muticus*.

2.4 Regeneration

Tabata et al. (1972) reported that callus cultures of *S. parviflora* developed plantlets on LS medium with 10^{-5} M IAA and 10^{-7} to 10^{-5} M kinetin and 3 g/l sucrose under fluorescent light. Vitamins in the medium were essential to support the elongation of the roots. These plantlets developed into normal plants when they were transfered to soil. On medium with 10^{-6} or 10^{-7} M 2,4-D, callus developed many roots. IAA at concentrations ranging from 10^{-7} to 10^{-5} M also had a positive effect on root formation. Shimomura et al. (1981a, b) have published studies on the breeding of *Scopolia* and described the regeneration of plants from stem segments and the clonal multiplication through shoot tip culture.

2.5 Bioconversion

One approach to produce secondary products with plant cells is to explore the enzymatic potential of the cells for the conversion of relatively cheap precursors into the final product. Romeike (1975) tried to esterify the precursors tropine and tropic acid into hyoscyamine through feeding experiments with cell suspensions of, among others, *S. lurida*. In these experiments *Datura* species converted tropine into acetyltropine. In *S. lurida*, however, no reaction product could be found. Cheng and Cheng (1978) reported that feeding the precursor phenylalanine to callus of *S. acutangula* (200 mg/l) resulted in an increase of the total alkaloid content of the cells from 0.142 to 0.294% dry wt. In continuation of this work, Zheng (1989) reported that the bioconversion of hyoscyamine to scopolamine by cells of *Anisodus acutangulus* (= *S. acutangula*) is increased from 0 to 27.4% by a nutrition depletion of the cells. Cheng et al. (1987, 1989) achieved the bioconversion of hyoscyamine to scopolamine by suspension cultures of *A. tanguticus*. For the experiments, a cell line was used which was derived from seedlings and which did not contain tropane alkaloids. When 100 mg/l DL-hyoscyamine was fed to a suspension culture, after 4 weeks 10 mg/l 6-hydroxyhyoscyamine, an intermediate in the biosynthesis of hyoscyamine to scopolamine, was found. After 5 weeks, scopolamine could be measured in concentrations up to 4 mg/l. The authors showed a relationship between the disappearance of hyoscyamine and the increase in 6-hydroxyhyoscyamine content. Also, the decrease in 6-hydroxyhyoscyamine when fed to cell suspensions and the increase in scopolamine was related. However, when the hyoscyamine concentration was raised from 100 to 1000 mg/l, the 6-hydroxyhyoscyamine content increased from 7 to 40 mg/l, but the scopolamine concentration remained about 8 mg/l, indicating that the enzymatic conversion of 6-hydroxyhyoscyamine to scopolamine is affected by some unknown factors. It was very interesting also that alginate-entrapped cells showed a high bioconversion ability with 6-hydroxyhyoscyamine as a precursor, and the scopolamine was recovered from the medium (Cheng et al. 1989). These results suggest that the cell wall has been made permeable and medium components or essential compounds from inside the cell can reach the site of biosynthesis. Accumulation of scopolamine always occurred in the latter growth phase, regardless of the time of addition of the precursor to the cells. Furthermore, a time period of some weeks was needed before the synthesis of scopolamine was fully expressed. For the biotechnological production of scopolamine it is important that cells can be grown in a fermentor. Zheng (1989) reported that the growth rate of cells of *A. acutangulus* in a fermentor was three times as high as in suspension cultures, without loss of biosynthetic capacity of the cells.

2.6 Analysis of Extracts of Plants and Cells

Scopolia plants differ in alkaloid pattern in comparison with the cell cultures. Most cultures contain only traces of the tropane alkaloids. For screening the cell cultures for alkaloids, selective and sensitive analytical procedures are necessary. It is known that cell cultures contain compounds which are not present or only in very low concentrations in the intact plants, and which can interfere with procedures developed for the intact plants (e.g., a codeine-like compound in cultures of *Papaver*; Galewsky and Nessler 1986). Oksman-Caldentey (1986) reviewed the analytical techniques available for the determination of the scopolamine and hyoscyamine content of the plants and cell cultures. In our laboratory, thin layer chromatography (TLC), gas liquid chromatography (GLC), high performance liquid chromatography (HPLC) with UV detection or diode array detection (DAD), and mass spectrometry (GC/MS and LC/MS) were used. It was found that the extraction procedure of alkaloids according to the Dutch Pharmacopoea Ed VI gave the highest recovery (90–95%), both for the intact plants of *S. carniolica* and the cell cultures. Freeze-dried plant material was macerated and agitated overnight in a mixture of 96% alcohol, 30% acetic acid, and water (150:10:140). After removal of the cell debris, the liquid fraction was increased to pH 10 with ammonia (25%) and extracted three times with diethylether. The combined ether layers were evaporated and the residue was dissolved in methanol and analyzed. A representative TLC plate is shown in Fig. 2. The plant contains both scopolamine and

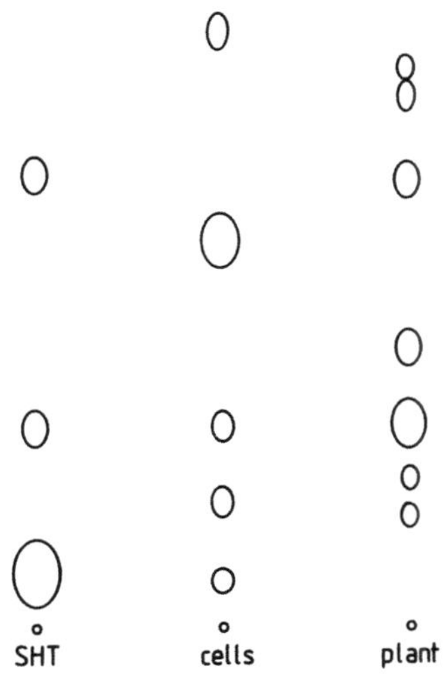

Fig. 2. TLC of extracts of plant leaves and of cell cultures. *SHT* is a mixture of scopolamine, hyoscyamine, and tropine. The Rf is 0.7, 0.3, and 0.1, respectively. Plate: silica F254 (Merck). Eluent: acetone, water, ammonia 25% (90:7:3). Detection: Dragendorff, (Stahl 1967)

hyoscyamine. In the cell suspension, only a small spot with the Rf of hyoscyamine was found. In highly concentrated samples, the individual alkaloid spots after immersion in Dragendorff reagent (Stahl 1967) cannot be recognized. In this case, the cobalt (II) thiocyanate reagent (Stahl 1967) offers an advantage; the alkaloid spots develop a bright blue color. From the color intensity and the detection limit of about 2 μg, a hyoscyamine content of the cells of about 0.01% was calculated.

Tabata et al. (1972) reported that yeast extract (1 g/l) in the medium increased the total alkaloid content of root cultures of *S. parviflora* from 0.033% to 0.072% dry wt. Undifferentiated cell cultures showed an average content of total alkaloids of about 0.01% dry wt. When grown in the light (7000 lx) as compared with cells grown in the dark, the alkaloid content of the cells was reduced 50–70%. Casamino acids (2 g/l) fed to cells of *S. parviflora* grown in the dark increased the total alkaloid content from 0.008 to 0.012%. With peptone or yeast extract the content was increased to 0.016 and 0.017% dry wt., respectively. Our results with *S. carniolica* confirm these results: in cells grown with 2 g/l casamino acids, the hyoscyamine concentration was increased by a factor of 1.5. Tabata et al. (1972) found that callus tissue of *S. parviflora* contained as much as seven Dragendorff-positive spots. Most of these spots were also present in the intact plant. Four of these were identified as hyoscyamine, scopolamine, apoatropine, and tropine. Cheng and Cheng (1978) studied extensively the biosynthesis of tropane alkaloids in cell cultures of *S. acutangula* and observed the production of considerable amounts of hyoscyamine and scopolamine. In stem callus, a scopolamine content as high as 0.77% dry wt. was measured (Zheng 1989). Light increased the alkaloid content; also, low 2,4-D concentrations had a positive effect. When cultured cells from a medium without kinetin were transferred to a medium with kinetin (1 mg/l), the scopolamine content increased from 0.056 to 0.495% dry wt. Lactalbumin hydrolysate (1 g/l) stimulated the growth of the cells and the scopolamine concentration increased from 0.033 to 0.149%. Valine and phenylalanine (200 mg/l) also influenced the biosynthesis of scopolamine: three times as much scopolamine was measured. From the organic acids tested, fumaric acid had some positive effect.

Unfortunately, the growth conditions and the procedures for identification and determination of the alkaloids were described in inaccessible Chinese literature. It was mentioned that newly induced calli were used. This could account for the high concentrations of scopolamine found in the cells as compared with the results of other authors. More confusing is the UV-absorption spectrum of scopolamine isolated from callus given in their report. This spectrum is considerably different from that given in handbooks.

In cells of *S. carniolica*, hyoscyamine was found in some cultures when using TLC (Fig. 2). The HPLC analysis confirmed these results (Fig. 3). When using TLC, all cell extracts gave a tropine spot. A number of our extracts have been analyzed with GC/MS and no other known alkaloids could be detected. Several experiments have been performed in order to increase the tropane alkaloid content of cell cultures of *S. carniolica*. We have tested a number of medium compositions, culture conditions, and incubation periods, all without any success. When we cultivated cell suspensions in the light (1500 lx) with 3×10^{-7} M IAA and 3% sucrose, the cells became greenish and the hyoscyamine content of the cells seemed to be higher with TLC. From these green cultures, a dark greenish-blue-colored line (code SC 2111) was selected and subcultured on B5 medium with 4% sucrose. The cells were transferred to a B5 medium with 3% sucrose and without hormones. One batch was grown under light of 2000 lx and one under 1200 lx. With TLC, no scopolamine could be detected, but hyoscyamine was present; the concentration was estimated to be 0.01% in the 2000-lx cultures and 0.03% in the 1200-lx cultures. Clearly, light has a positive effect on alkaloid biosynthesis, but higher light

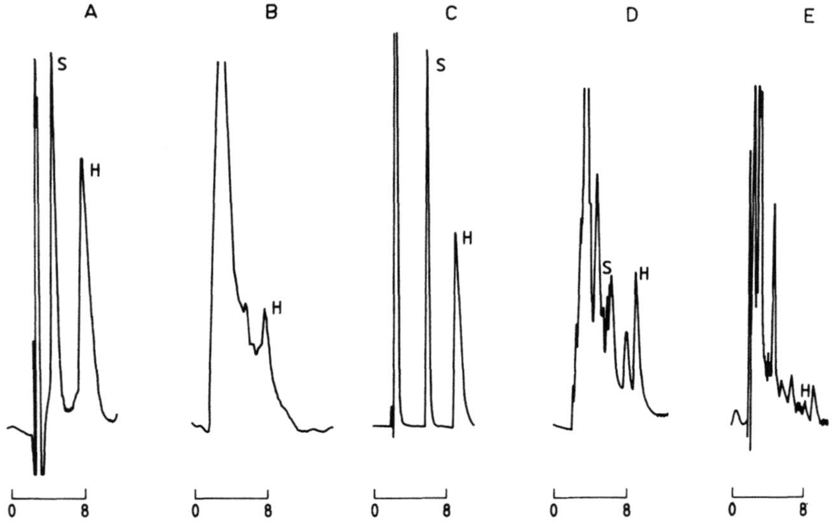

Fig. 3A–E. HPLC chromatograms of standards (**A**) and a cell extract (**B**) on a column of Nucleosil RP5 C18 cartridge (100 × 3 mm). Eluent: methanol, water, phosphoric acid 0.2% (10:20:10). Flow rate: 0.3 ml/min. Detection. UV at 204 nm. HPLC chromatograms of standards (**C**), extract of plant leaves (**D**), and of a cell culture (**E**) on a column of Spherisorb CN, cartridge (100 × 3 mm). Eluent: 25 mM ammonium acetate, 20% methanol pH·5 with acetic acid. Flow rate: 0.3 ml/min. Detection: UV at 220 nm. *H* Hyoscyamine; *S* Scopolamine intensities, e.g., 2000 lx, when using our cells, or 7000 lx in the case of *S. parviflora* (Tabata et al. 1972), have a negative effect. As was demonstrated by Cheng and Cheng (1978), our results also indicate that changes in the hormonal composition of the medium might be important to achieve an economically interesting production system with undifferentiated cells.

Both TLC and HPLC patterns often gave rise to misinterpretations. The cell cultures of *S. carniolica* contain a major compound, which has an Rf (0.75) close to scopolamine (Rf 0.8) when using TLC, and which is Dragendorff-positive. This compound is not specific for *Scopolia*, because it is also a major compound of *Agrobacterium tumefaciens* transformed calli of *Duboisia myoporoides* and regenerated plants from these calli (Scholten, unpubl. results). Attempts to identify this compound have not been successful to date. We do know that it is a low molecular weight, non-UV-absorbing compound. Addition of casamino acids to the medium induced higher concentrations of it in the cells. Another interesting aspect is that there is an inverse relationship between the presence and concentration of hyoscyamine and the compound. The coumarin scopoletin and the glucoside scopoline could not be detected in the cell extracts. However, another coumarin was found which showed chromatographic properties almost identical to scopoletin and which had the typical UV-spectrum of a coumarin in an analysis with HPLC-DAD.

In Russian literature, reports have been published on the formation of alkaloids in tissue culture of *S. stramonifolia* (Bereznegovskaya et al. 1971; Smordin et al. 1974). Zheng and Zheng (1976) and Zheng and Liang (1977) published a report on the production of hyoscyamine and scopolamine in callus cultures of *S. acutangula*. In addition to the accumulation of tropane alkaloids, *S. japonica* cells have been used for studies on the production of a plasmin inhibitory substance (Misawa et al. 1975) and proteinase inhibitors (Sakato et al. 1975).

For HPLC analysis of extracts of *S. carniolica* we developed several systems. Figure 3A, B shows the HPLC chromatograms of a mixture of standards and the cell extract shown in Fig. 2, on a Nucleosil RP5 C18 cartridge column. From the HPLC response a hyoscyamine concentration of 0.01% dry wt. was again calculated. HPLC-DAD proved to be suitable for plant extracts with high concentrations of alkaloids. Because of the low alkaloid content of cell extracts, highly concentrated extracts have to be used in the analyses. In such extracts other major compounds are present in excess and interfere with each other. Therefore, in these extracts no pure spectra could be obtained for the identification and verification of the compositions of the hyoscyamine peak. Plant extracts could be diluted in order to obtain separated peaks. A Lichrosorb RP18 cartridge column (Chrompack, The Netherlands) gave identical results when an eluent was used consisting of acetonitril and 20 mM triethylamine in water (pH 7.3 with H_3PO_4, 80:20). More separated peaks of scopolamine and hyoscyamine were obtained with a Spherisorb CN cartridge column and as a mobile phase, 25 mM ammonium acetate with 20% methanol, pH 5 with acetic acid (Fig. 3C, D, E).

Both TLC and HPLC patterns often gave rise to misinterpretations. Cell cultures of *S. carniolica* contain compounds which show all the chromatographic characteristics of hyoscyamine and scopolamine. In several extracts, a scopolamine peak with HPLC was found, but the presence could not be confirmed with mass spectrometry. Coupling of HPLC and mass spectrometry (LC/MS) proved to be

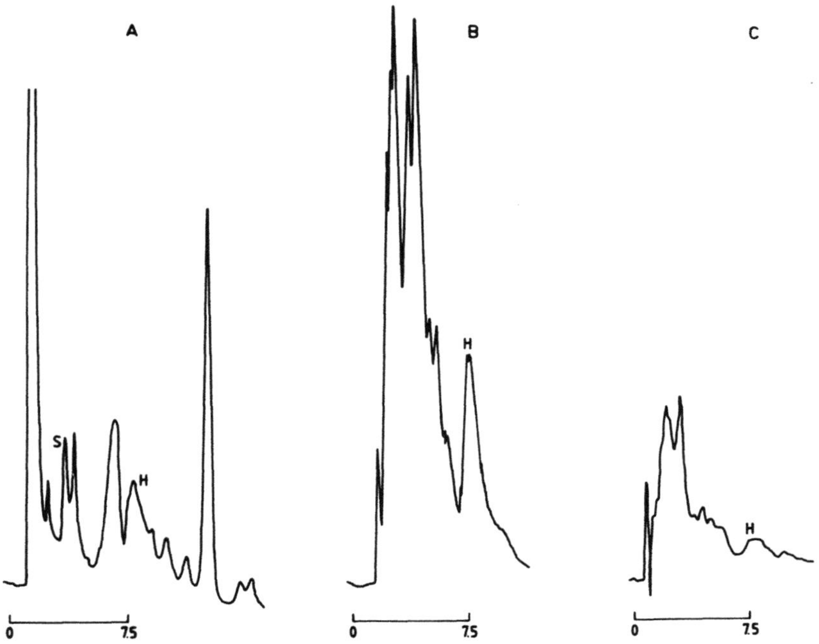

Fig. 4A–C. HPLC chromatograms of an extract of plant leaves (**A**) and of cell line SC2111 grown at 1200 lx (**B**) or at 2000 lx (**C**). Column: Nucleosil RP5 C18 (100 × 3 mm). Eluent: 12.5% methanol in water, pH 3.5 with acetic acid. Flow rate: 0.3 ml/min. Detection: UV at 215 nm. *H* Hyoscyamine; *S* Scopolamine

an especially useful tool for the identification of the presumed compounds. The system used positive chemical ionization with ammonia and an ion-spray technique which required a low ion strength of the eluent. An eluent consisting of 12.5% methanol in water (pH 3.5 with acetic acid) in combination with a Nucleosil cartridge column resulted in chromatograms of extracts of the authentic plant material and of cell line SC2111, as shown in Fig. 4. With LC/MS (Fig. 5) the presence of hyoscyamine in line SC2111 was confirmed. Figure 5A shows the HPLC peaks for hyoscyamine (M + 1 = 290) and scopolamine (M + 1 = 304). With the MS system, the eluate was screened for the presence of ions with the molecular weight (+1) of hyoscyamine and scopolamine. With a second MS coupled to the first MS (LC/MS/MS), fragments of the contents of the peaks can also be analyzed. In the cell line SC2111, no scopolamine could be detected. The concentration of hyoscyamine in cell cultures under 1200 lx appeared to be 0.032% dry wt. (Fig. 5B).

Fig. 5A, B. LC/MS chromatograms of plant material (**A**) and cell line SC2111 grown at 1200 lx (**B**). RIC = 290 and RIC = 304 show the peaks in the eluate of ions with molecular weight M + 1 = 290 (hyoscyamine) and M + 1 = 304 (scopolamine). Column: Nucleosil RR5 C18 (100 × 3 mm). Eluent: 12.5% methanol in water, pH 3.5 with acetic acid

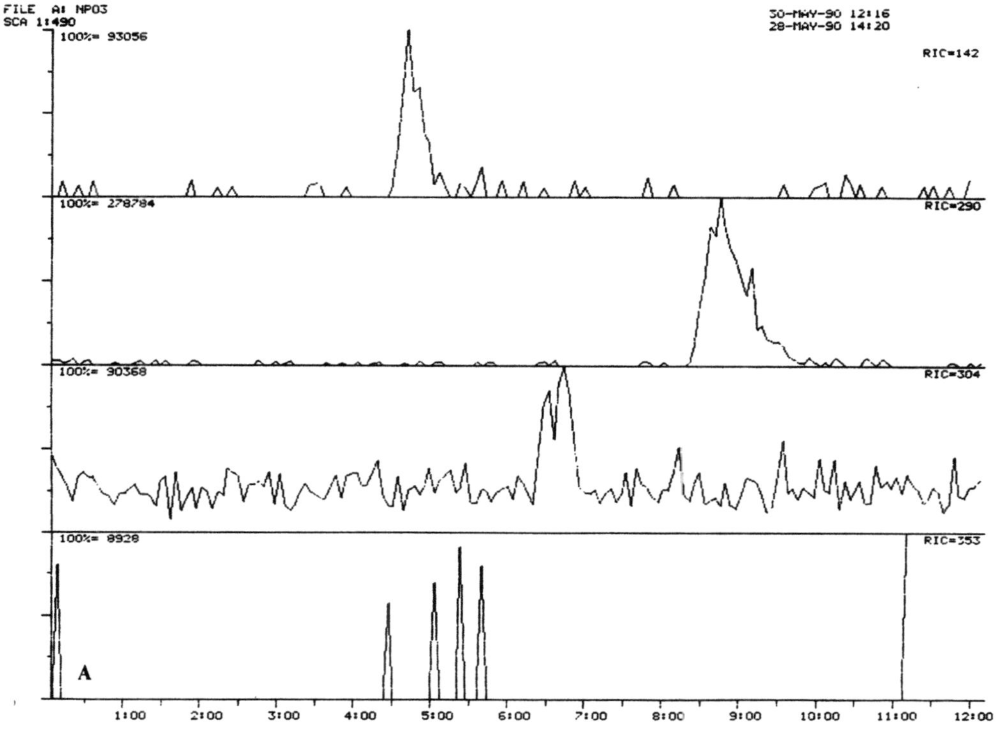
FILE A: NP03
SCA 1:490
30-MAY-90 12:16
28-MAY-90 14:20
100%= 93056
RIC=142
100%= 278784
RIC=290
100%= 90368
RIC=304
100%= 8928
RIC=353
A
1:00 2:00 3:00 4:00 5:00 6:00 7:00 8:00 9:00 10:00 11:00 12:00

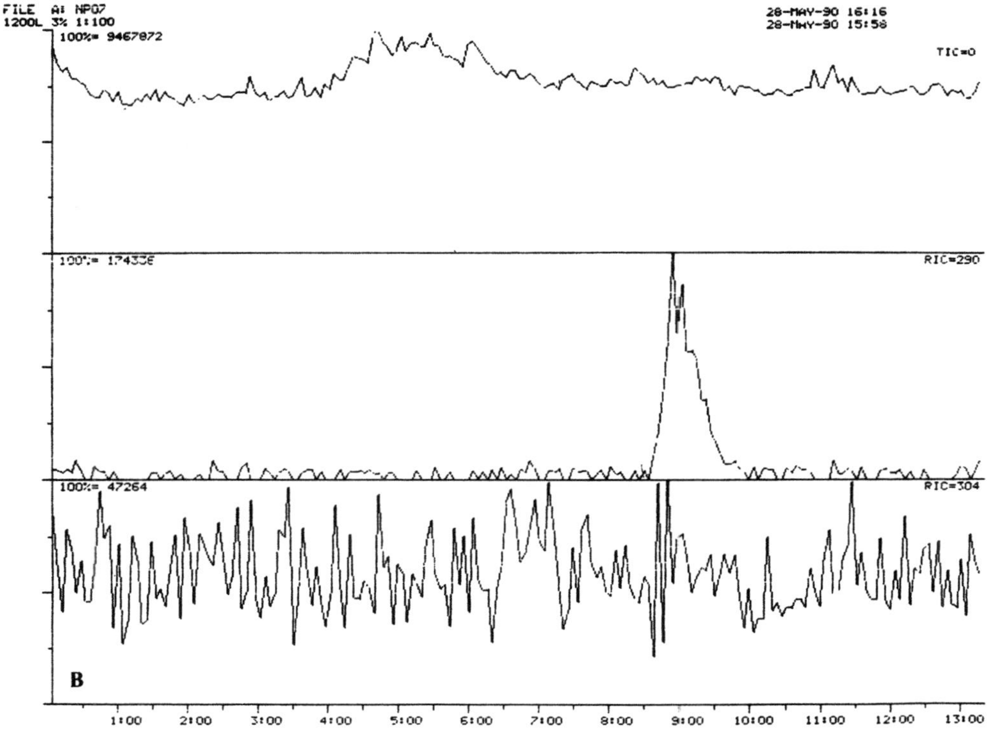
FILE A: NP07
1200L 3% 1:100
28-MAY-90 16:16
28-MAY-90 15:58
100%= 9467872
TIC=0
100%= 174336
RIC=290
100%= 47264
RIC=304
B
1:00 2:00 3:00 4:00 5:00 6:00 7:00 8:00 9:00 10:00 11:00 12:00 13:00

3 Protocol

Explant.

1. To initiate cell cultures, every part of the plant can in principle be used. Leaf strips are often used.

2. Outside the growing season of the plant, which is relatively short, seeds form a convenient starting material. Sterilized seeds are germinated on sterile perlite, moistened with a nutrient solution, or on a solidified MS medium without growth regulators. Any part of the seedling can be used as explant.

Callus Induction.

1. For callus induction a standard basal medium (e.g., MS, LS, or B5) with 3 or 4% sucrose can be used. As growth regulator 2,4-D (10^{-6} to 10^{-5} M) is effective. Kinetin or BA at a concentration of 5×10^{-6} M may be beneficial.

2. For suspension cultures identical media are used.

3. The cultures are grown at 25 °C in the dark or at low light intensities.

Alkaloid Production.

1. The alkaloid content of both *Agrobacterium* transformed roots and undifferentiated cells is influenced by the medium composition.

2. At the end of the growth cycle, the highest alkaloid content can be measured. In the stationary phase, the highest bioconversion capacity is also reached.

3. Organic supplements like yeast extract (1 g/l), casamino acids (2 g/l), or lactalbumin (1 g/l) stimulate the biosynthesis of tropane alkaloids.

4. A two-stage culture has a positive effect. Cells can be transferred from a medium with hormones to a medium without, or the balance of hormones can be changed.

5. Light at low intensities (below 2000 lx) increases the alkaloid content.

4 Conclusions and Prospects

Cell cultures of *Scopolia* show all the benefits of tissue culture of solanaceous plants: cell cultures can be easily handled and regeneration of plants has been described. For the production of tropane alkaloids with root or cell cultures, the concentrations of hyoscyamine and scopolamine are not of economic value. Cell cultures have to compete with high producing plants like *H. muticus* with over 4% dry wt. scopolamine (Oksman-Caldentey et al. 1987). However, if high cellular concentrations of tropane alkaloids can be reached, the controlled production of biomass has clear advantages. To reach such high levels, selection of high producing cell lines and genetic manipulation are inevitable. Some basic level of alkaloids in cells is favorable, and *Scopolia* meets all the requirements. Perhaps more promising are the results of Cheng et al. (1987, 1989) and Zheng (1989) with bioconversion of hyoscyamine to scopolamine. Apparently, in *Scopolia* cells all biosynthetic systems are active, and therefore these cells are very suitable for studies which focus on the key problem in undifferentiated tissues: the esterification of tropine and tropic acid to hyoscyamine.

References

Bereznegovskaya LN, Dmitruk SE, Doshchinskaya NV, Trofimova NA (1971) The formation of alkaloids in tissue culture of *Scopolia stramonifolia* D. Rastit Resur 7: 511–517

Cheng K, Cheng L (1978) Callus cultures of the three well-known Chinese herbs and their medicinal contents. In: Proc Symp on Plant tissue culture, Science Press, Peking, pp 469–497

Cheng KD, Zhu WH, Li XL, Meng C, Sun ZM, Yang DH (1987) Biotransformation of hyoscyamine by suspension cultures of *Anisodus tanguticus*. Planta Med 53: 211–213

Cheng KD, Fang HJ, Zhu WH, Mang C, Yang DH, Li L (1989) The biotransformation of 6-hydroxy-hyoscyamine to scopolamine by suspension and immobilized cells of *Anisodus tanguticus*. Planta Med 55: 391–392

Evans WC (1989) Trease and Evans's pharmacognosy. Baillière Tindall, London

Frohne D, Jensen U (1979) Systematik des Pflanzenreichs. Gustav Fischer, Stuttgart

Galewsky S, Nessler CL (1986) Synthesis of morphinane alkaloids during opium poppy somatic embryogenesis. Plant Sci 45: 215–222

Hegnauer R (1973) Chemotaxonomie der Pflanzen. Birkhäuser, Basel Stuttgart

Iida H, Watanabe Y, Kibayashi C (1985) A new synthetic route to tropane alkaloids based on [4 + 2] nitroso cycloaddition to 1,3-cycloheptadienes. J Org Chem 50: 1818–1825

Kitamura Y, Miura H, Sugii M (1985) Change of alkaloid distribution in the regenerated plants of *Duboisia myoporoides* R. Br. Shoyakugaku Zasshi 39: 85–87

Mano Y, Nabeshima S, Matsui C, Ohkawa H (1986) Production of tropane alkaloids by hairy root cultures of *Scopolia japonica*. Agric Biol Chem 50: 2715–2722

Misawa M, Tanaka H, Chiyo O, Mukai N (1975) Production of a plasmin inhibitory substance by *Scopolia japonica* suspension cultures. Biotechnol Bioeng 17: 305–314

Oksman-Caldentey K-M (1986) Production of scopolamine and hyoscyamine in some solanaceous plants and cell cultures. Acta Pharm Fenn 95: 49–58

Oksman-Caldentey K-M, Vuorela H, Isenegger M, Strauss A, Hiltunen R (1987) Selection for high tropane alkaloid content in *Hyoscyamus muticus* plants. Plant Breed 99: 318–326

Romeike A (1975) Versuche zur Veresterung von Tropin und Tropasäure an pflanzlichen Zellkulturen. Biochem Physiol Pflanz 168: 87–92

Sakato K, Tanaka H, Misawa M (1975) Broad specificity proteinase inhibitors in *Scopolia japonica* (Solanaceae) cultured cells. Isolation, physiochemical properties, and inhibition kinetics. Eur J Biochem 55: 211–219

Shimomura K, Shoyama Y, Nishioka I (1981a) Studies on the breeding of *Scopolia*. I. The production of plants regenerated from stem segments in vitro. Jpn J Pharmacogn 35: 115–121

Shimomura K, Shoyama Y, Nishioka I (1981b) Studies on the breeding of *Scopolia*. II. Clonal multiplication through shoot tip culture. Jpn J Pharmacogn 35: 122–127

Smordin AV, Smordin VV, Bereznegovskaya LN (1974) Use of a 2 factor analysis of variance for optimizing the conditions of alkaloid production by tissue cultures of the seeds of Hindu *Datura innoxia* and roots of Himalayan *Scopolia stramonifolia*. Khim Farm Zh 8: 46–49

Stahl E (1967) Dünnschicht-Chromatographie. Springer, Berlin Heidelberg New York

Tabata M, Yamamoto H, Hiraoka N, Konoshima (1972) Organisation and alkaloid production in tissue culture of *Scopolia parviflora*. Phytochemistry 11: 949–955

Wernicke W, Kohlenbach HW (1975) Antherenkulturen bei *Scopolia*. Z. Pflanzenphysiol 77: 89–93

Zheng GZ (1989) *Anisodus acutangulus*: production of scopolamine and hyoscyamine in cell cultures. In: Bajaj YPS (ed) Biotechnology in agriculture and forestry, vol 7. Medicinal and aromatic plants II. Springer, Berlin Heidelberg New York, pp 23–45

Zheng GZ, Liang Z (1977) Studies on tissue culture of medicinal plants part 2. Chemical control of callus growth and synthesis of hyoscyamine and scopolamine by *Scopolia acutangula* callus. Acta Bot Sin 18: 163–169

Zheng GZ, Zheng L (1976) Studies on tissue culture of medicinal plants part 1. Callus cultures of *Scopolia acutangula* for the production of hyoscyamine and scopolamine. Acta Bot Sin 18: 163–169

XXII *Silene alba* (White Campion): In Vitro Culture and Benzylisoquinoline Alkaloids Biotransformation

C. VIEL[1], F. BISTER-MIEL[2], and J.L. GUIGNARD[2]

1 Botany and Phytochemistry of *Silene alba*

(Bonnier 1911–1935; Tutin et al. 1964; Hutchinson 1973; Cronquist 1981; Bonnier and Douin 1990)

Silene alba (Miller) E.H.L. Krause, also known as *Lychnis dioica* L. or *Melandrium album* (Miller) Garcke, of the family Caryophyllaceae (Fig. 1) has many common names: compagnon blanc, robinet, floquet, saponaire blanche, oeil de Dieu, bourbonnaise (in French); white campion, white bachelor's buttons, white bottle, white cuckoo flower (in English); weipe Lichtnelke, Sommetrose, Wiederstock, Junggesellenknöpfe, Je-länger-je-freundlicher (in German); fischi da fischiare, lichnide, gittone bianco, erba nocca, violina di macchia (in Italian) (Bonnier 1911–1935). It is a dioecious herb of varied habit, usually growing as a weed on plowed or cultivated ground throughout the whole of Europe.

A short-lived perennial (sometimes annual) growing up to 80 cm in height, *S. alba* usually has rather thick and soft hairs, usually glandular on the upper epidermis. The leaves are ovate or ovate-lanceolate, the cauline is sessile. The inflorescence consists of loosely bound dichasium of large flowers, which open in the evening and are slightly scented. The flowers are pentamerous. The limb of the petal is distinct from the claw, with ten stamens. The calyx of the male flowers is 15–22 mm in length, with ten veins, while that of the female is 20–30 mm, 20-veined, usually inflated and acrescent in the fruit. The calyx teeth are narrowly triangular. Petals are usually white with five styles. The capsule is 10–15 mm in length, usually ovoid, and dehiscing, with ten teeth.

The plants produce anthocyanins but not betalains, and are not tanniferous, lacking both proanthocyanins and ellagic acid, although commonly accumulating triterpenoid saponins.

Silene alba, one of the most widespread of the European species, belongs to a family that is unique in its chemical composition: it contains triterpenic saponins, but no biosynthesizing alkaloids.

We studied the biotransformations of papaverine and other 1-benzylisoquinolines in *Silene alba* cell cultures. The choice of this species was justified by the fact that *S. alba*, like other Caryophyllaceae, does not biosynthesize alkaloids. Consequently, it is possible without great difficulty to characterize and isolate all the

[1] University of Tours, Faculty of Pharmacy, 2 bis boulevard Tonnellé, 37042 Tours Cedex, France
[2] University of Paris-Sud, Faculty of Pharmacy, 3 rue J.B. Clément, 92296 Châtenay-Malabry Cedex, France

Biotechnology in Agriculture and Forestry, Vol. 21
Medicinal and Aromatic Plants IV (ed. by Y.P.S. Bajaj)

Fig. 1. *Silene alba* (Miller) E.H.L. Krause

molecules of the basic structure, such as substrates and their biotransformation products. It was also of interest to ascertain the reaction of this type of plant cell toward alkaloid xenobiotics, and as a result, if biotransformation takes place, to determine the type of reactions which occur. Finally, the choice of benzylisoquinoline molecules as substrates was interesting due to the diversity and number that were expected in this type of structures: oxidations (benzyl group, phenolic ethers, aromatic carbons), N-oxidation, reductions, glycosylations after O-dealkylation or O-oxidation, etc. It was also of interest in 1-benzyl and 4-benzyl derivatives to compare the importance of the isomerism of benzyl group, and also the influence of the stericity of this same group with the respective accessibility of the 6- and 7-methoxy groups if the O-dealkylation reaction occurs.

2 In Vitro Approaches

There are two groups working on cell suspensions of *Silene alba*, as far as we know, one of them in Professor Bouriquet's laboratory at the University of Lille and the other under Professor Guignard at the University of Paris-Sud.

The first cell suspension was the subject of studies on the nutrition of cells in vitro (nitrogen, carbohydrates, calcium, gibberellic acid) (Dubois 1975; Dubois et al. 1976; Brassart et al. 1977a, b, 1982), including in the fermentor (Dubois 1975; Brassart et al. 1976), and on the cell's physicochemical properties and enzymatic activities (nucleic acids, pectic acid polymers, nitrate reductase, glutamate dehydrogenase, glutamine synthetase) (Legrand and Dubois 1977, 1978; Rambour et al. 1977, 1983; Brassart et al. 1984; Marie-Alsabek et al. 1984; Poulle and Rambour 1985; Morvan et al. 1987a, b).

The second cell suspension was also the subject of studies on the nutrition of cells (naphthoxyacetic acid, macroelement concentration, phosphorus, boron, glutamine) (Henry et al. 1978; Agier et al. 1982, 1983; Bister-Miel et al. 1985), but served primarily as material for research on biotechnology: the biotransformation by cell suspensions in vitro of exogenous isoquinoleic substrates (Verdeil et al. 1985, 1986; Bister-Miel et al. 1986, 1988; Christinaki et al. 1987; Agier et al. 1988).

2.1 Establishment of Tissue Cultures and Cell Suspensions of *S. alba*

Cell suspensions of *Silene alba* were obtained in two successive stages. First, cultures were obtained on solid medium, then calli were transferred into a liquid suspension.

The process of culturing plant cells involves the induction of undifferentiated wound growth on sterile explants from stems, leaves, or flowers stalks, by 2,4-D (Table 1). Flower stalk explants proved best for these studies.

The callus material is then removed from the explant and either maintained on solid medium supplemented with the appropriate organic and inorganic nutrients and phytohormones: 2,4-D and kinetin (Table 2). The choice of 2,4-D concentration must be determined by a stepwise assay: the first culture was grown with NOA (naphthoxyacetic acid), but all the various concentrations of NOA were insufficient. Better results were attained by adding a low concentration of 2,4-D. Subsequent studies with NOA and 2,4-D showed 2,4-D to be optimal (Table 2). It was then possible to transfer the callus into liquid suspension with the same nutrients and

Table 1. Medium for growth of *S. alba* explants

Organic and inorganic nutrients: Nitsch medium (Nitsch and Nitsch 1956)
Adenine 2 mg/l 2,4-D 0.1 mg/l Sucrose 20 g/l

Table 2. Medium for culture of *S. alba* cells

Organic and inorganic nutrients: MS medium (Murashige and Skoog 1962)
Adenine 2 mg/l 2,4-D 0.1 mg/l Kinetin 1 mg/l Sucrose 30 g/l

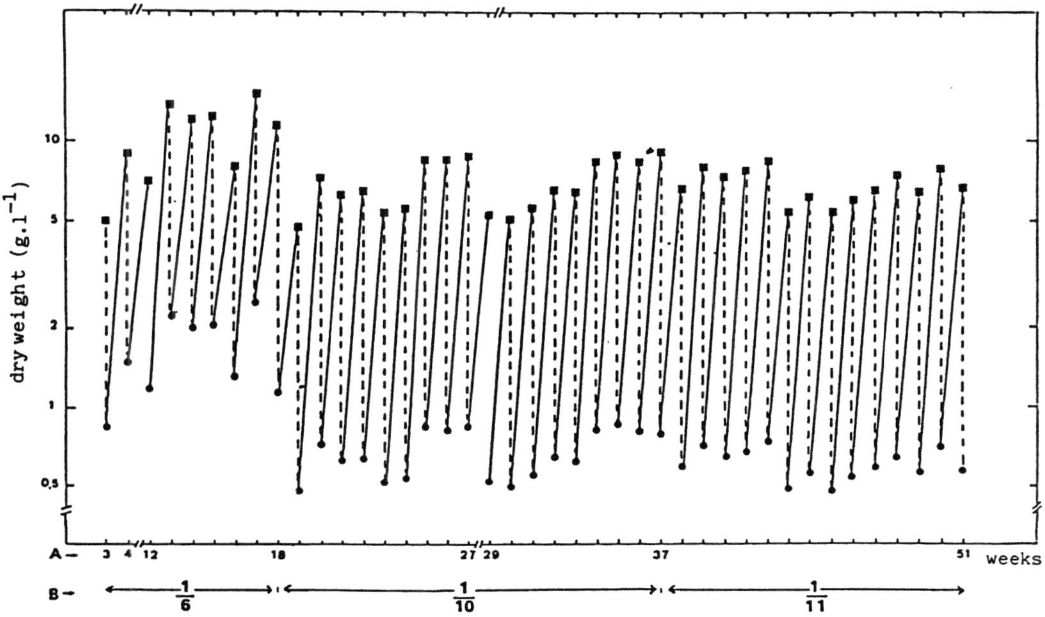

Fig. 2A, B. Evolution of suspension growth during the first 52 weeks. ((from Bister-Miel 1987). **A** Weeks. **B** Subculture dilutions. ● Dry weight of inoculum; ■ Dry weight after 7 days of culture

phytohormones (Bister-Miel et al. 1985). A hormone-free suspension of *S. alba* was also assayed, but was not suitable for biotransformation studies.

For optimal growth, the liquid suspension was subcultured every week. The growth rate increased gradually so that, using the same volume, the cell density became higher and higher. It was then necessary to increase the previous dilution rate to maintain a constant density in the inoculum. More than 50 weeks of study were necessary to obtain the optimal dilution for subcultures (Fig. 2), which was the only one that produced metabolites.

The suspension culture of *S. alba* was maintained for 10 years in a medium containing MS minerals and vitamins (Murashige and Skoog 1962), sucrose 30 g/l, 2,4-D 1 mg/l, adenine 2 mg/l, kinetin 1 mg/l, in flasks shaken at 180 rpm on a Gyrotory C 10. Cell suspensions were illuminated for 12 days and subcultured every 7 days onto a 1:11 dilution (Bister-Miel 1987). Cell suspensions were incubated at 23 ± 1 °C. Cell density was doubled after 48-h culture (Fig. 3).

2.2 Biotransformation Experiments; Structure of Metabolites

(Verdeil et al. 1985, 1986; Bister-Miel et al. 1986, 1988; Bister-Miel 1987; Christinaki et al. 1987; Agier et al. 1988)

Three molecules were selected for biotransformation experiments: a 1-benzylisoquinoline alkaloid, papaverine **1**, its 4-benzyl isomer, isopapaverine **2** and a 4-(4′-chlorobenzyl) analog, PV2, **3** (Fig. 4); the latter two were previously

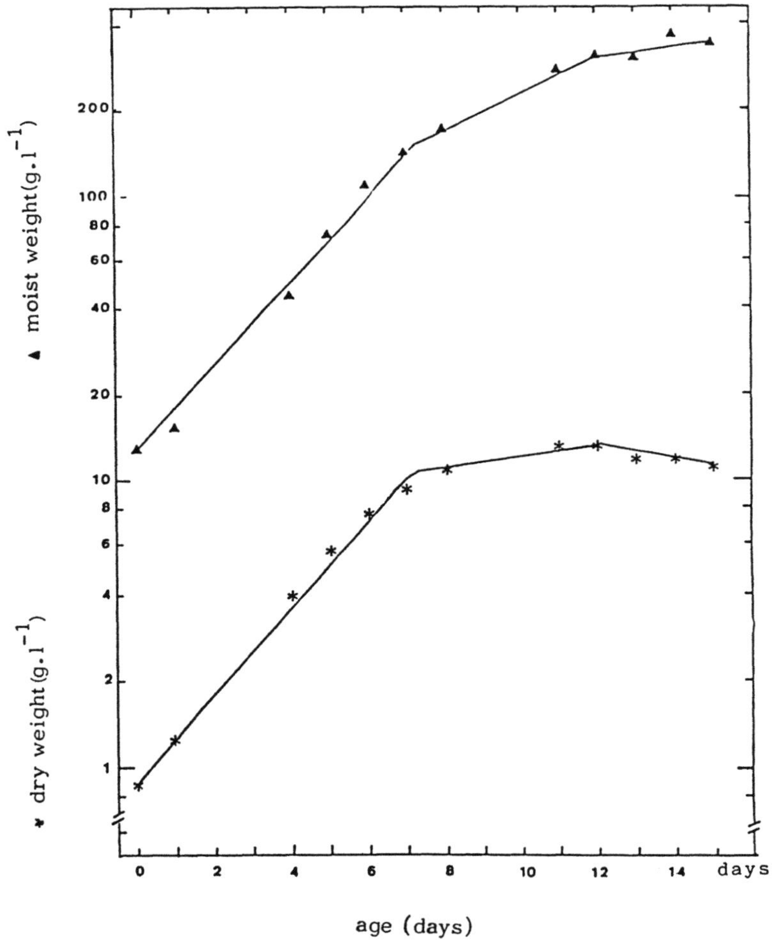

Fig. 3. Growth curves of *Silene alba* cell suspension. (from Bister-Miel 1987)

synthesized pharmacochemically (Prudhommeaux et al. 1971; Bouvier et al. 1976). These substrates were used in the form of water-soluble salts: hydrochloride for **1**, methane sulfonate for **2** and **3**.

The toxicity of the substrate must be estimated before selecting a suitable concentration for biotransformation.

2.2.1 Estimation of Toxicity

The toxicity of substrates in *S. alba* cell suspensions has been studied in media containing:

- 0 to 1000 mg/l papaverine hydrochloride
- 0 to 700 mg/l isopapaverine methane sulfonate
- 0 to 80 mg/l PV2.

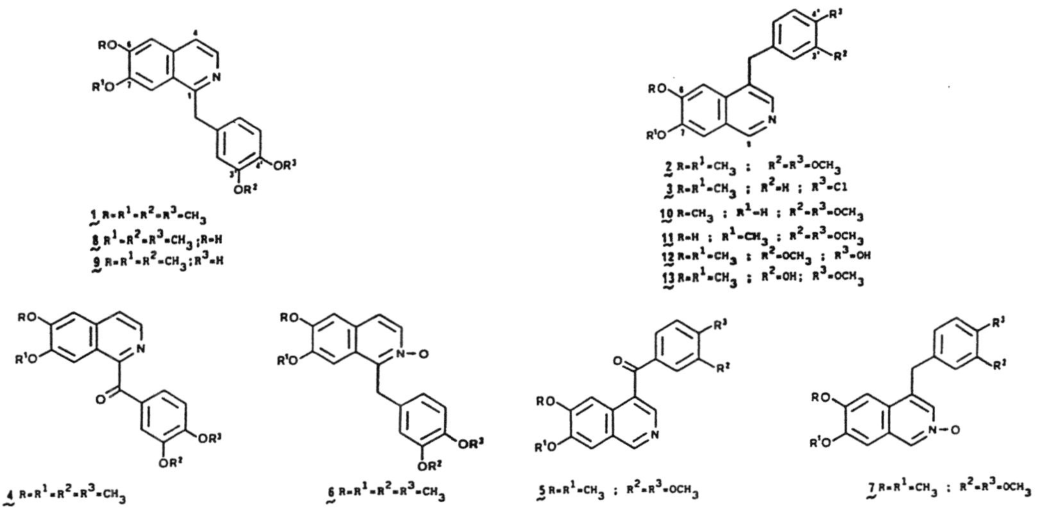

Fig. 4. Substrates and their biotransformation metabolisms. (from Bister-Miel et al. 1988)

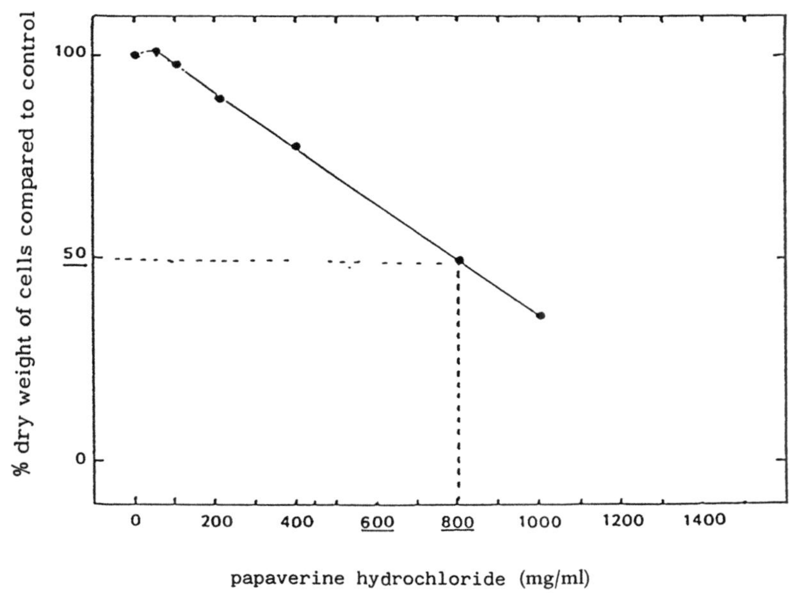

Fig. 5. Toxicity of papaverine hydrochloride in *S. alba* cell suspension. (from Bister-Miel 1987)

After inoculation and 7 days of growth in these media, dry weight (DW) evaluations were performed: cell samples were collected by reduced pressure sieving, and dried at 60 °C for 48 h. The dry weight at day 7 is given as a function of the substrate concentration in Figs. 5, 6, and 7. The toxicity data have been confirmed in three experiments.

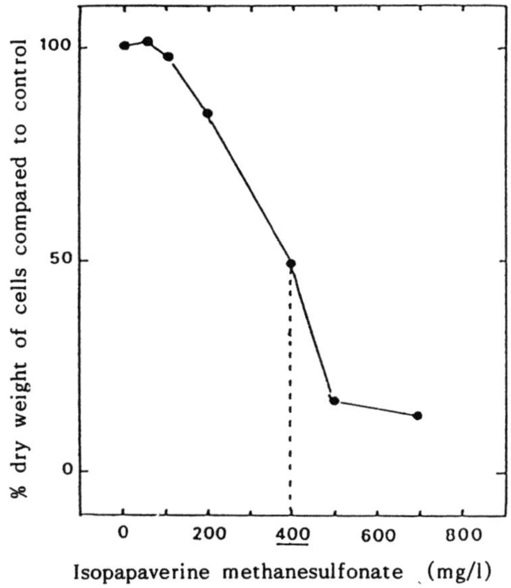

Fig. 6. Toxicity of isopapaverine methane sulfonate in *S. alba* cell suspension. (from Bister-Miel 1987)

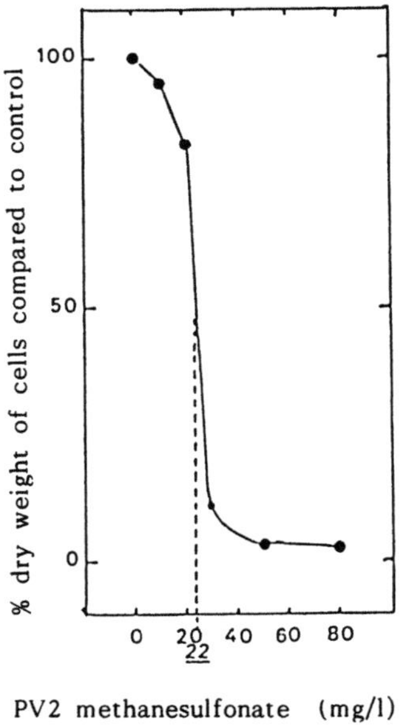

Fig. 7. Toxicity of PV2 methane sulfonate in *S. alba* cell suspension. (from Bister-Miel 1987)

A slight stimulation of growth can be observed with low amounts of papaverine hydrochloride and isopapaverine methane sulfonate (50 mg/l). At concentrations higher than 100 mg/l, growth decreased. Growth is reduced to 50 wt% at 800 mg/l for papaverine hydrochloride and 400 mg/l for isopapaverine methane sulfonate. In both, toxicity is light.

A considerable decrease in growth can be observed with low amounts of PV2. Growth is reduced to 50 wt% between 20 and 25 mg/l (22 mg/l on average). Toxicity is severe.

2.2.2 Estimation of Biotransformation Capacity

The substrates were used at 800 mg/l for papaverine **1**; 400 mg/l for isopapaverine **2**, and 20 mg/l for 4-(4'chlorobenzyl) isoquinoline **3**.

Prior to all biotransformation assays the stability of the three substrates was checked regarding the autoclaving conditions of the medium (120 °C), and lighting and extraction conditions. No degradation products were revealed by thin layer chromatography (TLC).

Cell suspensions with substrate were incubated at 23 ± 1 °C on a rotary shaker at 180 rpm under illumination of 25 $\mu E/m^2/s$ for 12 h/day. After inoculation and 7 days of growth, the biotransformation capacity of *S. alba* cell suspension cultures was evaluated qualitatively (at the end of the exponential growth phase). Three assays were performed with every substrate.

After biotransformation experiments (substrates in contact with *S. alba* cell suspension for 7 days), the cells and culture medium were extracted separately. The cells were frozen, ground, adjusted to pH 10 with ammonia, extracted first with ethyl acetate and then with methylene chloride. The medium was extracted in the same way. Subsequently, the cells and the medium were separately hydrolyzed in reflux with 5% sulfuric acid, then the acidic solution was alkalized with ammonia and extracted as above.

The biotransformation metabolites and residual substrates were examined on thin layer chromatographic plates by UV, light Dragendorff's and Gibb's reagents. Their characterization and identification were effected in different ways:

- comparison of Rf values and color characteristics in TLC with those of substrates and all compounds which have a strong probability of formation as metabolites and which we consequently synthesized;
- analysis of crude extracts by gas chromatography-mass spectra coupling (GC-MS) (70 eV, OV1 capillary column);
- extraction of alkaloid constituents from crude extracts, isolation on silica gel column chromatography, purification on preparative TLC silica gel, followed by proton NMR and mass spectrometries.

For quantitative appreciation, in the case of papaverine, densitometry was performed from the TLC plates after elution and immersion of the plates in Dragendorff's reagent; the color intensity of the alkaloid spots was estimated using a blue filter.

We observed only various metabolic oxidation reactions with all of the three substrates used with *S. alba* cell suspensions (Fig. 4). With papaverine **1** and isopapaverine **2**, the metabolites extracted from the culture medium result in benzylic oxidation (papaveraldine **4** and isopapaveraldine **5**), in N-oxidation (papaverine and isopapaverine N-oxides **6** and **7**), and in O-demethylation, in which the 6 and 4' monodemethyl derivatives (**8** and **9**) and 7, 3', and 4' monodemethyl compounds (**10**, **11**, and **12**) are predominantly formed from papaverine

1 and isopapaverine **2**, respectively; the 3′ monodemethylated compound is also formed from papaverine, but this separation is more difficult.

In the cells, the concentration of papaverine and isopapaverine metabolites is lower than in the medium. With papaverine, the papaveraldine and monohydroxylated derivatives are detectable and with isopapaverine, N-oxide and 7-demethylisopapaverine are especially evident.

Otherwise, with these two substrates, no metabolite was detected after acidic hydrolysis of cells and medium.

With 4-(4′-chlorobenzyl) isoquinoline **3** under the same experimental conditions, we showed the presence of two alkaloids in cell and culture medium extracts: non-metabolized isoquinoline **3** and a compound containing no chlorine, whose spectroscopic and chromatographic characteristics are in agreement with those of isopapaverine **2**. This derivative corresponds to 78% of the alkaloid fraction found in the cellular extract, but to only 38% of that found in the medium extract. By TLC, some traces of the 7-demethyl chlorobenzylisoquinoline **10** were identified. None of the complex metabolites were formed after acidic hydrolysis.

The yields of biotransformation of all three substrates studied are low. For example, the 6-monodemethyl papaverine **8**, the main compound of O-demethylation, represents less than 1% of papaverine **1** in *Silene* cultures (Verdeil et al. 1986; Bister-Miel et al. 1988). If the derivatives of monodemethylation are at a low concentration, they are preponderant in combined forms in a *Silene* suspension accustomed to papaverine (Bister-Miel et al. 1988); this suggests that the free monodemethyl form is transitional and that it combines when its concentration increases. For isopapaverine, the low concentrations of metabolites did not allow quantification.

An exception to these low yields is 4-benzyl isoquinoline **3** in isopapaverine **2** for which about 20% of the alkaloid substrate was converted (Agier et al. 1988).

Nevertheless, it should be noted that a concentration of only 22 mg/l of isoquinoline was used with the *Silene* cells, because of the toxicity of this alkaloid in cell suspensions, in contrast to the other two substrates, which have low toxicity, and for which concentrations of 400 to 800 mg/l can be used. In the case of isoquinoline **3**, a good yield results from a weak concentration of metabolite product **2**.

3 Summary and Conclusions

We have noted that for *Silene alba* cells, there are great differences in the toxicity of the alkaloid substrates, although they differ only minimally in their structure: isomerization of the benzyl group of position 1 to position 4, and replacement of a dimethoxy group by a chlorine atom. In this respect, it should be noted that in animals, the toxicity of chlorobenzylisoquinoline **3** is less than papaverine **1** and isopapaverine **2** (Prudhommeaux et al. 1971; Bouvier et al. 1976; Servin et al. 1985). The lower toxicity of isopapaverine than its chloro-derivative analog in *S. alba* cells shows that a detoxification process intervenes at the cellular level (Bister-Miel et al. 1988). This is also valid for the N-oxides and benzoylisoquinolines metabolites (Bister-Miel et al. 1988).

For oxidation reactions in biotransformation we observed that with papaverine and isopapaverine the major metabolites formed are O-demethylation derivatives. They are accompanied by N-oxides and benzoyl substrate derivatives. The benzyl secondary alcohol, papaverinol or isopapaverinol, was never evident as it was formed as a metabolite product in only one strain (*Thevetia nereifolia*) of the 19 strains that were chemically studied (Rideau et al. 1988), as was also the case with *Glycyrrhiza glabra* cells (Dorisse et al. 1988). Other strains of various species produced N-oxide and benzoyl derivatives of papaverine (Rideau et al. 1988), but to our knowledge, only *Silene alba* cells give O-demethylation metabolites with papaverine (and isopapaverine). These last reactions are analogous to those observed for papaverine and also to certain microorganisms (Rosazza et al. 1977), hepatic microsomia (Belpaire and Bogaert 1975), and in animal and human metabolism (Axelrod et al. 1958; Servin et al. 1985).

For the monohydroxylated metabolites, it can be stated that the accessibility of the 3′- and 4′-methoxy groups to enzymes is similar in papaverine **1** and isopapaverine **2**. In contrast, demethylation of the C-7 methoxy group of isopapaverine occurs predominantly, whereas in the case of papaverine it is the C-6 methoxy group. This fact can be explained by the conformation of the benzyl group in these two alkaloid molecules (Bouvier et al. 1976; Lejay et al. 1977; Christinaki et al. 1987). Thus, the possibility of rotation of a benzyl group around the isoquinoleic skeleton allows one to consider two conformers for papaverine

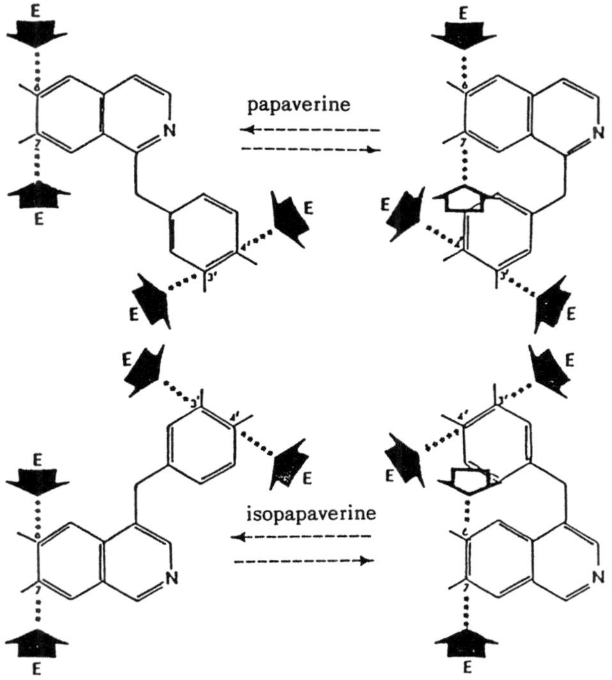

Fig. 8. Conformation scheme of the 1-benzyl- and 4-benzylisoquinolines. Scheme of the steric process which is liable to inhibit enzymatic attack in the 7 position of papaverine and in the 6 position of isopapaverine. ▲ Possible attack; △ difficult attack. (from Bister-Miel et al. 1988)

and isopapaverine (Fig. 8). In the one, the benzyl group is thrown to the nitrogen nucleus and the two methoxy in 6 and 7 are accessible to oxidative enzymes; in the other, the 7-methoxy in papaverine and the 6-methoxy in isopapaverine are hindered by benzyl at C-1 or at C-4 and are less accessible to oxidative enzymes.

The results obtained with papaverine and isopapaverine show that the oxidative enzymatic systems are not specific to a benzylisoquinolic molecular structure.

For 4-(4′-chlorobenzyl)isoquinoline **3** and its biotransformation, a few examples of aromatic nucleus deshalogenation have been reported in animals and plants (Garner 1976; Tissut and Severin 1984), and it appears to be probably the first example of a substrate deshalogenation by plant cells in vitro, this reaction being coupled with *ortho*-dimethoxylation of the aromatic nucleus. A mechanism was proposed which involves the formation of a transitory epoxide **14**, which leads either to an *ortho*-diphenol methylated in isopapaverine, or to a phenol consecutively methylated in isopapaverine **2** (Fig. 9).

The prospects offered by *Silene alba* cell cultures are interesting for the possible O-dealkylation of various alkaloid substrates, this reaction being possibly regio- and/or stereo-specific, according to the nature and configuration of the molecules studied.

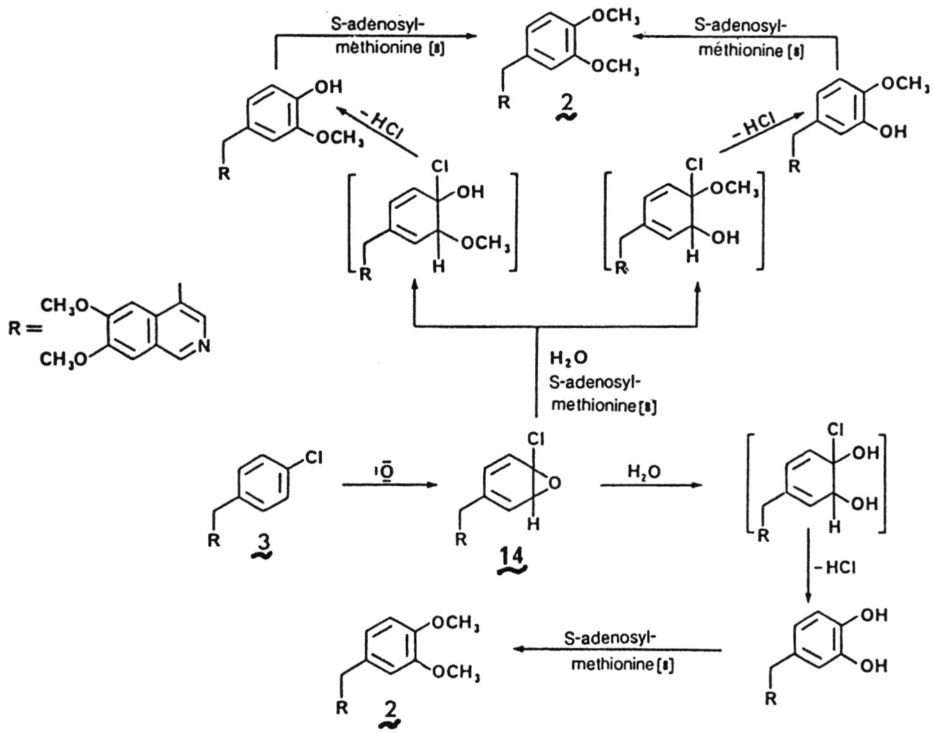

Fig. 9. Mechanisms proposed for the dehydrodechlorination of 4-(chlorobenzyl)isoquinoline **3**. (from Agier et al. 1988)

References

Agier C, Bister-Miel F, Guignard JL (1983) Study of the mean growth rate of *Silene alba* cells in suspension culture: effect of macroelement concentration in the medium. Bull Soc Bot Fr 130, Lett Bot (4–5): 291–300
Agier C, Bury M, El-Ech F, Cousson-Fonfraid S, Bister-Miel F, Guignard JL, Viel C (1988) Ortho-dimethoxylation and coupled dechlorination of an aromatic nucleus: a new reaction of plant cell suspension cultures. CR Acad Sci Paris Sér III, 306: 265–269
Agier C, Henry M, Bister-Miel F, Guignard JL (1982) Effect of boron on the starch catabolism in a cell suspension of *Silene alba* (Miller) E.H.L. Krause, CR Acad Sci Paris, Sér III, 294: 73–76
Axelrod J, Shofer R, Inscoe JK, King WM, Sjoerdsma A (1958) The fate of papaverine in man and other mammals. J Pharmacol Exp Ther 124: 9–15
Belpaire FM, Bogaert MG (1975) Metabolism of papaverine, II, Species differences. Xenobiotica 5: 421–429
Bister-Miel F (1987) Biotransformation of the papaverine, isopapaverine and analogs by non-alkaloid-producing vegetal cellular suspensions. PhD Thesis, University of Paris-Sud
Bister-Miel F, Agier C, Bury M, Postaire E, Guignard JL, Viel C (1986) Biotransformation of papaverine to papaveraldine by *Silene alba* cells. Effects of extraction techniques. Plant Med Phytothér 20: 3–7
Bister-Miel F, Agier C, Bury M, Viel C, Guignard JL (1988) Comparative biotransformations of benzylisoquinolines by non-alkaloid-producing cell suspensions. Bull Soc Bot Fr 135; Lett Bot (1): 57–70
Bister-Miel F, Guignard JL, Bury M, Agier C (1985) Compensatory glutamine effects on in vitro plant cell growth with phosphorus deficiency. Bull Soc Bot Fr 132; Lett Bot (1): 5–14
Bonnier G (1911–1935) Flore complète de France, Suisse et Belgique, vol 2. Librairie Générale de l'Enseignement, E Orlhac, Paris, p 39 and pl 78
Bonnier G, Douin R (1990) La grande flore en couleurs de Gaston Bonnier, Vol 1. Berlin, Paris, pl 78
Bouvier P, Branceni D, Prouteau M, Prudhommeaux E, Viel C (1976) Synthesis, pharmacological and conformational study of 4-benzyl isoquinoline analogs of papaverine. Eur J Med Chem 11: 271–278
Brassart C, Dubois J, Bouriquet R (1977) Carbohydrate nutrition of a cell suspension of *Silene alba* (Miller) E.H.L. Krause. CR Acad Sci Paris Sér D 285: 759–762
Brassart C, Dubois J, Bouriquet R (1977) Nitrogen nutrition of *Silene* cells in suspension culture (*Silene alba*) Miller E.H.L. Krause). CR Acad Sci Paris Sér D 287: 1373–1376
Brassart C, Dubois J, Bouriquet R (1982) Effects of gibberellic acid on the growth of *Silene alba*. Bull Soc Bot Fr 129; Lett Bot (4–5): 259–269
Brassart C, Eb P, Plichon B, Dubois J (1976) Use of antibiotics in cell cultures of *Silene* in the fermentor. CR Congr Natl Soc Savantes, Sect Sci 101: 455–463
Brassart C, Rambour S, Bouriquet R (1984) Glutamate dehydrogenase and glutamine synthetase activities in a cellular suspension of *Silene alba* grown in the presence of various nitrogen sources. Physiol Veg 22: 801–809
Christinaki H, Bister-Miel F, Hammoumi H, Bury M, Guignard JL, Viel C (1987) Biotransformation of isopapaverine by *Silene alba* cell suspensions. Phytochemistry 26: 2991–2994
Cronquist A (1981) An integrated system of classification of flowering plants. Columbia Univ Press, New York
Dorisse P, Gleye J, Loiseau P, Puig P, Edy AM, Henry M (1988) Comparison between two plant cell suspensions towards papaverine biotransformation. J Nat Prod (Lloydia) 51: 532–536
Dubois J (1975) Quantitative analysis of growth in a batch culture of *Silene alba* (Miller) E.H.L Krause. Bull Soc Bot Fr 122: 269–279
Dubois J, Rambour S, Vasseur J (1976) Growth of a cellular suspension of *Silene alba* (Miller) E.H.L. Krause and the variation of nucleic acids. CR Congr Natl Soc Savantes, Sect Sci 101: 441–454
Garner RC (1976) The role of epoxides in bioactivation and carcinogenesis. In: Bridges JW, Chasseaud LF (eds) Progress in drug metabolism, Vol 1. John Wiley, London, pp 77–128 (see pp 83–87)
Henry M, Benois M, Ly Thi B, Guignard JL (1978) The effect of naphtoxyacetic acid (NOA) on the growth of *Silene alba* (Miller) E.H.L. Krause cellular suspensions. Bull Soc Bot 125: 5–16
Hutchinson J (1973) The families of flowering plants. Clarendon Press, Oxford
Legrand B, Dubois J (1977) Development of peroxidases and auxin oxidases during the growth cycle of a *Silene alba* (Miller) suspension culture. CR Acad Sci Paris Sér D 285: 661–666

Legrand B, Dubois J (1978) Effect of growth regulators on proliferation, peroxidase activity and isoperoxidase in cell suspension of *Silene alba*. Biol Plant 20: 107–113

Lejay P, Viel C, Uchida-Ernouf G (1977) New analogs of papaverine: synthesis, pharmacological and conformational study of 4-benzylisoquinolines substituted in the 1 or 3 position. Farm Ed Sci 32: 3–24

Marie-Alsabek F, Morvan H, Dubois J, Bouriquet R (1984) Effect of pectic acid polymers released during plant cell culture on the growth of cell suspension cultures of *Silene*. CR Acad Sci Paris Sér III, 298: 301–304

Morvan C, Priem B, Verdus MC, Kwan JS, Morvan C (1987) Relationship between the acidity of the pectic acid polymers from suspension cultured cells of white campion and their promoting effects on the growth. Food Hydrocol 1: 501–503

Morvan C, Verdus MC, Feghali R, Morvan H (1987) Influence of culture age and calcium deficiency on the physicochemical properties and ultrastructural aspects of the cell wall of suspension cultured cells of white *Silene*. Bull Soc Bot Fr 134, Lett Bot (4–5): 323–336

Murashige T, Skoog F (1962) A revised medium for rapid growth and bioassays with tobacco tissue culture. Physiol Plant 15: 473–497

Nitsch JP, Nitsch C (1956) Auxin-dependent growth of excised *Helianthus tuberosus* tissues. Am J Bot 43: 839–851

Poulle M, Rambour S (1985) Regulation of nitrate reductase in suspension culture of *Silene alba*. Immuno-chemical approach. Plant Sci (Limerick, Irel) 40: 111–115

Prudhommeaux E, Viel C, Delbarre B (1971) Researches in isoquinoline series IX. Syntheses and properties of new 6,7-dimethoxy isoquinolines V. Benzyl derivatives substituted in position 2 or in position 4. Chim Thér 6: 358–366

Rambour S, Poulle M, Deniau JL, Brassart C (1983) Regulation of nitrate reductase activity in a cell suspension culture of *Silene alba*. Can J Bot 61: 3523–3528

Rambour S, Vasseur J, Dubois J (1977) Demonstration of polyadenylic RNA in *Silene alba* cell suspensions. CR Acad Sci Paris Sér D 284: 1063–1066

Rideau M, Morard P, Gansser C, Chénieux JC, Viel C (1988) Biotransformation and inductive behaviour of papaverine on various vegetal strains. Structure of compounds obtained. Pharmazie 43: 332–334

Rosazza JP, Kammer M, Youel L, Smith RW, Erhardt PW, Truong DH, Leslie SW (1977) Microbial models of mammalian metabolism O-demethylation of papaverine. Xenobiotica 7: 133–134

Servin A, Jacquot C, Lussiana JP, Viel C (1985) Metabolism of 6,7-dimethoxy 4-(*para*-chlorobenzyl) isoquinoline, I. Single dose pharmacokinetics in the rat and mouse. Xenobiotica 15: 553–566

Tissut M, Severin F (eds) (1984) Plantes, herbicides et desherbage. Association de coordination agricole, Paris, pp 115–119

Tutin TG, Heywood VH, Burges NA, Valentine DH, Walters SM, Webb DA (1964) Flora Europaea, vol 1. University Press, Cambridge

Verdeil JL, Bister-Miel F, Agier C, Plat M, Guignard JL, Viel C (1985) Difference in behavior of two vegetal cellular suspensions towards papaverine biotransformation. CR Acad Sci Paris Sér III, 300: 305–308

Verdeil JL, Bister-Miel F, Guignard JL, Viel C (1986) Papaverine biotransformation by *Silene alba* cell suspension. Planta Med 52: 1–3

XXIII *Solanum dulcamara* L. (Bittersweet): Accumulation of Steroidal Alkaloids in the Plant and in Different In Vitro Systems

A. EHMKE and U. EILERT[1]

1 Introduction

1.1 Biology and Distribution

Solanum dulcamara L. (= *Dulcamara flexuosa* Moench) (Fig. 1), known as dogwood or bittersweet (Solanaceae), is a clambering or prostrate, perennial shrub which may grow to a height of 2 m (Hegi 1927). Its stem is angular and woody with the exception of the herbaceous top and ranges in diameter between 0.25 and 2 cm, rarely up to 5–6 cm. The leaves are alternate, long-stalked, sparsely pubescent on both sides, and quite variable in shape. The oval- to egg-shaped leaf blade is pointed at the tip. Its base, however, may also be cordate, arrow-shaped, or may consist of one or two lobes. Different leaf forms may be found on the same plant. The flowers emerge axillary in panicle-like loose clusters. The calyx bears five narrow teeth; the five joint petals are bright purple and their tips are somewhat reflexed when fully expanded. The five stamens have yellow anthers which form a conspicuous column. The fruit is a round- to egg-shaped berry, green when young and becoming bright red when mature. In Europe, the flowering season is May to September. It is distributed throughout Europe and is also a native to North Africa, West Asia, India, the USSR, China, and Japan. It is not clear whether its occurrence in North America is of native origin or whether it was introduced from Europe. Growth habitats of the plant are wet meadows, ditches, and poplar-alder forests, but it may also grow on dry and sandy soils, waste areas, between gravel or rubble, and in clefts of old walls. The variability in morphological appearance, habitat, and its wide distribution gave rise to the description of several subspecies and varieties. Their taxonomic significance and position, however, is not quite clear and is further complicated by the occurrence of chemovarieties.

1.2 Steroidal Alkaloid Patterns

The three chemovarieties of *S. dulcamara* contain either solasodine, soladulcidine, or tomatidenol glycosides (Fig. 2; Table 1) in leaves and stems (Sander 1963; Willuhn 1966).

[1] Institut für Pharmazeutische Biologie der Technischen Universität Braunschweig, Mendelssohnstraße 1, W-3300 Braunschweig, FRG

Biotechnology in Agriculture and Forestry, Vol. 21
Medicinal and Aromatic Plants IV (ed. by Y.P.S. Bajaj)

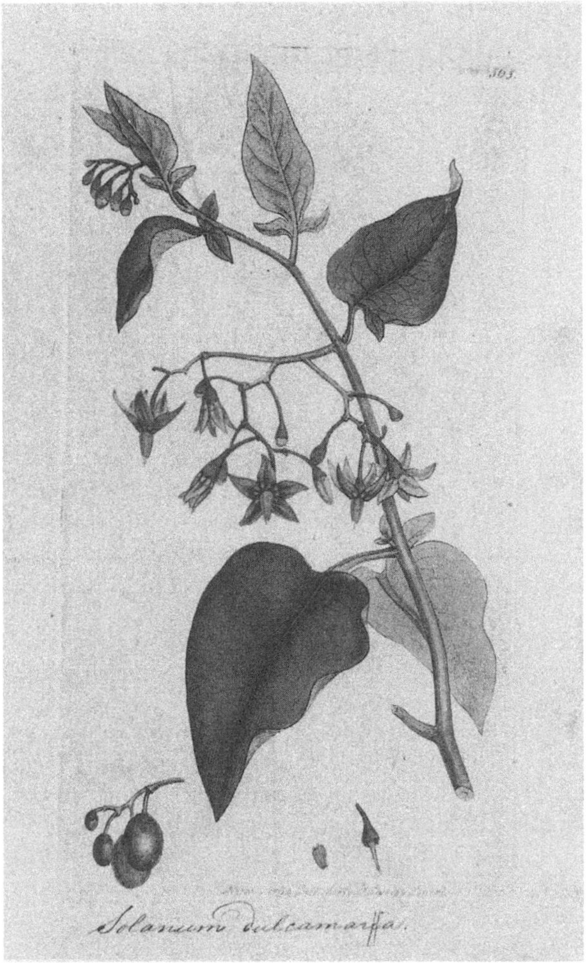

Fig. 1. *Solanum dulcamara* L. Flowering plant and berries. (English Botany, Sowerby London 1799)

The difference in alkaloid composition is a genetically fixed character of the plant (Sander 1963; Willuhn 1966; Mathé and Mathé 1979). The differences in alkaloid composition are restricted to the chlorophyllous vegetative organs. Fruits and flowers always contain solasodine glycosides in amounts comparable to the content of the type-specific alkaloid (Sander and Willuhn 1961; Willuhn 1967, 1968). Roots additionally contain the aglycones tomatidine and/or 15α-hydroxylated derivatives of solasodine, soladulcidine, tomatidenol, and tomatidine (Rönsch and Schreiber 1966; Willuhn and Kun-Anake 1970). The steroidal alkaloids are always accompanied by the biosynthetically closely related neutral sapogenins diosgenin, tigogenin, and yamogenin, which are nitrogen-free oxygen-analogous compounds.

The geographical distribution of the chemovarieties indicates a spatial separation: the tomatidenol-producing taxa are found in the humid Atlantic climate of western Europe, the soladulcidine type occurs in drier continental climates

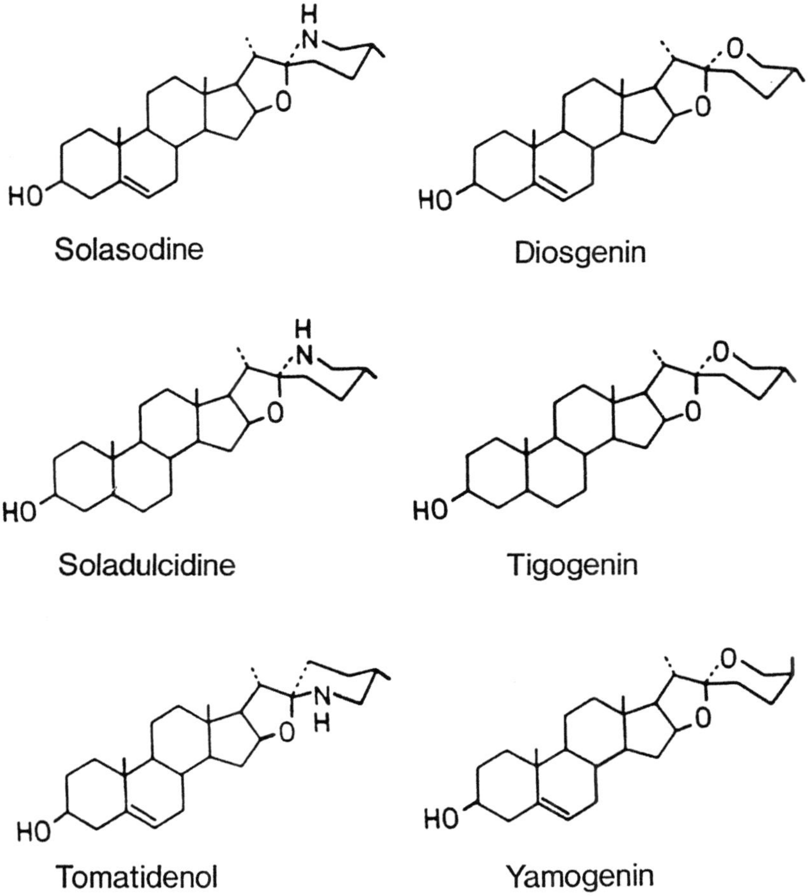

Fig. 2. The spirosolane alkaloids and the corresponding spirostanes in the green vegetative organs of the three chemovarieties of *Solanum dulcamara*

Table 1. Steroid alkaloid glycosides in *S. dulcamara*

Aglycon	Glycoside	Sugar moiety
Solasodine	Solasonine	Solatriose
	Solamargine	Chacotriose
Soladulcidine	Soladulcidine-tetraoside	Lycotetraose? D-Gal, 2 D-Glu, D-Xyl
Tomatidenol	α-Solamarine	Solatriose
	β-Solamarine	Chacotriose
	γ-Solamarine	Chacobiose L-Rha-(α 1 → 4) D-Glu
	γ^1-Solamarine	L-Rha-(α 1 → 2) D-Glu
	δ-Solamarine	D-Glu-(β 1 → 3) D-Gal

(Sander 1963; Mathé and Mathé 1978). Plants of the solasodine variety are comparatively rare. Along the boundary of the areals, populations of *S. dulcamara* exist which contain all three steroidal alkaloids (Willuhn 1966).

1.3 Medicinal Uses

Solanum dulcamara plays a role in folk medicine and as a drug of limited importance in Europe, China, and Japan. In European folk medicine, the dried stems are used separately or in combination with other drugs for the preparation of teas for "blood purification" (List and Hörhammer 1979). The drug is used against bronchitis and asthma. Decoctions are applied externally to treat eczema, ulcers, and to improve healing of wounds.

In India, *S. dulcamara* (*kakmachi, ruba-barik*) is used for the treatment of tumors, warts, chronic rheumatism, and skin affections. Infusions made from the dried stem are credited with sedative and analgesic properties (Ambasta 1986).

In Chinese medicine, *S. dulcamara* is known under the name *ku qie* and infusions of the dried stem are used for similar purposes as in Europe and India, e.g., to treat asthma and rheumatism. Vinegar-marinated berries and decoctions of the root are applied externally to cancerous sores and other swellings (Duke and Ayensu 1985). Hsu (1986) lists in his *Materia Medica* the drug Solani lyrati herba which is the entire dried plant of *S. lyratum* Thunb. (pai-mao-teng) or of *S. dulcamara* L. Therapeutical applications of the drug are used for malaria, jaundice, rheumatalgia, ascites, edema, furuncle, and erysipelas. Both *Solanum* species look very similar, and in the *Index Kewensis* of 1885 they are considered to be identical. In the *Flora Reipublicae Popularis Sinicae* of 1978 it is listed with the synonyms *S. dulcamara* L. var. *pubescens* Blume and *S. dulcamara* L. var. *lyratum* (Thunb.) Sieb et Zucc. ex Bonati in Lecte. A reassessment of its position, however, seems to be necessary.

1.4 Pharmacological and Biological Activities of Steroidal Alkaloids

Attention should be paid to the fact that the external application of herbal medicines containing steroidal alkaloids against skin disorders is common throughout all cultures. A patent application by Spencer et al. (1982) shows that this application proves true in modern medical therapy. Steroidal alkaloids are considered to be the active compounds in the drug. As early as 1965, Kupchan et al. demonstrated a tumor-inhibiting activity of the glycoside β-solamarine. Cham et al. (1987) report antitumor effects of the glycoalkaloid fraction of *S. sodomaeum*, which consists mostly of solasonine and solamargine. The experiments were performed with mice (Herston Whites) and with rats (Sprague Dawley) using Sarcoma 180 cells. Bähr and Hänsel (1982) found immunomodulating properties of the aglykone solasodine in an in vitro assay using spleen cell cultures. A patent application on preparations with antitumor activity of *S. lyratum* by a Japanese research group (Ushio et al. 1988) indicates the interest and expectations set in this activity. Strong antiviral activity of the steroidal glycoalkaloids α-chaconine, α-solasonine, and α-tomatine was shown by Thorne et al. (1985) using herpes-simplex infection of in vitro cultured cells as a test system.

Toxic effects of steroidal alkaloids on livestock and man were repeatedly reported and experimentally confirmed. A teratogenic activity was assigned to these compounds (Keeler 1975). In laboratory experiments, a single dose of solasodine (1.2 g/kg) given to pregnant hamsters produced spina bifida, exencephaly, and cranial bleb. Tomatidine and diosgenin were not teratogenic (Keeler et al. 1976). Steroidal alkaloids bearing a basic nitrogen atom in ring F, shared or unshared with ring E, with bonding capabilities in α-position to the steroid plane may be teratogenic (Gaffield and Keeler 1984).

Glycoalkaloids are long known to be inhibitors of acetylcholin-esterase (Roddick 1989). In this investigation, α-chaconine and α-solanine were the strongest inhibitors, tomatine was already less active, and solasonine and solamargine only showed a weak inhibitory activity. The aglycones were basically inactive.

Beside the pharmacological effects listed, steroidal alkaloids possess a number of additional biological activities such as antifungal, moluscicidal, and insecticidal properties. All these effects are most likely based on the saponin-like structure, in particular of the glycoalkaloids which will allow strong interactions with membranes, thus inflicting membrane damage (Roddick 1987). The membrane-disrupting properties of solasonine and solamargine were studied thoroughly in artificially prepared phosphatidylcholine/cholesterol liposomes (Roddick et al. 1990).

In the 1960s, tomatine gained interest as an externally applied antimycotic in the USA, but it did not prove to be effective (Wolters 1965a). Wolters investigated the antifungal activities of aqueous leaf and fruit extracts (1964a, 1965a), isolated steroidal alkaloids (1964b), and neutral saponins (1965b) from *S. dulcamara* and studied structure activity relationships. Steroidal alkaloids inhibited the growth of a number of different fungi. Against some fungi the aglycones were more active than the glycosides, but generally the spirosolanole glycosides were more active than the respective aglycones. The aglycones soladulcidine, solasodine, tomatidine, and Δ^5-tomatidenol showed some differences in their spectrum of activity. Compared to the steroidal alkaloids, the corresponding O-analogous neutral steroid sapogenins exhibited a similar spectrum of activity, but were less active when the minimal inhibitory activity was compared. A recent screening of the antifungal activity of a number of steroidal alkaloids showed strong differences between their biological activity. Minor changes in the side chain strongly affect the activity (Kusano et al. 1987a). The mode of action is still insufficiently elucidated, but complex formation with membrane sterols plays an important role in it (Roddick 1987). Structure activity studies, however, do not correlate with the sterol-binding activity (Roddick and Rijnenberg 1986). When investigating the effect of steroidal alkaloids (including solasodine) on cholesterol biosynthesis, Kusano et al. (1987b) measured a significant inhibition of the enzymatic conversion of dihydrolanosterol into cholesterol. They discuss an inhibition of sterol synthesis as a possible mechanism for fungitoxicity.

A role as a disease-resistance factor of plants is discussed for the steroidal alkaloids. The data available, however, lead to the conclusion that they are not key factors in the defence against fungal attacks (Roddick 1987).

1.5 Field Production

The growing interest in plants containing solasodine for commercial solasodine production led to the exploitation of many *Solanum* species in this respect. Various aspects of steroidal alkaloid production of *Solanum laciniatum* and *Solanum aviculare* in the field and in vitro were reviewed by Macek (1989). Also see Chapter XXIV (this Vol.) for *Solanum glaucophyllum*. Solasodine (and the C-22 epimer tomatidenol) can easily be converted to key intermediates for the synthesis of corticosteroids and contraceptive hormones. Mathé et al. (1986) suggested the cultivation of the tomatidenol/solasodine variety of *S. dulcamara* in temperate climates as an alternative to the subtropical species *S. aviculare*/*S. laciniatum* or *S. khasianum*. *S. dulcamara* is easily cultivated and produces tomatidenol in yields (15–45 kg/ha) comparable to or even better than the subtropical species under the same conditions in Hungary.

2 In Vitro Culture

There are only few reports on in vitro cultivation of *Solanum dulcamara*. The main interest was the production of steroids in different in vitro systems (Willuhn and May 1982; Bhatt et al. 1986; Ehmke and Eilert 1986). In addition in vitro experiments aiming at regeneration of plants from anthers (Zenkteler 1972), experiments with cultures from leaf disks and cell colonies (Bhatt et al. 1979, 1986) and from protoplasts (Binding and Nehls 1977) were conducted. Chand et al. (1989) used protoplasts for studies on electroporation aided transformation with *Agrobacterium tumefaciens*.

2.1 Initiation of Callus Cultures

Callus cultures were initiated from various explants like cultured shoots (Binding and Nehls 1977), leaves and stems of garden-grown plants, or hypocotyls of sterile seedlings (Willuhn and May 1982; Ehmke and Eilert 1986). Basal media employed were MS (Murashige and Skoog 1962) and Eriksson medium (Wetter and Constabel 1982). An auxin supplement of either 2,4-D (1 mg/l) or NAA (1 mg/l) with or without kinetin as cytokinin supported good callus formation. Willuhn and May (1982) reported the necessity of an addition of coconut milk (12%) to initiate and sustain callus growth. In our experiments, in contrast, no further requirements were noted.

We easily obtained suspension cultures by transfer of friable callus to liquid growth medium and incubation on a gyratory shaker (Ehmke and Eilert 1986). A heterotrophic and a mixotrophic callus cell line gave rise to respective suspension cultures. The heterotrophic one showed vigorous growth and consisted of single cells and small, pale yellow cell aggregates. The mixotrophic line grew slowly and formed compact bead-like aggregates consisting of parenchymatous cells and tracheidal elements. Chlorophyll formation varied, increasing with the size of the aggregates. Both lines have maintained their characteristics over a period of years.

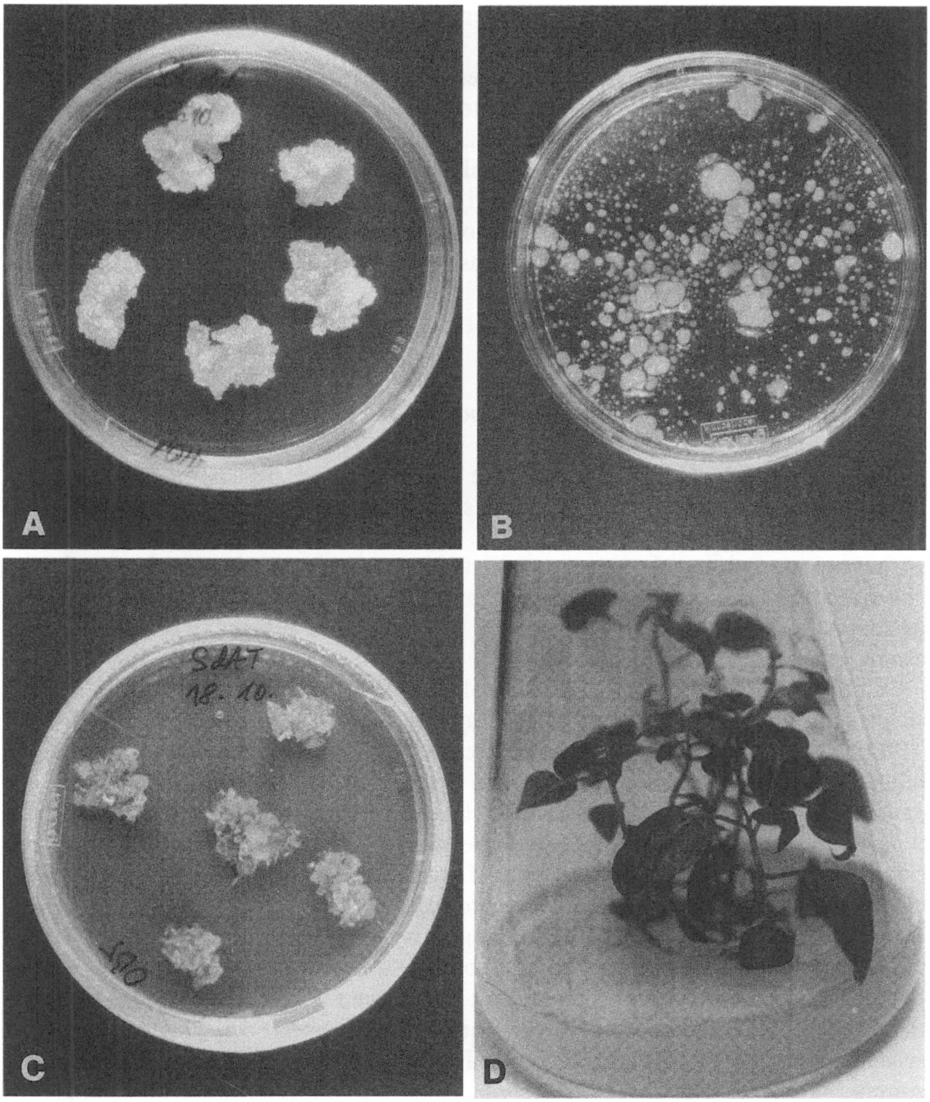

Fig. 3A–D. *Solanum dulcamara* L. **A** 1-week-old amorphous friable callus. **B** Mixotrophic suspension culture (transferred to a Petri dish for better demonstration of the bead-like structures). **C** 2-week-old differentiating tumor cultures transformed by *A. tumefaciens*. **D** Regenerated plants on a hormone-free medium

2.2 Plant Regeneration

Solanum dulcamara has a high potential for regeneration. Plants were regenerated from seedling leaves on growth medium containing 1 mg/l IAA and 1 mg/l kinetin. This hormone concentration effected the induction of leaf buds (Zenkteler 1972). Bhatt et al. (1979) obtained regeneration of leaf disks with a 100% yield on various media supplemented with 1 or 10 μM benzylaminopurin (BA) and with or without IAA.

Suspension-cultured cells were successfully regenerated on media with 0.1 or 1 μM BA (Bhatt et al. 1986) and protoplasts on media with 2.5 or 5 μM BA (Binding and Nehls 1977). Chand et al. (1988) observed a stimulation of plant regeneration after electroporation of cell suspension protoplasts of *S. dulcamara*. This resulted in better tissue growth and increased morphogenesis of such tissues. A tendency of spontaneous regeneration of roots, and eventually plants, was observed for callus cultures when transferred to MS medium without growth hormones (Ehmke and Eilert 1986).

2.3 In Vitro Production of Steroidal Saponins and Steroidal Alkaloids

Willuhn and May (1982) reported the occurrence of the sapogenin tigogenin accompanied by traces of the alkaloid soladulcidine in callus cultures of *S. dulcamara* which were grown in the dark. The presence of the sterols cholesterol, 24-methylenecholesterol, campesterol, stigmasterol, isofucosterol, sitosterol, and 4,4-dimethylsterols was verified by GC and GC-MS analysis. Sitosterol proved to be the main compound. As in the plant, the sterols are present in the free form as esters, glucosides, and acylglycosides.

Changes in the contents of free sterols, steryl esters, and steroidal alkaloids in suspension cultures as compared to plant organs were detected by Bhatt et al. (1986). The contents of steryl esters increased 15 days after transfer and remained high during the stationary phase. The concentration of steroidal alkaloids increased during the growth phase but started to decline at day 9 after transfer. Solasodine, only a minor alkaloid in the plant, became the main alkaloid in cell suspension cultures. Alkaloid contents of regenerated plants from this culture were compared with those of plants being propagated from seeds of the source plant. The regenerants showed a strong heterogeneity in alkaloid level but their average alkaloid level (5.4 mg/g dry wt.) was about 50% higher.

Our experiments (Ehmke and Eilert 1986) were directed at investigating the relationship between alkaloid formation and differentiation. Different in vitro systems of a soladulcidine-producing source plant were analyzed for their alkaloid content (Table 2). The fine heterotrophic suspension formed only the sapogenins diosgenin and tigogenin, but no alkaloids. The green mixotrophic culture, in addition, produced the 25-R alkaloids soladulcidine and solasodine, the latter not being present in the leaves of the source plant (Table 2). A positive correlation existed between aggregate size, chlorophyll content, and alkaloid level. In contrast to the results by Bhatt et al. (1986), the alkaloid level remained fairly constant throughout the growth cycle (Fig. 4). With increasing differentiation the alkaloid content also increased. Regenerated plantlets accumulated alkaloids in the same amount and pattern as the source plant. As shoot formation seems to be of primary importance for alkaloid production, we started to develop shoot cultures by transformation of leaf disks with *Agrobacterium tumefaciens* according to the method of Horsch et al. (1985) to study this effect more closely. Tumors and teratoma cultures were obtained; preliminary investigations show quite a complex

Table 2. The effect of culture conditions and differentiation on steroidal alkaloid accumulation; mean values of at least three replicate determinations, each consisting of ≥ 8 bulked samples.

Culture	Culture conditions	Alkaloid concentration (mg/g dry wt.)	Alkaloids found
Fine yellow suspension culture	Dark	–	–
Fine yellow suspension culture	Light (16 h)	–	–
Green suspension culture	Light (16 h)	0.095	Soladulcidine Solasodine
Green callus	Light (16 h)	0.18	Soladulcidine Solasodine
Green callus with rootlets	Light (16 h)	0.35	Soladulcidine Solasodine
Regenerated plants on hormone-free medium (shoots)	Light (16 h)	1.05	Soladulcidine (Solasodine)
Potted regenerated plants (shoots)	Light (16 h)	1.6	Soladulcidine

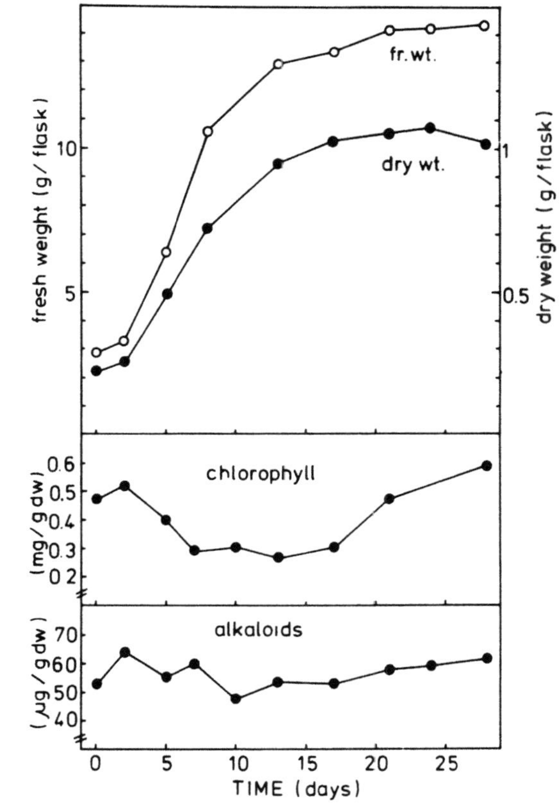

Fig. 4. Growth, chlorophyll and steroidal alkaloid content in green cell suspension cultures as a function of culture time (Ehmke and Eilert 1986)

alkaloid profile. Additionally, the alkaloid concentration was generally higher in these transformed cultures.

2.4 Extraction and Identification of Saponins and Steroidal Alkaloids

In plants and in cell cultures, the steroidal alkaloids occur as glycosides. For the isolation of glycoalkaloids and neutral saponins, freeze-dried tissue is extracted with 3% acetic acid (or 0.2 N HCl). The crude glycoalkaloids can be precipitated by the addition of NH_4OH; extraction of the supernatant with butanol yields the saponin fraction.

The direct identification of glycoalkaloids (e.g., solasonine, solamargine) is possible by HPLC (Elridge and Hockridge 1983; Hunter and Heftmann 1983; Cham and Wilson 1987) in combination with fast atom bombardment (FAB) or thermospray mass spectrometry (Price et al. 1985; Ehmke et al. 1987). For the sensitive and specific determination of solasodine glycosides in minute samples, Weiler et al. (1980) established a radioimmunoassay. This can be a valuable method for the screening for solasodine-rich cell culture strains, and regenerated plants, but due to the high specificity, it is not possible to detect the accompanying sapogenins.

Most methods, however, determine the aglycones after hydrolysis of the glycosides. Here again, HPLC is useful for the separation and quantification of the steroidal alkaloids and sapogenins (Crabbe and Fryer 1980; Ehmke and Eilert 1986). One problem of the HPLC method is the relatively low sensitivity due to the lack of chromophores. The sensitivity problem is overcome by the use of high-resolution GC (van Gelder 1985), but here the relatively low volatility of the compounds is disadvantageous.

For a rapid preliminary investigation of the alkaloid composition in plants and cell cultures, the well-established TLC methods prove to be valuable (Rönsch and Schreiber 1967; Hunter et al. 1976; Willuhn and May 1982). Estimation of solasodine concentration is often done by colorimetric procedures with bromthymol blue (Coxon et al. 1979) or methyl orange (Pandeya et al. 1981; Crabbe and Fryer 1982) after direct hydrolysis of plant material.

3 Conclusion

The importance of *Solanum dulcamara* as a source of steroidal alkaloids for the production of steroid hormones is limited. Only the tomatidenol and solasodine varieties may be of some interest; in general, the yield will not compare to the one obtainable with some other *Solanum* species, e.g., *Solanum laciniatum*. If its steroidal alkaloid production could be boosted by any specific manipulation, *Solanum dulcamara* may become more attractive for commercial growing due to its other qualities, fair cold hardiness, good growth on poor soils and perennial life cycle. These properties may also be valuable for breeding efforts aiming at the production of hybrid *Solanum* with unique qualities. Here, *Solanum dulcamara* may be of particular interest, as it shows a high potential for plant regeneration. The most fascinating aspects in working with *S. dulcamara*, however, are not in the application, but concern basic research questions. The developmentally regulated biosynthesis of steroidal alkaloids in *Solanum dulcamara* represents an exciting system for studies on differential gene expression in secondary metabolism.

4 Protocol

4.1 Establishment of Tissue Culture

Seed or stem segments are surface sterilized in 10% sodium hypochlorite for 10 min and washed three times with sterile distilled water. Seeds are germinated in the dark in Petri dishes on moist filter paper under aseptic conditions. Seedling hypocotyl explants or stem segments are placed on MS medium with 1 mg/l 2,4-D, 0.4 mg/l kinetin and 3% sucrose, solidified with 1% agar in Petri dishes and cultured either in the dark or with 16-h photoperiods at 25–26 °C. After about 5 weeks, calli are isolated and subcultured every 4 weeks while maintaining the conditions for callus initiation.

4.2 Establishment and Maintenance of Suspension Cultures

Take soft, friable callus and transfer about 2 g into 500 ml MS medium with 0.1 mg/l 2,4-D, 0.25 mg/l kinetin and 3% sucrose in 250-ml flasks. Shake at 100–130 rpm on a gyratory shaker. It is possible to select fast-growing heterotrophic cell lines and slower-growing green mixotrophic cell lines.

Subculture the suspensions every 12 days (heterotrophic culture) or 16–18 days (mixotrophic culture).

4.3 Regeneration of Plants

To induce plantlet development, place pieces of callus on MS medium with 1 to 2% sucrose without growth hormones.

Regeneration of protoplasts and suspension cells is achieved by transfer to media with 0.1 to 5 µmol/l benzylaminopurin (Binding and Nehls 1977; Bhatt et al. 1986).

4.4 Isolation and Detection of Steroidal Alkaloids and Sapogenins

Freeze-dried tissue is finely ground and extracted three times with 0.2 N HCl. The combined extracts are hydrolyzed with 1 N HCl (final concentration) at 95 °C for 2 h under reflux. After alkalinization (pH 9) with 25% NH_4OH, extract three times with chloroform. The concentrated chloroform phase can be used for qualitative and quantitative determination of the aglycones.

An estimation of the aglycone composition can be done by TLC on silica gel with the following solvent system: $CHCl_3$/MeOH 95:5 (v/v), but notice that the Δ^5-saturated and unsaturated compounds cannot be separated. Rf values: solasodine 0.17, soladulcidine 0.18, tomatidine 0.29, tomatidenol 0.29, diosgenin 0.5, tigogenin 0.5. Detection: anisaldehyde/H_2SO_4.

References

Ambasta SSP (1986) The useful plants of India. Publication and Information Directorate, CSIR, New Delhi

Bähr V, Hänsel R (1982) Immunomodulation properties of 5,20α(R)-dihydroxy-6α,7α-epoxy-1-oxo-(a5α)-witha-2,24-dienolide and solasodine. Planta Med 44: 32–33

Bhatt PN, Bhatt DP (1984) Regulation of sterol biosynthesis in *Solanum* species. J Exp Bot 35: 890–896

Bhatt PN, Bhatt DP, Sussex IM (1979) Organ regeneration from leaf disks of *Solanum nigrum*, *S. dulcamara* and *S. khasianum*. Z Pflanzenphysiol 95: 35–362

Bhatt DP, Bhatt PN, Mehta AR (1986) Steroid analysis and plant regeneration from cell suspension cultures of *Solanum dulcamara* L. Beitr Biol Pflanz 61: 203–213

Binding H, Nehls R (1977) Regeneration of isolated protoplasts to plants in *Solanum dulcamara* L.. Z Pflanzenphysiol 85: 279–280

Cham BE, Wilson L (1987) HPLC of glycoalkaloids from *Solanum sodomaeum*. Planta Med, pp 59–62

Cham BE, Gilliver M, Wilson L (1987) Antitumour effects of glycoalkaloids isolated from *Solanum sodomaeum*. Planta Med, pp 34–36

Chand PK, Ochatt SJ, Rech EL, Power JB, Davey MR (1988) Electroporation stimulates plant regeneration from protoplasts of the woody medicinal species *Solanum dulcamara* L. J Exp Bot 39: 1267–1274

Chand PK, Rech EL, Golds TJ, Power JB, Davey MR (1989) Electroporation stimulates transformation of freshly isolated cell suspension protoplasts of *Solanum dulcamara* by *Agrobacterium*. Plant Cell Rep 8: 86–89

Coxon DT, Price KR, Jones PG (1979) A simplified method for the determination of total glycoalkaloids in potato tubers. J Sci Food Agric 30: 1043–1049

Crabbe PG, Fryer C (1980) Rapid quantitative analysis of solasodine, solasodine glycosides and solasodiene by high-pressure liquid chromatography. J Chromatogr 187: 87–100

Crabbe PG, Fryer C (1982) Evaluation of chemical analysis for the determination of solasodine in *Solanum laciniatum*. J Pharm Sci 71: 1356–1362

Duke JA, Ayensu ES (1985) Medicinal plants of China, vol 2. Reference Publ Inc, Algonac, p 611

Ehmke A, Eilert U (1986) Steroidal alkaloids in tissue cultures and regenerated plants of *Solanum dulcamara*. Plant Cell Rep 5: 31–34

Ehmke A, Schiebel H-M, McDowell M (1987) Liquid chromatography mass spectrometry. On-line and off-line coupling for the analysis of *Solanum* glycoalkaloids. Pharmaceut Weekbl Sci Ed 9: 232 (Abstr)

Elridge AC, Hockridge ME (1983) High-performance liquid chromatographic separation of eastern black nightshade (*Solanum ptycanthum*) glycoalkaloids. J Agric Food Chem 31: 1218–1220

Gaffield W, Keeler RF (1984) Structure and stereochemistry of steroidal amine teratogens. Adv Exp Med Biol 177: 214

Hegi G (1927) Illustrierte Flora von Mitteleuropa V (4). JF Lehmanns Verlag, München, pp 2589–2592

Horsch RB, Fry JE, Hoffmann NL, Eichholtz D, Rogers SG, Fraley RT (1985) A simple and general method for transferring genes into plants. Science 228: 1229–1231

Hsu HY (1986) Oriental materia medica, a concise guide. Oriental Healing Arts Inst, Long Beach, p 437

Hunter IR, Heftmann E (1983) Isolation of glycoalkaloids with the chromatotron and their determination by high-performance liquid chromatography. J Liquid Chromatogr 6: 281–289

Hunter IR, Walden MK, Wagner J, Heftmann E (1976) Thin-layer chromatography of steroidal alkaloids. J Chromatogr 118: 259–262

Hunter IR, Walden MK, Heftmann E (1980) High-performance liquid chromatography of *Solanum* and *Veratrum* alkaloids. J Chromatogr 198: 363–366

Keeler RF (1975) Toxins and teratogens of higher plants. J Nat Prod 38: 56–86

Keeler RF, Young S, Brown D (1976) Spina bifida, exencephaly, and cranial bleb produced in hamsters by the *Solanum* alkaloid solasodine. Res Commun Chem Pathol Pharmacol 13: 723–730

Kupchan SM, Barboutis SJ, Know JR, Lau Cam CA (1965) Beta-solamarine: tumor inhibitor isolated from *Solanum dulcamara*. Science 150: 1827–1828

Kusano G, Takahashi A, Sugiyama K, Nozoe S (1987a) Antifungal properties of *Solanum* alkaloids. Chem Pharm Bull 35: 4862–4867

Kusano G, Takahashi A, Nozoe S, Sonoda Y, Sato Y (1987b) *Solanum* alkaloids as inhibitors of enzymatic conversion of dihydrolanosterol into cholesterol. Chem Pharm Bull 35: 4321–4323

List P, Hörhammer L (1979) Hagers Handbuch der pharmazeutischen Praxis, VI b. Springer, Berlin Heidelberg New York, pp 437–439

Macek TE (1989) *Solanum aviculare* Forst., *Solanum laciniatum* Ait (poroporo): in vitro culture and the production of solasodine. In: Bajaj YPS (ed) Biotechnology in agriculture and forestry, vol 7. Medicinal and aromatic plants II. Springer, Berlin Heidelberg New York Tokyo, pp 443–467

Mathé I Jr, Mathé I (1978) Steroids of *Solanum dulcamara* L. and their variability. Herba Hung 1: 39–55

Mathé I Jr, Mathé I (1979) Variation in alkaloids in *Solanum dulcamara* L. In: Hawkes JG, Lester RN, Skelding AD (eds) The biology and taxonomy of the solanaceae. Acad Press, London, pp 211–222

Mathé I Jr, Mathé I, Botz L, Koch L (1986) Possibilities of *Solanum* alkaloid production in european temperate zone. Proc 5th Int Symp Medicinal, aromatic and spice plants. Mungpp/Darjeeling. Acta Hortic 188: 193–201

Murashige T, Skoog F (1962) A revised medium for rapid growth and bioassays with tobacco tissue cultures. Physiol Plant 15: 473–497

Pandeya SC, Saratbabu GV, Bhatt AB (1981) A quick method for the estimation of solasodine in leaves and berries of *Solanum sysimbrifolium* Lamk. Indian J Exp Biol 19: 1207–1208

Price KR, Mellon FA, Self R, Fenwick GR, Osman SF (1985): Fast atom bombardment mass spectrometry of *Solanum* glycoalkaloids and its potential for mixture analysis. Biomed Mass Spectrom 12: 79–85

Roddick JG (1987) Antifungal activity of plant steroids. In: ACS Symp Ser 325. Ecology and metabolism of plant lipids. pp 286–303

Roddick JG (1989) The acetylcholinesterase-inhibitory activity of steroidal glycoalkaloids and their aglycones. Phytochemistry 28: 2631–2643

Roddick JG, Rijnenberg AL (1986) Effect of steroidal glycoalkaloids of potato on the permeability of liposome membranes. Physiol Plant 68: 436–440

Roddick JG, Rijnenberg AL, Weissenberg M (1990) Membrane-disrupting properties of the steroidal glycoalkaloids solasonine and solamargine. Phytochemistry 29: 1513–1518

Rönsch H, Schreiber K (1966) *Solanum* Alkaloide LXV. Über vier neue 15α-Hydroxy-spirosolan-Alkaloide aus Wurzeln von *Solanum dulcamara* L.. Liebigs Ann Chem 694: 169–182

Rönsch H, Schreiber K (1967) LXXXIII. Analytische und präparative dünnschichtchromatographische Trennung von 5α-gesättigten bzw. Δ^5-ungesättigten Steroidalkaloiden und -sapogeninen an $AgNO_3$-haltigen Adsorptionsschichten. J Chromatogr 30: 149–154

Sander H (1963) Chemische Differenzierung innerhalb der Art *Solanum dulcamara* L. Planta Med 11: 303–316

Sander H, Willuhn G (1961) Über *Solanum dulcamara* L. 5. Mitt. Isolierung von Tigogenin neben Soladulcidin aus reproduktiven Organen. Flora 151: 150–154

Spencer J, Coutler MF, Coutler P (1982) Topical salve for skin growth removal. Brit UK Pat Appl GB 2,0066,066

Thorne HV, Clarke GF, Suce R (1985) The inactivation of herpes simplex virus by some Solanaceae glycoalkaloids. Antiviral Res 5: 335–343

Ushio R, Murakami K, Sogawa K (1988) Antitumor extracts (steroidal saponins) of *Solanum lyratum*. Jpn Kokai Tokkyo Koho JP Chem Abstr 63,202,357

Van Gelder WMG (1985) Determination of the total C-27-steroidal alkaloid composition of *Solanum* species by high-resolution gas chromatography. J Chromatogr 331: 285–293

Weiler EW, Krüger H, Zenk MH (1980) Radioimmunoassay for the determination of the steroidal alkaloid solasodine and related compounds in living plants and herbarium specimens. Planta Med 39: 112–124

Wetter LR, Constabel F (1982) Plant tissue culture methods, 2nd edn. NRCC Prairie Regional Laboratory, Saskatoon

Willuhn G (1966) Untersuchungen zur chemischen Differenzierung bei *Solanum dulcamara* L. I. Genetische Fixierung der unterschiedlichen Steroidalkaloidführung. Planta Med 14: 408–420

Willuhn G (1967) Untersuchungen zur chemischen Differenzierung bei *Solanum dulcamara* L. II. Der Steroidgehalt in Früchten verschiedener Entwicklungsstadien der Tomatidenol- und Soladulcidin-Sippe. Planta Med 15: 58–73

Willuhn G (1968) Untersuchungen zur chemischen Differenzierung bei *Solanum dulcamara* L. III. Der Steroidalkaloidgehalt in den Früchten der Sosasodin-Sippe. Planta Med 16: 462–466

Willuhn G, Kun-Anake A (1970) Untersuchungen zur chemischen Differenzierung bei *Solanum dulcamara* L. V. Isolierung von Tomatidin aus den Wurzeln der Solasodin-Sippe. Planta Med 18: 354–360

Willuhn G, May S (1982) Triterpene und Steroide in Kalluskulturen von *Solanum dulcamara*. Planta Med 46: 153–158

Wolters B (1964a) Antibiotische Wirkung von *Solanum dulcamara* gegen Pilze. Naturwissenschaften 51: 111–112

Wolters B (1964b) Beziehungen zwischen Struktur und antibiotischer Wirkung bei einigen Steroidalkaloiden. Arch Pharm 297: 748–754

Wolters B (1965a) Die antibiotische Wirkung von *Solanum dulcamara* L. Pharm Z 110: 1718–1720

Wolters B (1965b) Der Anteil der Steroidsaponine an der antibiotischen Wirkung von *Solanum dulcamara*. Planta Med 18: 189–193

Zenkteler M (1972) In vitro formation of plants from leaves of several species of the Solanaceae family. Biochem Physiol Pflanz 163: 509–512

XXIV *Solanum glaucophyllum* Desf. (Duraznillo Blanco): In Vitro Culture and the Production of Steroidal Secondary Metabolites

M. WEISSENBERG[1], A. LEVY[2], and R.H. WASSERMAN[3]

1 Introduction

Solanum glaucophyllum Desf. is a calcinogenic plant growing wild mainly in South America, where its ingestion causes intoxication of grazing animals. Early work suggested that glycosides of various vitamin D_3 sterols occur in minute concentrations in the plant leaves and are responsible for a calcinotic disease of cattle prevalent in Argentina and Brazil, with symptoms resembling vitamin-D intoxication. The vast literature on this topic describing chemical, toxicological, and physiological aspects has been extensively reviewed in the last decade (Wasserman and Nobel 1980; Morris 1982; Boland 1986; Weissenberg 1989). A few communications reporting the occurrence of small amounts of other steroids such as β-sitosterol, diosgenin and solasodine, and also of phenolic derivatives (*vide infra*), have already been summarized (Weissenberg 1989). Since this chapter is concerned with the in vitro production of metabolites in *S. glaucophyllum*, the naturally occurring steroidal derivatives which have been found to occur also in vitro will be treated preferentially in the following survey.

1.1 Distribution and Importance of *S. glaucophyllum*

S. glaucophyllum is a deciduous, rhizomatous shrub (Fig. 1) growing wild typically in low-lying, wet land in the poorly drained eastern and central parts of the Buenos Aires province and in the northeastern provinces of Argentina (Okada et al. 1977), as well as in the Matto Grosso region of Brazil, in Uruguay and Paraguay (Cabrera 1965), and in India (Deb 1979). Accordingly, the incidence of the toxicose is greatest in flooded areas having pastures of higher feeding value than drier areas (Okada et al. 1977). The taxonomic keys available to identify *S. glaucophyllum* are summarized in Table 1. The plant has an extensive and deep underground root system (Fig. 2) which apparently confers considerable resistance to weed killers. Dispersal is probably by animals and birds that consume the berries or seeds. When a seedling is formed, vegetative propagation develops and a large array of

[1] Department of Chemistry of Natural Products and Pesticides, Agricultural Research Organization, The Volcani Center, Bet Dagan 50250, Israel
[2] Department of Medicinal, Spice and Aromatic Plants, Agricultural Research Organization, The Volcani Center, Bet Dagan 50250, Israel
[3] New York State College of Veterinary Medicine, Cornell University, Ithaca, NY 14853, USA

Biotechnology in Agriculture and Forestry, Vol. 21
Medicinal and Aromatic Plants IV (ed. by Y.P.S. Bajaj)

Fig. 1. Inflorescence of a young *S. glaucophyllum* plant cultivated in the experimental fields of the Agricultural Research Organization, The Volcani Center, Bet Dagan, Israel

S. glaucophyllum can thus be established. Fallen leaves retaining their toxicity for several months mix intimately with neighboring herbage on the underlying pasture (thus improving palatability) and are likely to be eaten accidentally by the grazing animal (Okada et al. 1977). Extensive information on the growth habit, distribution, and ecology of *S. glaucophyllum* has been presented in detail (Okada et al. 1977).

The economic relevance of *S. glaucophyllum* might be assessed in two ways: as the causal factor of the toxicosis of grazing cattle and, implicitly, as a potentially valuable source of vitamin D_3 sterols and other steroids, the latter aspect being the main subject of this survey. Conversely, the calcinotic disease of livestock induced by *S. glaucophyllum* ingestion causes considerable economic losses to the cattle

Fig. 2. Shoots and part of the root system of young *S. glaucophyllum* plants cultivated in the experimental fields of the Volcani Center, A.R.O., Bet Dagan, Israel

industry in Argentina and in Brazil (Worker and Carrillo 1967; Okada et al. 1977) and measures for field control of the toxicosis have been studied (INTA-FAO 1970), including the possibility of removal of *S. glaucophyllum* from pastures, mechanical eradication, treatment with herbicides, or better drainage of the area (Okada et al. 1977).

Table 1. Identification of *S. glaucophyllum*. (After Weissenberg 1989)

Character	*S. glaucophyllum* Desfontaines
Conspecific names	*S. malacoxylon* Sendtner
	S. glaucum Dunal
	S. glaucum Bertolini
	S. glaucescens Baile
	S. glaucumfrutescens Larrañaga
Common name	Duraznillo blanco (white little peach)
Height (m)	1–3
Leaf	
Size (cm)	(9–13) × (1–2)
Flower	
Corolla color	Blue – purple
Corolla size (mm)	20–30, across
Corolla lobes	Short
Fruit	
Color	Purplish – black
Size (mm)	7–10, across
Seed	Numerous, 5 mm long
Chromosome number (2n)	24

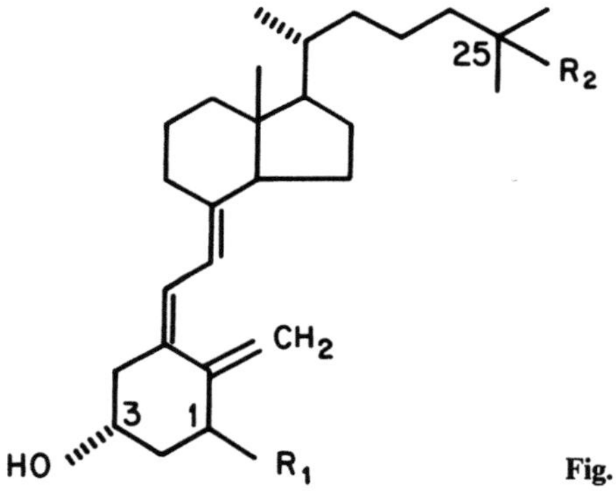

1a $R_1 = R_2 = H$

1b $R_1 = H$, $R_2 = OH$

1c $R_1 = R_2 = OH$

Fig. 3. Basic structure of vitamin D_3 and related sterols

1.2 Distribution of 1α,25-$(OH)_2D_3$ Activity in *S. glaucophyllum*

1.2.1 Composition and Content

Early research demonstrated that glycoside conjugates of 1α,25-$(OH)_2D_3$ (1α,25-dihydroxycholecalciferol, calcitriol) (Fig. 3, *1c*) occur in low concentrations in the plant leaves (Wasserman et al. 1976). Further investigation revealed the presence of three fructoglucosides of 1α,25-$(OH)_2D_3$ (Vidal et al. 1985), and also of vitamin D_3 (*1a*, cholecalciferol) and 25-OHD_3 (*1b*) in the aglycone portion of the glycosides (Esparza et al. 1982), thus suggesting that different vitamin D_3 sterols might be conjugated to various carbohydrate moieties or sequences, affording a series of naturally occurring calcinogenic glycosides (Haussler et al. 1976). The presence in this plant of enzymes involved in the conversion of vitamin D_3 into $1\alpha,25\text{-}(OH)_2D_3$ has been reported, inferring a hydroxylation pattern similar to that which is operative in animals (Esparza et al. 1982). Variable contents of 0.1–140 mg of $1\alpha,25\text{-}(OH)_2D_3$ per kg dry leaves were recorded from a large array of bioassays (see Weissenberg 1989 for a comprehensive review), along with a radioimmunoassay (RIA) employing a commercially available antibody (Weissenberg et al. 1988). The clinically useful hormone $1\alpha,25\text{-}(OH)_2D_3$ itself is a secosterol important in the regulation of vitamin D metabolism, calcium and mineral homeostasis, control of the immune system, bone remodeling and mineralization, action on cell proliferation and differentiation, and possibly anticancer action (Haussler 1986; Pike 1987).

1.2.2 1α,25-(OH)₂D₃ Activity Distribution and Extraction

The highest concentration of the active principle, as indicated by the induction of the synthesis of calcium-binding protein (CaBP), is found in the berries as glycosides, and their value is significantly greater than that of glycosides in leaves, stems, and roots (Table 2). The leaves have the lowest content of aglycones of the various extracts and significantly less than the parent glycosides level and than the aglycones in berries, whereas all other plant organs contain substantial amounts of the aglycone form of the active principle. Significant amounts of the glycoside derivatives of $1\alpha,25\text{-}(OH)_2D_3$ are present in each of the plant parts, and most of the $1\alpha,25\text{-}(OH)_2D_3$-like material in leaves and berries is in the form of the glycoside (Weissenberg et al. 1989). The exceedingly high ratio of glycosides to aglycones in leaves, which reportedly are a site of hydroxylation of the aglycones *1a* and *1b* (Esparza et al. 1982), might suggest that the glycosylation of the aglycones occurs in leaves in the presence of the appropriate glycosylating enzymes and/or glycosyl transferases; the resultant glycosides are subsequently translocated to the other organs for storage, the berries being apparently a preferred site of accumulation (Weissenberg et al. 1989). Substantial activity was also detected in the plant cell cultures by both CaBP and RIA (Weissenberg et al. 1988, 1989; see Sect. 2.3.1).

Extraction of plant glycosides with water at room temperature was found preferable to reflux with alcohols, and neither prolonged periods of reflux, nor reextraction, improved the yield (Weissenberg et al. 1988). For larger operations, the optimized extraction process was carried out with 70% aqueous ethanol (Weissenberg et al. 1990). Further purification of $1\alpha,25\text{-}(OH)_2D_3$ glycosides was done by ion-exchange chromatography of their borate complexes on Amberlite CG-400, which revealed the presence of three fructoglucosides (a nona-, penta-, and trisaccharide); yet their chemical or spectral characterization was not presented at the time (Vidal et al. 1985). The identification of the aglycones *1a–c* following hydrolysis of the glycosides fraction was reviewed in detail (Weissenberg 1989).

1.2.3 Conventional Cultivation of S. glaucophyllum

Optimal practices have been developed for large-scale cultivation of *S. glaucophyllum* as a potential vitamin D crop, based on studies of germination, vegetative

Table 2. Distribution of $1\alpha,25\text{-}(OH)_2D_3$ activity[a] in different parts of *S. glaucophyllum*. (Weissenberg et al. 1989)

Plant part[b]	Contribution to total dry plant weight	Crude CH_2Cl_2 extract (aglycones)[c]			Crude EtOH extract (glycosides)[d]			Ratio of glycosides/ aglycones extracts		Total extract (aglycones and glycosides)	
		Yield		CaBP[e]	Yield		CaBP[e]	Weight	CaBP	CaBP	
	(%)	(mg)	(%)	(μg/mg protein)	(mg)	(%)	(μg/mg protein)			(μg/mg protein)	(%)
Leaves	18.1	172	8.6	0.2 ± 0.2	269	13.4	2.5 ± 0.8	1.6	12.50	2.7	11.1
Berries	3.9	262	13.1	4.0 ± 1.4	372	18.6	9.0 ± 2.0	1.4	2.25	13.0	53.5
Stems	44.8	14	0.7	3.2 ± 1.1	252	12.6	1.2 ± 0.5	18	0.38	4.4	18.1
Roots	33.2	11	0.5	2.2 ± 0.8	215	10.7	2.0 ± 0.8	19.5	0.91	4.2	17.3

[a] Expressed as concentration of duodenal CaBP (calbindin-D_{28K}).
[b] The mean ± SE of four samples was 18.2 ± 1 μg CaBP/mg protein for the $1\alpha,25\text{-}(OH)_2D_3$-treated group and 0.61 ± 0.1 μg CaBP mg protein for the propylene glycol control.
[c] Originating from 2 g dry plant material.
[d] Originating from 2 g dry plant material previously extracted with CH_2Cl_2.
[e] Values represent the mean ± SE of eight samples.

propagation, accessions performance, and environmental factors (Weissenberg et al. 1987). The findings on activity distribution within the plant (Weissenberg et al. 1989) proved important for developing the appropriate cultivation practices needed to maximize the yield. In fact, although the berries possess the highest activity, their contribution to the total dry biomass yield is only 3.9% (see Table 2). The low berry yield of the species is a combined result of low flowering rate and fruit set on the inflorescences (0–40 berries per plant), and low dry weight of the berries (0.1–0.3 g/berry). In contrast, the stems and roots contribute substantially to the total biomass yield and plant activity (Table 2) and, in addition, are easier to process than the leaf material which is contaminated with pigments. Production conditions should therefore be directed toward the exploitation of the total biomass of *S. glaucophyllum* plants rather than their leaves alone, as done currently (the dry leaf yield accounts for ca. 3.3 t/ha at two seasonal harvests) (Weissenberg et al. 1989).

1.2.4 Clinical Uses and Market Demand

The pharmacological properties of the calcinogenic glycosides of *S. glaucophyllum* are presumably related to their main aglycone $1\alpha,25\text{-}(OH)_2D_3$. Under the trademark Rocaltrol and in doses of 0.5–2.5 μg/day, $1\alpha,25\text{-}(OH)_2D_3$ itself has proved effective in treatment of several disorders of bone and calcium metabolism in humans, such as hypocalcemia of hypoparathyroidism, hypophosphatemic osteomalacia, vitamin D-dependent rickets, and renal osteodystrophy (Meier 1981; Haussler and Cordy 1982; Simpson and Wishaar 1986). The as yet limited application of pharmaceutical preparations of unprocessed *S. glaucophyllum* leaves in human medicine, veterinary medicine, and animal husbandry has also been reviewed in considerable detail (Weissenberg 1989).

Steadily increasing use of $1\alpha,25\text{-}(OH)_2D_3$ in clinical and biochemical studies has led to the need for this compound in larger quantities and at reasonable prices. So far, $1\alpha,25\text{-}(OH)_2D_3$ and its derivatives have been prepared by expensive multistep syntheses (Pardo and Santelli 1985). However, the likelihood of using *S. glaucophyllum* as a field crop of vitamin D_3 sterols seems remote at this stage, although at the current sale price of about $900 per ,25 mg of synthetic $1\alpha,25\text{-}(OH)_2D_3$, it may provide a fair alternative, considering the aforementioned plant material yields (Weissenberg 1989). Furthermore, the reported plant content (20–140 mg of $1\alpha,25\text{-}(OH)_2D_3$ per kg dry leaves; Weissenberg et al. 1988) might conceivably be improved by appropriate cultivation practices, tissue culture techniques, and breeding studies. On the other hand, a concurrent approach was recently devised involving work with synthetic analogues of plant calcinogenic glycosides (Fürst et al. 1983), which would be of interest both for preparative and for physiological purposes (Londowski et al. 1986; Rambeck et al. 1986).

Considering the laborious and difficult isolation and purification of *S. glaucophyllum* calcinogenic glycosides, an alternative means of therapeutic application mainly in veterinary uses might arguably by offered by dose-controlled administration of the plant in powdered form or as crude extracts. In that fashion, *S. glaucophyllum* may represent a fairly inexpensive source of $1\alpha,25\text{-}(OH)_2D_3$ (Weissenberg 1989).

1.3 Distribution of Sterols, Spirostanols, and Spirosolanols in *S. glaucophyllum*

1.3.1 β-Sitosterol

β-Sitosterol (Fig. 4, *2*) was identified in vivo in seeds of *S. glaucophyllum* at a very low concentration (about 0.02%) in glycoside form (Sahoo and Jain 1987). Isolation was done by hydrolysis of seed samples with 5% solution of hydrochloric acid in ethanol, followed by extraction with ethyl acetate and analytical and/or preparative thin layer chromatography (TLC) of the resulting extracts (Sahoo and Jain 1987). Identification of the isolated sitosterol was achieved by comparing physical, chromatographic, and spectral data with an authentic sample. Assays were carried out by a colorimetric procedure (Das and Banerjee 1980). So far, β-sitosterol was not found in other organs of *S. glaucophyllum*, but it occurs in vitro in still lower amounts (approx. 0.01%; see Sect. 2.3.2.1).

1.3.2 Diosgenin

The spirostane derivative diosgenin (Fig. 4, *3*) was also identified in vivo in seeds of *S. glaucophyllum*, in trace amounts (ca. 0.01%) and in glycoside form (Sahoo and Jain 1987). Isolation and identification of diosgenin were performed as described previously for sitosterol (see Sect. 1.3.1), by hydrolysis and TLC separation (Sahoo and Jain 1987), and assays were carried out by a colorimetric method (Sanchez et

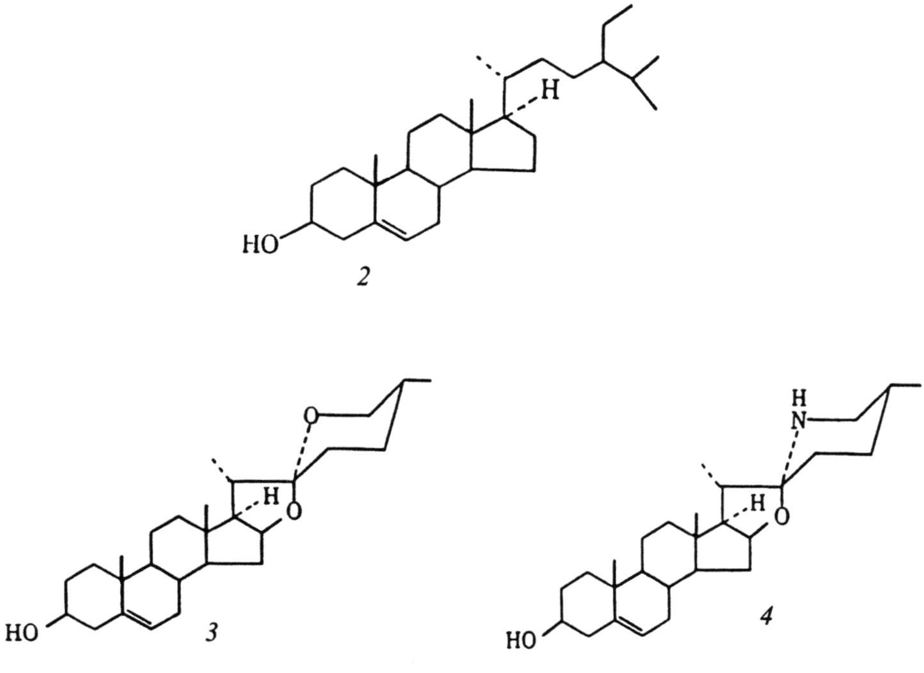

Fig. 4. Structures of β-sitosterol (*2*), diosgenin (*3*) and solasodine (*4*)

al. 1972). Like sitosterol, diosgenin was also not found in other organs of *S. glaucophyllum*, but it does occur in tissue cultures in higher amounts (ca. 0.03%; see Sect. 2.3.2.2).

1.3.3 Solasodine

The spirosolane derivative solasodine (Fig. 4, *4*) was identified in glycoside form in *S. glaucophyllum* seeds (0.015%; Sahoo and Jain 1987), leaves (<0.1%; Weiler et al. 1980; Weissenberg unpubl.), berries (ca. 0.01%; Weissenberg unpubl.), stems, and roots (trace amounts; Weissenberg unpubl.). Isolation from seeds was done by extraction with a 2% ethanolic solution of acetic acid, and crude glycoalkaloids were precipitated with ammonium hydroxide. Following hydrolysis of the glycosides fraction with an alcoholic solution of hydrochloric acid, free solasodine was extracted with chloroform, purified by preparative TLC, identified by comparing physical, chromatographic, and spectral data with an authentic sample (Sahoo and Jain 1987), and assayed by a spectrophotometric procedure (Birner 1969). Solasodine is also synthesized in tissue cultures of *S. glaucophyllum* at much lower levels (0.004%; see Sect. 2.3.2.3).

1.3.4 Occurrence and Importance

β-Sitosterol, diosgenin and solasodine were found in small amounts in seeds of *S. glaucophyllum*, while solasodine itself occurs in trace amounts in other plant organs as well. All three compounds are starting materials of choice for manufacture of steroidal drugs including male and female hormones, contraceptive, diuretics, antiinflammatory agents, and anabolic compounds, which represent a huge market requiring large amounts of raw steroids. Diosgenin was regarded during the 1950s and 1960s as the most versatile raw steroidal material, occurring in 4–6% yield in *Dioscorea* species grown mainly in Mexico, and being easily cleaved to 16-dehydropregnenolone acetate (16-DPA) which, in turn, may be converted to most steroidal drugs. However, political aspects associated with increasing control by the Mexican government, combined with overharvesting and rising production costs, led by the mid-1970s to a decline in the use of diosgenin, and to a search for alternatives (Djerassi 1976). One of them was believed to be solasodine, the nitrogen analogue of diosgenin, which can also be easily degraded to the common intermediate 16-DPA and occurs in 2–5% yield in various *Solanum* species (Mann 1978) from whom it could be readily isolated (Weissenberg 1979). It was accordingly suggested that solasodine might become "the diosgenin of the 1980s" (Djerassi 1976) and indeed, at the time, Glaxo, Diosynth, and Schering AG produced it commercially in India, New Zealand, and Ecuador, respectively (Coppen 1979). Yet, development of microbiological methods for side-chain cleavage (Martin 1984) pointed to sitosterol as a convenient source of steroidal drugs, being readily available in important quantities as by-product in the processing of soya bean oil and of tall oil, and thus replacing diosgenin and solasodine as raw materials. Nonetheless, the meagre content of diosgenin, solasodine, and sitosterol in *S. glaucophyllum*, as compared with other plant species, clearly disqualifies it for any practical use in this respect.

1.4 Distribution of Phenolic Metabolites in *S. glaucophyllum*

Extraction of *S. glaucophyllum* leaves with water, followed by concentration, extraction of the aqueous solution with n-butanol, solvent removal, and chromatography on a polyamide column afforded hydroquinone, kaempferol, quercetin and eight phenolic glycosides (O-methyl arbutin, arbutin, kaemferol-3-O-rutinoside, isorhamnetin-3-O-rutinoside, quercetin-3-O-(2^G-β-D-apiosyl rutinoside), rutin, avicularin, and isoquercitrin (Rappaportt et al. 1977). Addition of lead acetate to *S. glaucophyllum* leaf extracts for precipitation of the phenolic derivatives which interfere with the isolation process of the calcinogenic glycosides, revealed unexpected distribution of vitamin D activity between both saponin and phenolic crude fractions. The activity displayed by the phenolic fraction has been attributed to partial coprecipitation, and/or interaction of saponins with lead acetate (Weissenberg et al. 1990).

Since these phenolic metabolites were not found in tissue cultures of *S. glaucophyllum* and their potential value is limited, they will not be referred to further in this chapter.

2 In Vitro Approaches

2.1 Review of Tissue Culture Studies on *S. glaucophyllum*

A few preliminary reports pointed out the ability of this plant to synthesize in vitro seco-sterols, sterols, and spirostane and spirosolane derivatives. Thus, our team at the Volcani Center and Cornell University studied the establishment of tissue cultures (Weissenberg et al. 1987) and demonstrated the presence of substantial $1\alpha,25\text{-}(OH)_2D_3$ activity in cell suspensions (Weissenberg et al. 1987, 1988, 1989). A second group, at the Universities of Rajasthan and Utkal in India, reported the occurrence of β-sitosterol, diosgenin and solasodine in tissue cultures of this species (Jain and Sahoo 1986; Sahoo and Jain 1987). The work published so far in that area was summarized briefly within the framework of a survey on calcinogenic plants (Weissenberg 1989). While $1\alpha,25\text{-}(OH)_2D_3$ activity has not yet been encountered in tissue cultures of other plants, the in-vitro formation of solasodine in various *Solanum* species has been reviewed extensively (Ripperger and Schreiber 1981; Roddick 1986; Ellis 1988; Gröger 1988; Macek 1989), as has that of diosgenin in *Dioscorea* species (Ellis 1988; Furmanowa and Guzewska 1989), and of β-sitosterol in a series of plant genera (Ellis 1988).

2.2 Establishment of Tissue Culture

2.2.1 $1\alpha,25\text{-}(OH)_2D_3$ Studies

Various media were investigated for their efficiency to induce calli from hypocotyls of young seedlings (2-week-old) of *S. glaucophyllum* (Weissenberg et al. 1987).

Murashige and Skoog's medium (1962) (MS) supplemented either with 2,4-D (2 mg/l) and zeatin (0.1 mg/l), or with 2,4-D (1 mg/l) and kinetin (0.3 mg/l), and solidified with agar (0.8%), was the most efficient in inducing callus formation, and 80–90% of the hypocotyls developed a white friable callus on these media following 2 weeks' incubation in the dark at 25 °C.

Suspension cultures were initiated by dispersing small fragments of the callus in the medium without agar, and maintaining the cells in suspension on a gyratory shaker at 100 rpm in the dark at 25 °C. The cells were subcultured every 3 weeks on the same medium and under the same environmental conditions, then separated by filtration 18 days after inoculation, lyophilized, and radioimmunoassayed (Weissenberg et al. 1988) or bioassayed (Weissenberg et al. 1989) for $1\alpha,25\text{-}(OH)_2D_3$ activity. The cell dry weight was found to be significantly higher after 10 days in medium containing zeatin (0.44 g cell per 50 ml of medium) than in the presence of kinetin (0.32 g cell per 50 ml of medium) (Weissenberg et al. 1987).

2.2.2 Sterols, Spirostanols, and Spirosolanols Studies

Seeds of *S. glaucophyllum* were treated with an aqueous solution containing gibberellic acid (1000 ppm) and various amino acids (glutamic acid, histidine, leucine, lysine, methionine, proline, tryptophan, and valine, 0.5 ppm each) and vitamins (nicotinic acid and thiamine hydrochloride, 0.5 ppm each) during 12–16 h for quick germination (Sahoo and Jain 1987). Unorganized callus tissue cultures of the plant were established from sterilized seeds inoculated in MS medium (1962) supplemented with 2,4-D (1 ppm) and agar (1%). The seeds germinated within 7 days and callus initiated from the radicle and cotyledonary portions of the seedling. The undifferentiated tissue was dark brown, irregularly lobed, flat, and frägile. The calli were maintained for 10 months by periodic subculturing every 6–8 weeks onto fresh medium. The static cultures were grown at 25 ± 2 °C and 55% relative humidity under room-light conditions. The maintained tissues were harvested at 2, 4, 6, and 8 weeks of age. Each tissue sample was oven-dried at 100 °C for 15 min in order to inactivate the enzymatic processes, and then at 60 °C to constant weight. The samples, as well as the stale media from the static cultures, were analyzed for steroids and steroidal alkaloids (Jain and Sahoo 1986; Sahoo and Jain 1987).

2.3 Production of Metabolites by *S. glaucophyllum* Cultures

2.3.1 $1\alpha,25\text{-}(OH)_2D_3$ Production

Reports on UV-light-dependent formation of vitamin D_3 in *Trisetum flavescens* (L.) Beauv. (yellow oat grass) (Zucker et al. 1980), *Medicago sativa* L. (alfalfa) (Horst et al. 1984), and phytoplankton (Holick et al. 1982) had suggested the presence of the natural provitamin 7-dehydrocholesterol. Hence the inference was drawn (Esparza et al. 1982) that vitamin D_3 is synthesized in plants by photolytic activation of a precursor, in a manner similar to that which occurs in the skin of vertebrates

(Holick et al. 1982). A very low intensity of UV-irradiation was required for the photoconversion of the precursor into vitamin D_3, and brief exposure to daylight of extracts of *T. flavescens* grown without UV light was sufficient to obtain activation (Rambeck et al. 1981). Growth, as judged by yield and appearance, did not differ markedly in *T. flavescens* and *S. glaucophyllum* between plants grown with and without UV light (Rambeck et al. 1981).

In order to evaluate the potential use of cell cultures for the production of the plant active principle, and to find out whether its in vitro synthesis can take place without exposure to daylight, preliminary experiments with *S. glaucophyllum* cell suspensions were performed in the dark (Weissenberg et al. 1989). The results indeed point to appreciable activity formed under these conditions, as indicated either by RIA (Weissenberg et al. 1988) or by a CaBP bioassay (Weissenberg et al. 1989). The latter test is based on the observation that feeding of a high-strontium, low-calcium, vitamin D-replete diet to chicks can depress the formation of intestinal CaBP (Corradino et al. 1971), due to the inhibition of the renal formation of $1\alpha,25\text{-(OH)}_2D_3$ from 25-OH D_3 by this diet (Omdahl and DeLuca 1972). Accordingly, administration of a source of $1\alpha,25\text{-(OH)}_2D_3$ to strontium-inhibited chicks restores the capacity of the intestine to synthesize CaBP, and this thereby serves as an indicator of the presence of a $1\alpha,25\text{-(OH)}_2D_3$-like derivative in the test sample. The ability of the cell suspensions of *S. glaucophyllum* to stimulate the synthesis of CaBP in strontium-inhibited chicks provides definitive evidence for the presence of compounds with $1\alpha,25\text{-(OH)}_2D_3$-like activity therein.

Some of the cell suspensions of *S. glaucophyllum* synthesized significant amounts of the active principle in vitro (Table 3, entries 2 and 4, the former significantly greater than the latter; see also Weissenberg et al. 1988). The wide variability of $1\alpha,25\text{-(OH)}_2D_3$ activity among the different cell suspensions might reflect genotypic differences between hypocotyls from which the cell cultures originated, and can be used to develop high-yielding cell clones. A comparable variability in activity was also observed at the plant level (Weissenberg et al. 1989).

Table 3. $1\alpha,25\text{-(OH)}_2D_3$ activity[a] of cell suspensions from *S. glaucophyllum* (Weissenberg et al. 1989)

Cell suspension No.	Cells dry weight (g/flask)	CaBP[b] (μg/mg protein)
1	0.362	0.4 ± 0.3
2	0.306	9.3 ± 0.8[c]
3	0.424	1.5 ± 0.3
4	0.473	2.8 ± 1.4
5	0.410	1.5 ± 0.3

[a] Expressed as concentration of duodenal CaBP (calbindin-D_{28K}).
[b] Values represent mean ± SE of four samples per cell suspension. The $1\alpha,25\text{-(OH)}_2D_3$ control values were 18.2 ± 1 μg/mg protein. The propylene glycol control values were 0.61 ± 0.1 μg/mg protein.
[c] Significantly greater than cell suspensions 1, 3, 4 and 5, at $p < 0.05$.

2.3.2 Sterols, Spirostanols and Spirosolanols Production

Ten-month-old callus cultures of *S. glaucophyllum* were used in studies for isolation of steroids and steroidal alkaloids (Table 4; Jain and Sahoo 1986). The highest growth value was observed with the increased growth period up to the sixth week (2.32) and the lowest at the second week, with a decline during the eight week (1.85). Under the conditions of these studies, the tissue cultures had the ability to biosynthesize the following compounds as glycosides.

2.3.2.1 β-Sitosterol

A linear increase in sitosterol content with the growth index was observed with the increased growth period up to the sixth week (0.163 mg/g dry wt.), but declined during the eighth week (Table 4; Jain and Sahoo 1986; Sahoo and Jain 1987). Sitosterol was identified and isolated by hydrolysis of dried tissue samples with 5% solution of hydrochloric acid in ethanol, followed by extraction with ethyl acetate and analytical and/or preparative TLC of the resulting extracts (Sahoo and Jain 1987). The isolated sitosterol was identified by comparing physical, chromatographic, and spectral properties with an authentic sample, and its quantitative estimation was carried out by a colorimetric procedure (Das and Banerjee 1980).

2.3.2.2 Diosgenin

The maximum amount of diosgenin has been reported for 6-week-old tissue (0.317 mg/g dry wt.), with a minimum in 2 weeks (Table 4; Jain and Sahoo 1986; Sahoo and Jain 1987). The correlation of the diosgenin content with the growth index was similar to that noted above for sitosterol. Isolation and identification of diosgenin were carried out as for sitosterol (Sahoo and Jain 1987; see Sect. 2.3.2.1) by hydrolysis and TLC separation, and its quantitation was performed by using a colorimetric method (Sanchez et al. 1972).

Table 4. Growth index and production of diosgenin, β-sitosterol and solasodine in *S. glaucophyllum* tissue cultures[a]. (Jain and Sahoo 1986)

Age at harvest (weeks)	GI[b]	Diosgenin	β-Sitosterol	Solasodine
		(mg/g dry wt.)		
2	1.25	0.224	0.120	0.012
4	2.10	0.242	0.143	0.020
6	2.32	0.317	0.163	0.028
8	1.85	0.300	0.114	0.037

[a] Values represent the mean of five replicates.

[b] Growth index (final dry wt. of tissue – initial dry wt. of tissue/initial dry wt. of tissue).

2.3.2.3 Solasodine

A gradual increase in solasodine content was noted up to 8 weeks (0.037 mg/g dry wt.) whereas the growth index and the diosgenin and β-sitosterol content decreased (Table 4). Isolation was done by macerating dried tissue samples with a 2% solution of acetic acid in 90% ethanol. The resulting solution was concentrated, processed with 10% acetic acid to precipitate accompanying impurities, defatted with benzene-ether and treated with ammonium hydroxide. The precipitated crude glycoalkaloids were purified by successive dissolution in 5% acetic acid and precipitation with ammonium hydroxide. The glycosides fraction was hydrolyzed with an alcoholic solution of hydrochloric acid and, following alkalinization with ammonium hydroxide, free solasodine was extracted with chloroform, purified by preparative TLC, identified by comparing physical, chromatographic and spectral data with an authentic sample (Sahoo and Jain 1987), and estimated by a spectrophotometric procedure (Birner 1969).

2.3.2.4 Occurrence of Sterols, Spirostanols, and Spirosolanols

Examination of Table 4 suggests that the in-vitro biosynthesis of diosgenin seems to be enhanced with the reduced production of β-sitosterol and solasodine (Jain and Sahoo 1986). In seeds, however, diosgenin was detected in trace amounts only, whereas β-sitosterol and solasodine were synthesized in higher amounts than in tissue cultures (Sahoo and Jain 1987).

The presence and content of spirostanols and/or spirosolanols may be dependent on the ratio of plant growth substances such as 2,4-D, kinetin, and gibberellic acid (Heble et al. 1971). The simultaneous occurrence of solasodine, diosgenin, and β-sitosterol in tissue cultures of various *Solanum* species has been reviewed (Gröger 1988; Macek 1989). The occurrence of solasodine with diosgenin in different plant organs of *Solanum* species is also well documented (Schreiber 1968; Mann 1978), while β-sitosterol was often identified in plants of the genus *Solanum* (Puri and Bhatnagar 1974). Admittedly, cholesterol or a biogenetically equivalent sterol might be the precursor of both C_{27} spirostane and spirosolane steroids occurring together in plants, and hence their biosynthesis and metabolism should be closely related (Heftmann 1983).

2.4 Exploitation Potential

2.4.1 $1\alpha,25\text{-}(OH)_2D_3$

Preliminary results reported so far indicate that the activity present in some of the cell cultures (Table 3, entry 2) is higher than that of the leaves, stems, and roots of the plant, and comparable to that of the berries (Weissenberg et al. 1989). It is certainly possible that exposure of the cell cultures to daylight or UV-irradiation, and optimization of the culture media and growth conditions would enhance activity. Moreover, the large variability in activity observed both at the cell and plant level, is most promising for the development of high-yielding cell clones. At any rate, the formation of $1\alpha,25\text{-}(OH)_2D_3$-like derivatives by cell suspensions of

S. glaucophyllum might provide a potentially useful technology for an economically feasible biosynthetic approach. Its ecological safety is another asset, since the plant is considered to be a serious hazard to animal health in those areas where it grows naturally, or would eventually be cultivated. However, more work is needed in order to assess whether the tissue culture production of $1\alpha,25\text{-}(OH)_2D_3$ may compete with the various synthetic procedures which were used so far for its preparation, albeit at high production and sale prices.

2.4.2 Sterols, Spirostanols and Spirosolanols

Data presented in Table 4 show that the potential of *S. glaucophyllum* cell cultures to synthesize diosgenin is better than that for β-sitosterol, and certainly more so than that for solasodine (Jain and Sahoo 1986). However, a comparison with other plant species points to a low capacity of *S. glaucophyllum* for in-vitro synthesis of spirostane and/or spirosolane derivatives, which parallels that assessed in vivo for the latter. Thus, *S. laciniatum* Ait. can accumulate as much as 0.4% solasodine in callus (Macek 1989) and up to 1.5% in clonal selected strains (Zenk 1978) as compared with *S. glaucophyllum* (0.004%; see Table 4), whereas other *Solanum* species could produce it at a level of 5% (Ellis 1988). Likewise, cell suspensions of certain *Dioscorea* species can synthesize up to 8% diosgenin, that is, as much as in whole plants (Ellis 1988; Furmanowa and Guzewska 1989), *vs.* 0.03% for *S. glaucophyllum* (Table 4). On the evidence of that data, *S. glaucophyllum* does not belong to those plant species which accumulate in vitro solasodine, diosgenin, and β-sitosterol in high amounts, and certainly cannot compete with the corresponding whole-plants cultivation for potential commercial application.

3 Conclusions and Prospects

S. glaucophyllum is regarded as the main plant source of vitamin D_3 sterols. Cultured in vitro, it affords calli and suspensions that display substantial $1\alpha,25\text{-}(OH)_2D_3$ activity, and have also the ability to synthesize β-sitosterol, diosgenin and solasodine, although at low concentrations. Besides providing a potentially useful biotechnological approach at least for $1\alpha,25\text{-}(OH)_2D_3$ production, this plant might constitute a highly interesting model for studying the in vitro metabolism of quite a large array of steroidal derivatives, and for evaluating the chemotaxonomic significance of their occurrence and distribution, both in vitro and in vivo. These studies are still in a very early phase and much work has yet to be done.

4 Protocol

$1\alpha,25\text{-}(OH)_2D_3$ Studies. Seeds of *S. glaucophyllum* were surface-sterilized with commercial 20% Chlorox solution and germinated on 1% agar at 28 °C in the dark. For callus induction, hypocotyls

from young seedlings (2-week-old) were transferred to Petri dishes containing Murashige and Skoog medium supplemented with 1 mg/l of 2,4-D, 0.3 mg/l of kinetin, and solidified with 0.8% agar. Following 2 weeks' incubation in the dark at 25 °C, friable white callus formed on the hypocotyls. Suspension cultures were initiated by dispersing small fragments of the callus in 100-ml flasks filled with 20 ml of medium without agar, and placed in the dark on a gyratory shaker (100 rpm) at 25 °C. Subculturing was done every 3 weeks by placing 10 ml cells in 250-ml Delong flasks filled with 50 ml of the medium solution. The cells from different suspensions, each originating from a different hypocotyl, were collected by filtration 18 days after inoculation, lyophilized, and radioimmunoassayed (Weissenberg et al. 1988) or bioassayed (Weissenberg et al. 1989) for 1α,25-$(OH)_2D_3$ activity.

Sterols, Spirostanols and Spirosolanols Studies. Unorganized callus tissue cultures of the plant were established from the sterilized seeds on revised Murashige and Skoog medium supplemented with 1 ppm of 2,4-D and 1% agar. The tissue was maintained for 10 months by subculturing every 6–8 weeks onto fresh medium, then harvested at 2-weeks intervals up to the eight week and analyzed for steroids and steroidal alkaloids. Each tissue sample was dried and the growth indices were calculated for each harvest (Jain and Sahoo 1986).

Acknowledgments. Research in the authors' laboratories quoted in this work was supported by grant No. I-389-81 from the United States-Israel Binational Agricultural Research and Development Fund (BARD). One of us (R.H.W.) also acknowledges partial support by U.S. National Institutes of Health grant No. DK-04652. This is contribution No. 3087-E, 1990 series, from the Agricultural Research Organization, The Volcani Center, Bet Dagan, Israel.

References

Birner J (1969) Determination of total steroid bases in *Solanum* species. J Pharm Sci 58: 258–260

Boland RL (1986) Plants as a source of vitamin D_3 metabolites. Nutr Rev 44: 1–8

Cabrera AL (1965) Solanaceae in Flora de la Provincia de Buenos Aires. In: Colección Cientifica del INTA, vol 4, part 5a. Instituto Nacional de Tecnologia Agropecuaria, Buenos Aires, pp 190–250

Coppen JJW (1979) Steroids: from plants to pills – the changing picture. Trop Sci 21: 125–141

Corradino RA, Ebel JG, Craig PH, Taylor AN, Wasserman RH (1971) Calcium absorption and the vitamin D_3-dependent calcium-binding protein. Calcif Tissues Res 7: 81–92

Das AK, Banerjee AB (1980) A rapid method for quantification of sterols after thin-layer chromatography. Indian J Exp Biol 18: 969–971

Deb DB (1979) Solanaceae in India. In: Hawkes JG, Lester RN (eds) The biology and taxonomy of the Solanaceae. Linnean Soc Symp Ser 7. Academic Press, London, pp 87–112

Djerassi C (1976) The manufacture of steroidal contraceptives: technical versus political aspects. Proc R Soc Lond Sect B 195: 175–186

Ellis BE (1988) Natural products from plant tissue culture. Nat Prod Rep 5: 581–612

Esparza MS, Vega M, Boland RL (1982) Synthesis and composition of vitamin D_3 metabolites in *Solanum malacoxylon*. Biochim Biophys Acta 719: 633–640

Furmanowa M, Guzewska J (1989) *Dioscorea*: in vitro culture and the micropropagation of diosgenin-containing species. In: Bajaj YPS (ed) Biotechnology in agriculture and forestry, vol 7. Medicinal and aromatic plants II. Springer, Berlin Heidelberg New York, pp 162–184

Fürst A, Labler L, Meier W (1983) Synthese von β-D-glucopyranosiden einiger hydroxylierter Vitamin D-Verbindungen. Helv Chim Acta 66: 2093–2102

Gröger D (1988) Terpenoids and steroid alkaloids. In: Constabel F, Vasil IK (eds) Cell culture and somatic cell genetics of plants, vol 5. Academic Press, New York, pp 435–448

Haussler MR (1986) Vitamin D receptors: nature and function. Annu Rev Nutr 6: 527–562

Haussler MR, Cordy PE (1982) Metabolites and analogues of vitamin D. J Am Med Assoc 247: 841–844

Haussler MR, Wasserman RH, McCain TA, Peterlik M, Bursak KM, Hughes MR (1976) 1,25-Dihydroxyvitamin D_3-glycoside: identification of a calcinogenic principle of *Solanum malacoxylon*. Life Sci 18: 1049–1056

Heble MR, Narayanaswamy S, Chadha MS (1971) Hormonal control of steroid synthesis is *Solanum xanthocarpum* tissue cultures. Phytochemistry 10: 2393–2394
Heftmann E (1983) Biogenesis of steroids in Solanaceae. Phytochemistry 22: 1843–1860
Holick MF, Adams JS, Clemens TL, MacLaughlin J, Horiuchi N, Smith E, Holick SA, Nolan J, Hannifan N (1982) Photoendocrinology of vitamin D: the past, present and future. In: Norman AW, Schaefer K, von Herrath D, Grigoleit HG (eds) Proc 5th Workshop Vitamin D (Vitamin D, Chemical, Biochemical and Clinical Endocrinology of Calcium Metabolism). Walter de Gruyter, Berlin, pp 1151–1156
Horst RL, Reinhardt TA, Russell JR, Napoli JL (1984) The isolation and identification of vitamin D_2 and vitamin D_3 from *Medicago sativa* (alfalfa plant). Arch Biochem Biophys 231: 67–71
INTA-FAO (1970) Estudio de las enfermedades y deficiencias nutricionales del ganado vacuno en la Argentina. Final Report, UN Special Fund Project, FAO, Rome
Jain SC, Sahoo S (1986) Isolation of steroids and glycoalkaloids from *Solanum glaucophyllum* Desf. Pharmazie 41: 820–821
Londowski JM, Kost SB, Meier W, Labler L, Kumar R (1986) Biological activity of the C-1, C-3, C-25, β-D-glucopyranosides of 1,25-dihydroxyvitamin D_3. J Pharmacol Exp Ther 237: 837–840
Macek TE (1989) *Solanum aviculare* Forst., *Solanum laciniatum* Ait. (Poroporo): in vitro culture and the production of solasodine. In: Bajaj YPS (ed) Biotechnology in agriculture and forestry, vol 7. Medicinal and aromatic plants II. Springer, Berlin Heidelberg New York, pp 443–467
Mann JD (1978) Production of solasodine for the pharmaceutical industry. In: Brady NC (ed) Advances in agronomy, vol 30. Academic Press, New York, pp 207–243
Martin KA (1984) Sterols. In: Rehm H-J, Reed G (eds) Biotechnology, vol 6a. Verlag Chemie, Weinheim, pp 80–95
Meier W (1981) Die chemische Entwicklung von Rocaltrol. Nieren Hochdruckkrankh 10: 175–179
Morris KML (1982) Plant induced calcinosis: a review. Vet Hum Toxicol 24: 34–48
Murashige T, Skoog F (1962) A revised medium for rapid growth and bioassays with tobacco tissue cultures. Physiol Plant 15: 473–497
Okada KA, Carrillo BJ, Tilley M (1977) *Solanum malacoxylon* Sendtner: a toxic plant in Argentina. Econ Bot 31: 225–236
Omdahl JL, DeLuca HF (1972) Rachitogenic activity of dietary strontium: inhibition of intestinal calcium absorption and 1,25 dihydroxycholecalciferol synthesis. J Biol Chem 247: 5520–5526
Pardo R, Santelli M (1985) Synthèse des métabolites de la vitamine D. Bull Soc Chim Fr (II): 98–114
Pike JW (1987) Emerging concepts of the biologic role and mechanism of action of 1,25-dihydroxyvitamin D_3. Steroids 49: 3–27
Puri RK, Bhatnagar JK (1974) Studies on the genus *Solanum* – a review. Pharmacos 19: 7–24
Rambeck WA, Kreutzerg O, Bruns-Droste C, Zucker H (1981) Vitamin D_3 in the grass *Trisetum flavescens*. Z Pflanzenphysiol 104: 9–16
Rambeck WA, Weiser H, Zucker H (1986) Antirachitische Aktivität von Glukosiden des 1,25-Dihydroxyvitamin D_3 und des 1α-Hydroxyvitamin D_3 beim Hühnerküken. Wien Tieraerztl Monatsschr 73: 169–171
Rappaportt I, Giacopello D, Seldes AM, Blanco MC, Deulofeu V (1977) Phenolic glycosides from *Solanum glaucophyllum*: a new quercetin triglycoside containing D-apiose. Phytochemistry 16: 1115–1116
Ripperger H, Schreiber K (1981) *Solanum* steroid alkaloids. In: Manske RHF, Rodrigo RGA (eds) The alkaloids: chemistry and physiology, vol 19. Academic Press, New York, pp 81–192
Roddick JG (1986) Steroidal alkaloids of the Solanaceae. In: D'Arcy WG (ed) Solanaceae: biology and systematics, chapter 16. Columbia Univ Press, New York, pp 201–222
Sahoo SL, Jain SC (1987) Isolation of steroids and glycoalkaloid from seeds and seedling callus of *Solanum glaucophyllum* Desf. tissue culture. Cell Chromosome Res 10: 104–109
Sanchez GL, Medina Acevedo JC, Soto RR (1972) Spectrophotometric determination of diosgenin in *Dioscorea composita* following thin-layer chromatography Analyst 97: 973–978
Schreiber K (1968) Steroid alkaloids: the *Solanum* group. In: Manske RHF (ed) THE alkaloids: chemistry and physiology, vol 10. Academic Press, New York, pp 1–192
Simpson RV, Wishaar R (1986) Newly discovered activities for calcitriol (1,25-dihydroxyvitamin D_3): implications for future pharmacological use. BioAssays 4: 65–70
Vidal MC, Lescano W, Avdolov R, Puche RC (1985) Partial structure elucidation of the carbohydrate moiety of 1,25-dihydroxycholecalciferol glycoside isolated from *Solanum glaucophyllum* leaves. Turrialba 35: 65–70

Wasserman RH, Nobel TA (1980) Vitamin D-related compounds as the toxic principle in calcinogenic plants. In: Norman AW (ed) Basic clinical nutrition, vol 2. Vitamin D: molecular biology and clinical nutrition. Marcel Dekker, New York, pp 455–487

Wasserman RH, Henion JD, Haussler MR, McCain TA (1976) Calcinogenic factor in *Solanum malacoxylon*: evidence that it is 1,25-dihydroxyvitamin D_3-glycoside. Science 194: 853–855

Weiler EW, Kruger H, Zenk MH (1980) Radioimmunoassay for the determination of the steroidal alkaloid solasodine and related compounds in living plants and herbarium specimens. Planta Med 39: 112–124

Weissenberg M (1979) Isolation of alkaloidal steroids from plants. Israeli Patent 58227, 30 July 1982; appl. 11 Sept 1979. Chem Abstr 99: 28002, 1983

Weissenberg M (1989) Calcinogenic glycosides. In: Cheeke PR (ed) Toxicants of plant origin, vol 2. Glycosides. CRC Press, Boca Raton, pp 201–238

Weissenberg M, Levy A, Wasserman RH (1987) Studies on calcinogenic Solanaceae plants of potential value as source of active vitamin D-like derivatives. Report I-389-81, United States-Israel Binational Agricultural Research and Development Fund (BARD), Bet Dagan, Israel

Weissenberg M, Maoz A, Levy A, Wasserman RH (1988) Radioimmunoassay for rapid estimation of vitamin D derivatives in calcinogenic plants. Planta Med 54: 63–65

Weissenberg M, Levy A, Wasserman RH (1989) Distribution of calcitriol activity in *Solanum glaucophyllum* plants and cell cultures. Phytochemistry 28: 795–798

Weissenberg M, Schaeffler I, Maoz A, Levy A, Wasserman RH (1990) The fractionation of *Solanum glaucophyllum* extracts with lead acetate revisited. Fitoterapia 61: 258–262

Worker NA, Carrillo BJ (1967) "Enteque seco", calcification and wasting in grazing animals in the Argentine. Nature 215: 72–74

Zenk MH (1978) The impact of plant cell culture on industry. In: Thorpe TA (ed) Frontiers of plant tissue culture, Proc 4th Int Congr Plant tissue and cell culture, Univ Calgary. Offset Printing Service, Calgary, Alberta, pp 1–13

Zucker H, Stark H, Rambeck WA (1980) Light-dependent synthesis of cholecalciferol in a green plant. Nature 283: 68–69

XXV *Strophanthus* Species (Members of the Dogbane Family): In Vitro Culture and the Production of Cardenolides

K. Kawaguchi[1], M. Hirotani[2], and T. Furuya[2]

1 Introduction

The cardiac glycosides are pharmaceutically potential drug groups that are available to medicine today for the treatment of congestive heart failure. Clinically availabilities are derived from the leaves and seeds of plant in the genera *Digitalis* and *Strophanthus*. In vitro culture, regeneration, and production of *Digitalis* (foxglove) cardenolides and other secondary metabolites were reviewed in detail (Rücker 1988).

Strophanthus, belonging to the family Apocynaceae (the dogbane family), is from the Greek meaning "a turn or twist" and "a flower" and refers to the twisted lobes of the corolla. About 40 species of *Strophanthus* native to Africa and Asia, chiefly tropical, are perennial trees, shrubs or climbers up to 3 m tall. The leaves are feathery or leathery and opposite. The cymose inflorescence is terminal. Members of this genus have a variety of fragrant flowers, ranging in color from white, through the yellows and reds, to purple. The calyxes is glandular, the corolla funnel-shaped, with five lobes tapering into attenuated, long tails. The two carpeled ovaries develop into capsular fruits having two diverging free follicles, which enclose the hairy seeds. *Strophanthus* seeds have long been used by the native Africans in the preparation of arrow poisons (Claus et al. 1970; Morton 1977; Evans 1989).

The flowering of *Strophanthus gratus* grown in the greenhouse can be seen in Fig. 1. A large number of cardiac glycosides have been isolated from *Strophanthus* plants (Table 1), and highly oxygenated aglycones in comparison with *Digitalis* cardenolides are characteristic of *Strophanthus* cardenolides (Fig. 2) (Hegnauer 1964; Reichstein 1967; Singh and Rastogi 1970). *S. gratus*, found in western tropical Africa from Sierra Leone to the Congo, contains 4–8% G-strophanthin (ouabain) in the seeds. Its chemical structure is 3β-(α-L-rhamnopyranosyloxy)-1β,5,11α,14,19-pentahydroxy-5β,14β-card-20(22)-enolide octahydrate (Fig. 3) and this compound is the most polar of the commonly used cardiac glycosides. The seeds of *S. kombé*, native to central and eastern Africa, mainly Mozambique, Northern Rhodesia, Nyasaland, and Tanzania, provide the precursor for semisynthetic compound acetylstrophanthidin. Both of these compounds are of clinical interest

[1] Medicinal Plant Garden, School of Pharmaceutical Sciences, Kitasato University, Sagamihara, Kanagawa 228, Japan
[2] School of Pharmaceutical Sciences, Kitasato University, Minato-ku, Tokyo 108, Japan

Biotechnology in Agriculture and Forestry, Vol. 21
Medicinal and Aromatic Plants IV (ed. by Y.P.S. Bajaj)

Fig. 1. Flowering *Strophanthus gratus* plants cultivated in the greenhouse of the Medicinal Plant Garden, School of Pharmaceutical Sciences, Kitasato University, Japan

because of their rapid onset of action and the relatively rapid offset of effect when they are administered intravenously (Antmann and Smith 1986). In biochemical studies, ouabain is well known as a specific inhibitor of Na^+, K^+-ATPase.

G-Strophanthin (ouabain) obtained from the seeds of *S. gratus* or from the wood of *Acokanthera schimperi* or *A. ouabaio* (Apocynaceae) is official in the pharmacopoeias of many countries; Argentina, Austria, Belgium, Czechoslovakia, Egypt, France, West Germany, Italy, Japan, Mexico, The Netherlands, Scandinavia, Poland, Romania, Spain, Switzerland, and Turkey (all 8 H_2O); Portugal (9 H_2O). Additionally, proprietary names and manufacturers for commerical use are as follows; g-Strofantin (DAK, Denmark, Sweden); Ouabaïne Aguettant (France); Ouabaine Arnaud (Gamaprod, Australia; Nativelle, France; Lipomed, United Kingdom, The Netherlands, Spain); Purostrophan (KaliChemie, West Germany); Strodival (Herbert, West Germany); Strophoperm (West Germany) (Reynolds 1989).

In order to obtain *Strophanthus* cardenolides, we continue fundamental studies using plant cell and tissue cultures. No cardenolides were detected by TLC analysis

Table 1. Principal cardiac glycosides of *Strophanthus* species

Aglycones	Glycosides	Sugars[a]
G-Strophanthidin (Ouabagenin)	G-Strophanthin (Ouabain)	-Rha
Periplogenin	Emicymarin	-Dgt
Sarmentogenin	Sarmentocymarin	-Sar
	Divaricoside	-Ole
	Divostroside	-Dgn
Sarmutogenin	Amboside	-Dgn
Sarverogenin	Intermedioside	-Dgn
	Sarveroside	-Sar
	Interoside	-Dgn-Glc
	Isostrophanthoside	-Dgn-Glc-Glc
Strophanthidin	Cymarin	-Cym
	k-Strophanthin-β	-Cym-Glc
	k-Strophanthoside	-Cym-Glc-Glc
Strophanthidol	Cymarol	-Cym
	k-Strophanthol-γ	-Cym-Glc-Glc

[a] Cym = cymarose; Dgn = diginose; Dgt = digitalose; Glc = glucose; Ole = oleandrose; Rha = rhamnose; Sar = sarmentose.

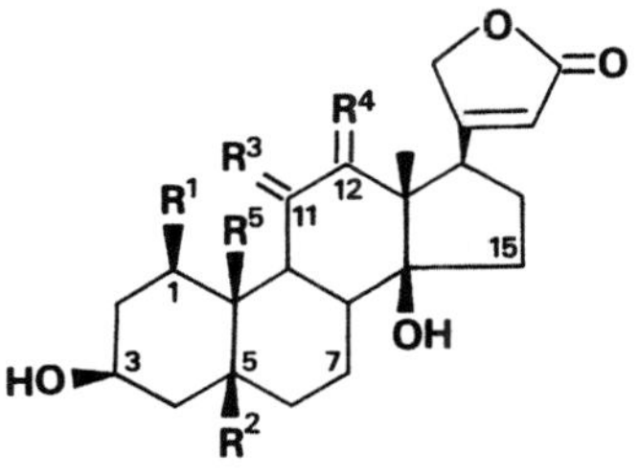

Aglycones	R^1	R^2	R^3	R^4	R^5	
Digitoxigenin	H	H	2H	2H	CH_3	
Periplogenin	H	OH	2H	2H	CH_3	
Sarmentogenin	H	H	···OH, ◄H	2H	CH_3	
Bipindogenin	H	OH	···OH, ◄H	2H	CH_3	
Sarmutogenin	H	H	=O	◄OH, ···H	CH_3	
Sarverogenin	H	H	···OH, ◄H	=O	CH_3	C-7···O···C-15
Strophanthidin	H	OH	2H	2H	CHO	
Strophanthidol	H	OH	2H	2H	CH_2OH	
Sarmentologenin	H	OH	···OH, ◄H	2H	CH_2OH	
G-Strophanthidin (Ouabagenin)	OH	OH	···OH, ◄H	2H	CH_2OH	

Fig. 2. Chemical structures of the principal aglycones of *Strophanthus* cardenolides

of extracts of the calli derived from *S. gratus*, *S. amboensis*, *S. intermedius*, and *S. divaricatus*. Moreover, cardenolide production could not be observed by biotransformation of progesterone, a precursor of cardenolides, with these suspension cultures, similarly using cultured cells of *Digitalis purpurea* (Furuya et al. 1973).

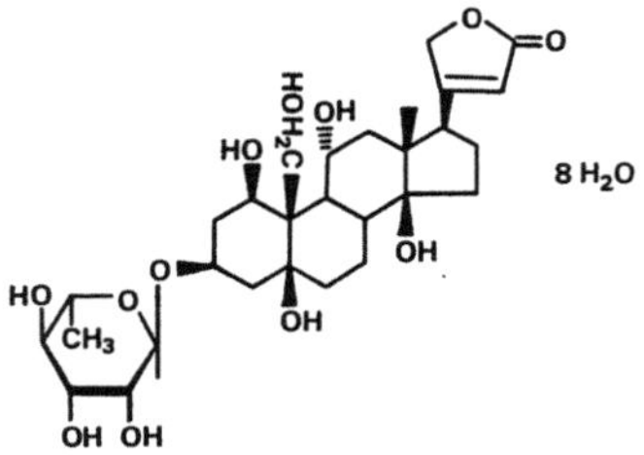

G-Strophanthin (Ouabain)

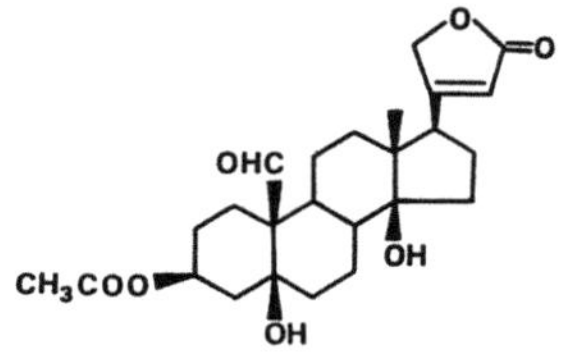

Acetylstrophanthidin

Fig. 3. Chemical structures of clinically useful *Strophanthus* cardenolides

In this chapter, we describe the biotransformation of digitoxigenin (**1**), a precursor of cardiac glycosides, by cell suspension cultures of *Strophanthus gratus* (Furuya et al. 1988), *S. amboensis* (Kawaguchi et al. 1988), *S. intermedius* (Kawaguchi et al. 1989), and *S. divaricatus* (Kawaguchi et al. 1991) and cardiac glycosides from the regenerated plants of *S. divaricatus* through the calli under controlled culture conditions (Kawaguchi et al., unpubl.).

2 In Vitro Approaches

There has been only one report on the use of tissue cultures of *Strophanthus* plants as far as we know. Dohnal et al. (1981) have reported on the growth and phytochemical analysis of *S. amboensis* callus tissue culture, and have identified phytosterols and ursolic acid, but cardenolides could not be detected.

2.1 Establishment of Tissue Cultures

2.1.1 Establishing Callus Cultures

In January 1983, the leaves, the volubile stems, and the flower buds of *Strophanthus gratus* Baill., cultivated in the greenhouse of the Medicinal Plant Garden, Kitasato University, were sterilized by treatment with 70% EtOH for 1 min and a saturated solution of bleaching powder for 20 min and then rinsed with sterile H_2O and cut into ca. 5-mm segments. These segments were placed on MS basal medium (Murashige and Skoog 1962) containing 1 ppm 2,4-dichlorophenoxyacetic acid (D), 0.1 ppm kinetin (K), 3% sucrose and 0.9% agar (abbreviated as DK medium) at 25 °C in the dark. The Sg-1 callus (obtained from the leaves), the Sg-2 callus (from the volubile stems), and the Sg-3 callus (from the flower buds) were subcultured at 25, 30, and 35 °C in the dark and/or under illumination of 3000–5000 lx for 16 h every 4 weeks.

The Sa-1 callus from the leaves of *Strophanthus amboensis* Engl. and Pax and the Si-2 callus from the stems of *S. intermedius* Pax, collected at Kyoto Takeda Herbal Garden in January 1983, were obtained in the same manner. After sterilization, the seeds of *S. divaricatus* Hook and Arn., collected at Botanical Gardens, Faculty of Science, University of Tokyo in February 1983, were placed on MS basal medium containing 3% sucrose and 0.9% agar (abbreviated as B medium) at 25 °C in the dark. After 3 weeks, the cotyledons of seedling were cut into ca. 3-mm

Fig. 4A–D. Tissue cultures of *Strophanthus divaricatus*. **A** Calli cultured on MS medium containing 1 ppm 2,4-D and 0.1 ppm kinetin in the dark. **B** Embryoids and shoots derived from the calli on MS medium containing 0.5 ppm kinetin in the light. **C** Rooting from the shoots cultured on MS medium without plant hormone in the light. **D** The regenerated plants grown in pots segments and then these segments were placed on the DK medium at 25 °C in the dark. The calli were subcultured every 4 weeks.

Nevertheless, callus induction from the leaves and stems of *S. kombé* Oliv., *S. hispidus* DC., and *S. petersianus* Klotz. could not be established.

2.1.2 Regeneration of Plants from Callus Cultures

The shoots through the embryoids were regenerated from the calli by transferring the calli (cultured on DK medium) onto the B medium containing 0.5 ppm K at 25 °C in 3000–5000 lx for 16 h, or on the B medium without plant growth regulator at 15 °C in 300–500 lx for 16 h. The regenerated plants were obtained after root formation by transferring the shoots onto the B medium. After that the regenerated plants were cultivated on the soil in pots (Fig. 4A–D). The regenerated plants were obtained only with *Strophanthus divaricatus* calli, not with *S. gratus*, *S. amboensis*, and *S. intermedius* calli.

Although no cardenolides were detected in the embryoids and shoots, six Kedde-positive spots (R_f 0.37, 0.23, 0.19, 0.11, 0.08, 0.04; $CHCl_3$—EtOH 10:1) were detected on TLC in the $CHCl_3$ and $CHCl_3$—*iso*-PrOH (3:2) extracts from the aerial parts of the regenerated plants (340 g fresh wt., 50–130 cm height) grown in pots.

2.2 Biotransformation of Digitoxigenin by Cell Suspension Cultures of *Strophanthus* Species

In the biotransformation experiments, digitoxigenin (**1**) (30 mg) suspended with Tween 80 (5%) was added to each flask (250 ml medium/1-l flask) and then the calli (30–50 g fresh wt. per flask) from 4-week-old static cultures were incubated in a shaker (90 spm = stroke per minute) for 16–18 days.

After incubation, the calli were harvested with nylon cloth and homogenized with MeOH in a Waring blender. The $CHCl_3$ and the $CHCl_3$—MeOH (2:1) extracts were obtained from the calli, and also the medium was extracted with the same method. These extracts were examined on TLC with Kedde's reagent and 10% H_2SO_4 to detect the biotransformation products.

Purification of the products was achieved by repeated chromatography on a silica gel column (Wako gel C-200 and C-300) using $CHCl_3$—MeOH solvent system, HPLC (Nucleosil 5C18: 10ϕ × 300 mm column, coupled to a UV detector and a differential refractometer using 60–80% MeOH in H_2O solvent system) and DCCC [droplet countercurrent chromatography, $CHCl_3$—MeOH—H_2O (5:6:4); descending method].

The elucidation of their chemical structures was accomplished using mainly ^1H and ^{13}C NMR (300, 400 MHz; 75, 100 MHz) and mass spectral data. The assignment of each signal in the NMR spectra of the products was performed using the DEPT pulse sequence, two-dimensional proton-proton, and proton-carbon shift correlation data. Mass spectra were run using a direct inlet system at 20 eV.

The biotransformation percentage yield of these products was calculated from the ratio of the amount of the isolated product to the starting amount of **1**, taking into consideration their M_rs, and are summarized in Table 2.

Table 2. Biotransformation products of digitoxigenin (**1**) by cell suspension cultures of *Strophanthus* species

Products		*S. gratus* Yield (%)	*S. amboensis*	*S. intermedius*	*S. divaricatus*
17βH-Digitoxigenin	(**2**)	4.7			9.7
3-*epi*-17βH-Digitoxigenin	(**3**)	6.7		0.8	17.9
17βH-Acovenosigenin-A	(**4**)	0.4			
Periplogenin	(**5**)		5.3	0.9	
17βH-Periplogenin	(**6**)	2.8			0.7
3-*epi*-17βH-Periplogenin	(**7**)	2.3			1.9
3-*epi*-4β-hydroxy-17βH-Digitoxigenin	(**8**)	0.2			
3-Epidigitoxigenin	(**9**)		4.2	8.8	
3-Epiperiplogenin	(**10**)		1.4	6.4	
Digitoxigenin β-D-glucoside	(**11**)		0.7	0.3	
Periplogenin β-D-glucoside	(**13**)		3		
Gitoxigenin	(**14**)			0.5	1.1
3-Epigitoxigenin	(**15**)			0.8	1.3
17βH-Gitoxigenin	(**19**)				2.9
3-*epi*-17βH-Gitoxigenin	(**20**)				2.2

2.2.1 *Strophanthus gratus*

After incubation of digitoxigenin (**1**) (960 mg) with the cell suspension cultures of *S. gratus* (1.5 kg fresh wt.) for 16 days, the biotransformation products were isolated as crystals and their structures established as 17βH-digitoxigenin (**2**); 3-*epi*-17βH-digitoxigenin (**3**); 17βH-acovenosigenin-A (**4**); 17βH-periplogenin (**6**); 3-*epi*-17βH-periplogenin (**7**); and 3-*epi*-4β-hydroxy-17βH-digitoxigenin (**8**), respectively. Furthermore, periplogenin (**5**) was identified by TLC and HPLC. Products **7** and **8** are new compounds in nature and product **3** was main product in this experiment.

In addition, products **2**–**6** were detected on TLC, but **7** and **8** were not observed after incubation for 6 days. Therefore the biotransformation pathway of **1** by cell suspension cultures of *S. gratus* was presumed as shown in Fig. 5.

The biotransformation by isomerization of the 17β-lactone ring (**1** to **2** and **5** to **6**) has been demonstrated for the first time by a plant cell suspension culture. However, the reaction did not occur with ten cardenolides and bufadienolides when incubated with tissue slices from the livers of ten animal species, including man (Herrman and Repke 1964), and in microbiological conversion of cardenolides (Charney and Herzog 1967; Iizuka and Naito 1967). Since several 17βH-cardenolides have been found from natural sources, some plants are believed to possess an isomerization enzyme system for conversion of 17αH- to 17βH-cardenolides (Singh and Rastogi 1970; Abe and Yamauchi 1977). Thus, it is very interesting that the high isomerization ability has been demonstrated in the *S. gratus* cells.

The epimerization of 3β-hydroxyl to 3α-hydroxyl (**2** to **3** and **6** to **7**, probably via 3-keto) and 5β-hydroxylation (**1** to **5**, **2** to **6** and **3** to **7**) performed by the *S. gratus* cells have been previously observed with cell cultures of other plants (Jones et al. 1978; Hirotani and Furuya 1980a; Reinhard and Alfermann 1980). The

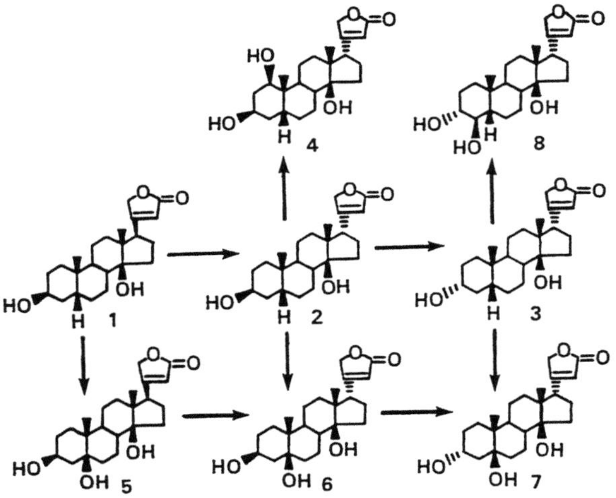

Fig. 5. Biotransformation of digitoxigenin (**1**) by cell suspension cultures of *Strophanthus gratus*. (Furuya et al. 1988)

1β- and 4β-hydroxylations in the *S. gratus* cells are notable, especially the 4β-hydroxylation which was not previously known even in the microbial transformation of cardenolides.

As 1β- and 5β-hydroxylations are indispensable reactions for the production of G-strophanthin, it is noteworthy that the reactions have been achieved not only with the whole plant, but also with the cultured cells.

2.2.2 *Strophanthus amboensis*

Six biotransformation products of digitoxigenin (**1**: 600 mg) by cell suspension cultures of *S. amboensis* (1.0 kg fresh wt.) were obtained after 18 days of incubation. Their structures were confirmed as periplogenin (**5**), 3-epidigitoxigenin (**9**), 3-epiperiplogenin (**10**), digitoxigenin β-D-glucoside (**11**), 3-epidigitoxigenin β-D-glucoside (**12**), and periplogenin β-D-glucoside (**13**). The formation of **10** and **13**, which have been isolated from *Adonis vernalis* (Junior and Wichtl 1980) and *Strophanthus vanderijstii* (Brenneisen et al. 1964), respectively, was first demonstrated in plant cell cultures.

The possible biotransformation pathway of **1** by cell suspension cultures of *S. amboensis* is shown in Fig. 6. 5β-Hydroxylation (**1** to **5**, **9** to **10** and **11** to **13**) and glucosylation (**1** to **11**, **5** to **13** and **9** to **12**) were distinct biotransformation abilities in this cells. The formation of **11** and **12** has been observed using *Digitalis purpurea* cells, with production of **12** predominating over **11** (Hirotani and Furuya 1980b). This fact is consistent with the results of enzymatic studies which showed that digitoxigenin (**1**) has a very low affinity for sterol: UDPG glucosyltransferase purified from the cells and plant (Yoshikawa and Furuya 1979). From these results, it is interesting to note the different extents of glycosylation, the axial 3β-hydroxyl group (**11**) was glucosylated preferentially compared to the equatorial 3α-hydroxyl group (**12**), as was observed with the *S. amboensis* cells.

The epimerization of 3β-hydroxyl to 3α-hydroxyl (**1** to **9** and **5** to **10**, probably via 3-keto) has been observed, but isomerization of the 17β-lactone ring and 1β-

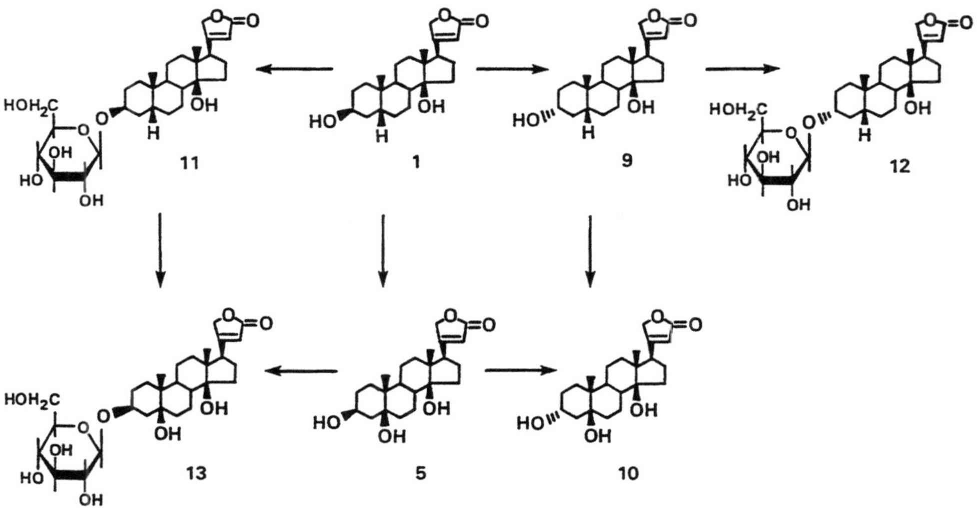

Fig. 6. Biotransformation of digitoxigenin (**1**) by cell suspension cultures of *Strophanthus amboensis*. (Kawaguchi et al. 1988)

and 4β-hydroxylations performed by the *S. gratus* cells were not demonstrated in the *S. amboensis* cells.

2.2.3 *Strophanthus intermedius*

As shown in Fig. 7, 11 biotransformation products of digitoxigenin (**1**; 810 mg) by cell suspension cultures of *S. intermedius* (1.4 kg fresh wt.) were obtained after 18 days of incubation. The formation of gitoxigenin (**14**) and 3-epigitoxigenin (**15**), which were obtained with these cells, were presumed to be produced by 16β-hydroxylation of **1** and **9**, respectively. 16β-Hydroxylation was observed only in the biotransformation of digitoxin by *Digitalis purpurea* cells (Furuya et al. 1970). In addition, the sole product was detected on TLC and identified with authentic digitoxigenone (**16**) after 3 days of incubation. The 3β-hydroxyl was oxidized first to the 3-keto before the other reactions proceeded in the *S. intermedius* cells. This oxidation had been also observed with cell cultures of *Digitalis lanata* (Stohs and Rosenberg 1975), *D. purpurea* (Stohs and Rosenberg 1975; Hirotani and Furuya 1980b) and *Thevetia neriifolia* (Döller and Reinhard 1979). So, epimerization of the 3β-hydroxyl to the 3α-hydroxyl (**5** to **10** and **14** to **15**) proceeded probably via 3-keto compounds such as **17** and **18**, which we could not isolate in this experiment.

2.2.4 *Strophanthus divaricatus*

A new compound [3-*epi*-17βH-gitoxigenin (**20**)], a new product [17βH-gitoxigenin (**19**)], and six known compounds (**2**, **3**, **6**, **7**, **14**, and **15**) were isolated as biotransformation products of digitoxigenin (**1**; 725 mg) by cell suspension cultures of

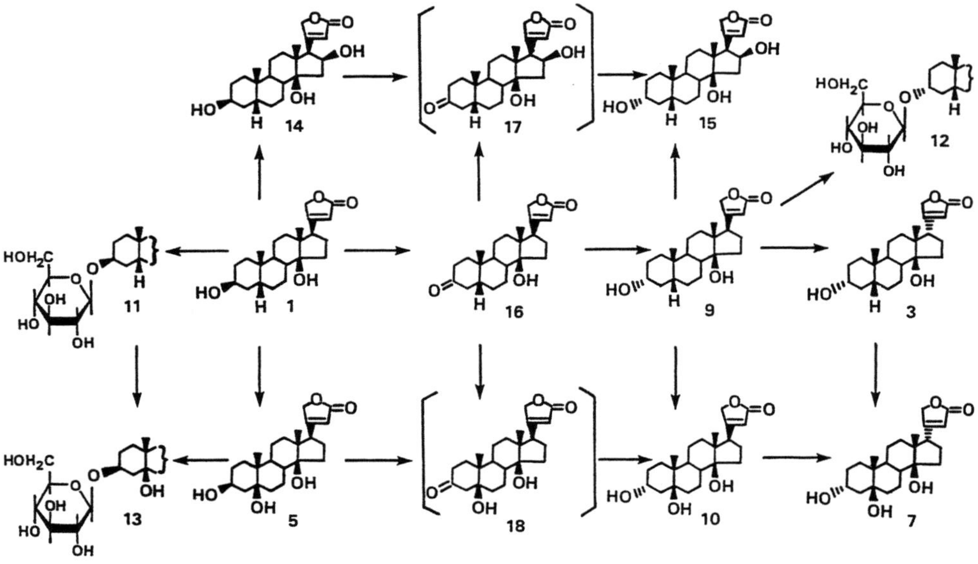

Fig. 7. Biotransformation of digitoxigenin (1) by cell suspension cultures of *Strophanthus intermedius*. (Kawaguchi et al. 1989)

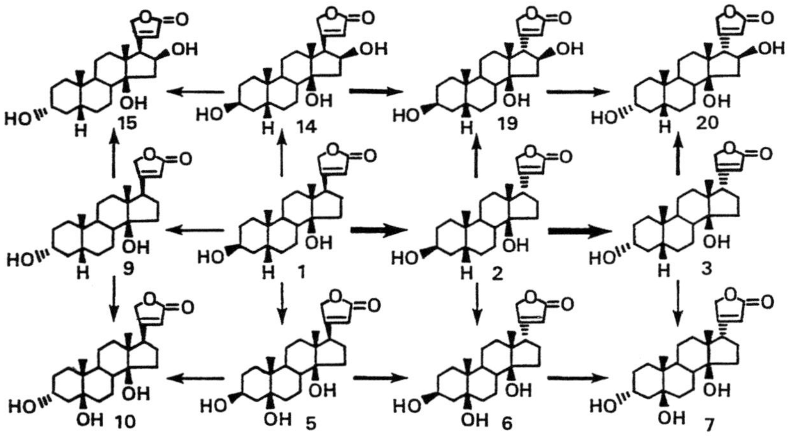

Fig. 8. Biotransformation of digitoxigenin (1) by cell suspension cultures of *Strophanthus divaricatus*. (Kawaguchi et al. 1991)

S. divaricatus (540 g fresh wt.) after 18 days of incubation. At the same time, three products (**5**, **9**, and **10**) were identified with authentic samples by TLC and HPLC. Product **3**, 3-*epi*-17βH-digitoxigenin, was the main product in this experiment (yield 17.9%). Although product **19** was synthesized chemically (Russel et al. 1960), product **20** has not been reported previously as either a naturally occurring or chemically synthesized substance.

The possible biotransformation pathway of **1** by cell suspension cultures of *S. divaricatus* is shown in Fig. 8. It was demonstrated that isomerization of the 17β-lactone ring (**1** to **2**, **5** to **6**, **14** to **19**, **9** to **3**, **10** to **7** and **15** to **20**); epimerization of the 3β-hydroxyl to the 3α-hydroxyl (**1** to **9**, **2** to **3**, **5** to **10**, **6** to **7**, **14** to **15** and **19** to **20**); 5β-hydroxylation (**1** to **5**, **2** to **6**, **3** to **7** and **9** to **10**); and 16β-hydroxylation (**1** to **14**, **2** to **19**, **3** to **20** and **9** to **15**) proceed in these cells. Isomerization of the 17β-lactone ring in **1**, the highest activity (ca. 95% in total yield of the products) in these cells, was performed for the first time with *S. gratus* cells to a great extent, and slightly with *S. intermedius* cells.

Divostroside and divaricoside, whose aglycone is sarmentogenin, having a 11α-hydroxyl group, were isolated from the regenerated plants of *S. divaricatus* through the calli (see Sect. 2.3). On the other hand, although 16β-hydroxylation of the digitoxigenin (**1**) molecule was demonstrated distinctly, 11α-hydroxylation was not found in these cells. Moreover, glycosylation could not be demonstrated with glucose, oleandrose, or diginose in these suspension cultures.

2.3 Cardenolides from the Regenerated Plants of *Strophanthus divaricatus*

Four cardenolides were isolated from the $CHCl_3$ extract of the aerial parts of the regenerated plants grown in pots. These cardenolides were confirmed as 17βH-divostroside, divostroside, 17βH-divaricoside, and divaricoside (Fig. 9). Moreover, the structural determination of another cardenolide is now in progress. Chen et al. (1987) reported on the isolation and structures of 12 cardenolide glycosides including divostroside and divaricoside from the leaves, stems, and roots of *S. divaricatus*. On the other hand, 17βH-divostroside and 17βH-divaricoside were isolated as new compounds in our experiment.

There is interest in the relationship between the existence of 17βH-cardenolides in the regenerated plants and the highest activity of isomerization of the 17β-lactone ring in cell suspension cultures of *S. divaricatus*. Moreover, it is noteworthy for the expression of totipotency of plant cells in vitro that 11α-hydroxylation and glycosylation with oleandrose or diginose are observed distinctly in the regenerated plants as well as the original plants.

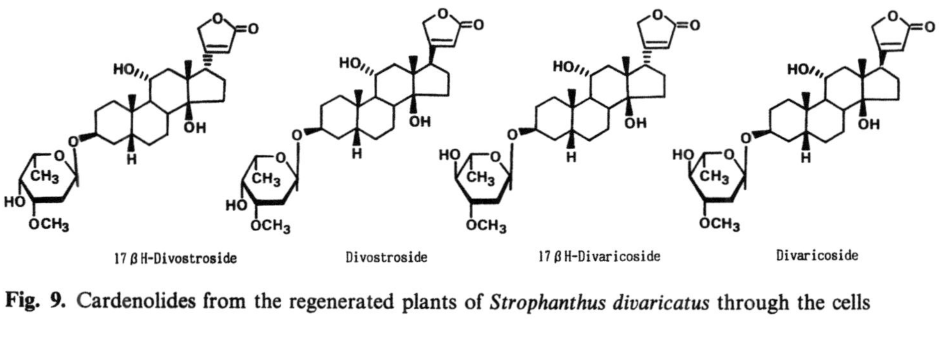

Fig. 9. Cardenolides from the regenerated plants of *Strophanthus divaricatus* through the cells

3 Conclusions and Prospects

Three new compounds, nine new biotransformation products, and five known products were obtained by the biotransformation of digitoxigenin (**1**) using four species of *Strophanthus* cultured cells. Isomerization of the lactone ring and 1β- and 4β-hydroxylation in *S. gratus* cells, glucosylation in *S. amboensis* cells, 16β-hydroxylation in *S. intermedius* cells and isomerization of the lactone ring and 16β-hydroxylation in *S. divaricatus* cells were observed as characteristic biotransformation reactions of different origin. However, 11α- and 19-hydroxylations, the indispensable reactions for G-strophanthin production, could not be demonstrated with these *Strophanthus* cultured cells (Fig. 10). Moreover, we are interested in the relation between the enzymes for 1β-, 4β-, 5β-, and 16β-hydroxylation of **1** in these cells and the digitoxin 12β-hydroxylase from cell cultures of *Digitalis lanata* (Petersen and Seitz 1988).

On the other hand, it seems that epimerization of the 3β-hydroxyl to the 3α-hydroxyl (probably via 3-keto) and 5β-hydroxylation of **1** are common reactions in cell and tissue cultures of plants with and without cardenolides (Table 3).

Recently, five new compounds were obtained as biotransformation products of digitoxigenin (**1**) by ginseng hairy root cultures (Kawaguchi et al. 1990). The hairy roots showed high glycosylation and esterification abilities to the digitoxigenin (**1**) molecule. Therefore, the various biotransformation abilities of plant cell and tissue cultures from different origins will possibly contribute to the provision of new and more effective and safer cardiac glycosides, some of which may prove useful in the pharmaceutical industry (Spieler et al. 1985; Kreis and Reinhard 1990). For example, we observed that **15**-diacetate possessed the highest inotropic activity in the biotransformation products obtained with the *Strophanthus* cells, in a preliminary experiment using a guinea-pig left-atrial preparation (Furuya et al., unpubl.). Similarly, Humber et al. (1985) had already reported on the synthesis and biological activity of the four diastereomeric 3,16-diacetates related to gitoxigenin (**14**) and found high inotropic activity of **15**-diacetate, 7.5 times as potent as digoxin.

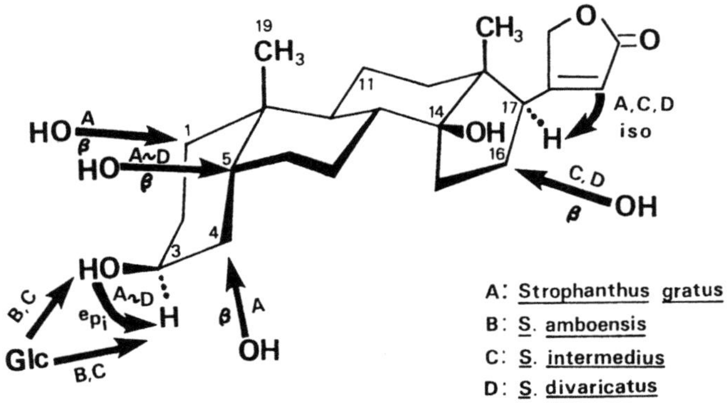

Fig. 10. Biotransformation reactions of the digitoxigenin (**1**) molecule of *Strophanthus* cultured cells

Table 3. Biotransformation reactions of the digitoxigenin (**1**) molecule by plant cell cultures

Plant species	Oxidation (3β-OH → 3-keto)	Epimerization (3β-OH → 3α-OH)	Hydroxylation (1β-	4β-	5β-	12β-	16β-)	Glycosylation	Isomerization (17β- → 17α-lactone ring)	Esterification
Digitalis lanata	+					+		+		
D. purpurea	+	+			+			+		
Thevetia neriifolia	+							+		
Daucus carota					+					
Strophanthus gratus		+	+	+	+				+	
S. amboensis		+			+			+		
S. intermedius	+	+			+		+	+	+	
S. divaricatus		+			+		+		+	
Panax ginseng		+			+			+ (Gentiobioside, sophoroside)		+ (Stearate, palmitate, myristate, laurate)

Cardenolide production could not be observed in these calli, which were not only from different species (*S. gratus*, *S. amboensis*, *S. intermedius*, and *S. divaricatus*) but also from different parts (leaves, stems, and flower buds). On the other hand, cardenolide production was demonstrated in the regenerated plants of *S. divaricatus* through the calli. So, cardenolide production in *Strophanthus* species seems to correlate strictly with morphological differentiation, as in that of *Digitalis purpurea*, reported by Hirotani and Furuya (1977). Also, Hagimori et al. (1984) reported on digitoxin production by jar fermenter culture of shoot-forming cultures of *D. purpurea* using a revised medium. More recently, Saito et al. (1990) investigated production of cardenolides in the green hairy roots transformed using an *Agrobacterium*-Ri plasmid binary vector in *D. purpurea*. In the future, it is expected that utilization of the shoot-forming cultures or the green hairy-root cultures will be profitable in the production of *Strophanthus* cardenolides.

4 Protocol

Explants from the leaves, the volubile stems, and the flower buds of *S. gratus* from calli within 2 weeks on solid MS medium (1 ppm 2,4-D; 0.1 ppm K) at 25 °C in the dark. Subcultured at 4-week intervals, the Sg-3 callus obtained from the flower buds grows the most vigorously at 30 °C in the dark. Similarly, the Sa-1 callus from the leaves of *S. amboensis* and the Si-2 callus from the stems of *S. intermedius* can be induced and maintained at 30 °C in the dark or under light (3000–5000 lx, 16 h). Callus cultures of *S. divaricatus* can be obtained by germinating surface-sterilized seeds under aseptic conditions and setting ca. 3-mm hypocotyl segments on the same medium at 25 °C in the dark. Suspension cultures for the biotransformation experiments are readily established by placing calli in liquid MS medium (1 ppm, 2,4-D; 0.1 ppm K) and using reciprocal shaking (90 spm = stroke per minute). No cardenolides can be detected by TLC analysis of extracts of these calli.

MS medium (0.5 ppm K) is suitable for the shoot formation through the embryoids regenerated from the callus of *S. divaricatus* (at 25 °C under light). The shoot can also be obtained by transferring the callus into MS medium without plant growth regulator at 15 °C at 300–500 lx for 16 h. When complete plantlets are needed, these shoots are grown on fresh MS medium at 25 °C under light. After 4–5 weeks, rooted plantlets are moved to small pots containing vermiculite. Maintain plantlets for about 2 weeks under a plastic cover in order to avoid desiccation and then in a plant growth chamber at 25 ± 3 °C under light. Two months later, transfer the regenerated plants to large pots with soil. The production of cardenolides is observed in the aerial parts of the regenerated plants (50–130 cm height, about 6 months later) as described in Sects. 2.2 and 2.3.

References

Abe F, Yamauchi T (1977) Studies on *Cerbera*. I. Cardiac glycosides in the seeds, bark, and leaves of *Cerbera manghas* L. Chem Pharm Bull 25: 2744–2748

Antman EM, Smith TW (1986) Pharmacokinetics of digitalis glycosides. In: Smith TW (ed) Digitalis glycosides. Grune & Stratton, Orlando, pp 45–60

Brenneisen K, Euw JV, Tamm Ch, Reichstein T (1964) Die Cardenolide der Samen von *Strophanthus vanderijstii* Staner 3. Mitteilung. Identifizierung und Konstitutionsermittlung der wasserlöslischen Glykoside. Helv Chim Acta 47: 814–827

Charney W, Herzog HL (eds) (1967) Microbial transformation of steroids. Acad Press, New York
Chen R-F, Abe F, Yamauchi T, Taki M (1987) Cardenolide glycosides of *Strophanthus divaricatus*. Phytochemistry 26: 2351–2355
Claus EP, Tyler VE, Brady LR (1970) Cardioactive glycosides. In: Pharmacognosy, 6th edn. Lea and Febiger, Philadelphia, pp 81–96
Dohnal B, Rozkrutowa B, Supniewska JH (1981) Growth and phytochemical analysis of *Strophanthus amboensis* tissue culture. Herba Hung 20: 83–90
Döller PC, Reinhard E (1979) Biotransformation of cardenolides. Comparative studies with cell cultures of *Thevetia neriifolia* and *Digitalis lanata*. Planta Med 37: 277–288
Evans WC (1989) Cardio-active drugs. In: Trease and Evan's pharmacognosy, 13th edn. Bailliere Tindall, London, pp 500–517
Furuya T, Hirotani M, Shinohara T (1970) Biotransformation of digitoxin by suspension callus culture of *Digitalis purpurea*. Chem Pharm Bull 18: 1080–1081
Furuya T, Kawaguchi K, Hirotani M (1973) Biotransformation of progesterone by suspension cultures of *Digitalis purpurea* cultured cells. Phytochemistry 12: 1621–1626
Furuya T, Kawaguchi K, Hirotani M (1988) Biotransformation of digitoxigenin by cell suspension cultures of *Strophanthus gratus*. Phytochemistry 27: 2129–2134
Hagimori M, Matsumoto T, Mikami Y (1984) Jar fermenter culture of shoot-forming cultures of *Digitalis purpurea* L. using a revised medium. Agric Biol Chem 48: 965–970
Hegnauer R (1964) Apocynaceae 2. Herzgifte. In: Chemotaxonomie der Pflanzen, Band 3. Birkhauser, Basel, pp 137–144
Herrmann I, Repke K (1964) Epimerisierung und Hydroxylierung von digitaloiden Steroidlactonen durch Fermente der Leber. Arch Exp Pathol Pharmakol 248: 351–369
Hirotani M, Furuya T (1977) Restoration of cardenolide-synthesis in redifferentiated shoots from callus cultures of *Digitalis purpurea*. Phytochemistry 16: 610–611
Hirotani M, Furuya T (1980a) Biotransformation of steroids and alkaloids by plant cell cultures. Shoyakugaku Zasshi 34: 1–7
Hirotani M, Furuya T (1980b) Biotransformation of digitoxigenin by cell suspension cultures of *Digitalis purpurea*. Phytochemistry 19: 531–534
Humber DC, Jones PS, Phillipps GH (1985) Synthesis and biological activity of some cardiotonic compounds related to gitoxigenin. Steroids 45: 19–31
Iizuka H, Naito A (1967) Microbial transformation of steroids and alkaloids. Univ Tokyo Press and Univ Park Press, Tokyo
Jones A, Veliky IA, Ozubko RS (1978) Biotransformation of cardenolides by plant cell suspension cultures. I. Isolation and identification of periplogenin from cultures of *Daucus carota* Ca68 incubated with digitoxigenin. Lloydia 41: 476–487
Junior P, Wichtl M (1980) 3-Epi-periplogenin: ein neues Cardenolid aus *Adonis vernalis*. Phytochemistry 19: 2193–2197
Kawaguchi K, Hirotani M, Furuya T (1988) Biotransformation of digitoxigenin by cell suspension cultures of *Strophanthus amboensis*. Phytochemistry 27: 3475–3479
Kawaguchi K, Hirotani M, Furuya T (1989) Biotransformation of digitoxigenin by cell suspension cultures of *Strophanthus intermedius*. Phytochemistry 28: 1093–1097
Kawaguchi K, Hirotani M, Yoshikawa T, Furuya T (1990) Biotransformation of digitoxigenin by ginseng hairy root cultures. Phytochemistry 29: 837–843
Kawaguchi K, Hirotani M, Furuya T (1991) Biotransformation of digitoxigenin by cell suspension cultures of *Strophanthus divaricatus*. Phytochemistry 30: 1503–1506
Kreis W, Reinhard E (1990) Production of deacetyllanatoside C by *Digitalis lanata* cell cultures. In: Nijkamp HJJ, van der Plas LHW, van Aartrijk J (eds) Progress in plant cellular and molecular biology. Kluwer Acad Publishers, Dordrecht, pp 706–711
Morton JF (1977) Smooth *Strophanthus* and green *Strophanthus*. In: Major medicinal plants. Botany, culture and uses. Charles C Thomas, Springfield, Illinois, pp 258–267
Murashige T, Skoog F (1962) A revised medium for rapid growth and bioassays with tobacco tissue cultures. Physiol Plant 15: 473–497
Petersen M, Seitz HU (1988) Reconstitution of cytochrome *P*-450-dependent digitoxin 12β-hydroxylase from cell cultures of foxglove (*Digitalis lanata* EHRH.). Biochem J 252: 537–543
Reichstein T (1967) Cardenolid- und Pregnanglykoside. Naturwissenschaften 54: 53–67
Reinhard E, Alfermann AW (1980) Biotransformation by plant cell cultures. In: Fiechter A (ed) Advances in biochemical engineering, vol 16. Springer, Berlin Heidelberg New York, pp. 49–83

Reynolds JEF (ed) (1989) Cardiac inotropic agents. In: Martindale – the extra pharmacopoeia, 29th edn. The Pharmaceutical Press, London, pp 822–834

Rücker W (1988) *Digitalis* spp.: in vitro culture, regeneration, and the production of cardenolides and other secondary products. In: Bajaj YPS (ed) Biotechnology in agriculture and forestry. vol. 4. Medicinal and aromatic plants I. Springer, Berlin Heidelberg New York, pp 388–418

Russel JH, Schindler O, Reichstein T (1960) 17α-Corotoxigenin, 17α-Coroglaucigenin, 17α-Gitoxigenin sowie vereinfachte Methode zur Herstellung von 16-Anhydro-gitoxigenin. Helv Chim Acta 43: 167–176

Saito K, Yamazaki M, Shimomura K, Yoshimatsu K, Murakoshi I (1990) Genetic transformation of foxglove (*Digitalis purpurea*) by chimeric foreign genes and production of cardioactive glycosides. Plant Cell Rep 9: 121–124

Singh B, Rastogi RP (1970) Cardenolides-glycosides and genins. Phytochemistry 9: 315–331

Spieler H, Alfermann AW, Reinhard E (1985) Biotransformation of β-methyldigitoxin by cell cultures of *Digitalis lanata* in airlift and stirred tank reactors. Appl Microbiol Biotechnol 23: 1–4

Stohs SJ, Rosenberg H (1975) Steroids and steroid metabolism in plant tissue cultures. Lloydia 38: 181–194

Yoshikawa T, Furuya T (1979) Purification and properties of sterol: UDPG glucosyltransferase in cell culture of *Digitalis purpurea*. Phytochemistry 18: 239–241

XXVI *Tagetes* spp. (Marigolds): In Vitro Culture and the Production of Thiophenes

H. BRETELER and D.H. KETEL[1]

1 Introduction

Tagetes species were used by ancient civilizations like the Aztecs for various purposes (Neher 1968). The pigments of the flowers were used as a dye and in chicken feed, oil was extracted from the leaves and used as an ingredient of perfumes, and the roots were also assumed to have interesting properties. Field tests in the USA in the 1930s showed that larvae of a root-knot nematode entered the roots of marigolds, but usually failed to develop and neither reached the adult stage nor produced eggs (Steiner 1941). In 1953, a Dutch bulb breeder (Van de Berg-Smit 1953) reported the biological activity of common garden marigolds (*Tagetes patula*) against root rot in *Narcissus* caused by free-living nematodes. The latter finding was an incentive for a scientific analysis of the effect of *Tagetes* plants by the crop protection industry and the academic world. A few years after the initial report by Van de Berg-Smit (1953), Uhlenbroek and Bijloo (1958, 1959, 1960) isolated and described some active principles from *Tagetes* plants. These chemicals belonged to a group of heterocyclic sulphur-containing compounds, the thiophenes. The thiophene α-terthienyl, which occurs in *Tagetes* and related species, was first synthesized in 1941 (Steinkopf et al. 1941) and isolated from plants in 1947 (Zechmeister and Sease 1947).

In the past three decades, much more has been discovered about these bioactive molecules. Recently (Lam et al. 1988), a conference was held on the chemistry and biology of naturally occurring acetylenes and related compounds, with an emphasis on thiophenes. The proceedings of this conference report on the various aspects of thiophenes, e.g., their ecological significance, their biological activity, and their potential applications. Although various patents have been granted, no large-scale commercial production and application of thiophenes has been achieved. The 1988 conference proceedings make clear that many fundamental hurdles are still to be crossed. Therefore, the area is still a subject of active interdisciplinary research in Europe, North America, and Asia.

The biocidal activity of thiophenes has also prompted the study of thiophene metabolism in cell and tissue culture: on the one hand, to have access to a relatively simple and reproducible biological model system and on the other, to eventually incorporate these compounds in a process based on modern plant biotechnology, either to enhance the resistance against plant diseases or to produce specific

[1] Centre for Agrobiological Research (CABO), P.O. Box 14, 6700 AA Wageningen, The Netherlands

Biotechnology in Agriculture and Forestry, Vol. 21
Medicinal and Aromatic Plants IV (ed. by Y.P.S. Bajaj)

compounds by cultured plant cells. Although species related to *Tagetes* and other compounds than thiophenes were part of these studies, emphasis seems to be on thiophene accumulation in a few *Tagetes* species. In this chapter we will summarize the work on cell and tissue culture of *Tagetes* spp. and where appropriate we will refer to studies with intact plants.

2 Characteristics of *Tagetes* Plants

Tagetes species belong to the Asteraceae family. This family comprises ca. 10% of all flowering plant species and contains many tribes with significant levels of polyacetylenes and related compounds, such as thiophenes (Bohlmann et al. 1973; Bohlmann and Zdero 1985). Especially the subtribe Tageteae (tribe Helenieae) is rich in thiophenes. Probably because of its widespread use as an ornamental plant, the research on natural thiophenes has focused on a few of the ca. 20 *Tagetes* species, mainly *T. patula* (Fig. 1) and *T. erecta.*

Fig. 1. An adult marigold plant (*Tagetes patula* L.)

Tagetes spp. are strongly scented herbaceous or woody annual and perannual plants, with South-central Mexico as the area of the greatest diversity. Within the genus, there is a great variation in plant size, from a few decimeters in cultivated marigolds to many meters in the bushes of *T. minuta* in Central America. In a greenhouse in Holland, the latter species could attain a height of more than 5 m.

In small-scale horticulture, marigolds are used for crop protection. Their use has been reported as a source of insecticides in Mexico (von Reis Altschul 1973) and for medicinal purposes in the Philippines (Quisumbing 1951), China (Duke and Ayensu 1985), India and Pakistan (Watt 1889–1896), Malaysia (Burkill 1966), and South America (Mihalik 1978/1979). In a review on the ethnobotany of *Tagetes*, Neher (1968) reports that these species are used in folk medicine as an antiseptic, analgesic, carminative, diuretic, expellant, stimulant, vermifuge, vermin repellant, and aphrodisiac. Apart from agriculture and medicine, the plants are used for a variety of mystical and magical ceremonies in Latin America, India, and East Africa. It may well be that the alleged properties, already exploited in history, relate to the same specific phytochemicals which are studied nowadays in cell and tissue culture.

3 Thiophenes

Basically, thiophenes are heterocyclic, sulfur-containing compounds with one S-atom and four C-atoms per pentagon as the building block:

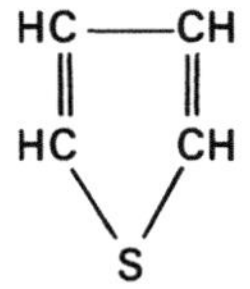

In *Tagetes* monothienyls are seldom found. Almost all naturally occurring thiophenes are di- and terthienyls as α-polymers (Table 1). α-Terthienyl is sometimes found in a mono(-α-)methylated form. The natural dithienyls commonly have one acetylenic α-side chain with four C-atoms. This side chain may contain a terminal double bond (BBT), contain OH-groups to form a mono- (BBTOH) or a dialcohol ($BBT(OH)_2$), or an acetate group in mono- (BBTOAc) or diacetate ($BBT(OAc)_2$). In the case of a dialcohol, partial acetate substitution yields a mono-alcohol-monoacetate [BBT(OH)(OAc)]. In other Asteraceae, and in at least one *Tagetes* sp. (*T. pauciloba*), monothienyl derivatives also occur. Cyclic dithio-polyacetylenes like thiarubrines are also found in Asteraceae but not in *Tagetes* (Bohlmann et al. 1973). The total number of thiophene compounds in the Asteraceae exceeds 150.

The biological activity of thiophenes depends on irradiation of the molecule by UVA light (Gommers et al. 1982). In the presence of triplet oxygen, the excited state of thiophenes generates highly reactive singlet oxygen molecules (Bakker et al. 1979), which interact with amino acids in proteins and other biomolecules in

Table 1. Molecular structures of the prevailing thiophenes in *Tagetes* species. (Croes et al. 1989b)

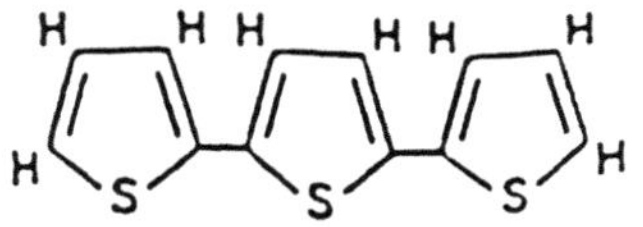	α-T α-terthienyl
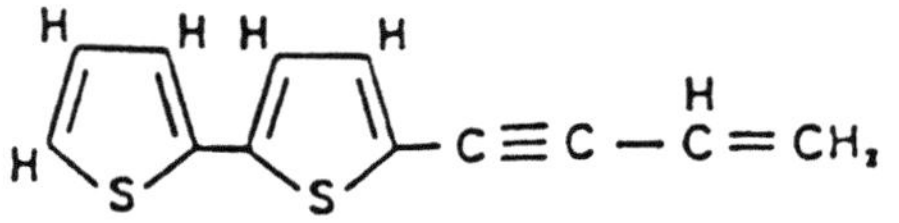	BBT butenylbithiophene
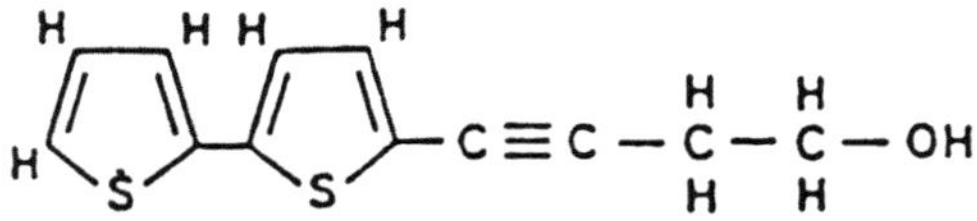	BBTOH hydroxybutenylbithiophene
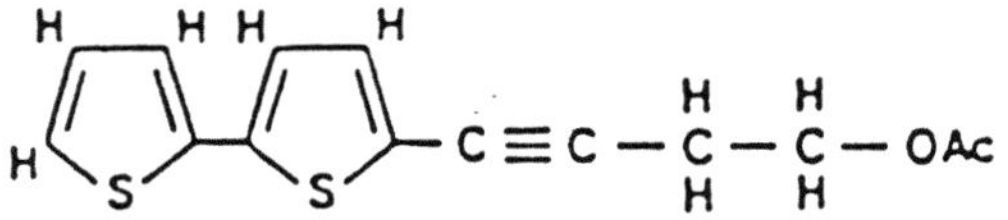	BBTOAc acetoxybutenylbithiophene
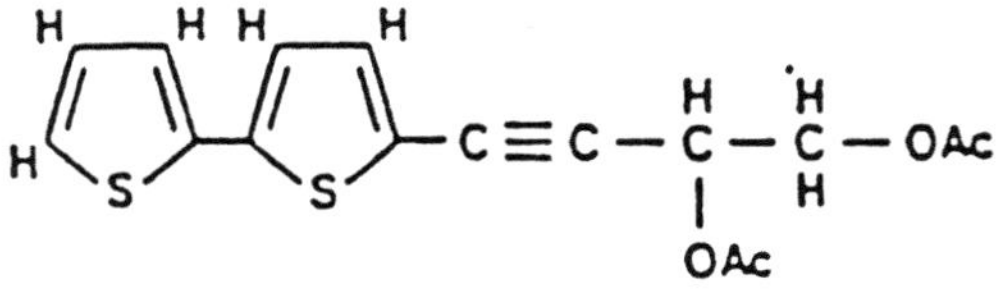	$BBT(OAc)_2$ diacetoxybutenylbithiophene

nematodes and other organisms (Gommers et al. 1980). The classification as photosensitizers, however, poses the question how di- or terthienyls in plant roots can be the active nematicidal principle in a soil environment devoid of light. Recently Wijnberg (1988) proposed a process tentatively called "photochemistry without light". This process requires the presence of plant peroxidases and the subsequent conversion of unsaturated compounds (fats, oils, terpenes) to hydroperoxide and 1,2-dioxetanes. During the decomposition of the latter compounds, excited-state carboxyl compounds are formed and transfer their energy to thiophene molecules. It is unknown what kind of in vivo mechanism operates in plant roots in the dark.

The photodynamic nature of thiophenes has an important bearing on the possible application of thiophenes.

1. Since photoactivation of thiophenes was unknown in the late 1950s, soil application of thiophenes was unsuccessful.
2. The cascade of events which may be required for a desired effect of thiophenes may explain many erratic results obtained so far.
3. The rapid degradation of thiophenes in normal daylight (Philogène et al. 1985; Sütfeld and Breteler 1988) may be an advantage in some and a disadvantage in

other aspects of application. Recently, Lam and Thomasen (1988) have shown that the degradation of thiophenes in the light can be considerably retarded by simple compounds like caffeine.

4. The intensity, the duration, and the quality of illumination may be an important factor in the control of thiophene metabolism in cell and tissue culture, and during the extraction and analysis of thiophenes from biological materials.

The biogenesis of the various thiophenes is largely unknown. There is circumstantial evidence for the involvement of acetylenic precursors, such as dehydromatricaria ester and crepenynic acid. A scheme proposed by Jente et al. (1981) supposes the progressive desaturation of fatty acids to C_{18} polyacetylenes and the incorporation of a suitable S-source into pentayneene (C_{13}). Jente's long-term experiments with tritiated precursors given to soil-grown *Tagetes* yielded the only data on the overall biosynthetic pathway known. The interest in cell and tissue culture of *Tagetes* has recently led to studies with ^{35}S to gain knowledge of the source of sulfur used in biosynthesis. Despite the scarcity of data, some schemes for biosynthesis have been proposed. A scheme is given in Fig. 2. Sütfeld and coworkers (Sütfeld and Towers 1982; Pensl and Sütfeld 1985; Sütfeld 1987, 1988) are the only investigators who have isolated and characterized enzymes, such as BBTOAc:acetate esterase and dithiophenol (acetyl-CoA):acetyltransferase (Metschulat and Sütfeld 1988), involved in thiophene metabolism. Acetate esterase activity was also found in callus and cell cultures, although to a much lesser extent than in the aerial parts of intact plants (Helsper et al. 1988a). These studies, however, are limited to the interconversion of some end products of the hypothetic pathway: the mono- and dialcohol, and the mono- and diacetate of BBT. Therefore, major unsolved questions remain about the mechanism by which polyacetylene chains condense with S-atoms to become thiophene molecules. The very low concentration of the short-lived and unstable polyacetylenes in plants, e.g., pentayneene, seriously hampers progress in this field.

Gradually, modifications of the conditions for the extraction of thiophenes from plant material and for their separation by HPLC yielded quick and reliable analyses, for example, those described by Sütfeld (1987, 1989), Sütfeld and Breteler (1988), and Croes et al. (1989a). The use of HPLC combined with multidiode array

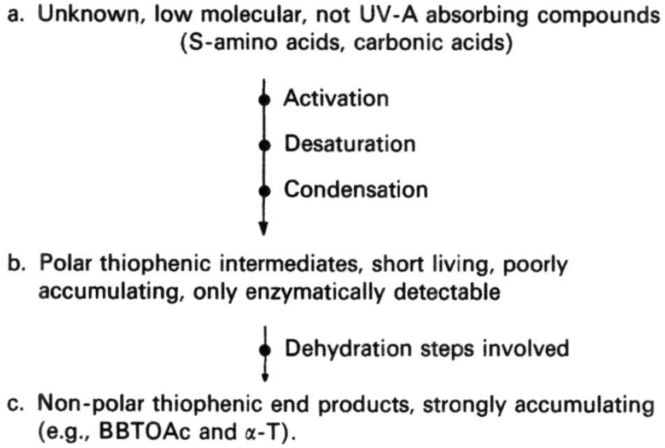

Fig. 2. Hypothetical scheme for the biosynthesis of thiophenes based on investigations by various authors. (Ketel 1987b)

UV-detection and computerized storage of spectra has shown that, in intact plants, thiophenes occur in all organs. The distribution of thiophenes, however, depends on species, plant age, and environmental conditions. The highest level of thiophenes in particular organs of the intact plants is in the range of some mg/g dry matter (DM). The relatively polar thiophenes BBTOH and $BBT(OH)_2$ are often most abundant in the roots of *Tagetes* (Sütfeld 1982). BBTOH is the prevailing thiophene in the rhizosphere of *T. patula* (Tang et al. 1987). Because of their relatively high solubility in aqueous media, much biotechnological work has focused on the production and the release of dithenylalcohols (see Sect. 4.2).

4 In Vitro Culture of *Tagetes*

The work reported in the literature on tissue culture is restricted to the three species *T. erecta*, *T. patula*, and *T. minuta*. We will briefly discuss the culture of plantlets in vivo and in vitro, the induction of callus, and the preparation of cell cultures and normal and transgenic root cultures. For all cultures, the results of thiophene accumulation and growth will be presented; results from fermentor studies will also be given.

4.1 Source of Explant and Callus Culture

Explants for the proliferation of callus can be obtained from various organs of plants grown in soil or nutrient solution, e.g., roots, achenes, stems, leaves, and flowers. We have successfully grown *T. erecta*, *T. patula*, and *T. minuta* in aerated nutrient solution with 5 mM $Ca(NO_3)_2$ as the N-source (Ketel et al. 1985). Seeds were germinated on paper moistened with demineralized water. The plants were grown at 24 °C, a 16-h photoperiod (30 W/m^2), and 75% RH.

Plants of all ages beyond 2 to 3 weeks could be used as a source of explants. The selected explants were sterilized for 2 s in 70% (v/v) EtOH and 14 min in 5% (w/v) $Ca(OCl)_2$, and subsequently washed for 3, 10, and 30 min in fresh aliquots of sterile demineralized water. Apical leaf and root sections of approximately similar size (length 1.5 cm) were excised and placed upright into a tube (16×2.5 cm ϕ) with 15 cm^3 MS medium (Murashige and Skoog 1962) supplemented with 2% (w/v) sucrose, 0.5 mg/dm^3 naphthylacetate, and 5 mg/dm^3 benzyladenine, and gelled with 0.2% (w/v) Gelrite (Costar Ltd., Badhoevedorp, The Netherlands). The pH was adjusted to 5.9 with KOH prior to autoclaving (20 min, 120 °C). Axenic calli developed at 24 ± 1 °C and a 16-h photoperiod (5 W/m^2; Philips TLW 33). After 3 weeks, primary (C_1) calli were transferred to fresh MS medium to obtain secondary (C_2) calli. To obtain C_3, C_4, etc. callus, the C_2, C_3, etc. calli were transferred every 2 weeks. When rapid multiplication of the biomass of callus was required, the calli were cut into two or three pieces and subcultured according to the standard method of callus culture, and Erlenmeyer flasks (150 or 250 cm^3) were used instead of tubes.

Due to differences in morphological structure and in the capacity to synthesize secondary metabolites, the selection of explants is of paramount importance to obtain productive callus and cell cultures (Wiermann 1981). In addition, the resistance of the explants against disinfection measures differs considerably between explant tissues. An explant was considered suitable when it yielded reasonable quantities of undifferentiated and friable callus, which was still capable of thiophene biosynthesis.

From plants grown in sandy soil, the leaves could be used as an explant with some success. The results could be drastically improved when plants grown on a regularly refreshed nutrient solution were used. This difference may be due to a lower level of microbial contamination in the leaves of such plants (Evans and Cocking 1977). In general, 95–100% of the leaf explants from the solution-grown plants survived sterilization and produced callus. In contrast, this percentage was below 50% for other organs such as roots and stems. Therefore, leaves from plants in nutrient solution grown under controlled conditions were used as the standard explant in our experiments. The best and most reproducible results were obtained with leaves from 5–7-week-old plants. Leaves of very young plants often failed to survive sterilization, while in leaves from older plants the infection rate rose gradually. The average weight of primary calli (C_1) of *T. minuta* obtained from leaves varying in age between 3 and 13 weeks was ca. 0.8 g fresh wt. in 3 weeks (Ketel et al. 1985). The growth rate of tertiary calli (C_3) obtained from the C_2-calli was between that of the C_1 (ca. 0.05 g/day) and C_2-calli (ca. 0.2 g/day) when originating from plants aged 3–8 weeks, and approximated 0.1 g/day. The growth rate of C_3 calli originating from plants older than 8 weeks increased drastically up until ca. 0.7 g/day in correspondence with the increase of the age of the original plants. We assume that cytoplasmic contacts between original leaf cells and the cells of the C_3 calli were gradually disrupted in the course of the C_1 and C_2 callus phases. Therefore, to distinguish them from well-known memory effects in cells (Thellier et al. 1982; Sanchez De Jimenez and Fernandez 1983), we suggested that such reproducible effects of gene expression be called "origin effects". In addition, the gradual increase of the growth rate of C_3 calli obtained from leaves of older *T. minuta* plants corresponded clearly with a gradual decrease of the dry matter content and an increase of the friability of the calli. While the C_1 and C_2 calli were hard and smooth, *T. minuta* C_3-calli had a friable structure and good suspensibility. Therefore, this origin effect was used to obtain fine cell suspensions of *T. minuta* .

In contrast with *T. minuta*, leaf explants of younger and older *T. erecta* plants produced friable C_1 calli. However, in agreement with the results of Kothari and Chandra (1985), the calli of *T. erecta* showed a high potential to form roots and/or shoots. Usually 40–60% of *T. erecta* C_1 and C_2 calli differentiated spontaneously under standard callus conditions (Ketel 1986). The observed differences between *T. minuta* and *T. erecta* calli were not significantly affected when the parent plants were grown in different nutrient solutions. Therefore, the potential of calli to differentiate was ascribed to a phytohormone effect mediated by differential gene activity. For that reason, *T. erecta* calli grown under standard callus conditions were unsuitable for obtaining fine cell suspensions. However, differentiation of calli of *T. erecta* could be suppressed by adding gibberellic acid, suggesting that GA_3 affects the expression of genes involved in callus differentiation (see p. 402).

Table 2. Fresh weight (FW, g per callus) and dry matter content (% DM) of primary (C_1), secondary (C_2) and tertiary (C_3) calli of *Tagetes minuta* and *T. patula* grown on gelled MS medium[a]

	Species	Phase of callus culture		
		C_1	C_2	C_3
Time (weeks)		1–3	4–5	6–7
FW	*T. minuta*	0.59 ± 0.26	3.80 ± 1.74	6.33 ± 1.60
% DM		11.4	10.4	9.4
FW	*T. patula*	1.63 ± 0.90	6.10 ± 1.63	11.80 ± 5.24
% DM		6.0	5.2	4.7

[a] FW data are given ± SD ($7 < n < 15$) (Ketel 1986).

With respect to their friability and the potency to differentiate, calli of *T. patula* were intermediate between the calli of *T. minuta* and *T. erecta*. The DM content of *T. patula* calli was generally half that of *T. minuta* calli (Table 2), which explains the higher friability of *T. patula* calli. In addition, the formation of few small roots and/or shoots on the cauliflower-like surface of *T. patula* calli in at most 20% of the calli (Ketel 1986) does not result in a negative judgement of *T. patula* calli as suitable source of liquid cultures. Minced calli of *T. patula*, however, put into liquid medium do not disintegrate into single cells or small groups of cells, but remain aggregates of considerable size (Fig. 6). After regular subculturing of these liquid cultures, the size of the different aggregates usually varied between ca. 0.5 and 15 mm. In conclusion, an appropriate selection of leaf explants of *Tagetes* species resulted in fine cell suspensions of *T. minuta* and aggregated liquid cultures of *T. patula*.

To prepare cell cultures for the production of secondary metabolites, an explant forming a friable and undifferentiated callus consisting of cells that do not only contain, but also produce the desired compounds is required. Based on the totipotency of plant cells (Muir et al. 1954), this combination of properties of an explant gives promise that newly formed callus cells retain the biosynthetic capacity of the cells of the original tissue. Memory effects may also play a role in plant cell cultures (Thellier et al. 1982; Sanchez De Jimenez and Fernandez 1983). Product formation in intact plants, however, can be considered as the result of a complex pattern of synthesis, transport, and storage processes (Wiermann 1981). Therefore, the presence of secondary compounds in explants gives no direct indication for the site of synthesis of these compounds in the plant organs. In addition, information about the occurrence of secondary metabolites in different plant organs is essential in comparing the level of production in explants and calli.

Leaves of young plants of *T. minuta* contain a higher level of thiophenes than the leaves of comparable plants of *T. patula* and *T. erecta* (Fig. 3). No thiophenes or only traces of these compounds were found in leaves from *T. erecta* plants older than 6 weeks. In contrast, the level of thiophenes in the roots of *T. patula* significantly exceeded the level of thiophenes in the roots of *T. minuta* and *T. erecta*

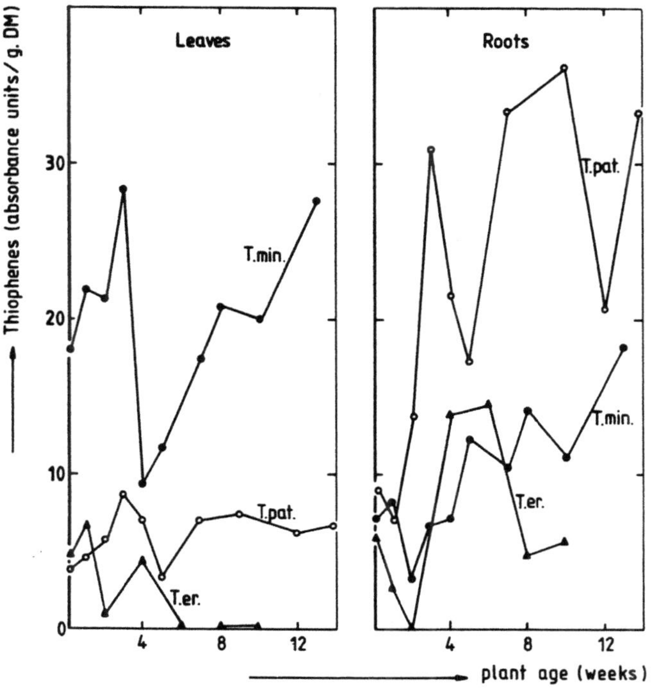

Fig. 3. Total amounts of thiophenes in leaves and roots of *Tagetes erecta* (▲; T. er), *T. patula* (○; T. pat), and *T. minuta* (●; T. min) plants. Data are presented in terms of absorbance units measured at UV peak wavelength: λ max 320 to 380 nm. (Ketel 1987b)

(Fig. 3). The occurrence of the major thiophenes (BBTOH, BBT, BBTOAc, α-T) in leaves and roots of *T. patula* corresponded with the data of Sütfeld (1982). A detailed analysis of the different thiophenes in the leaves and the roots of the different *Tagetes* species, however, made clear that increases and decreases of "leaf thiophenes" with age occurred independently of the variation in "root thiophenes". Some thiophenes occurred only in leaves and not in roots, and vice versa. These results suggest that thiophene metabolism in leaves and roots are regulated differentially or independently (Ketel 1987a).

It has been frequently observed that the production of secondary metabolites in calli neither qualitatively nor quantitatively mimicks the secondary metabolite production in the original explants (Thorpe 1978; Mulder-Krieger et al. 1982; Drawert and Berger 1982). In agreement with such reports, the occurrence of thiophenes in calli of the *Tagetes* species tested did not closely reflect the level of thiophenes in the different explants. In detail, the production of thiophenes in C_1 calli of *T. minuta* amounted to 2–3% of the total level of thiophenes found in the leaves of the plants (Table 3). In contrast with the occurrence of four or five different thiophenes in the leaves, only one or two different thiophenes were found in the calli. The total level of thiophenes in C_2 calli of *T. minuta* usually was one-half that in C_1 calli. This trend was continued in the subsequent passages of these calli, frequently resulting in C_3 calli of *T. minuta* devoid of thiophenes (Table 3).

Table 3. Thiophene concentration in primary (C_1), secondary (C_2), and tertiary (C_3) calli of *Tagetes erecta, T. patula*, and *T. minuta* as percentage of the total thiophene content in the original leaves (cf. Fig. 3)[a]

Species	Plant age (weeks)	$\frac{\text{Thiophene concentration in callus}}{\text{Thiophene concentration in leaves}}$ (× 100)		
		C_1	C_2	C_3
T. erecta	1	9	–[b]	–
	2	17	–	–
	4	3	–	–
	6		–	–
	8		–	–
	10	53	–	–
T. patula	2	16	0	9
	3	7	8	17
	4	–	–	32
	5	10	76	22
	7	2	9	8
	9	1	1	10
	12	1	3	2
	14	2	0	0
T. minuta	3	1	0	0
	4	1	1	0
	5	1	1	0
	7	2	3	0
	8	2	1	0
	10	2	1	0
	13	3	1	0

[a] Leaf explants were taken from different *Tagetes* plants varying in age (Ketel 1987b).
[b] No data available.

A reversed situation was found in differentiated calli of *T. erecta*. The total level of thiophenes in these calli was in the same order of magnitude as in *T. minuta* calli. However, it must be kept in mind, that the leaves of *T. erecta* contained only traces of thiophenes. So, the total level of accumulated thiophenes in C_1 calli of *T. erecta* sometimes exceeded that in the leaf explants (e.g., weeks 6 and 8 in Table 3). In similar experiments, C_1 calli of *T. patula* contained ca. 30 μg thiophenes per g DM. This amount corresponded with 1–16% of the total level of thiophenes found in the leaves of *T. patula* plants (Table 3) and with ca. 90% of the total level of thiophenes in C_1 calli of *T. minuta* and *T. erecta*. However, particularly when *T. patula* calli were obtained from leaves of 3- to 7-weeks-old plants, the total level of thiophenes in cauliflower-like C_2 calli was frequently found to be increased if compared with the C_1 calli. A decrease of the thiophene level in C_3 calli of *T. patula* relative to C_2 calli is also not unusual, but then the thiophene production in subsequent passages of the callus culture demonstrated a wavelike pattern (cf. Fig. 4).

The results from callus studies indicate that morphological and biochemical differentiation are closely related in *Tagetes* species. This relationship conforms to

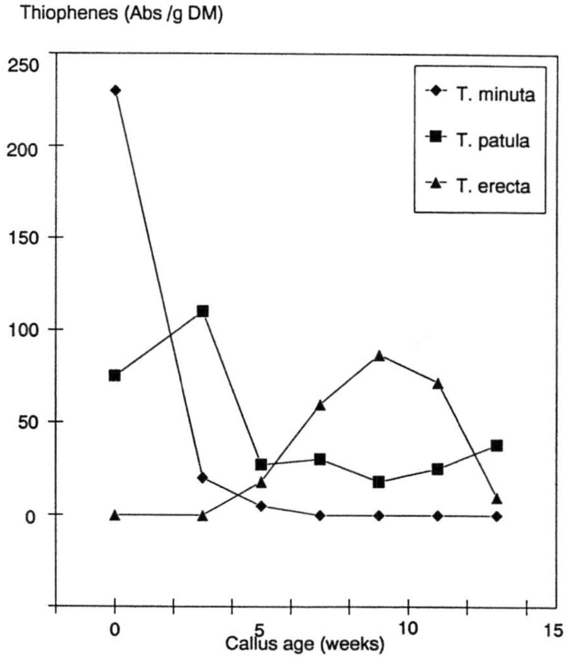

Fig. 4. Thiophene level in calli of varying age of three *Tagetes* species. (Breteler and Ketel, unpubl.)

the results obtained with secondary metabolite production in calli of several dicotyledonous plants (Yeoman et al. 1979) and emphasizes the importance of adequate callus analyses before cell suspensions are prepared (Ketel 1988).

Apart from morphological differentiation, up until now light was found to be the most important factor which influences the absolute levels of thiophenes in callus. In *T. minuta* callus even low light intensities (2 W/m^2) strongly inhibited thiophene accumulation (Croes et al. 1989a). The same light effect was found in cell cultures of *T. patula* where thiophene accumulation was reduced by at least 50%, 2 weeks after transfer of cultures previously grown in the dark to low light conditions (Adamse, pers. comm.). In root cultures of *T. erecta* and *T. patula*, it was established that in darkness the ratio of BBT/BBTOH was above unity. This ratio fell below unity in root cultures grown in light (Ketel, unpubl.). Therefore, light not only affects the absolute levels of thiophenes in cells and tissues of *Tagetes* spp., but also influences the pattern of thiophenes.

4.2 Cell Cultures

Minced *T. minuta* C_3 calli, with a friable and smooth structure, gradually formed single cells and small aggregates in continuously shaken liquid growth medium. This process eventually resulted in a dark-green cell suspension (Fig. 5), the biomass of which started to increase about 1 week after inoculation. Under standard conditions, the doubling time of the cells of a suspension in a 1-dm^3 Erlenmeyer flask was about 2–3 days (μ: ca. 0.30/day). The stock-suspensions were used to start different experiments and were regularly sampled to check the

Fig. 5. Fine granular liquid cultures of *Tagetes minuta*

thiophene content in the cells and the release of thiophenes into the medium. Cauliflower-like C_2 and C_3 calli of *T. patula* were similarly minced and used to start liquid cultures. The initial phase of these cultures generally coincided with an intense browning of the liquid medium. Frequent washing of the small aggregates is then necessary to cope with the browning of the medium. The browning of the medium is not an unusual process and can be considered as a result of the well-known release of polyphenolic compounds (Constabel 1965; Sahai and Shuler 1984). In contrast to minced calli of *T. minuta*, the aggregates of *T. patula* retained their structure and gradually formed bigger aggregates (Fig. 6). Young and friable aggregates, however, were disintegrated regularly, therefore the size of the different aggregates in *T. patula* liquid cultures usually varied between 0.5 and 15 mm.

When the packed cell volume of the aggregates in 1-dm^3 Erlenmeyer flasks was kept between ca. 15–70% of the total volume of liquid, the doubling time of the biomass was approximately 15 days ($\mu = 0.05$/day). Observations with the microscope indicated centers of cell differentiation in the big aggregates of *T. patula* (cf. Street et al. 1965, Helsper et al. 1988b). Such cell differentiation was not found in the small aggregates of *T. minuta* and the smallest aggregates of *T. patula*. We concluded therefore that the process of differentiation in liquid cultures of *T. patula* may considerably influence the growth rate of this type of biomass (Yeoman et al. 1979). In addition, we found that during a number of subcultures the DM content of the aggregates gradually increased (Fig. 7). This trend may result in an eventual cessation of growth of the aggregates, which we often observed. The results obtained with liquid cultures of *T. minuta* and *T. patula* indicate that differences in the potential for differentiation observed during the phase of callus culture continued during the phase of liquid culture. However, the kind of surrounding medium and the differences in gas environment between solid and liquid culture systems exert their effects on the expression of the potential to differentiate.

Cells of *T. minuta* cell suspensions generally contained four to seven different nonpolar compounds. These substances, e.g., sesquiterpenoids and sterols, were

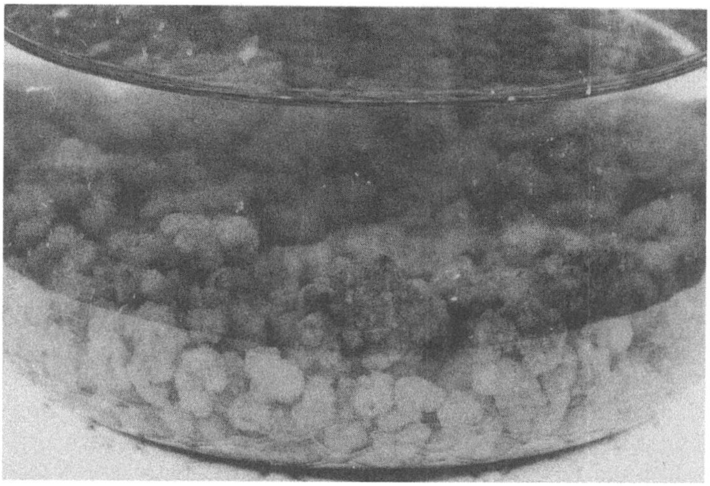

Fig. 6. Aggregated cell culture of *Tagetes patula*. (Ketel 1987b)

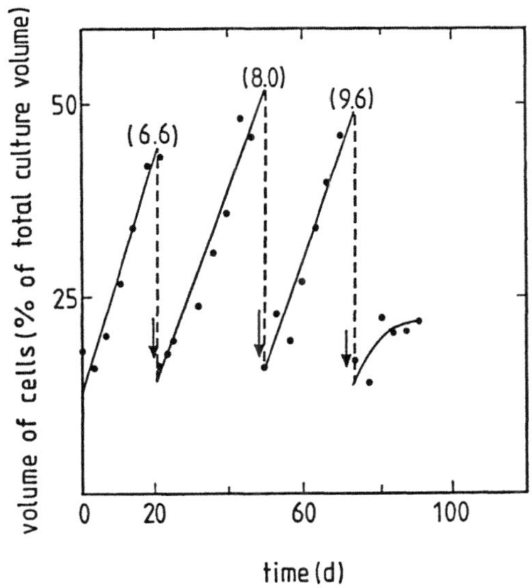

Fig. 7. Growth of a liquid culture of *Tagetes patula* in a 1-dm^3 Erlenmeyer flask. Dilution of cell aggregates in the Erlenmeyer flask to ca. 14% packed cell volume (PCV%) was done when the volume of the cell aggregates was ca. 0.25 dm^3. Measurements of the PCV were done twice a week. DM content of the cell aggregates, in percent of fresh weight (in parentheses) was determined at the end of each growth cycle. (Ketel 1987b)

identified by GC-MS (Groneman et al. 1984). The liquid media of the cell suspensions, however, were generally devoid of these compounds indicating that most of the cellular compounds were tightly associated with cell compartments. Furthermore, analysis over several years and at all phases of the suspension cultures of different subcultures indicated that thiophenes were pertinaciously absent. Also, when cultured *T. minuta* cells were entrapped into different supports (e.g., alginate, agar, K-carrageenan, or agarose), treated at 10 °C, N_2-gassed, or depleted of sugars, and when treated with combinations of entrapment and the mentioned alterations of environmental conditions, no thiophenes could be detected (Ketel et al. 1987).

Entrapment of cells and other stress conditions, however, considerably influenced the accumulation and release of a number of other nonpolar secondary metabolites.

The failure of fine cell suspensions of *T. minuta* to produce thiophenes underlines that the relationship between morphological and biochemical dedifferentiation is apparently strong in this as in many other cases. In essence, the differential effect of entrapment and environmental stress conditions on the production and release of nonthiophenic and thiophenic hexane-soluble metabolites in *T. minuta* cells illustrates that these treatments differently control the production of different secondary metabolites. At the level of gene expression, the differential sensitivity for methylation of cytosine residues in the DNA of the cells may be related to this phenomenon (van Slogteren et al. 1983).

During the various subcultures of liquid aggregate cultures of *T. patula*, the cells and the medium contained varying amounts of different thiophenes. No relationship was found between the occurrence of thiophenes in the cells and in the medium. The major thiophenes found in the aggregates were BBT and BBTOAc, whereas BBTOH was the major thiophene that occurred in the liquid media. Differences in solubility of various thiophenes in an aqueous solution may explain

Fig. 8. Airlift bioreactor of 165 dm^3 with a *Tagetes patula* culture. (Hulst et al. 1987)

this observation. Low amounts of the relatively polar compound BBTOH synthesized in the cells may be easily transported through cell membranes and cell walls, released into the medium, and thus cause a continuous endogenous supply by the cells' biosynthetic activity. The production of thiophenes was essentially maintained when *T. patula* aggregate cultures were grown in fermenters. This observation holds for airlift and stirred tank reactors up to a few liters, and also for a 165-dm^3 airlift fermentor with external loop (Hulst et al. 1987; Fig. 8). However, the pattern of thiophenes released into the medium (Fig. 9) illustrates that the production of thiophenes in liquid cultures of *T. patula* is unstable. This instability of thiophene production in liquid cultures may be due to the increase of the DM content of the cell aggregates (Fig. 7). This process coincided with the secretion of mucilaginous substances around the cells (cf. Street et al. 1965) and may hamper the continuation of the proliferation of young and metabolically active cells upon aging. The eventual effect may be that production of thiophenes gradually slows down. More research is necessary to exploit the observed cell differentiation in the bigger aggregates to develop methods keeping cell aggregates viable and controlling their size. Results obtained with defined aggregate fractions from liquid cultures of *T. patula*, differing in size from 0.1–13 mm (Fig. 10) show that the production of thiophenes is low in small aggregates and increases with the size of the aggregates (Hulst et al. 1989). The production of thiophenes, however, drops already before the maximum size of the aggregates is reached. This relationship may also explain the instability of the production of thiophenes in aggregated

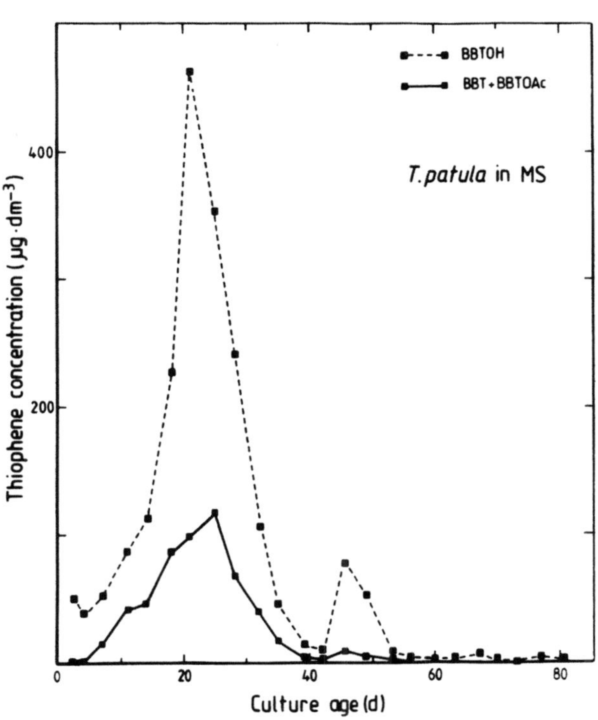

Fig. 9. Release of thiophenes into the medium by aggregated *Tagetes patula* cultures in the course of time. (Ketel, unpubl.)

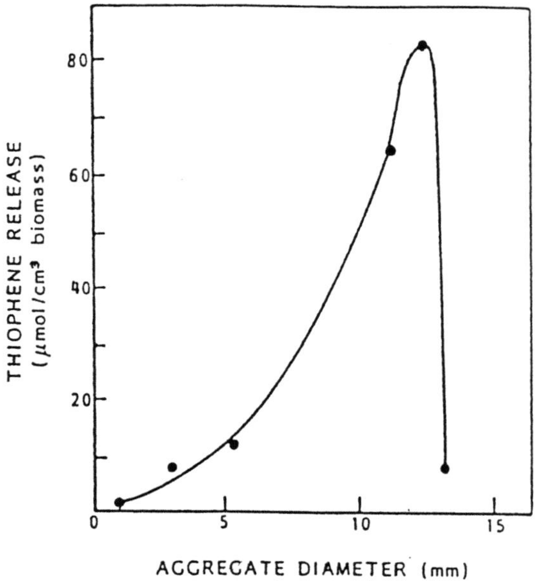

Fig. 10. Thiophene secretion by cell aggregates of *Tagetes patula* of defined diameter ranges. (Helsper et al. 1988b)

liquid cultures of *T. patula* and may also be related to the lack of thiophene production in fine cell suspensions of *T. minuta*.

A fluorescence microscope analysis of protoplasts derived from cultured cells of *T. patula* indicated that the intensity of blue fluorescence per protoplast varies considerably (Adamse 1990). This variation, which is related to differences in thiophene content per cell, may be used in a procedure to select high-yielding cell lines after selection and sorting by means of a flow cytometer (Brown 1984). The main bottle neck in this procedure for *Tagetes patula* appears to be the low frequency and slow rate of protoplast regeneration of this species (Adamse 1990).

The differentiation of *T. erecta* (see p. 393) is strongly sensitive to gibberellic acid (cf. Kothari and Chandra 1985; Ketel unpubl.). Calli treated with low levels of gibberellic acid (ca. 2 mg/dm^3) formed homogeneous and friable cell aggregates, sized 4–6 mm in liquid medium, without root formation. Similar to aggregates of *T. patula*, the aggregates of *T. erecta* released thiophenes, mainly BBTOH, into the medium. However, the level of BBTOH in the medium was low and variable. The level of thiophenes in the aggregates of *T. erecta* (0–11 μg/g DM) was negatively related to the exogenous concentration of gibberellic acid. In the calli of *T. erecta* the same relationship was found, but in these tissues the level of accumulated thiophenes was also (positively) related to the degree of root formation. Exogenous gibberellic acid in the aggregate cultures may hamper the differentiation essential for thiophene production, analogous to cell differentiation in aggregates of *T. patula* (Ketel and Breteler 1988). Gibberellic acid probably does not directly affect the biosynthesis and accumulation of thiophenes in *T. erecta* aggregates, because there was significant discrepancy between the level of thiophenes in (differentiated) root cells and (undifferentiated) callus cells. Gibberellic acid can thus be considered to be an effective component of the growth medium of *T. erecta* aggregates to suppress organ differentiation. However, more research is necessary to establish the merits of gibberellic acid as a stabilizing factor to obtain stable thiophene production.

4.3 Root Cultures

Although research in the past decades elucidated the principles of the mode of action of phytohormones and the use of some of these compounds has been adopted in plant cell cultures, the growth rate of isolated organs remained low when compared to the formation of biomass by fine cell suspensions. A considerable improvement of root cultures was obtained by using transformed roots. Transformed roots can be considered as neoplastic tissues which appear on plants (generally only on dicotyledonous plants) infected with the virulent pathogen *Agrobacterium rhizogenes*. The syndrome on host plants has been described by De Cleene and De Ley (1981), whereas the interaction between the pathogen *Agrobacterium* and its plant host has been subject of intensive studies (reviews by Nester et al. 1984, Birot et al. 1987). The genetic transformation, caused by incorporation of T-DNA from the bacterial plasmid into the plant genome is witnessed by the occurrence of opines (Otten and Schilperoort 1978; Petit et al. 1983). Isolated transformed roots can be aseptically grown on gelled MS-medium without the supply of phytohormones. This autotrophy results from the effect of auxin and cytokinin loci which are located on the incorporated T-DNA. Transformed roots generally develop root hairs at a very high density, so that the plant disease was called "hairy roots" by phytopathologists.

In vitro cultures of hairy roots of *Tagetes* and other Asteraceae have been described by Flores and coworkers (Flores 1987; Flores et al. 1987, 1988). The roots developed high growth rates, accumulated high levels of thiophenes and showed stable growth and thiophene content over a long period of time. In addition, the pattern of thiophenes in transformed roots reflects the pattern of thiophenes in normal roots, as has been described for other plant species (Rhodes et al. 1987). We have used transformed root cultures as a model system to investigate the relationship between the differentiation of roots and the production of thiophenes.

We compared different types of transformed roots which were formed on the different *Tagetes* species (Fig. 11). The different types of transformed roots were obtained from axenically grown plants infected with *A. rhizogenes* and with mutant strains of *A. tumefaciens* (Table 4). Infection with mutant strains of *A. tumefaciens* generally yielded crown galls. Depending on the *Tagetes* species used, some of the crown galls cultured in vitro gradually formed roots with a decreasing intensity in the order *T. erecta*, *T. patula*, *T. minuta* (Table 5). This observation lends support to the hypothesis that the response of plant species to transformation mirrors the differentiation of untransformed tissues in vitro (Offringa et al. 1986; Nester et al. 1984). In agreement with the tendency of the calli of the three species to form roots, transformed callus of *T. minuta* showed higher growth rates than transformed callus of *T. erecta* (Table 6). In contrast, the transformed roots of *T. erecta* grew many times faster than transformed *T. minuta* roots. Both transformed callus and transformed roots of *T. patula* had a growth rate in between that of *T. erecta* and *T. minuta*, similar to untransformed tissues (see Sects. 4.1, 4.2). The growth rate of transformed roots of *T. erecta* was in the same order of magnitude as the growth

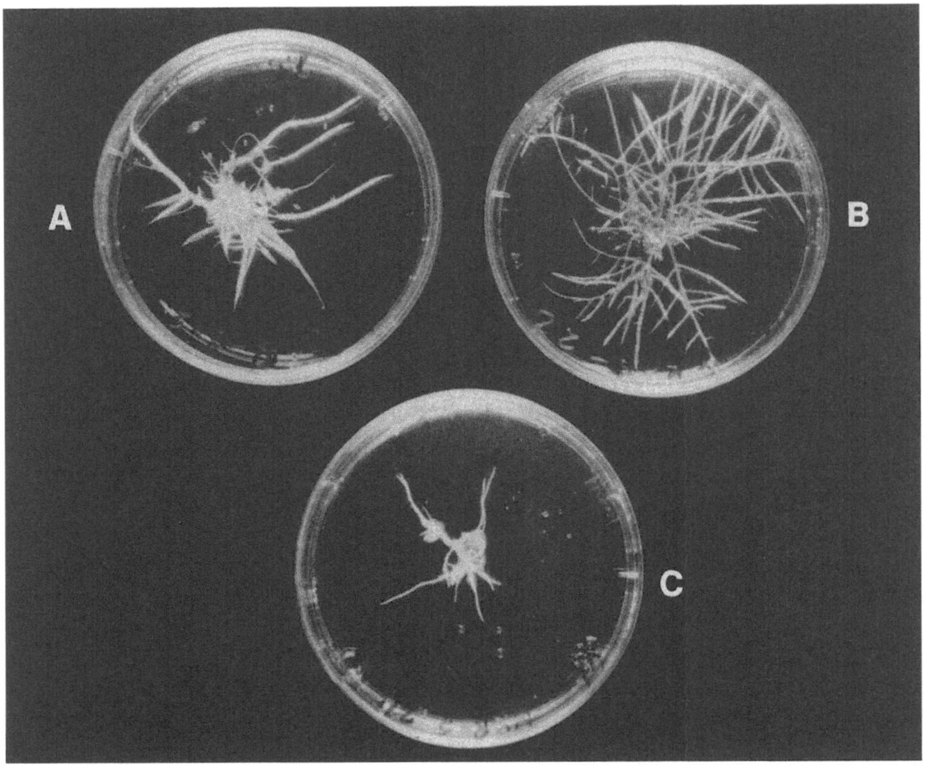

Fig. 11A–C. Root cultures of **A** *Tagetes erecta*; **B** *T. patula*; and **C** *T. minuta*. Plants were axenically grown and infected with *Agrobacterium rhizogenes* LBA 9402. Roots were excised from the infection site and cultured in vitro. Root cultures of *T. erecta* produced hard and short main roots with few branches. Roots of *T. patula* were long and highly branched, whereas *T. minuta* root cultures did not grow appreciably. (Ketel, unpubl.)

Table 4. Strains of *Agrobacterium* used to infect *Tagetes* plants. (Ketel and Croes, unpubl.; Croes et al. 1989b)

Strain	Plasmid	Genes	Characteristic opine[a]
A. tumefaciens			
LBA4001	pTiAch5	aux1, aux2, cyt	Octopine
LBA4060	pTiAch5::IS60	aux1, cyt	Octopine
LBA4210	pTiAch5::Tn904	aux1, aux2	Octopine
LBA1501	pTiB6::TN1831	aux1, cyt	Octopine
LBA1538	pTiB6::TN1831	aux2, cyt	Octopine
LBA8370	pTiT37	aux1, aux2, cyt	Nopaline
LBA288	–	–	–
A. rhizogenes			
LBA9402	pRi1855	aux1, aux2, rol A, B, C, D	Agropine

[a] Detected in the transformed *Tagetes* tissues.

Table 5. Gall and root formation on plants of three *Tagetes* species infected with wild-type and mutant strains of *Agrobacterium tumefaciens* and *A. rhizogenes* (cf. Table 4)

Strain of *Agrobacterium*	Plant species		
	T. erecta	*T. patula*	*T. minuta*
LBA4001	Crown gall and roots	Crown gall	Crown gall
LBA4060	Crown gall and roots	Crown gall	Crown gall
LBA1501	Crown gall	Crown gall	ND
LBA1538	Crown gall	Crown gall	ND
LBA8370	Crown gall and roots	Crown gall and roots	Crown gall
LBA4210	Crown gall and roots	Crown gall and roots	Crown gall and roots
LBA288	Roots[a]		
LBA9402	Roots	Roots	Roots

[a] Incidentally occurring (nontransformed), ND = not determined. The response of the plant was scored 6 weeks after infection (Ketel and Croes, unpubl.).

Table 6. Growth of transformed calli and roots of three *Tagetes* species cultured for 3 weeks on MS medium

Agrobacterium strain	Relative growth rate (per day)[b]		
	T. erecta	*T. patula*	*T. minuta*
Calli			
LBA4001	0.06 ± 0.02	0.11 ± 0.02	0.09 ± 0.01
LBA4060	ND[a]	ND	0.13 ± 0.03
LBA8370	ND	ND	0.11 ± 0.03
Roots			
LBA4210	0.06 ± 0.02	0.08 ± 0.02	− 0.03 ± 0.01
LBA4060	0.10 + 0.02	ND	ND
LBA8370	ND	0.08 ± 0.01	ND
LBA9402	0.10 ± 0.02	0.08 ± 0.03	0.05 ± 0.02
Untransformed[c]	0.09 ± 0.01	0.10 + 0.01	ND

[a] Not determined.
[b] Data are presented as mean with SD, $n \geq 4$ (Ketel and Croes, unpubl.).
[c] Roots on plants grown in hydroculture.

Table 7. Accumulation of thiophenes in crown galls and adventitious roots grown on stems of *Tagetes patula* plants 6 weeks after infection (Croes, unpubl.)

Agrobacterium strain	Neoplastic growth	Thiophene content (μg/g dry wt.)
LBA4001	Callus	577
LBA4060	Callus	900
LBA4210	Callus plus roots	1019
LBA9402	Roots	1492
Control	–	114

Table 8. Summary of the protocol to obtain cell cultures and root cultures of *Tagetes* spp. and the production of thiophenes by these cultures

Material	Medium	Cultivation period	Environmental conditions	Special treatments	Best results
1. Plants	Nutrient solution	5–7 weeks	Temp. 24 °C; RH 75 ± 5% Light: 16 h (30 W m^{-2})[a]	Basal mineral medium with 5 mM $Ca(NO_3)_2$ as N-source	Medium and conditions suitable for all *Tagetes* spp. Optimal explants after 5–7 weeks
2. Explants	–	–	–	Sterilization: 70% EtOH (2 s), 5% $Ca(OCl)_2$ (14 min), flushes of sterile water for 3, 10, and 30 min.	Upper part of fully expanded leaves placed upright in tubes (16 × ϕ 2.5 cm)[b] filled with 10–15 cm^3 gelled MS medium (see 3).
3. Callus	4.72 g MS medium powder + 20 g sucrose + 2 g Gelrite + 5 mg BA + 0.5 mg NAA (per dm^3). Final pH 5.9	C_1 callus 3 weeks C_2 callus 2 weeks C_3 callus 2 weeks etc. with 2-week intervals	Temp. 24 °C Light: 24 h (5 W m^{-2})[a]	Subculturing in tubes or Erlenmeyer flasks[b] (100–250 cm^3)	*T. erecta*: differentiates strongly and generally dies after a few subcultures, unless GA added. Growth rate C_3 calli: ca. 0.02 g/day. *T. patula*: differentiates gently, forms friable, yellow cauliflowerlike C_2 and C_3 calli. Growth rate C_3 calli: ca. 0.8 g/day. *T. minuta*: no differentiation. Calli smooth, and dark green. In general, only C_3 calli friable. Growth rate C_3 calli: ca. 0.5 g/day.
4. Cell cultures	4.72 g MS medium powder + 20 g sucrose + 0.1 g *myo*-inositol + 5 mg BA + 0.5 mg NAA + 1 g soluble PVP-10000 (per dm^3). Final pH 5.8	*T. erecta* limited *T. patula* 6–12 months *T. minuta* unlimited	Temp. 24 °C Light 24 h (5 W m^{-2})[a]	Regular subculturing in Erlenmeyer flasks (1 dm^3)[b]. Flasks continuously shaken: 120 rpm, 2.5 cm amplitude. Inoculum: 15% (75 cm^3 cells in 425 cm^3 liquid medium. Subculturing at 50 (*T. erecta*) and (*T. patula*) and 70% (*T. minuta*) PCV (packed cell volume).	Types of aggregates, doubling times (dt) and thiophene production *T. erecta* (unless GA added): brown aggregates (ϕ 4–6 mm), dt 10 days, 11 μg/g DM and release of BBTOH. *T. patula*: yellow-brown-green aggregates (ϕ 0.5–15 mm), dt 15 d, 50–380 μg/g DM and release of BBTOH into the medium (up to 10 mg/dm^3). *T. minuta*: fine granular suspensions, green, dt 1.8 days, no production of thiophenes

5. Root cultures	4.72 g MS medium powder + 20 g sucrose per dm^3, pH 5.8.	Plant phase: 3 weeks Solid medium phase: ca. 2 months Liquid medium phase: unlimited	Temp. 24 °C Light: 24 h (5 $W m^{-2}$)[a] or dark	*Plant phase*: growth of sterile plants in tubes (16 × 2.5 cm ϕ) filled with gelled (2 g gelrite/dm^3) MS medium + 2% sucrose. Growth of neoplastic tissues after infection with *Agrobacterium* strains *Solid medium phase*: growth of isolated transformed roots in Petri dishes with gelled MS medium with an antibiotic *Liquid medium phase*: regular subculturing of well-growing root cultures in liquid MS medium Cultures shaken as all cultures or slower	*Agrobacterium rhizogenes* LBA 9402 in combination with (1) *T. erecta*, (2) *T. patula*, and (3) *T. minuta* Relative growth rates: (1) 0.10/day; (2) 0.08/day; (3) 0.05/day Thiophene content (in the light): *T. erecta*: 1.9 mg/g DM; *T. patula* 3.6 mg/g DM; *T. minuta*; 1.8 mg/g DM General remarks: stable growth and thiophene production; highest thiophene contents in root tips and in the midexponential phase; no spontaneous release of thiophenes Thiophene release in medium containing organic solvents

[a] Philips TLW 33.
[b] Tubes and Erlenmeyer flasks covered with perforated metal closure foil.

rate of normal roots on intact plants (Table 6), whereas the growth rate of transformed roots of *T. minuta* was comparable with normal roots cultured in vitro in the absence of phytohormones. All of our transformed roots produced opines (Table 4).

Neoplastic tissues, either grown as crown gall or as roots or as crown gall with roots, contained different levels of thiophenes. In particular, the level of accumulated thiophenes in hairy roots obtained after infection of *T. patula* plants with *A. rhizogenes* (strain LBA 9402; Table 7) was about two to three times higher than in undifferentiated crown galls. The level of thiophenes in normal roots of *T. patula* (control) was more than ten times lower than in hairy roots. No evidence exists that the high level of thiophenes in the transformed roots is the consequence of elicitation or a direct effect of the hairy root disease. Therefore, the biosynthesis of thiophenes in transformed tissues may be quantitatively regulated by specific DNA sequences transferred by the infecting bacteria into the plant genome (Norton et al. 1985).

Croes et al. (1989b) reported that the auxin IAA affects the morphology and thiophene level of normal roots and roots transformed by *A. rhizogenes* of *T. patula*. Thiophene levels are reduced at high IAA concentrations and after transformation. The latter treatment usually stimulates branching. The highest thiophene levels were found in the tips of the roots. Arroo et al. (1990) related the thiophene concentration and the total amount of thiophenes in root cultures of *T. patula* to the growth phase. The concentration was highest in the midexponential phase (ca. 5 days after subculturing), while the total amount per culture was highest at the beginning of the stationary phase (ca. 25 days after subculturing). The same authors inferred from experiments with ^{35}S-feeding that in root cultures no thiophene degradation occurs during late exponential growth, and that in the early growth phases, synthesis is the only factor that determines the rate of thiophene accumulation.

Transformed and untransformed root cultures do not spontaneously release thiophenes into the growth medium. Organic solvents like hexadecane and FC 40, however, when used in a two-liquid-phase bioreactor trigger considerable excretion of thiophenes, without major effect on the growth rate (Buitelaar, pers. comm.).

The highest growth rates and thiophene levels in roots cultures have been summarized in Table 8.

5 Conclusions and Prospects

It may be assumed that the use of *Tagetes* plants in ancient civilizations was partly based on their content of acetylenic compounds like thiophenes. Many centuries later in the 1940s, α-terthienyl was synthesized and isolated from plant tissue. In the 1930s and 1950s, the biological activity of *Tagetes* against nematodes was (re?)discovered, followed by chemical analysis and unsuccessful nematicidal trials by the crop protection industry in the late 1950s.

The ascent of plant biotechnology and the need to use less hazardous biocides in agriculture has resulted in a new wave of interest for *Tagetes* and thiophenes in the 1980s. At present, various laboratories use *Tagetes* cell and tissue cultures to study thiophene accumulation for various reasons, ranging from potential application to academic attractiveness. Time will show whether fermentor-produced thiophenes will be competitive with chemically synthesized thiophenes and with other natural biocides. In Table 8, the best protocols and results from our work in terms of growth and thiophene content have been summarized.

The in vitro culture of *Tagetes* species has proven to be an interesting and manageable alternative for intact plants to further the understanding of the regulation of secondary metabolism in general and that of naturally occurring acetylenes and thiophenes in particular. Hitherto most of the research has been rather descriptive and empirical. It is encouraging to note that during the last years, the first enzymes, related to thiophene biosynthesis have been characterized, and that the study of the biosynthetic pathway of thiophenes and their acetylenic precursors is reinforced. Very recently the effects of genetic transformation of *Tagetes* by *Agrobactericum* species have prompted a molecular analysis of the transformation process and its consequences for growth, morphology and thiophene metabolism. It is to be expected that the methods currently available to study and influence gene expression will be directed toward the understanding and exploitation of thiophene accumulation by in vitro cultures.

The success of the molecular approach, however, may be curtailed by the problems of instability of growth and metabolite production detailed in the present review, and the lack of knowledge of biochemical key processes, key enzymes, and their regulatory controls. Also, more basic knowledge of the relationship between thiophene metabolism and the metabolism of related classes of compounds, e.g., polyunsaturated fatty acids, other acetylenes and other polar and nonpolar S-compounds, has to be generated before the altered expression of genes involved in one or a few steps of thiophene metabolism will result in higher rates of metabolite production. Finally, it may well be that the knowledge of thiophene metabolism in vitro may in the next century be applicable to improve the defense system of crops against their predators, rather than to use chemicals from in vitro cultures for practical purposes.

Acknowledgment. We are grateful to all our NOVAPLANT colleagues in Nijmegen and Wageningen for their stimulating cooperation in the past years.

References

Adamse P (1990) Selection of high-yielding cell lines of *Tagetes* using flow cytometry. In: Nijkamp HJJ, Plas LHW van der, Aartrijk J van (eds) Progress in plant cellular and molecular biology. Kluwer, Dordrecht, pp 726–731

Arroo RRJ, Kenkel H, Meijers H, Croes AF, Wullems GJ (1990) Cellular aspects of thiophene accumulation in relation to root differentiation. In: Nijkamp HJJ, Plas LHW van der, Aartrijk J van (eds) Progress in plant cellular and molecular biology. Kluwer, Dordrecht, pp 587–591

Bakker J, Gommers FJ, Nieuwenhuis J, Wijnberg H (1979) Photoactivation of the nematicidal compound α-terthienyl from roots of marigolds (*Tagetes* species). J Biol Chem 254: 1841–1844

Berg-Smit J van de (1953) Over het wortelrot bij Narcissen. Weekbl Bollencultuur 93/94: 437

Birot AM, Bouchez D, Casse-Delbart F, Durand-Tardif M, Jouanin L, Pautot V, Robaglia C, Tepfer D, Tepfer M, Tourneur J, Vilaine F (1987) Studies and uses of the Ri plasmids of *Agrobacterium rhizogenes*. Plant Physiol Biochem 25: 323–335

Bohlmann F, Zdero C (1985) Naturally occurring thiophenes. In: Weissberger A, Taylor EC (eds) Chemistry of heterocyclic compounds, part 1. Gronowitz S (ed) Thiophene and its derivatives. Wiley & Sons, New York, pp 221–323

Bohlmann F, Burkhardt T, Zdero C (1973) Naturally occurring acetylenes. Academic Press, London

Brown S (1984) Analysis and sorting of plant material by flow cytometry. Physiol Vég 22: 341–349

Burkill JH (1966) A dictionary of the economic products of the Malay peninsula. Ministry of Agriculture and Cooperation, Kuala Lumpur, Malaysia, 2444 pp

Cleene M de, Ley J de (1981) The host range of infectious hairy-root. Bot Rev 47: 147–194

Constabel F (1965) Phenolics in tissue cultures derived from *Juniperus communis* L.: studies on tannin synthesis. In: White PR, Grove AR (eds) Proc Int Conf Plant tissue Culture, Berkeley McCutcham, Berkeley, pp 183–190

Croes AF, Aarts AM, Bosveld M, Breteler H, Wullems GJ (1989a) Control of thiophene accumulation in calli of two *Tagetes* species. Physiol Plant 76: 205–210

Croes AF, Berg AJR van den, Bosveld M, Breteler H, Wullems GJ (1989b) Thiophene accumulation in relation to morphology in roots of *Tagetes patula*; effects of auxin and transformation by *Agrobacterium*. Planta 179: 43–50

Drawert F, Berger RG (1982) Uber die Biogenese von Aromastoffen bei Pflanzen und Früchten. XIX Mitt: Vergleich der Biosynthese von Aromastoffen in Segment-, Kallus- und Suspensionskulturen von Citrusarten. Chem Mikrobiol Technol Lebensm 7: 143–147

Duke J, Ayensu ES (1985) Medicinal plants of China, vol 1. Reference Publications, Michigan

Evans, PK, Cocking, EC (1977) Isolated plant protoplasts. In: Street HE (ed) Plant tissue and cell culture. Botanical monographs, vol 11, 2nd edn. Blackwell, Oxford, pp 103–135

Flores HE (1987) Use of plant cell and organ cultures in the production of biological chemicals. In: Mumma RO, Lebaron H, Honeycutt RC, Duesing JH (eds) Application of biotechnology to agricultural chemistry. Am Chem Soc Symp Ser Washington DC 334: 67–86

Flores HE, Hoy MW, Pickard JJ (1987) Secondary metabolites from root cultures. Trends Biotechnol 5: 64–69

Flores HE, Pickard JJ, Hoy MW (1988) Production of polyacetylenes and thiophenes in heterotrophic and photosynthetic root cultures of Asteraceae. In: Lam J, Breteler H, Arnason T, Hansen L (eds) Chemistry and biology of naturally-occurring acetylenes and related compounds (NOARC), Elsevier, Amsterdam, pp 233–254

Gommers FJ, Bakker J, Smits L (1980) Effects of singlet oxygen generated by the nematicidal compound α-terthienyl from *Tagetes* on the nematode *Aphelenchus avenae*. Nematologica 26: 369–375

Gommers FJ, Bakker J, Wijnberg H (1982) Dithiophenes as singlet oxygen sensitizers. Photochem Photobiol 35: 615–619

Groneman AF, Posthumus MA, Tuinstra LGM Th, Traag WA (1984) Identification and determination of metabolites in plant cell biotechnology by gas chromatography and gas chromatography/mass spectrometry: application to nonpolar products of *Chrysanthemum cinerariaefolium* and *Tagetes* species. Anal Chim Acta 163: 43–54

Helsper JPFG, Prins T, Lutke Willink D, Breteler H (1988a) 5-(4-Acetoxy-1-butinyl)-2,2-bithiophene (BBTOAc): acetate esterase in intact plants, calli and cell cultures of *Tagetes patula*, cv. *Nana Furia*. Plant Physiol 86 (Suppl): 143

Helsper JPFG, Ketel DH, Hulst AC, Breteler H (1988b) Production and secretion of thiophenes by differentiated cell cultures of *Tagetes*. In: Lam J. Breteler H, Arnason T, Hansen L (eds) Chemistry and biology of naturally-occurring acetylenes and related compounds (NOARC). Elsevier, Amsterdam, pp 279–286

Hulst AC, Verlaan P, Breteler H, Ketel DH (1987) Thiophene production by *Tagetes patula* in a pilot plant airlift-loopreactor (ALR). In: Neyssel OM, Meer RR van der, Luyben KCAM (eds) Proc 4th Eur Congr Biotechnol. Elsevier, Amsterdam, pp 401–404

Hulst AC, Meyer MMT, Breteler H, Tramper J (1989) Effect of aggregate size in cell cultures of *T. patula* on thiophene production and cell growth. Appl Microbiol Biotechnol 30: 18–25

Jente R, Olatunji GA, Bosold F (1981) Formation of natural thiophene derivatives from acetylenes by *Tagetes patula*. Phytochemistry 20: 2169–2175

Ketel DH (1986) Morphological differentiation and occurrence of thiophenes in leaf callus cultures from *Tagetes* species: relation to the growth medium of the plants. Physiol Plant 66: 392–396

Ketel DH (1987a) Distribution and accumulation of thiophenes in plants and calli of different *Tagetes* species. J Exp Bot 38: 322–330

Ketel DH (1987b) Callus and cell culture of *Tagetes* species in relation to production of thiophenes. Thesis, Agric Univ Wageningen, The Netherlands

Ketel DH (1988) Selection of thiophene-producing calli of *Tagetes patula* in relation to the production of thiophenes in liquid cultures. In: Robins RJ, Rhodes MJC (eds) Manipulating secondary metabolism in culture. Cambridge Univ Press, Cambridge, pp 225–232

Ketel DH, Breteler H, Groot B de (1985) Effect of explant origin on growth and differentiation of calli from *Tagetes* species. J Plant Physiol 118: 327–333

Ketel DH, Hulst AC, Gruppen H, Breteler H, Tramper J (1987) Effects of immobilization and environmental stress on growth and production of non-polar metabolites of *Tagetes minuta* cells. Enzyme Microb Technol 9: 303–307

Ketel DH, Breteler H (1988) Morphogenesis and thiophene production in cell cultures of *Tagetes* species. In: Lam J, Breteler H, Arnason T, Hansen L (eds) Chemistry and biology of naturally-occurring acetylenes and related compounds (NOARC). Elsevier, Amsterdam, pp 267–278

Kothari SL, Chandra N (1985) Plant regeneration in callus and suspension cultures of *Tagetes erecta* L. (African marigold) J Plant Physiol 122: 235–241

Lam J, Thomasen T (1988) Complexing agents for protection of highly conjugated compounds against photodegradation. In: Lam J, Breteler H, Arnason T, Hansen L (eds) Chemistry and biology of naturally-occurring acetylenes and related compounds (NOARC). Elsevier, Amsterdam, pp 47–60

Lam J, Breteler H, Arnason T, Hansen L (1988) Chemistry and biology of naturally occurring acetylenes and related compounds (NOARC). Proc NOARC-Conf Aarhus-Denmark. Elsevier, Amsterdam, 366 pp

Metschulat G, Sütfeld R (1988) Acetyl-CoA: 4-hydroxy butinyl bithiophene o-acetyltransferase isoenzymes from *Tagetes patula* seedlings. Z Naturforsch 42c: 885–890

Mihalik GJ (1978/79) Guyanese ethnomedical botany. A folk pharmacopoeia. Ethnomedicine 1/2: 83–96

Muir WH, Hildebrandt AC, Riker AJ (1954) Plant tissue cultures produced from single isolated cells. Science 119: 877–878

Mulder-Krieger T, Verpoorte R, Water A de, Gekel M van, Oeveren BCJA van, Baerheim Svendsen A (1982) Identification of the alkaloids and anthraquinones in *Cinchona ledgeriana* callus cultures. Planta Med 46: 19–24

Murashige T, Skoog F (1962) A revised medium for rapid growth and bioassays with tobacco cultures. Physiol Plant 15: 473–497

Neher RT (1968) The ethnobotany of *Tagetes*. Econ Bot 22: 317–325

Nester EW, Gordon MP, Amasino RM, Yanofsky MF (1984) Crown gall: a molecular and physiological analysis. Annu Rev Plant Physiol 35: 387–413

Norton RA, Finlayson AJ, Towers GHN (1985) Thiophene production by crown galls and callus tissues of *Tagetes patula*. Phytochemistry 24: 719–722

Offringa JA, Melchers LS, Regenburg-Tuink AJG, Constantino P, Schilperoort RA, Hooykaas PJJ (1986) Complementation of *Agrobacterium tumefaciens* tumor-inducing *aux* mutants by genes from the Tr-region of the Ri plasmid of *Agrobacterium rhizogenes*. Proc Natl Acad Sci 83: 6935–6939

Otten LABM, Schilperoort RA (1978) A rapid microscale method for the detection of lysopine and nopaline dehydrogenase activities. Biochim Biophys Acta 527: 497–500

Pensl R, Sütfeld R (1985) Occurrence of 3,4-diacetoxybutinylbithiophene in *Tagetes patula* and its enzymatic conversion. Z Naturforsch 40c: 3–7

Petit A, David C, Dahl GA, Ellis JG, Guyon P, Casse-Delbart F, Tempé J (1983) Further extension of the opine concept: plasmids in *Agrobacterium rhizogenes* cooperate for opine degradation. Mol Gen Genet 190: 204–214

Philogène BJR, Arnason JT, Berg CW, Duval F, Champagne D, Taylor RG, Leitch LC, Morand P (1985) Synthesis and evaluation of the naturally occurring phototoxin, Alpha-terthienyl as a control agent for larvae of *Aedes intrudens*, *Aedes atropalpus* (Diptera: Culicidae) and *Simulium verecundum* (Diptera: Simuliidae). J Econ Entomol 78: 121–126

Quisumbing E (1951) Medicinal plants of the Philippines. Bureau of printing, Manila, 100 pp

Reis Altschul S von (1973) Drugs and foods from little-known plants. Harvard Univ Press, Cambridge, MA, 366 pp

Rhodes MJC, Robins RJ, Kamill JD, Parr AJ, Walton NJ (1987) Secondary product formation using *Agrobacterium rhizogenes* transformed "hairy root" cultures. IAPTC Newslett 53: 2–15

Sahai OP, Shuler ML (1984) Environmental parameters influencing phenolics production by batch cultures of *Nicotiana tabacum*. Biotech Bioeng 26: 111–120

Sanchez de Jimenez E, Fernandez L (1983) Biochemical parameters to assess cell differentiation of *Bouvardia termifolia* Schecht callus. Planta 158: 377–383

Slogteren GMS van, Hoge JHC, Hooykaas PJJ, Schilperoort RA (1983) Clonal analysis of heterogeneous crown gall tumor tissues induced by wild-type and shooter mutant strains of *Agrobacterium tumefaciens* expression of T-DNA genes. Plant Mol Biol 2: 321–333

Steiner G (1941) Nematodes parasitic on and associated with roots of marigolds (*Tagetes* hybrids). Proc Biol Soc Wash 54: 31–34

Steinkopf W, Leitsmann R, Hofmann KH (1941) Studien in der Thiophenreihe. LVII. Über α-Polythienyle. Liebigs Ann Chem 546: 180–199

Street HE, Henshaw GG, Buiatti MC (1965) The culture of isolated plant cells. Chem Ind 2 (January): 27–33

Sütfeld R (1982) Distribution of thiophene derivatives in different organs of *Tagetes patula* seedlings grown under various conditions. Planta 156: 536–540

Sütfeld R (1987) HPLC of thiophenes for phytochemical and biochemical research. In: Linskens HF, Jackson JF (eds) High performance liquid chromatography in plant sciences. Springer, Berlin Heidelberg New York, pp 104–113

Sütfeld R (1988) Enzymological investigations into the metabolism of bithiophene derivatives. In: Lam J, Breteler H, Arnason T, Hansen L (eds) Chemistry and biology of naturally occurring acetylenes and related compounds (NOARC), vol 7. Elsevier, Amsterdam, pp 201–208

Sütfeld R (1989) Preparative liquid chromatography with analytical separation quality. Interval injection/displacement reversed-phase high performance liquid chromatography. J Chromatogr 464: 103–115

Sütfeld R, Breteler H (1988) Effects of plant material and extract treatment on the yield of natural products from *Tagetes*. In: Lam J, Breteler H, Arnason T, Hansen L (eds) Chemistry and biology of naturally occurring acetylenes and related compounds (NOARC). Elesevier, Amsterdam, pp 101–106

Sütfeld R, Towers GHN (1982) 5-(4-Acetoxy-1-butinyl)-2,2′-bithiopheneacetate esterase from *Tagetes patula*. Phytochemistry 21: 277–279

Tang CS, Wat CK, Towers GHN (1987) Thiophenes and benzofurans in the undisturbed rhizosphere of *Tagetes patula* L. Plant Soil 98: 93–97

Thellier M, Desbiez MO, Champagnat P, Kergosien Y (1982) Do memory processes occur also in plants? Physiol Plant 56: 281–284

Thorpe A (1978) Regulation of organogenesis in vitro. In: Hughes KW, Henke R, Constantin M (eds) Propagation of higher plants through tissue culture. A bridge between research and application. Univ Tennessee Symp Proc Conf 7804111, 16–19 April, USD of Energy, Washington

Uhlenbroek JH, Bijloo JD (1958) Investigations on nematicides I. Isolation and structure of a nematicidal principle occurring in *Tagetes* roots. Rec Trav Chim 77: 1004–1009

Uhlenbroek JH, Bijloo JD (1959) Investigations on nematicides II. Structure of a second nematicidal principle isolated from *Tagetes* roots. Rec Trav Chim 78: 382–390

Uhlenbroek JH, Bijloo JD (1960) Investigations on nematicides III. Polythienyls and related compounds. Rec Trav Chim 79: 1181–1198

Watt G (1889–1896) A dictionary of the economic products of India, 7 Volumes. Periodical Experts, Delhi, India

Wiermann R (1981) Secondary plant products and cell and tissue differentiation. In: Stumpf PK, Conn EE (eds) The biochemistry of plants. A comprehensive treatise, vol 7. Academic Press, New York, pp 86–116

Wijnberg H (1988) The chemistry of α-terthienyl. In: Lam J, Breteler H, Arnason T, Hansen L (eds) Chemistry and biology of naturally-occurring acetylenes and related compounds (NOARC). Elsevier, Amsterdam, pp 21–28

Yeoman MM, Miedzybrodzka MB, Lindsey K, McLauchlan WR (1979) The synthetic potential of cultured plant cells. In: Sala F, Parasi B, Cella R, Ciferri O (eds) Plant cell cultures: results and perspectives. Developments in plant biology, vol 5. Elsevier, Amsterdam, pp 327–348

Zechmeister L, Sease JW (1947) A blue-fluorescing compound, terthienyl, isolated from marigolds. J Am Chem Soc 69: 273–275

XXVII *Thymus vulgaris* L. (Thyme): In Vitro Culture and the Production of Secondary Metabolites

H. Tamura, T. Takebayashi, and H. Sugisawa[1]

1 Introduction

Thymus is a perennial herb of the Labiatae family (Fig. 1A) and can grow to a height of some 30 cm. There are more than 100 varieties of *Thymus* distributed throughout Europe. It has dark green leaves and light mauve-pink flowers in early summer. Thyme is a good rock-garden plant, as it spreads easily. It originates on the coast of the Mediterranean Sea and in the region of Asia Minor. Now it is cultivated in numerous European countries, Russia, Canada, America, and some of the North African countries. Thyme is one of the great culinary and medicinal herbs, and is a classic constituent of a bouquet garni (Fleming 1990), but only a few species are used in the kitchen, e.g., common thyme (*T. vulgaris*) and lemon thyme (*T. citriodorus*). The essential oil (Marhuenda et al. 1988; Blazquez and Zafra-Polo 1989) of thyme has also been reported to be effective in food preservatives, antioxidants (Farag et al. 1990), antimicrobials (Morris et al. 1979), breath-freshening dentifrices (Ishikawa and Shibuya 1989), medical disinfectants, etc. In general, the dried leaves contain 0.7–2.5% dry wt. of the essential oil. The major components are thymol (15–62%), carvacrol (2–44%), p-cymene, and γ-terpinene. The typical chemical structure is shown in Fig. 2.

In this chapter, the main discussion is focused on the selection of the callus cells and management of cell environments for production of flavor metabolites. Various aspects of tissue culture and micropropagation have been recently reviewed (Furmanowa and Olszowska 1992).

One of our experiments involved the selection of strains according to visual color differences in the cells. Green, yellow, and white calli, those colors which are independent of the desired metabolite, were determined as the representative colors of thyme callus. Furthermore, we examined the correlation between color difference and the production of characteristic monoterpenes such as thymol and carvacrol. Another experiment was designed to trap the volatiles before these compounds can act as toxic substances. To promote the production of volatiles from tissue culture of the leaves of *Thymus vulgaris* L., special attention was given to the cell oil, because it has been found that volatiles can sometimes act as toxic substances (Cormier, pers. comm. 1986). The cell has to have some organs, e.g., oil globules, in order to store the toxic materials or to carry the toxic and volatile

[1]Department of Bioresource Science, Faculty of Agriculture, Kagawa University, Miki-cho, Kagawa-ken, Japan 761-07

Biotechnology in Agriculture and Forestry, Vol. 21
Medicinal and Aromatic Plants IV (ed. by Y.P.S. Bajaj)

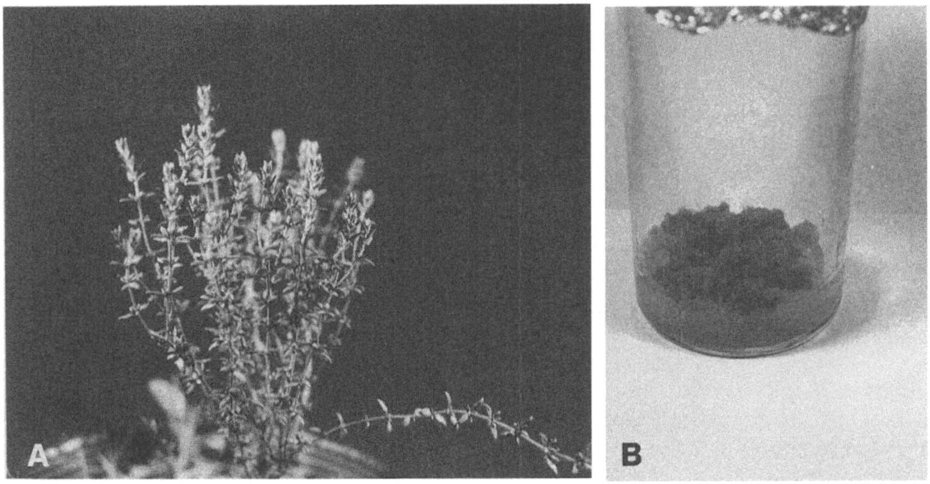

Fig. 1A, B. Photographs of the intact plant (**A**) and the callus tissue (**B**) of *Thymus vulgaris* L.

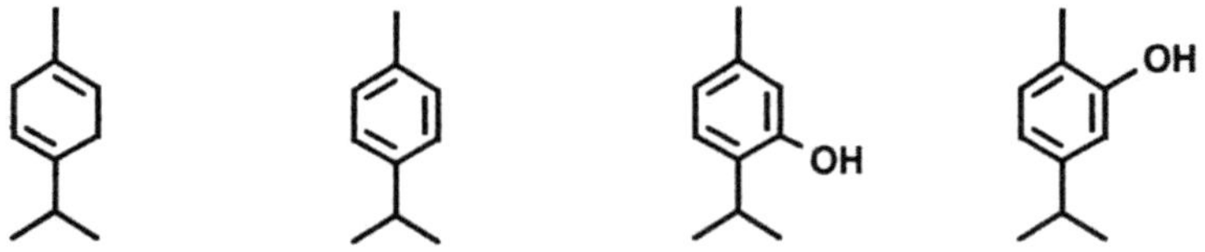

γ-terpinene p-cymene thymol carvacrol

Fig. 2. Typical compounds in *Thymus vulgaris* L.

compounds to the outside of the cells. The effect of several adsorbents on cell growth and the production of secondary metabolites was examined.

2 Production of Secondary Metabolites in *T. vulgaris* L.

2.1 Intact Plant

2.1.1 Volatile Components

Thymus vulgaris L. was cultivated on the farm of Kagawa University in 1985 and 1986. The aerial parts, leaves, and stems were collected in the flowering stage. The climate of our area, Seto-uchi, is moderate and somewhat similar to that of the Mediterranean. The volatiles of thyme leaves and stems were extracted by simultaneous distillation extraction (SDE) (Nickerson and Linkers 1966; Sugisawa et al. 1987). Leaves and stems (106 g) were homogenized with an equal volume of water for 3 min at 0 °C. The volatiles were extracted with 50 ml of ether by SDE for

60 min. After treatment with anhydrous Na_2SO_4 and Whatman 1-PS, the extract was concentrated at room temperature under purified N_2 (30 ml/min). The yield of concentrates (oil) was 0.34% of the plant fresh weight.

The oil obtained by SDE was fractionated into seven fractions containing hydrocarbons and oxygenated compounds using silica gel column chromatography with a gradient elution of n-pentane and ether as solvents. The flavor components in these fractions were identified by the retention index (RI) of GC (Kovats Index, Sugisawa et al. 1988; Tamura and Sugisawa 1989; Tamura et al. 1990) and GC/MS. As shown in Tables 1 and 2, a total of 29 hydrocarbons and 44 oxygenated compounds was identified. In the hydrocarbon fraction, hexane, heptane, m-xylene, 3-ethyltoluene, m-cymene, dimethylstyrene, α-copaene, β-elemene, β-bourbonene, β-cedrene, and β-farnesene were identified. The fraction of oxygenated compounds comprised 61.1% of the total oil with thymol (46.1%) as the major

Table 1. Compounds identified in the hydrocarbon fraction of thyme oil.[a] (Sugisawa et al. 1987)

Peak no.	Compound	Content(%)[b]	Kovats index Ref.[c]	Found
5	Hexane[d]	2.6	600	603
6	Heptane[d]	t	700	700
17	m-Xylene[d]	t	864	863
19	3-Ethyltoluene[d]	t	920	942
20	α-Thujene	1.2	938	931
21	α-Pinene	0.5	941	941
22	Camphene	0.2	956	955
25	Sabinene	0.2	975	974
27	β-Pinene	0.1	980	981
28	Myrcene	1.7	984	986
30	α-Phellandrene	0.1	1005	1004
31	δ-3-Carene	t	1012	1008
32	α-Terpinene	1.1	1016	1017
33	m-Cymene[d]	t	1015	1021
34	p-Cymene	13.2	1020	1021
35	Limonene	0.5	1028	1030
37	Ocimene	0.1	1038	1042
38	γ-Terpinene	14.0	1055	1058
40	Dimethylstylene[d]	t	–	1082
41	Terpinolene	t	1081	1082
62	α-Copaene[d]	t	1384	1383
64	β-Elemene[d]	t	1386	1390
65	β-Bourbonene[d]	t	1388	1391
66	β-Caryophyllene	1.5	1428	1426
67	β-Cedrene[d]	t	1429	1431
68	β-Farnesene[d]	0.1	1449	1453
69	α-Caryophyllene	0.1	1459	1458
71	Aromadendrene	0.4	1475	1476
72	δ-Cadinene	0.1	1520	1519

[a] OV 101 column (50 m × 0.25 mm), 80–200 °C, 2 °C/min, N_2 0.71 ml/min.
[b] t, Trace amount < 0.1%.
[c] Our data base.
[d] The compounds were newly identified.

Table 2. Compounds identified in the oxygenated fraction of thyme oil.[a] (Sugisawa et al. 1987)

Peak no.	Compound	Content(%)[b]	Kovats index Ref.[c]	Found
1	Acetaldehyde[d]	t	363	
2	Ethanol[d]	0.1	500	
3	Ethyl methyl ketone[d]	t	553	567
4	Ethyl acetate[d]	t	583	587
7	3-Methyl-3-buten-1-ol[d]	t	711	714
8	(Z)-2-Pentenol[d]	0.1	746	747
9	Methyl 2-methylbutyrate	0.1	761	761
10	Hexanal[d]	0.1	778	778
11	Butyl acetate[d]	0.1	789	785
12	(Z)-2-Hexenal	t	–	822
13	(E)-2-Hexenal[d]	2.5	830	830
14	(Z)-3-Hexenol[d]	0.4	833	838
15	(E)-2-Hexenol[d]	0.1	844	846
16	1-Hexanol[d]	t	848	849
18	(2E, 4E)-2, 4-Hexadienal[d]	t	881	883
23	1-Octen-3-ol[d]	1.2	962	964
24	Ethyl pentyl ketone[d]	0.1	969	968
26	3-Octanol	0.1	977	974
29	2-Octanol[d]	t	985	986
36	1, 8-Cineole	0.4	1030	1030
39	trans-Sabinene hydrate	1.1	1069	1062
42	(Z)-3-Hexenyl propionate[d]	t	1083	1082
43	Linalool	1.7	1087	1087
44	Fenchyl alcohol	t	1108	1114
45	Camphor	0.2	1130	1131
46	Borneol	0.5	1159	1158
47	Terpinen-4-ol	0.7	1170	1169
48	Dihydrocarvone[d]	t	1177	1176
49	α-Terpineol	0.2	1178	1179
50	Myrtenal[d]	t	1185	1186
51	Carveol[d]	t	1202	1204
52	Citronellol[d]	t	1214	1214
53	2-Isopropyl-5-methylanisole	0.1	1222	1215
54	5-Isopropyl-2-methylanisole	0.2	1233	1230
55	Geranial	0.1	1253	1252
56	Decanol[d]	t	1263	1261
57	Thymol	46.1	1269	1274
58	Bornyl acetate	3.4	1277	1280
59	Carvacrol		1279	1280
60	Thymol acetate[d]	t	1330	1329
61	Eugenol[d]	t	1335	1339
63	iso-Eugenol[d]	t	1384	1385
70	1-Dodecanol	t	1468	1466
73	Caryophyllene oxide[d]	0.2	1578	1577

[a] OV 101 column (50 m × 0.25 mm), 80–200 °C, 2 °C/min, N_2 0.71 ml/min.
[b] t, Trace amount < 0.1%.
[c] Our data base.
[d] The compounds were newly identified.

constituent. Twenty-six compounds such as 1-octen-3-ol, thymol acetate, and caryophyllene oxide were identified in this fraction, as shown in Table 2.

2.1.2 The Characteristic Impact Compounds in Thyme Oil

The odor quality of the concentrated volatiles and the dried leaves of intact plants were compared by a duo-trio test and paired comparison test with six panel members. In the test the flavor qualities of both samples were judged to be similar.

From the pair test of the hydrocarbon and oxygenated compound fractions, oxygenated compound fractions were found to exhibit the characteristic aroma of thyme oil. In the oxygenated fractions, separated into six fractions through a silica gel column, fraction 3 (the pentane fraction containing 2% ether), showing the most characteristic aroma of thyme was subjected to the GC sniff test using the flavor wheel. The GC sniff device was designed according to the method of Acree et al (1984).

For a flavor wheel for thyme oil, four main terms, such as floral-fruity, herbaceous, woody, and spicy were selected, and 32 secondary terms were used to form the wheel. Each component, which was eluted from the GC sniff device, was classified by aroma quality and its intensity according to the flavor wheel. Ten components selected by this wheel were subjected to an omission test to determine the importance of each compound as a characteristic component of thyme oil. In this test, bornyl acetate was omitted. Nine compounds, thymol (46.1%), carvacrol (3.4%), linalool (1.7%), 1-octen-3-ol (1.2%), trans-sabinene hydrate (1.1%), terpinen-4-ol (0.7%), borneol (0.5%), camphor (0.2%), and 3-octanol (0.1%) were found to be the most significant volatiles of thyme oil aroma.

2.2 Callus Tissue Cultures

2.2.1 Callus Cultures

Callus culture and the suspension culture of thyme were prepared according to the method described in previous papers (Sugisawa et al. 1987, 1988, 1990). Callus tissues were induced from the explants of leaves on Murashige and Skoog (MS) medium at 25°C for 8 weeks at a light intensity of 2000 1x. The medium (100 ml) contained 3 g of sucrose and 0.9 g of Difco Bactoagar and various concentrations of the growth regulators, auxin and cytokinin, naphthaleneacetic acid (NAA) and kinetin, indoleacetic acid (IAA) and kinetin, and 2,4-dichlorophenoxyacetic acid (2,4-D) and kinetin. The undifferentiated, induced callus (a typical example is shown in Fig. 1B) was transferred every 30 days to new MS agar media containing various concentrations of the growth regulators when rapid growth was observed. The concentrations of auxin and cytokinin for the subculture affected the growth factor and productivity of volatiles from the callus. The subculture conditions were determined from the growth to be 1 ppm NAA and 1 ppm kinetin or 1 ppm 2,4-D and 1 ppm kinetin, with 3% sucrose and 0.9% agar on MS medium at 25 °C and 2000 1x.

2.2.2 *Identification of Volatiles*

In the study on volatile production from callus tissues or suspension cells, sample preparation for GC and GC/MS represents a special problem because the growth of cultured cells in vitro may be slow and the productivity of volatiles may be low. SDE, modified SDE (MSDE), and simultaneous distillation absorption (SDA) (Sugisawa et al. 1987) were adapted to separate volatiles from callus tissues at 0.01–0.001% fresh weight. SDE can deal with about 50–100 g of callus; MSDE and SDA approximately 1–5 g of callus.

Callus tissues (5 g fresh weight) cultured on MS medium with 1 ppm NAA and 1 ppm kinetin were harvested from the 39th subcultures 30 days after transference. Volatiles were separated by MSDE and analyzed by the RI of GC and GC/MS. Recovery yield by MSDE was improved 1.3-fold more than the original SDE. The quantitative analysis of components in the oil obtained from callus was carried out using undecane as an internal standard. Unfortunately, the oil yield was very low, corresponding to 1/500–1/1000 of that of the entire plant. The identification of compounds was carried out by GC/MS with our RI data base for identification, as previously mentioned. Furthermore, aliphatic compounds were predicted from the 53 equations for RI values, prepared for the identification of aliphatic compounds (Tamura et al. 1990). Figure 3 shows a total ion monitoring (TIM) chromatogram of volatiles from the callus. As shown in Table 3, 61 components, including 25 aldehydes, 7 alcohols, 4 ketones, 8 terpenes, and others, were identified. A remarkable feature of the essential oil obtained from the callus was the presence of a variety of C15-C17 aldehydes and sesquiterpenes. (Z)-7-Heptadecenal, (Z, Z)-7,11-heptadecadienal, (Z, Z, Z)-7,11,14-heptadecatrienal, aliphatic aldehydes, which were detected as the major components in cucumber (Kemp 1975) and seaweed, such as *Ulva pertusa* and *Enteromorpha clathrata* (Kajiwara et al. 1987; Sugisawa et al. 1990), were found in *in vitro* cultured cells. The enzymatic formation of these C17 aldehydes from C18 fatty acids such as oleate, linoleate, and linolenate was recently reported by Kajiwara et al. (1988). Small amounts of thymol (0.3 μg/100 g fresh wt. callus) and β-elemene (0.7 μg/100 g fresh wt. callus) were also found in the volatiles from the callus tissue as well as in those of the intact plant. However, γ-terpinene and p-cymene were not found. Geranylacetone (1.8), β-chamigrene (0.7), nerolidol (46), β-guaiene (0), cuparene (0), γ-cuprenene (0), which were not detected in the intact plant, were found in this callus (μg/100 g fresh wt. callus of the eighth generation). Typical terpenoids found in callus tissue are listed in Fig. 4. From these sesquiterpenes, cuparene and β-chamigrene were also detected in the volatiles from the callus of *Perilla* sp. (Labiatae) (Nabeta and Sugisawa 1983). Sesquiterpenes, instead of the characteristic monoterpenes, were commonly produced as the major terpenoids from callus tissues of the Labiatae family (both *Thymus* and *Perilla*).

2.2.3 *Sensory Evaluation of the Volatiles from Thyme Callus*

The sensory evaluation of volatiles from the callus was examined by the GC-sniff test, as previously mentioned. The aromatic quality of the callus volatiles was generally judged to resemble that of the overripe, sweet oriental persimmon fruit (*Diospyros kaki* L.), completely lacking any spicy odor.

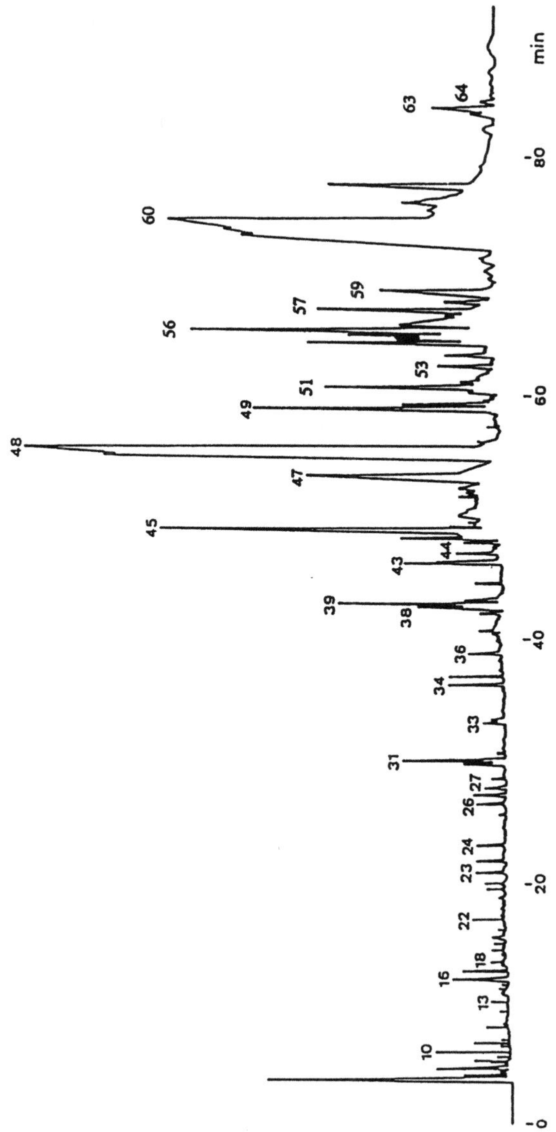

Fig. 3. Total ion monitor chromatogram of volatiles from the callus. Analytical condition; OV 101 column (50 m × 0.25 mm), 80–200 °C, 2 °C/min, N_2 0.71 ml/min

Table 3. The volatile components identified in thyme callus.[a] (Sugisawa et al. 1987)

Peak no.	Compound	Kovats index Ref.	Found
1	Acetaldehyde	363	+
2	Ethanol	500	+
3	Biacetyl	558	+
4	Ethyl acetate	583	+
5	Chloroform[b]	599	+
6	Benzene	657	656
7	Pentanal	671	671
8	3-Pentanone	677	676
9	1-Pentanol	742	751
10	Hexanal	778	777
11	Furfural	809	805
12	(E)-2-Hexenal	830	830
13	Benzaldehyde	934	934
14	2,4-Heptadienal	–	970
15	2-Pentylfuran	983	979
16	(2E, 4E)-2,4-Heptadienal	984	985
17	Benzyl alcohol	1010	1008
18	Phenylacetaldehyde	1015	1014
19	3,5-Octadien-2-one	1040	1044
20	1-Octanol	1054	1057
21	(3E, 5E)-3,5-Octadien-2-one	1064	1069
22	Nonanal	1087	1087
23	2,4-Dichlorophenol	1147	1155
24	Decanal	1188	1188
25	(2E,4E)-2,4-Nonadienal	1191	1191
26	(E)-2-Decenal	1240	1242
27	1H-Indole	1255	1260
28	2,4-Decadienal	–	1268
29	Thymol	1269	1268
30	4-Vinylguaiacol	1282	1286
31	(2E,4E)-2,4-Decadienal	1290	1290
32	Tridecane	1300	1300
33	(E)-2-Undecenal	1339	1345
34	β-Elemene	1394	1394
35	Tetradecane	1400	1404
36	Geranylacetone	1432	1432
37	β-Chamigrene	1480	1481
38	β-Guaiene	1499	1493
39	2,6-Di-tert-butyl-p-cresol[b]	1495	1495
40	Tridecanal	1495	1496
41	Cuparene	1506	1507
42	γ-Cuprenene	1530	1530
43	Nerolidol	1550	1551
44	1-Tridecanol	–	1563
45	Tetradecanal	1589	1597
46	Hexadecane	1600	1600
47	1-Tetradecanol	1661	1663
48	Pentadecanal	1691	1707
49	1-PentadecanOl	1760	1763
50	(7Z)-7-Hexadecenal	–	1768
51	Hexadecanal	1791	1796
52	Octadecane	1800	1798

Table 3. Continued

Peak no.	Compound	Kovats index Ref.	Found
53	Diisobutyl phthalate[b]	1823	1821
54	(8Z,11Z)-8,11-Heptadecadienal	1846	1855
55	(8Z,11Z,14Z)-8,11,14-Heptadecatrienal	1864	1860
56	(8Z)-8-Heptadecenal	1871	1867
57	Heptadecanal	1892	1900
58	Methyl hexadecanoate	1909	1903
59	Dibutyl phthalate[b]	1915	1919
60	Hexadecanoic acid	1946	1945
61	Ethyl hexadecanoate	1982	1975
62	Ethyl linoleate	2144	2136
63	Ethyl linolenate	2145	2141
64	Ethyl oleate	2150	2145
65	Ethyl octadecanoate	2180	2174

[a] These data were collected from 9th generation callus cultured on the Murashige and Skoog agar medium with 2,4-D (1 ppm), kinetin 1 ppm at 25 °C.
[b] Artifact.
[c] + Could not be calculated.

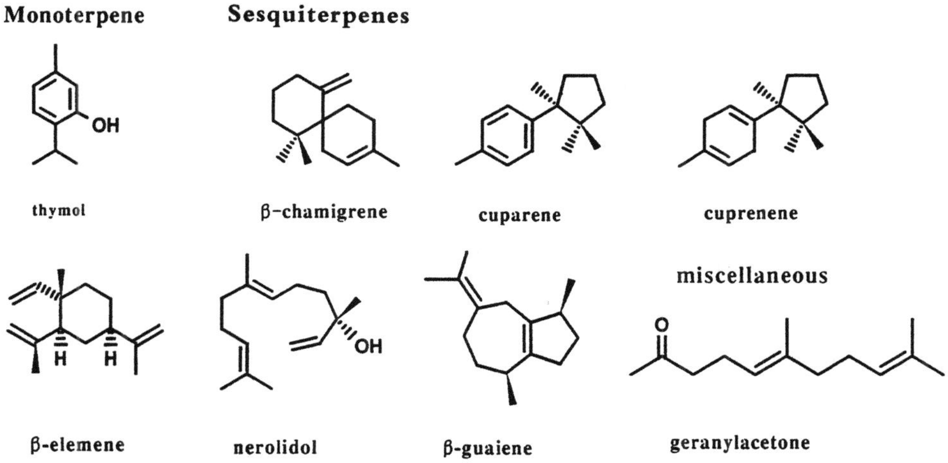

Fig. 4. Terpenoids identified in thyme callus tissue

2.3 Control and Regulation of the Production of Secondary Metabolites

2.3.1 Transformation of Some Precursors of Thymol

At present, it is assumed that thymol is biosynthesized by the aromatization of γ-terpinene leading to p-cymene followed by hydroxylation. Thus, it has been reported that p-cymene might be the direct precursor of oxygenated derivatives

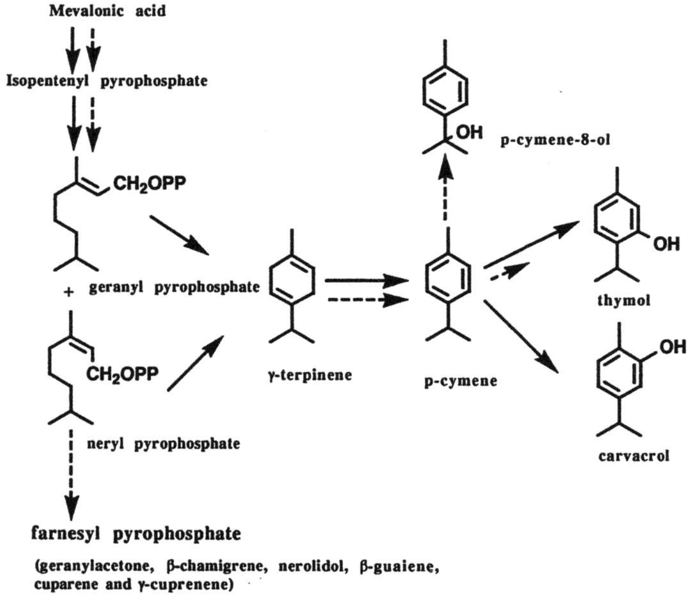

Fig. 5. Biosynthetic pathway of monoterpenes, which are the main metabolites from the intact plant, and sesquiterpenes, which are metabolites from cultured cells. ←— intact plant; ←----- cultured cell with or without precursors (modified after Croteau 1981)

(Poulose and Croteau 1978; Croteau 1981). γ-Terpinene and p-cymene as precursors were added to callus tissue on agar medium at 25 days to determine the enzyme activity in the thyme callus; the callus was then grown for 10 more days. The result of the transformation of the precursor materials is shown in Fig. 5 and Table 4. The enzyme activity of aromatization or dehydrogenation of γ-terpinene was weakly observed. Furthermore, p-cymen-8-ol could also be detected in the extracts. The transformation yield of γ-terpinene to p-cymene, including p-cymen-8-ol, was about 20% (Table 4). The biogenetic pathway from γ-terpinene to aromatic monoterpenes in thyme callus could be suggested by the detection of p-cymene and p-cymen-8-ol as the main products. Hydroxylation (oxygenation) of p-cymene occurred to form p-cymen-8-ol as well as thymol and carvacrol when p-cymene was added to the callus on the 25th day after transference. Although the production of p-cymen-8-ol was the main metabolic pathway for p-cymene, the production of thymol and carvacrol increased tenfold more when p-cymene was added, rather than γ-terpinene or no additive. Linalool was also transformed oxidatively to linalool oxides 1 and 2 by this callus enzyme system (data not shown). It seems to have a high oxygenase activity in this callus. The thyme callus is potentially valuable as a flavoring or medicine-making enzyme of the hydrocarbon monoterpene by hydroxylation or oxygenation.

Addition of mevalonic acid (1.09 μg/10 ml) enhanced the amount of the volatiles 2.1-fold more than that of the reference sample (subcultured callus, MS medium with 1 ppm NAA, 1 ppm kinetin). Terpene syntheses were weakly activated and

Table 4. Transformation of the precursors of thymol biosynthesis

Precursor	Products γ-Terpinene	p-Cymene	Thymol	Carvacrol	p-Cymene-8-ol
None	–[a]	–	0.02	–	–
γ-Terpinene[b]	10.25	1.52	0.02	0.02	0.91
p-Cymene	–	3.71	0.26	0.58	4.52

[a] ppm.
[b] Each 5 μl precursor was added to the top of the callus after 25 days of subculture.

produced mono- and sesquiterpene such as thymol (1.3 μg/100 g fresh wt. callus), β-elemene (2), geranylacetone (11.9), β-chamigrene (5.3), and β-guaiene (1.4).

Thus, it was proved that the enzyme activities of the monoterpene syntheses of thyme were latent in this callus. Although most of the monoterpenes which were found in intact plants could not be detected, except for thymol, new sesquiterpenes were produced in the calli. In this callus system, geranyl pyrrophosphate, which might be an intermediate precursor of thymol and sesquiterpene synthesis, was assumed to affect the production of sesquiterpene rather than monoterpene, as shown in Fig. 5 (dotted arrow). It was found that a flow control of geranyl pyrophosphate should be conducted for monoterpene synthesis. Since p-cymen-8-ol was not detected in intact plants nor in the callus without the addition of the precursors, the production of p-cymen-8-ol might be explained by 8-hydroxylase, which only acted against an excess amount of p-cymene in the callus.

2.3.2 The Difference in Volatiles in Green, Yellow, and White Calli

It was pointed out that cell selection is very important for plant cell culture (Yamamoto et al. 1982). However, most useful chemicals are colorless; therefore, it is difficult to distinguish highly productive cells. However, cell lines of thyme were visually selected according to color differences. Callus (31st generation), which was grown with 1 ppm NAA and 1 ppm kinetin on MS agar medium, was divided into three fractions according to color, e.g., green, yellow, and white, and then inoculated on each new medium as done previously. After 23 days, the colored fractions (green, yellow, and white portions) of each callus were subcultured in the same way for 38 days to promote color differences in each selected callus. The volatile components in these calli were not the same. On agar medium, green and yellow calli produced trace amounts of thymol (1 μg/100 g fresh wt.) as well as geranylacetone, β-chamigrene, γ-cuprenene, and nerolidol. White callus produced ethanol (11 μg/100 g fresh wt.) but not thymol. In the suspension culture, green and yellow calli produced thymol (2 μg/100 g FW), twofold more than that in agar culture. White callus of the suspension culture enhanced the production of ethanol (570 μg/100 g fresh wt.). However, sesquiterpenes such as geranylacetone, β-chamigrene, γ-cuprenene, and nerolidol were not detected or only detected in a small amount in every colored callus of the suspension culture.

2.3.3 Trapping Volatiles from Callus Tissues by Some Adsorbents and by the Double-Liquid Layer System

Charcoal and octadecyl silane (ODS) together with silica gel were used as adsorbents to trap toxic volatiles. The 22nd callus was subcultured on MS agar medium (1 ppm NAA, 1 ppm kinetin) and 1.7% absorbent. Five to 10% of the volatiles was adsorbed. However, a fair amount of the volatiles was retained in the callus. These adsorbents did not enhance the growth and production of volatiles in the callus. In the double-liquid layer system, i.e., suspension culture with an immiscible liquid, Migliol 812 (trigylceride) and hexadecane were used as the polar and nonpolar liquid fractions instead of charcoal and ODS. Many volatiles were still observed in each callus. Migliol and hexadecane trapped 349 and 159 μg of volatiles per gram callus fresh weight, respectively. The amount of volatiles accumulated in Migliol 812 was tenfold more than that in the callus cells. However, ethanol composed more than 90% of the volatiles. Using both liquid adsorbents, an increase in monoterpene could not be observed. Ethanol accumulation seemed to be related to the splitting and then the reduction of acetyl CoA or related compounds. Therefore, the activation of the biosynthesis pathway of terpenoids and fatty acid derivatives was inhibited at the initial stage by the polar polymer solvent, Migliol.

3 Conclusions and Prospects

The volatiles of thyme oil were identified by using retention indices (RI) of GC and GC/MS. The characteristic compounds were determined to be thymol (46.1%), carvacrol (3.4%), linalool (1.7%), 1-octen-3-ol (1.2%), trans-sabinene hydrate (1.1%), terpinen-4-ol (0.7%), borneol (0.5%), camphor (0.2%), and 3-octanol (0.1%) from the sensory evaluations.

Thyme callus was induced and subcultured every 30 days on MS medium with 1 ppm NAA and 1 ppm kinetin or 1 ppm 2,4-D and 1 ppm kinetin with 3% sucrose and 0.9% agar at 25 °C under 2000 1x.

In the callus tissue of *Thymus vulgaris* L., eight terpenes, including β-chamigrene, nerolidol, geranylacetone, β-guaiene, cuparene, γ-cuprenene, β-elemene, and thymol, were found. Only thymol and β-elemene were commonly identified in the intact plant and callus tissue. Sesquiterpene synthesis was considered to be superior to monoterpene synthesis due to the detection of seven sesquiterpenes in the volatiles of the callus tissue. This callus tissue produced a flavor similar to overripe, sweet oriental persimmon fruit, therefore, callus tissues are potentially capable of producing new flavors. It is important that the flavor regulates the biosynthetic pathway, thus, leading to monoterpene synthesis.

Some possible methods to regulate the productivity of metabolites in order to change the flavor quality are summarized as follows:

1. The concentrations of auxin and cytokinin in subculture affect the growth factor and productivity of volatiles in the callus. In the previous paper, it was reported that the concentration of β-chamigrene was enhanced 400-fold more than that of subcultured callus by changing the type and concentration of auxin and

cytokinin from 1 ppm NAA and 1 ppm kinetin to 10 ppm IAA and 10 ppm kinetin (Sugisawa et al. 1987).
2. From the results of transformation of some precursors of thymol, γ-terpinene could be biotransferred to p-cymene and then thymol. Therefore, enzymes necessary for the production of thymol were found in the callus systems. In particular, oxidative activity was found to be high. Therefore, the addition of some precursors to the callus tissue may make it possible to use thyme flavors or other chemicals. for industrial purposes.
3. Cell selection is very important for plant cell culture. However, it is difficult to differentiate the high productive cells from the others because of the colorless chemicals. Nevertheless, cell lines of thyme were visually selected according to color differences. Green and yellow calli contained somewhat more thymol than the nonselected subcultured callus, especially in suspension culture. Visual selection of cells seemed to be effective for the production of secondary metabolites.
4. Finally, under two-layer culture conditions, a large amount of the produced ethanol was trapped in organic solution. It might be concluded that since primitive materials (primary metabolites) necessary for the production of secondary metabolites were metabolized and then trapped as ethanol, an increase in the flavor monoterpenes and sesquiterpenes could not be observed. If two-layer culture had been conducted at the final stage of culture to trap the final products, inhibition of toxic activity might have been possible.

To produce thyme flavors by plant tissue culture, yellow or green cells have to be prepared in suspension culture with MS medium with 1 ppm NAA, 1 ppm kinetin, and 3% sucrose. A combination of two-layer culture and additional precursors might be effective at the final stage, to increase the amount of the secondary metabolites.

Acknowledgements. We gratefully thank our coworkers, K. Miwa and T. Matsuo, for their technical assistance. This work was supported by a grant from the Yamasaki Spice Research and Promotion Foundation, Tokyo, Japan.

References

Acree TE, Barnard J, Cunningham DG (1984) The analysis of odor-active volatiles in gas chromatographic effluents. In: Schreier P (ed) Analysis of volatiles, methods. application. Walter de Gruyter, Berlin, pp 251–267

Blazquez MA, Zafra-Polo MC (1989) Essential oil analysis of *Thymus godayanus*, an endemic species growing in northeastern Spain. Pharmazie 44: 651

Croteau R (1981) Biosynthesis of monoterpenes. In: Porter JW, Spurgeon SL (eds) Biosynthesis of isoprenoid compounds, vol 1. Wiley & Sons, New York, pp 225–282

Farag RS, Ali MN, Taha SH (1990) Use of some essential oils as natural preservatives for butter. J Am Oil Chem Soc 67: 188–191

Fleming S (ed) (1990) Herbs. Mitchell Beazley, New York

Furmanowa M, Olszowska O (1992) Micropropagation of thyme (*Thymus vulgaris* L.). In: Bajaj YPS (ed) Biotechnology in agriculture and forestry, vol 19. High-Tech and Micropropagation III. Springer, Berlin Heidelberg New York, pp 230–243

Ishikawa M, Shibuya K (1989) Breath-freshening dentifrices containing copper gluconate, a fluorine compound, and alkyl sulfates. Eur Pat Appl 16

Kajiwara T, Hatanaka A, Kawai T, Ishihara M, Tuneya T (1987) Long chain aldehydes in the green marine algae Ulvaceae. Nippon Suisan Gakkaishi 53: 1901

Kajiwara T, Yoshikawa H, Saruwatari T, Hatanaka A, Kawai T, Ishihara M, Tsuneya T (1988) Enzymatic formation of unsaturated long chain fatty aldehydes in *Ulva pertusa*. Phytochemistry 27: 1643–1645

Kemp TR (1975) Characterization of some new C16 and C17 unsaturated fatty aldehydes. J. Am Oil Chem Soc 52: 300–302

Marhuenda E, Menendez M, Lastra CA (1988) Constituents of essential oil of *Thymus carnosus* Boiss. J Chromatogr 436: 103–106

Morris JA, Khettry A, Seitz EW (1979) Antimicrobial activity of aroma chemicals and essential oils. J Am Oil Chem Soc 56: 595–603

Nabeta K, Sugisawa H (1983) Volatile components produced by callus tissues from three *Perilla* plants. In: Charalambous G, Inglett G (eds) Instrumental analysis of foods. Academic Press, New York, pp 65–84

Nickerson GB, Linkers ST (1966) Gas chromatographic evidence for the occurrence of hop oil components in beer. J Chromatogr 21: 1–5

Poulose AJ, Croteau R (1978) Biosynthesis of aromatic monoterpenes. Arch Biochem Biophys 187: 307–314

Sugisawa H, Miwa K, Matsuo T, Tamura H (1987) Volatile compounds produced from the cultured cells of thyme (*Thymus vulgaris* L.). In: Schreier P (ed) Bioflavour '87. Walter de Gruyter, Berlin, pp 327–340.

Sugisawa H, Tamura H, Nakahara K (1988) The calculation program of retention index in gas chromatography. Nippon Shokuhin Kogyo Gakaishi 35: 709–712

Sugisawa H, Nakahara K, Tamura H (1990) The aroma profile of the volatiles in marine green algae (*Ulva pertusa*). Food Rev Int. 6: 573–589

Tamura H, Sugisawa H (1989) Calculation program of Kovats retention index values based on those of known compounds. Nippon Shokuhin Kogyo Gakaishi 36: 148–151

Tamura H, Kihara S, Sugisawa H (1990) A new identification method for aliphatic compounds using linear equations of the GC retention index value. Agric Biol Chem 54: 3171–3176

Yamamoto Y, Mizuguchi R, Yamada Y (1982) Selection of a high and stable pigment-producing strain in cultured *Euphorbia millii* cells. Theor Appl Genet 61: 113–116

XXVIII *Uncaria elliptica*: In Vitro Culture and the Production of Epicatechin and Rutin

N.P. Das and K. H. Law[1]

1 Introduction

1.1 Distribution and Importance of the Plant

At least 34 species of the genus *Uncaria* (tribe Cinchonea, subtribe Mitragyninae Havil, family Rubiaceae) are widely distributed in Africa, tropical America, tropical and subtropical Asia (Ridsdale 1978). About 14 species have been reported in Malaysia. *Uncaria* species are characterized by the presence of hooks on the nodes lying opposite leaves with interpetiolar stipules and a head-shaped inflorescence. It is a woody climber the stems of which are square-shaped, bearing ovate leaves with five to seven lateral veins. The midribs are typically pubescent with long straight hairs, although sometimes only finely pubescent with glabrous laminae.

A typical young and an adult *Uncaria elliptica* plant (woody climber) are shown in Figs. 1 and 2, respectively. This plant was originally found growing among *Hevea* trees in a large rubber plantation near Kuala Lumpur, Malaysia. The identity was determined with the help of Dr. C.E. Ridsdale and Dr B.A. Krukoff, Botanists of Malesian Botany, Leiden.

In open land, the *Uncaria* plant grows as a bush, also thriving well in the forest or cleared jungles. Some of the species commonly found in Malaysia, Sri Lanka, India, Burma, Thailand, Cambodia, and Indonesia include *U. gambier* (Hunt) Roxb., *U. elliptica* R. Br. ex G. Don, *U. longiflora* (Poir) Merr., and *U.* callophylla Bl. ex North.

1.2 Flavonoids and Alkaloids in *Uncaria*

The *Uncaria elliptica* plant has been reported to contain several flavonoids such as the glycoside rutin, the aglycone (−)-epicatechin (Balz and Das 1979; Law and Das 1986, 1987), and (+)-catechin was reported to be present in *U. gambir* (Das and Griffiths 1967).

Various *Uncaria* species also contain several types of alkaloids. The alkaloids isolated from the different species were mainly the heteroyohimbines and oxindoles (Goh and Aishah 1985). The species *U. borneesis*, *U. longiflora* var. *longiflora*, and

[1] Laboratory of Flavonoid Research, Department of Biochemistry, Faculty of Medicine, National University of Singapore, Singapore 0511

Biotechnology in Agriculture and Forestry, Vol. 21
Medicinal and Aromatic Plants IV (ed. by Y.P.S. Bajaj)

Fig. 1. Young *Uncaria elliptica* plant

Fig. 2. Adult *Uncaria elliptica* plant (a 3-year-old plant grown in a tropical garden)

U. longiflora var. *pteropoda* are particularly rich in alkaloids of the oxindole type, namely, corynoxeine, rhyncophylline, and pteropodine (Phillipson et al. 1978). The yohimbine alkaloids (e.g., pseudoyohimbine) were reported to be found in *U. borneesis* and *U. callophylla* (Goh and Aishah 1985; Goh et al. 1986). *U. callaphylla* is also unique in that it contains the dimeric indole alkaloid, callophylline, which is made up of the gambirine and pseudoyohimbine moieties (Goh and Aishah 1985; Goh et al. 1986). *U. elliptica* of Thailand has been reported to contain several types of indole alkaloids such as rauniticine pseudoindoxyl, 19-epi-ajmalicine, and 3-iso-rauniticine (Herath et al. 1979; Phillipson and Supavita 1981, 1983). The *U. gambir* of Malaysia contained the indole alkaloid gambirine (Merlini et al. 1967). The *U. thwaitessi* of Sri Lanka was reported to contain triterpenes such as uncaric acid, diacetyl uncaric acid, and ursolic acid (Herath et al. 1978).

1.3 Pharmacological and Medicinal Use

Several species of *Uncaria* have been used as folk medicine and as a source of tannins (Ridsdale 1978). Recently, several reports have described the use of the *Uncaria* as a traditional medicine for the treatment of hypertension (Goh et al. 1986, 1988; Chang et al. 1989). The alkaloid gambirine isolated from *Uncaria* gambier was reported to affect the vegetative nerves of the uterus (Rothlin and Raymond-Hamet 1935).

2 Plant Propagation

Most of the *Uncaria* species are not available commercially, therefore, the plant is conventionally propagated through stem cuttings. In this method, a portion of a stem, root, or leaf is cut from the parent plant. The cuttings are then placed under certain favorable environmental conditions and induced to form roots and shoots, thereby producing a new independent plant which is usually an exact reproduction of the parent plant. For species that can be propagated by cuttings, this method has numerous advantages. Many new plants can be started in a limited space from a few stock plants. The method is inexpensive, rapid, and simple. In addition, special techniques in grafting and budding have also been used for some species.

Many of those "hard-to-root" plant species will usually require the treatment of the cuttings with growth regulators in order to increase the percentage of cuttings that form good quality roots, and to increase uniformity of rooting (Hartmann and Kester 1983). Auxins such as indole-3-butyric acid (IBA) and α-naphthalene acetic acid (NAA) have been found to be reliable in stimulating adventitious root production in a large number of plant species.

We examined the effects of growth regulators and other physical factors such as light and dissolved air on the formation of root on the stem cutting, hence establishing a rooting protocol for the propagation of *U. elliptica.*

In general, there are three major methods for applying growth regulators to stem cuttings for the induction of root formation: (1) the quick dip or concentration dip; (2) prolonged soaking; and (3) powder methods. Preliminary attempts to utilize the quick dip and the powder methods were unsuccessful and not pursued further. Therefore, only the prolonged soaking method was utilized and investigated in greater detail in our present study.

Stem cuttings obtained from the *U. elliptica* plant were separated into five groups and placed in several test solutions (Table 1) for varying lengths of time (6, 24, 48, 72, 96, and 120 h) at ambient temperatures (26 ± 2 °C). After the incubation period, the stems were transferred to deionized water (about 5 cm depth) and the water level replenished daily. The container was covered with black paper. After 18 days of culture in deionized water, the number of visible roots formed were counted. The test substances, namely, IBA, NAA, and indole acetic acid (IAA) were dissolved in ethanol and the solutions diluted with deionized water. The final ethanol concentration was 0.1–0.4% and each test solution by itself had no significant effects on adventitious root formation. Each stem cutting had six fully developed uppermost leaves, unless otherwise stated. Various concentrations of auxins were added to the test solutions and their effect on root formation studied (Table 2).

Table 1. Influence of duration of auxin treatment on root number in *Uncaria elliptica* cuttings[a]

Duration of treatment (h)	No. of roots per cuttings produced in: 10 μM IAA	10 μM IBA	10 μM NAA
6	–	–	–
24	–	–	–
48	–	4 ± 2	2 ± 2
72	–	6 ± 1	3 ± 2
96	2 ± 1	9 ± 3	4 ± 2
120	2 ± 1	10 ± 2	3 ± 1

[a] The newly excised cuttings were placed in rooting solution containing the auxin after treatment for the time denoted. The values shown are the mean ± SD for five cuttings. Number of roots in untreated control is zero. At 144 h or longer, the leaves tend to turn yellow and results are therefore not shown (Law 1989).

Table 2. Rooting response of *Uncaria elliptica* cuttings to various concentration of auxins[a]

Concentration (μM)	No. of roots per cutting after treatment in: IAA	IBA	NAA
0.1	–	1 ± 1	–
1	–	4 ± 2	2 ± 1
10	2 ± 1	9 ± 3	4 ± 2
100	4 ± 1	6 ± 3	2 ± 2

[a] The compounds were present in the rooting solution during the first four days of the rooting period. The values shown are the mean ± SD of 5 cutting per treatment. Number of roots in untreated control ± 0 (Law 1989).

The rooting efficiency between lateral and terminal shoots was compared. The influence of the number of leaves on the rooting response was also tested. A series of cuttings was prepared as follows: (1) apex with two uppermost leaves; (2) apex with four uppermost leaves; and (3) apex with six uppermost leaves. The cuttings were treated with IBA (10 μM) and NAA (10 μM).

The effect of light on the formation of roots from the cutting was also investigated. A white fluorescent lamp of about 4000 lx (16-h daily exposure) was used. The black paper covering the test containers was removed to expose the submerged portion of the stem to light. The effect of dissolved air on the formation of the roots was studied by sparging air bubbles through the test solution using an aquarium aeration pump. The number of roots formed after 18 days of culture was counted.

The rooting response of the cuttings of *U. elliptica* was found to depend upon the auxin concentration and duration of the treatment. This phenomenon is also observed in other species, e.g., *Pinus radiata* (Smith and Thorpe 1975) and *Malus pumica* (James 1983). IBA and other auxins had to be present for several days in order to stimulate rootings (Table 1). Twenty-four-hour treatment with these auxins had no effect and resulting only in a rather small stimulation. The fact that synthetic auxins such as IBA and NAA are much more stable than IAA may explain the difference in response to IAA and similar concentrations of synthetic auxin.

Table 3. Effect of a combination of IBA and NAA on the number of roots formed[a]

Amount of IBA (μM)	No. of roots per cuttings produced in: without NAA	10 μM NAA
1	2 ± 1	12 ± 3
10	4 ± 2	16 ± 4
100	2 ± 2	10 ± 3

[a] The compounds were present in the rooting solution during the first 4 days of the rooting period. The values shown are the mean ± SD for five cuttings per treatment.

Table 4. Effect of light on root formation[a]

No. of leaves per shoot	No. of roots formed
2	3 ± 2 (5 ± 2)
4	3 ± 2 (12 ± 2)
6	10 ± 4 (16 ± 4)

[a] The newly excised cuttings were placed in rooting solution containing 10 μM each of NAA and IBA for 4 days and then transferred to deionized water. The values shown are the mean ± SD of five cuttings. The values in parentheses were obtained in the absence of light (Law 1989).

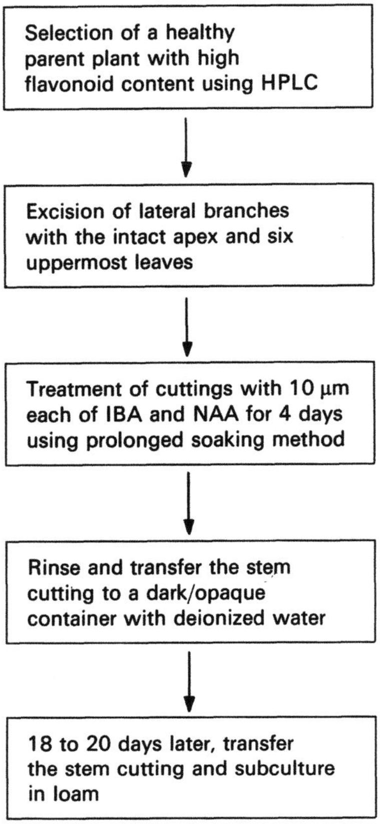

Fig. 3. Outline of rooting protocol for *Uncaria elliptica* (Law 1989)

The rooting response was maximum at 10 μM for both IBA and NAA when applied singly (Table 2). IBA is, however, more potent than NAA. The synergy between these two compounds was also maximal when they are applied together at 10 μM (Table 3).

Light was found to be unfavorable for *U. elliptica* root formation (Table 4).

Lack of dissolved air and attached leaves resulted in a slightly lower efficiency of root formation. However, though the availability of dissolved air did enhance root formation, it was not very important and therefore not implemented in the rooting protocol (Fig. 3) which was established for the future propagation of *U. elliptica* in plantations for the production of the flavonoids rutin and/or (−)-epicatechin.

3 Production of Rutin and (−)-Epicatechin

Uncaria elliptica r. Br. ex G Don (Figs. 1, 2) has received considerable interest from the pharmaceutical industry as an alternative source of the flavonoid, following reports of large amounts of rutin and moderate amounts of (−)-epicatechin present in the plant (Balz and Das 1979; Law and Das 1986, 1987). Rutin is a flavonol-3-glycoside, while (−)-epicatechin belongs to the group flavan-3-ol (Fig. 4). Rutin has been long known to be of therapeutic importance. Together with its

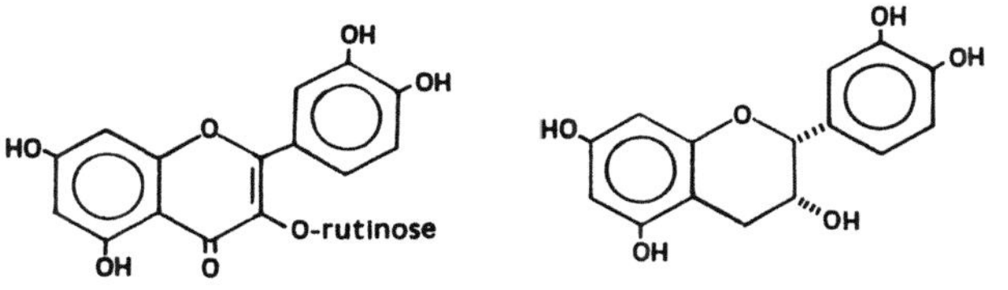

Rutin (−) - Epicatechin

Fig. 4. Structures of rutin and (−)-epicatechin

Table 5. Rutin and (−)-epicatechin contents in leaves and stems of different ages of the plant *Uncaria elliptica* by HPLC determination. (Law and Das 1987)[a]

Position and state of plant tissues	% Rutin contents/g dry wt.	% (−)-Epicatechin contents/g dry wt.
Young leaves next to apex	25.6 ± 0.3	0.9 ± 0.1
	16.8 ± 0.5[b]	1.5 ± 0.3[b]
Intermediate, mature leaves	21.0 ± 0.6	2.0 ± 0.1
	11.9 ± 0.2[b]	2.9 ± 0.2[b]
Basal, old leaves	11.8 ± 0.8	4.0 ± 0.3
	13.8 ± 1.1[b]	3.8 ± 0.1[b]
Young stems	11.3 + 0.4	1.2 + 0.1
	7.5 + 0.2[b]	3.3 + 0.2[b]
Intermediate, mature stems	2.7 + 0.4	1.1 + 0.3
	2.0 + 0.3[b]	1.3 + 0.1[b]
Bark (from mature stems)	6.9 ± 0.1	1.5 ± 0.03
	4.7 + 0.4[b]	3.0 + 0.1[b]
Cortical tissue (from mature stems)	0.6 + 0.003	0.5 + 0.04
	0.6 + 0.1[b]	0.4 + 0.1[b]
Pith (from mature stems)	1.9 + 0.04	0.6 + 0.01
	0.04 + 0.05[b]	0.5 + 0.1[b]
Basal, old stems	1.0 + 0.08	0.9 + 0.07
	2.5 + 0.3[b]	1.1 + 0.1
Bark (from old stems)	2.7 + 0.01	2.2 + 0.06
	2.5 + 0.3	1.6 + 0.2
Cortical tissue (from old stems)	0.5 + 0.05	0.4 + 0.03
	0.7 + 0.2[b]	0.3 + 0.1[b]
Pith (from old stems)	1.5 + 0.1	0.4 + 0.03
	1.5 + 0.1[b]	0.4 + 0.1

[a] Separation was carried out using HPLC. Detector conditions: 280 nm, 0.1 a.u.f.s. (absorption units full scale) and threshold = 5% f.s. (full scale). The values given are mean ± SD calculated from three measurements.

[b] Values obtained from plants grown in the shade.

Table 6. Rutin and (−)-epicatechin contents of *Uncaria elliptica* flower[a]

Part of blossom	Rutin content[b]	(−)-Epicatechin[b]
Petals, full bloom	13.2 + 0.4	1.2 + 0.3
Petals, unopened buds	14.3 + 0.8	1.3 + 0.1
Sepals, full bloom	18.9 + 1.0	3.0 + 0.3
Sepals, unopened buds	14.6 + 0.9	2.0 + 0.4
Stamen, full bloom	27.1 + 0.3	1.3 + 0.1

[a] The values shown are mean + SD of three determinations (Law and Das 1986).
[b] g% dry wt. of tissue.

hydroxyethyl derivatives, rutin has been marketed as the drugs "Venoruton" and "Paraven" for the treatment of vascular diseases in man (Gabor 1986). (−)-Epicatechin is a diastereoisomer of (+)-catechin, the latter had been marketed as the drug "Catergen" or "Cianidanol-3" (Balant 1981).

In view of the pharmacological importance of these flavonoids, several studies were carried out to establish the tissue distribution pattern of these flavonoids and tissue culture, in order to achieve a better understanding of flavonoid production, both in vivo and in vitro by the plant *U. elliptica.*

Rutin has been extracted from the leaves and blossoms of the buckwheat plant (*Fagopyrum esculentum*) (Couch et al. 1946) and several species of *Eucalyptus* leaves (Rodwell 1950; Humphreys, 1964) for pharmaceutical manufacturing. It is now mainly obtained from the flower buds of *Sophora japonica* (Couch et al. 1952) which have a rutin content of about 17.6 g% dry weight.

Tables 5 and 6 demonstrate that *U. elliptica* can be a very good alternative source of rutin, and to a smaller extent, of (−)-epicatechin. Due to its high content, the extraction and recovery of rutin from this plant are simple and efficient. The recovery is about 80–85% successful, using the extraction and purification methods reported earlier (Law and Das 1987). Alternatively, it can also be extracted using boiling water or isopropanol, adapting the extraction procedure of rutin from the buckwheat plant (Griffiths et al. 1955). The production of rutin from the leaves harvested from a 3-year-old plant using the method of Griffiths et al. (1955) was estimated to be 200–300 g/year, from about 1500 g of dried leaves of the *U. elliptica* plant. This plant has therefore a very promising commercial value for rutin production.

4 Establishment of Callus Culture and the Production of Rutin and Epicatechin

Buds and young twigs of *U. elliptica* were not a good explant source because they turned brown on all the media tested, probably due to rapid oxidation. They eventually died after isolation and culture. Soaking these explants in sterile deionized water (Rodriguez 1982) and 1% ascorbic acid (Mante and Tepper 1983) or transferring the explants to a fresh medium (Broome and Zimmerman 1978) after a short time interval (2 days) did not increase tissue viability or reduce tissue blackening. The slow to zero rate of callus formation and higher rate of contamination ruled out mature stems, mature leaves, old stems, and old leaves as the explant of choice. The most appropriate explant was therefore found to be the immature, young leaves which were used effectively in the subsequent experiments.

There was no apparent polarity involved in callus formation and explant orientation was not important, as opposed to that of *Gasteria verrucosa*, *Harworthia faciata* (Beyl and Sharma 1983), and *Camelia sinesis* (Frisch and Camper 1987). Seasons of the year are known to affect the efficiency of callus initiation and production in some plant species such as tea (Frisch and Camper 1987). However,

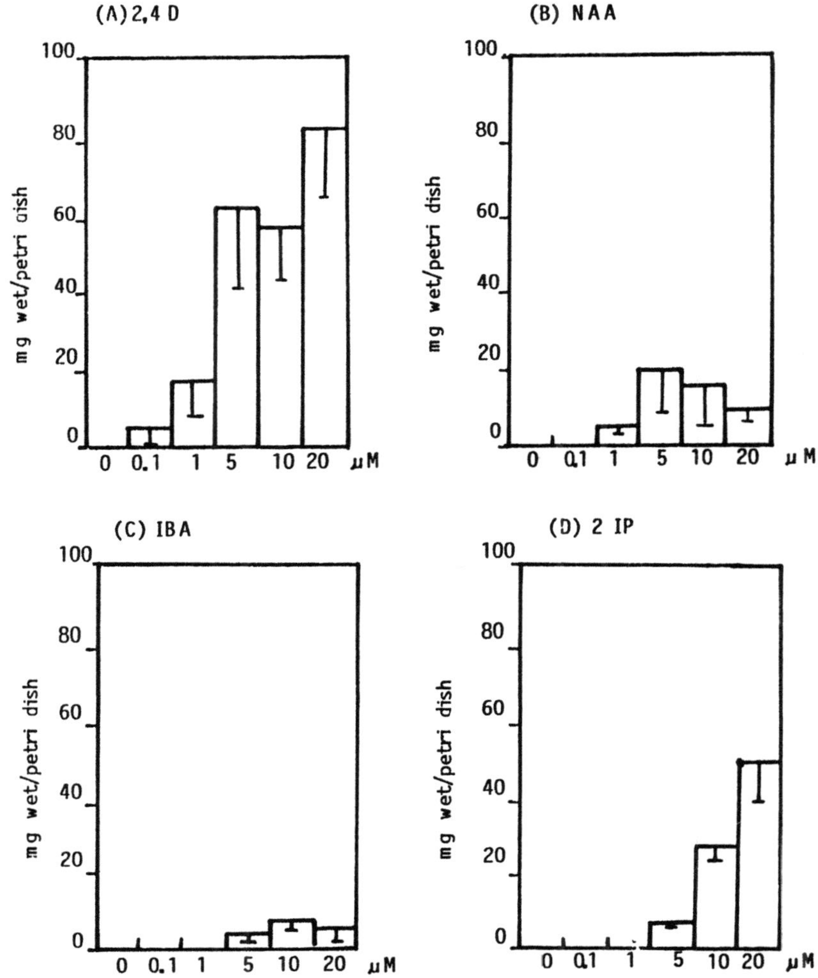

Fig. 5. Effect of different concentrations of **A** 2,4-D, **B** NAA, **C** IBA, and **D** 2-IP on the amount of callus formation. (Law and Das 1990)

there was no significant variation in the rate of callus formation from the young leaves of *U. elliptica* throughout the year. This may be due to the fact that this plant is found in the tropics where there are only wet and dry seasons. There was, however, more frequent contamination during the rainy period. A similar observation has been reported for the calli of tea plant (Frisch and Camper 1987). Washing with 70% ethanol reduced contamination by about 20%, probably through the removal of bacteria, and by providing better contact of the explant with calcium hypochlorite. This sterilant, although more difficult to handle, was preferred to sodium hypochlorite because it is cheaper and less toxic to plant material (Sweet and Bolton 1979).

Frequently there is an optimum size for explants used for the initiation of tissue culture. In this study, although explant size was not critical, 0.5 cm^2 of the explant had been found to be most effective. A two- to three-fold expansion was observed in the size of the explant after 4 weeks of incubation. Smaller explants were less viable, while larger explants had a higher rate of contamination.

The formation and growth of the *U. elliptica* tissue culture have been reported recently (Law and Das 1990) and the effect of growth regulators was studied.

2,4-D alone at a concentration of 0.1–20 µM in the medium could initiate callus formation (Fig. 5). 2-IP, NAA, and IBA also produced callus, but were less effective and higher amounts were required for callus formation. KIN and IBA, when present singly, were ineffective. It was observed that the degree of callus initiation was highest with auxins, the order being, 2,4-D > NAA > IBA. The cytokinins, on the other hand, were less effective in the order of 2-IP > KIN > BA (Law and Das 1990). In the absence of a growth regulator, there was no callus formation. The separate use of auxins, namely, NAA and IBA, resulted in rhizogenesis of negatively geotropic adventitious root structures (Fig. 6). Rhizogenesis was not desirable, as

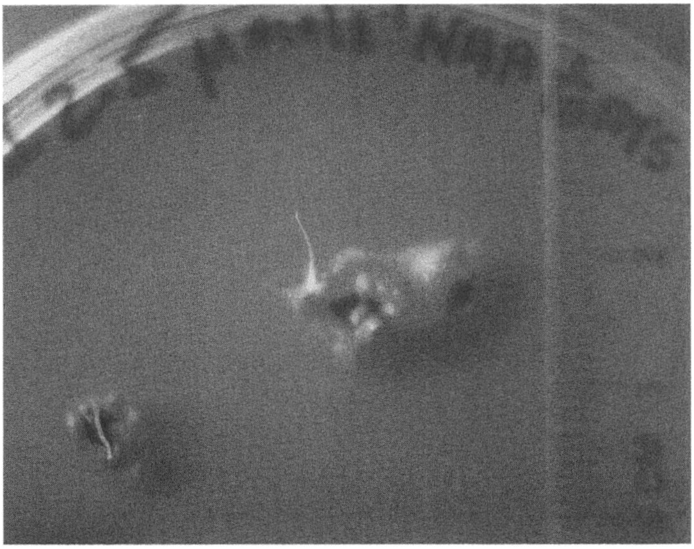

Fig. 6. Rhizogenesis during callus initiation. Explant used is a section of young leaf. Culture is 5 weeks old, on 1/5 medium and 1% sucrose. *Scale*: ⟼ 1 cm

it somehow decreased and/or arrested further growth and development of callus in the subsequent subculture. IBA and NAA appeared to stimulate rhizogenesis, while kinetin and 2-IP depressed it. The percentage of callus-forming root structure decreased as the number of passages increased.

Although a single growth regulator can result in callus initiation from an explant, a combination of auxin and cytokinin was required for the maintenance of

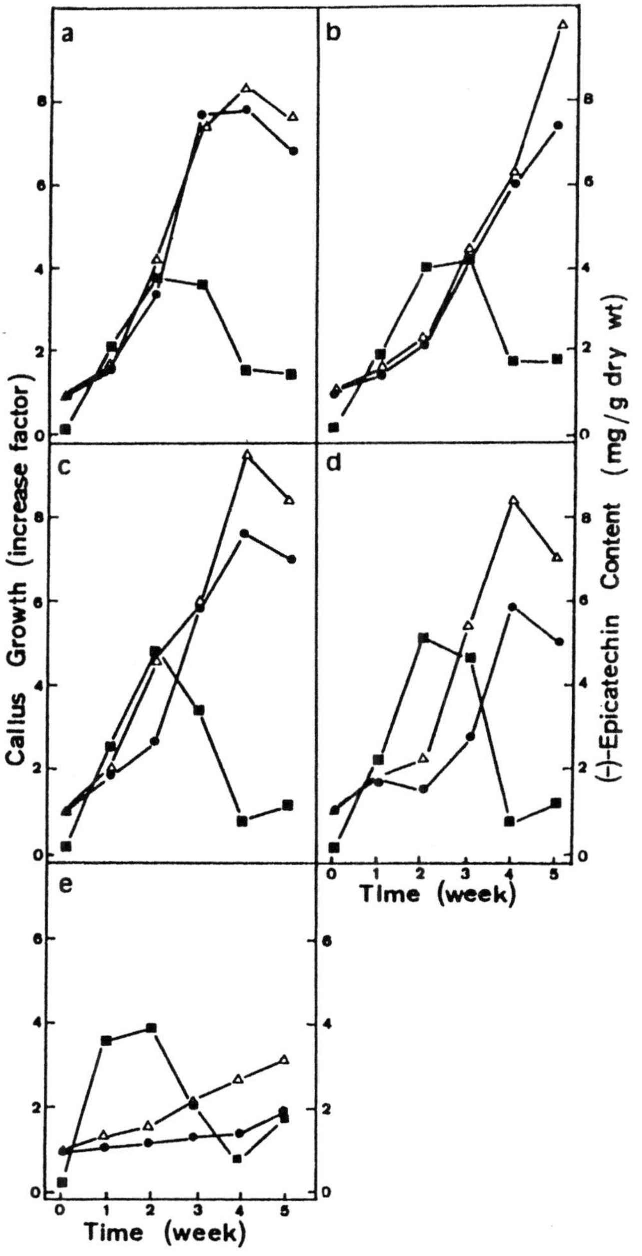

Fig. 7. Time course of growth and (−)-epicatechin content (■ —— ■) in LKH 1/1 callus of *Uncaria elliptica* grown under continuous light. The tissues were cultured on $\frac{1}{5}$ MS agar medium supplemented with 5 μM NAA, 0.1 μM 2-IP, and various sucrose concentrations: **a** 1%; **b** 2%; **c** 3%; **d** 4%; and **e** 6%. Growth was expressed as an increase factor defined as the g wt. of tissue at harvest per g wt. of the inoculum: (△ —— △) dry wt. increase; (● —— ●) fresh wt. increase

the culture. The concentration of growth regulators for the optimum maintenance of callus was found to be either 5 or 10 μM of IBA combined with 5 or 10 μM of kinetin. The calli obtained from this study varied from white/colorless to pale yellow and turned green on exposure to illumination (Fig. 8).

Increasing the amount of sucrose, fructose, and glucose did not exert a marked effect on the (−)-epicatechin level. Its content in the callus increased from 0.25

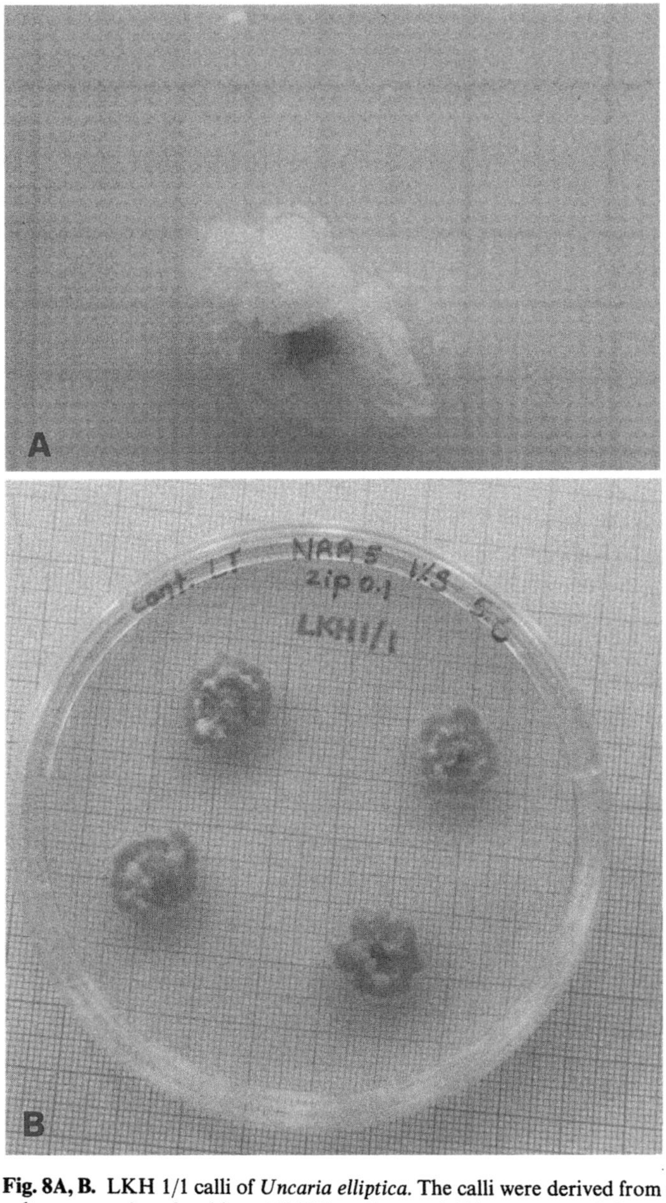

Fig. 8A, B. LKH 1/1 calli of *Uncaria elliptica*. The calli were derived from explants of a young leaf. The cultures were 4 weeks old, on 1/5 MS medium and 1% sucrose. **A** Culture grown in the dark, **B** Culture grown under continuous light. *Scale*: ⟼ 1 cm to 5.2 mg/g dry weight. The (−)-epicatechin production pattern was growth-associated (Fig. 7).

Cytokinins, especially kinetin were found to be stimulatory for (−)-epicatechin formation. Auxins, especially 2,4-D, suppressed its formation. A combination of 10 μM each of indole-3-butyric acid (IBA) and kinetin was found to be optimal for (−)-epicatechin formation while maintaining a satisfactory growth rate. The (−)-epicatechin content in the callus was comparable to its level in the major parts of the source plant.

5 Summary and Conclusion

A callus cell line was established using the natural variation of cells induced by different growth regulators. The identification or selection of a high-yielding cell line for both rutin and (−)-epicatechin have not been successful. Within the time frame of the present study, only (−)-epicatechin-producing calli were found. With this (−)-epicatechin-producing cell line, it was hoped that, by varying the different physical and culture conditions, the level of (−)-epicatechin could be enhanced and the production of rutin induced.

Rutin, the major flavonoid found in the source plant, was not detected under the present culture conditions. The ability of the callus to synthesize (−)-epicatechin but not rutin appeared to favor the concept that when more than one flavonoid derivative is simultaneously formed in a vegetative plant tissue, the accumulation of the simplest form in callus or cell cultures is preferred over the more highly oxidized flavonoids (Stafford et al. 1986).

References

Balant L (1981) Clinical pharmacology of (+)-cyanidanol-3: a synopsis with emphasis on pharmacokinetics. In: Conn HO (ed) Int Worksh on (+)-Cyanidanol-3 in a disease of the liver. The Royal Society of Medicine, Int Congr Symp Ser 47. Academic Press, London, pp 4954

Balz JP, Das NP (1979) *Uncaria elliptica* a major source of rutin. Planta Med 36: 174–177

Beyl CA, Sharma GC (1983) Picloram induced somatic embryogenesis in *Gasteria* and *Haworthis*. Plant Cell Tissue Organ Cult 2: 123–132

Broome OC, Zimmerman H (1978) In vitro propagation of Malling 7 apple rootstock. *HortScience* 15: 151–510

Chang P, Koh YP, Geh SL, Soepadmo E, Goh SH, Wong AK (1989) Cardiovascular effects in the rat of dihydrocorynantheine isolated from *Uncaria callophylla*. J Ethnopharmacol 25: 213–218

Couch JF, Naghski J, Krewson CF (1946) Buckwheat as a source of rutin. Science 103: 197–198

Couch JF, Naghski J, Krewson CF (1952) Rutin content of *Sophora japonicao*. J Am Chem Soc 74: 424–425

Das NP, Griffiths LA (1967) Biosynthesis of (+)-[^{14}C]-catechin by the plant *Uncaria gambir* Roxb. Biochem J 105: 73–77

Frisch CH, Camper ND (1987) Effect of synthetic auxins on callus induction from tea stem tissue. Plant Cell Tissue Organ Cult 8: 207–227

Gabor M (ed) (1986) The pharmacology of benzopyrone derivative and related compounds. Akademiai Kiado, Budapest

Goh SH, Aishah S (1985) Alkaloids of *Uncaria callophylla*. Phytochemistry 24: 880–886

Goh SH, Soepadmo E, Chang P, Aishah S, Koh YP, Nasrulhaq A, Taylor CE, Wong AK (1986) Malaysian medicinal plants, antihypertensive principles of *Uncaria callophylla*. Malays J Sci 8: 109–115

Goh SH, Liew SL, Soepadmo E (1988) Malaysian medicinal plants used for the treatment of hypertension. In: Seminar on Malaysian Traditional Medicine, University of Malaya, Kuala Lumpur, Malaysia, pp 503–508

Griffiths JQ, Krewson CF, Naghski J (eds) (1955) Rutin and related flavonoids chemistry, pharmacology, clinical applications. Mack, Easton, Pennysylvania

Hartmann HT, Kester DE (eds) (1983) Plant propagation. Principles and practices. Prentice Hall International, London

Herath WHMW, Sultanbawa MUS, Wannigama GP (1978) Three new ursene carboxylic acids from *Uncaria* thwaitessii. Phytochemistry 17: 1979–1982

Herath WHMW, Sultanbawa MUS, Wannigama GP, Cave A (1979) Alkaloidal and other constituents of *Uncaria elliptica* and *Canthium dicoccum*. Phytochemistry 18: 1385–1387

Humphreys FR (1964) The occurrence and industrial production of rutin in southeastern Australia. Econ Bot 18: 195–253

James DJ (1983) Adventitious root formation "in vitro" in apple rootstocks (*Malus pumica*). I. Factors affecting the length of the auxin-sensitive phase in M.P. Physiol Plant 57: 149–153

Law KH (1989) Studies on rutin and (−)-epicatechin production in *Uncaria elliptica* plant and tissue culture. PhD Thesis, National University of Singapore

Law KH, Das NP (1986) Quantitation of rutin and (−)-epicatechin in the flower of *Uncaria elliptica* plant using high performance liquid chromatography. In: Kon OL, Chung MCM, Hwang PLH, Leong SF, Loke KH, Thiyagarajah P, Wong PTH (eds) Contemporary themes in biochemistry. Cambridge University Press, London, pp 378–379

Law KH, Das NP (1987) Dual-wavelength absorbance ratios and spectrum scanning techniques for identification of flavonoids by HPLC. J Chromatogr 388: 225–233

Law KH, Das NP (1989) Growth of *Uncaria elliptica* callus and its (−)-epicatechin production. Phytochemistry 28: 1099–1100

Law KH, Das NP (1990) Studies on the formation and growth of *Uncaria elliptica* tissue culture. J Nat Prod 53: 125–130

Mante S, Tepper HB (1983) Propagation of *Musa textilis* Nee plants from apical meristem slices in vitro. Plant Cell Tissue Organ Cult 2: 151–159

Merlini L, Mendelli R, Nasini G, Hesse M (1967) Cambirine, a new indole alkaloid from *Uncaria gambier*. Tetrahedron Lett 16: 1571–1574

Phillipson JD, Supavita N (1981) Alkaloids of *Uncaria elliptica* from Thailand. J Pharm Pharmacol 33: 13 (Suppl)

Phillipson JD, Supavita N (1983) Alkaloids from *Uncaria* species Pt. 8. Alkaloids of *Uncaria elliptica*. Phytochemistry 22(8): 1809–1813

Phillipson JD, Hemingway SR, Ridsdale CE (1978) Alkaloids of *Uncaria*, Part V. Their occurrence and chemotaxonomy. J Nat Prod 41: 503–570

Ridsdale CE (1978) A revision of *Mitragyna* and *Uncaria* (Rubiaceae). Blumea 24: 43–100

Rodriguez R (1982) Callus initiation and root formation from in vitro culture of walnut cotyledons. HortScience 17: 195–196

Rodwell CN (1950) Rutin in two *eucalyptus*. Nature 165: 773–774

Rothlin E, Raymond-Hamet, A (1935) Effects of gambirine on the vegetative nerves of the uterus. C R Seances Soc Biol Fil 119: 37–40

Smith DR, Thorpe TA (1975) Root initiation in cuttings of *Pinus radiata*. II. Growth regulator interactions. J Exp Bot 26: 193—202

Stafford HA, Kreitlow KS, Lester HH (1986) Comparison of proanthocyanidins and related compounds in leaves and leaf-derived cell cultures of *Gingko bioloba L*, *Pseudotsuga menziesii* Franco and *Ribes sanguineum*. Plant Physiol 82: 1132–1138

Sweet HC, Bolton WE (1979) The surface decontamination of seeds to produce axenic seedlings. Am J Bot 66: 692–698

Subject Index

MIX
Papier aus verantwortungsvollen Quellen
Paper from responsible sources
FSC® C105338

FSC
www.fsc.org

If you have any concerns about our products,
you can contact us on
ProductSafety@springernature.com

In case Publisher is established outside the EU,
the EU authorized representative is:
Springer Nature Customer Service Center GmbH
Europaplatz 3, 69115 Heidelberg, Germany

Printed by Libri Plureos GmbH
in Hamburg, Germany